Novel Nanomaterials

Edited by Karthikeyan Krishnamoorthy

Published in London, United Kingdom

IntechOpen

Supporting open minds since 2005

Novel Nanomaterials
http://dx.doi.org/10.5772/intechopen.91600
Edited by Karthikeyan Krishnamoorthy

Contributors
Musa Yahaya Pudza, Zurina Zainal Abidin, Valerii Barbash, Ravi Kumar Chidrala, Ramachandran Rajendran, Zong-Xiang Xu, Fei Wang, Mihrican Muti, Kübra Gençdağ Şensoy, Juyoung Ha, Ana Rovisco, Pedro Barquinha, Rita Branquinho, Rodrigo Martins, Joana Vaz Pinto, Elvira Fortunato, Yaser E. Greish, Hesham F. F. El-Maghraby, Rudy Situmeang, Chuan-Pei Lee, Chi-Ang Tseng, Prasanta Kumar Sahoo, Yi-June Huang, Lebogang Katata-Seru, Hugues Kamdem Paumo, Tshepiso Moremedi, Soumen Sardar, Mpitloane Joseph Hato, Abhijit Bandyopadhyay, Jan Tkac, Lenka Lorenocova, Peter Kasak, Kishor Kumar Sadasivuni, Betul Gürünlü, Mahmut Bayramoğlu, Karthikeyan Krishnamoorthy, Jeyasubramanian Kadarkaraithangam, Gokul Raja T. S., Parani Bramma Nayagi S., Sivaji Mathivanan, Suresh Babu Naidu Krishna, Jerushka S. Moodley, Karen Pillay, Patrick Govender, Olga Yaschenko

First published in London, United Kingdom, 2021 by IntechOpen
IntechOpen is the global imprint of INTECHOPEN LIMITED, registered in England and Wales, registration number: 11086078, 5 Princes Gate Court, London, SW7 2QJ, United Kingdom
Printed in Croatia

British Library Cataloguing-in-Publication Data
A catalogue record for this book is available from the British Library

Additional hard and PDF copies can be obtained from orders@intechopen.com

Novel Nanomaterials
Edited by Karthikeyan Krishnamoorthy
p. cm.
Print ISBN 978-1-83881-025-2
Online ISBN 978-1-83881-026-9
eBook (PDF) ISBN 978-1-83881-027-6

For EU product safety concerns:
IN TECH d.o.o., Prolaz Marije Krucifikse Kozulić 3, 51000 Rijeka, Croatia, info@intechopen.com or visit our website at intechopen.com.

Meet the editor

Dr. Karthikeyan Krishnamoorthy is currently a contract professor in the Department of Mechatronics Engineering, Jeju National University, South Korea. He has more than eight years of teaching and research experience in the synthesis and characterization of 2D nanomaterials. His research works are focused on Raman spectroscopic analysis of low-dimensional materials, development of next-generation energy harvesting and storage devices using two-dimensional nanostructures beyond graphene, and advanced EES systems or Electric Vehicles (EVs). With a Hirsch index of 41, Dr. Karthikeyan's great research achievements have resulted in six patent applications and more than 100 research articles in international journals. To his credit, he was recognized as one of Stanford's top scientists (2%) in the world across all scientific (nano) disciplines in 2019.

Contents

Preface

Nanomaterials are a class of materials ranging in size from 1 to 100 nm. They have received much attention in academic research and industrial sectors due to their inherent distinct characteristics and properties. The smaller size and the high surface-to-volume ratio of these materials results in exciting and exceptional properties that are not available in their conventional macroscopic/bulk counterparts. Nanomaterials and nanotechnology are a key area of research and their properties are being explored widely in various laboratories and industrial sectors. Nanomaterials have potential applications in diverse fields such as nanoelectronics, memory devices, optoelectronics, energy storage devices (lithium-ion batteries and supercapacitors), catalysis (photo-, chemical-, surface-), sensors (chemical and biomedical), paints and coatings (anti-corrosive, antibacterial, self-cleaning), nanomedicine, and even agriculture.

The chapters of this book are categorized in a logical manner from preparation to the characterization of nanomaterials for varied applications in energy storage, harvesting, sensing, environmental and biomedical applications, coatings, and the construction industry. This book covers advanced preparation methods for obtaining novel nanomaterials (such as graphene, carbon dots, zinc tin oxide, nanocellulose) and structural nanomaterials (hollow nanomaterials). It also summarizes the utilization of nanostructured materials and their composites towards supercapacitive energy storage and triboelectric energy harvesting sectors. A section is dedicated to biorecognition of nanomaterials in building advanced biosensors (fundamentals, dopamine detection, recognizing cancer biomarkers) using emerging nanostructured materials (2D siloxene and 2D MXene). Another section demonstrates the biological and environmental applications of nanomaterials such as green synthesis routes, antimicrobial activity, drug delivery, bone cement, and polysaccharide-based adsorbents for dye removal. The final section discusses the application of nanostructured materials in self-cleaning coatings and the construction industry.

As the editor of this book, I would like to express my sincere gratitude to the chapter authors for submitting their valuable research findings and previously published and presented review-based articles. This book is a useful resource for scientists working in the fields of interdisciplinary science and technology, experts in a variety of disciplines, both academic and industrial, as well as students.

Karthikeyan Krishnamoorthy
Faculty of Applied Energy System,
Major of Mechatronics Engineering,
Jeju National University,
Jeju, South Korea

Section 1

Preparation of Novel Nanomaterials

Chapter 1

Hydrothermal Synthesis of Zinc Tin Oxide Nanostructures for Photocatalysis, Energy Harvesting and Electronics

Ana Isabel Bento Rovisco, Rita Branquinho, Joana Vaz Pinto, Rodrigo Martins, Elvira Fortunato and Pedro Barquinha

Abstract

The massification of Internet of Things (IoT) and Smart Surfaces has increased the demand for nanomaterials excelling at specific properties required for their target application, but also offering multifunctionality, conformal integration in multiple surfaces and sustainability, in line with the European Green Deal goals. Metal oxides have been key materials for this end, finding applications from flexible electronics to photocatalysis and energy harvesting, with multicomponent materials as zinc tin oxide (ZTO) emerging as some of the most promising possibilities. This chapter is dedicated to the hydrothermal synthesis of ZTO nanostructures, expanding the already wide potential of ZnO. A literature review on the latest progress on the synthesis of a multitude of ZTO nanostructures is provided (e.g., nanowires, nanoparticles, nanosheets), emphasizing the relevance of advanced nanoscale techniques for proper characterization of such materials. The multifunctionality of ZTO will also be covered, with special attention being given to their potential for photocatalysis, electronic devices and energy harvesters.

Keywords: hydrothermal synthesis, zinc tin oxide, nanostructures, nanowires, multifunctionality, sustainability

1. Introduction

Nanotechnology attracted wide attention over the last decades, leading to a very fast development of materials and processing routes. Different areas such as electronics, cosmetics, medicine/biology, optical systems, energy, and many others, have profited from this rapid growth. Having in mind the environmental issues that we are facing in the modern era, the importance of searching for environmentally friendly, recyclable and low cost nanomaterials and fabrication processes is essential [1].

This has been a concern in strategic areas as large area electronics (LAE), one of the fastest growing technologies in the world, with projected market growth from $31.7 billion in 2018 to $77.3 billion in 2029 [2]. LAE includes many segments (e.g., displays, sensors, logic, memory), which are desired to be seamlessly integrated on virtually any object to create smart surfaces. Due to their good electrical properties, transparency, large area uniformity and good mechanical flexibility, oxide thin films have been crucial materials to advance these concepts [3]. Depending on the metal cations (and on the metal to oxygen ratio), metal oxide thin films can be considered as dielectrics, semiconductors or even conductors [4–6]. Owing to their remarkable electrical properties, In-based materials, such as ITO (indium tin oxide) and IGZO (indium-gallium-zinc oxide) are currently the multicomponent oxide conductor and semiconductor thin films with larger market relevance in LAE [4, 7]. However, indium is an expensive material, due to its scarcity and high market value, appearing in the current list (2020) of the critical raw materials from the European Commission [8]. The same applies for gallium, another element of IGZO. Therefore, the replacement of these materials is imperative to assure long-term sustainability [1].

This quest for new oxide materials is naturally also transposed for nanostructures, as their fascinating properties will certainly boost even further the demand for oxide (nano)materials in a plethora of industries. Departing from critical cations, ZnO is perhaps the most widely studied oxide nanostructure. Its properties are nowadays well-known and useful for multiple applications, from photocatalysis, to solar cells or biosensors [9]. It can also be prepared by a multiplicity of methodologies, from vapor- to solution-based processes [10]. Multicomponent oxide nanostructures, particularly those based on sustainable materials, have been significantly less studied, but already show great potential to enhance properties and enlarge the range of applications of oxides. As in thin films, a great advantage of these multicomponent materials is the possibility of tuning their properties by adjusting the cationic ratio [11, 12]. Zinc tin oxide (ZTO) is one of the multicomponent oxides that has been explored and has shown very interesting properties when compared with its binary counterparts (ZnO and SnO_2). In fact, ZTO was already demonstrated to exhibit similar properties to IGZO in low-temperature thin film transistors (TFTs), while avoiding the use of critical raw materials [13].

This chapter provides a literature review on the hydrothermal synthesis of ZTO nanostructures, the main properties of this material, and its applications, highlighting its multifuncionality.

1.1 Hydrothermal synthesis of ZTO nanostructures

Hydrothermal methods have been widely explored and developed in the last years [14]. This method consists in a chemical reaction in an aqueous solution, under high pressure (> 1 atm) and at temperatures usually ranging between 100°C and 300°C. In case of using non-aqueous solvents, the method is called solvothermal. Typically, the solution is kept inside an autoclave and a conventional oven is used as heat source. The pressure inside the autoclave is dependent both on the temperature and the volume used. This allows for a high energy supply for the reactions even at relatively low temperatures. While the typical nucleation and growth mechanism of the oxide nanostructures in these reactions is thought to consist mainly in dissolution–reprecipitation, these mechanisms are often not well understood.

The synthesis of multicomponent oxide materials such as ZTO is usually easier and more efficient by vapor phase methods, such as chemical vapor deposition and thermal evaporation, than by solution processes, due to the higher temperatures of the former. However, vapor phase methods present drawbacks that are important to consider, such as high temperatures (>700°C) and high costs. On the other hand, while inexpensive and simple, the hydrothermal technique still allows for a well-controlled synthesis of the desired nanostructures' shape and structure with high reproducibility, thus presenting as an excellent alternative to the conventional physical methods [15, 16]. Additionally, while conventional ovens are typically used as the heat source, microwave-assisted synthesis started recently to be widely explored, enabling reduced synthesis duration due to its more efficient and more homogeneous heat transfer process [17].

1.2 Overview on ZTO nanostructures produced by hydrothermal synthesis

ZTO appears commonly in two main forms, a stable one, Zn_2SnO_4, and $ZnSnO_3$, a metastable phase. The stable Zn_2SnO_4 phase is an orthostannate with and inverse spinel structure and is a n-type semiconductor with a band gap of 3.6 eV [18]. $ZnSnO_3$, the metastable phase, can have a rhombohedral structure or a perovskite structure either orthorhombic (−orth) or ordered face centered structure (−fcc) [14, 19]. This phase is a well-known piezo/ferroelectric material and presents a band gap of 3.9 eV [18]. Several ZTO nanostructures such as nanoparticles, octahedrons, nanocubes, nanowires, and nanoflowers, have been produced by hydrothermal synthesis, appearing in both $ZnSnO_3$ and Zn_2SnO_4 phases. **Figure 1** shows examples of Zn_2SnO_4 and $ZnSnO_3$ nanostructures with different morphologies and dimensions (0D, 1D, 2D and 3D) produced by hydrothermal synthesis.

Lehnen *et al.*, for example, reported very small Zn_2SnO_4 quantum dots (with diameters below 30 nm), produced with a microwave-assisted hydrothermal synthesis, followed by high-temperature annealing [20]. Numerous other reports on Zn_2SnO_4 nanoparticles have been shown (**Figure 1a**), either using standard hydrothermal synthesis or solvothermal synthesis [16, 21–28]. Regarding $ZnSnO_3$ nanoparticles (**Figure 1b**), several hydro and solvothermal routes have been reported for its synthesis [29–32]. For instance, Beshkar *et al.* reported the use of the Pechini method at 80°C to synthesize fcc-$ZnSnO_3$ nanoparticles, followed by a calcination at 700°C for 2 h [33].

Concerning 1D structures, while several reports on Zn_2SnO_4 nanowires exist, these consist essentially in vapor phase methods, more specifically in thermal evaporation at high temperature (>750°C) [19], showing the difficulty in obtaining the stable phase of ZTO in the nanowire form [34]. This is emphasized by the fact that there are only a few reports for Zn_2SnO_4 nanowires from hydrothermal synthesis, mostly assisted by seed-layers. For example, Zn_2SnO_4 nanowires were grown on a stainless steel seed-layer and from Mn_3O_4 nanowires [35, 36]. Zn_2SnO_4 nanorods by hydrothermal synthesis were also reported by Chen *et al.* (**Figure 1c**), but only organized in 3D flowerlike superstructures [37]. Regarding $ZnSnO_3$, only a few reports for nanowires exist, also consisting typically in physical processes (carbon-thermal reaction, thermal evaporation or CVD processes) [38, 39]. For hydrothermal processing of $ZnSnO_3$ nanowires seed-layers are typically used. Lo *et al.* employed an FTO thin film as seed-layer for this end (**Figure 1d**) [40–42]. A different approach was reported by Men *et al.* who transformed ZnO nanowires into $ZnSnO_3$ nanowires by a hydrothermal synthesis [41]. Recently, our group demonstrated for the first time $ZnSnO_3$ nanowires obtained by an one-step hydrothermal synthesis without employing any seed-layer (**Figure 1e**) [32, 43].

2D structures of ZTO have also been reported. Joseph *et al.* synthesized fcc-$ZnSnO_3$ flakes by a hydrothermal method at only 100°C [44]. Guo *et al.* produced orth-$ZnSnO_3$ nanoplates (**Figure 1f**) by a hydrothermal process at 260°C for 24 h [45]. Chen *et al.* obtained orth-$ZnSnO_3$ nanosheets through a hydrothermal synthesis at 180°C for 12 h, where a precipitate of $ZnSn(OH)_6$ was achieved followed by a calcination at 600°C for 3 h [46]. Zn_2SnO_4 nanoplates have also been reported, for example, by Cherian *et al.* (**Figure 1g**) [34].

There are also several reports regarding 3D ZTO nanostructures. Gao *et al.* reported the synthesis of $ZnSnO_3$ hollow spheres (**Figure 1h**) by hydrothermal synthesis at 120°C for 3 h [47]. A commonly reported shape for ZTO nanostructures

Figure 1.
Multiple ZTO nanostructures obtained by hydrothermal synthesis, analyzed by scanning electron microscopy (SEM): (a) Zn_2SnO_4 nanoparticles, from reference [28]; (b) fcc-$ZnSnO_3$ nanoparticles produced by our group; (d) rhombohedral-$ZnSnO_3$ nanowires growth on FTO seed-layer, reprinted with permission from [42], copyright (2020) American Chemical Society; (e) orth-$ZnSnO_3$ nanowires synthesized without employing seed-layers (in form of powder), from reference [43]; (f) orth-$ZnSnO_3$ nanoplates, reprinted with permission from [45], copyright (2020) American Chemical Society; (h) orth-$ZnSnO_3$ hollow spheres, reprinted with permission from [47], copyright (2020) American Chemical Society; (i) orth-$ZnSnO_3$ nanocubes produced by our group; and by transmission electron microscopy (TEM): (c) Zn_2SnO_4 nanorods, reprinted with permission from [37], copyright (2020) American Chemical Society; and (g) Zn_2SnO_4 nanoplates, reprinted with permission from [34], copyright (2020) American Chemical Society.

is the nanocube shape (**Figure 1i**). For instance, Chen *et al.* reported a synthesis which could result in $ZnSnO_3$ nanocubes or $ZnSnO_3$ nanosheets, depending on the processing temperature [46]. The octahedron shape is also common, and octahedrons of Zn_2SnO_4 have been reported by several groups, being these identified as the most stable phase and shape for ZTO nanocrystals. Zn_2SnO_4 octahedrons constituted by nanoplates can also be formed [48].

While **Figure 1** shows the wide range of possibilities offered by hydrothermal synthesis within the ZTO system, it is challenging to obtain structures with a targeted phase ($ZnSnO_3$ or Zn_2SnO_4 in this case) and shape (e.g. nanosheet or nanowire) [40, 49]. For this end, a comprehensive tailoring of the synthesis parameters is required.

2. Research methods

Usually the hydrothermal synthesis of ZTO nanostructures is performed inside a teflon-lined stainless-steel autoclave using a conventional oven as heating source. Nevertheless, as previously shown, there are already a few examples of microwave-assisted hydrothermal synthesis of ZTO nanostructures. As an example of a typical method, our synthesis starts with the dissolution of the zinc and tin precursors separately in 7.5 mL of deionized water, followed by their mixture. Then a surfactant (ethylenediamine, EDA) is added, and the solution is magnetically stirred for 30 minutes. The last step is the addition of the mineralizer agent (NaOH). It is observed that milling the precursors before their dissolution in water leads to a more homogeneous result. After the solution preparation, it is transferred into the autoclave and kept in the oven for 24 h at 200°C. After the synthesis, the resultant precipitate (comprising the nanostructures) should be washed several times with deionized water and isopropyl alcohol, alternately, and centrifuged at each time. The nanostructures are usually dried at ≈60°C, in vacuum, for a at least 2 hours [32, 43].

3. Results and discussion

3.1 Growth mechanism of ZTO nanostructures

3.1.1 Tailoring the chemico-physical parameters of the hydrothermal synthesis

Understanding the influence of each synthesis parameter is a key step in achieving the desired structures. Specifically, considering seed-layer free processes allows evaluation of the intrinsic influence of each synthesis parameter on the nanostructures' growth. Moreover, the solvent also plays a major role in the process strongly determining the dissolution and diffusion of the species during the synthesis. When the precursors' solubility is not high enough, precluding an efficient reaction, mineralizer agents can be used (NaOH, KOH, etc.) to increase the solubility of the species [50, 51].

To understand the growth within Zn:Sn:O system it is essential to revise the main equations related with the chemical reactions behind each ZTO phases. The chemical reaction processes for the formation of $ZnSnO_3$ nanostructures have been represented in the literature by the following equations [52]:

$$Zn^{2+} + Sn^{4+} + 6OH^- \rightarrow ZnSn(OH)_6 \quad (1)$$

$$ZnSn(OH)_6 \rightarrow ZnSnO_3 + 3H_2O \quad (2)$$

Regarding Zn_2SnO_4, its formation has been described by different reactions depending on the precursors and solvents involved in the synthesis. For example, Li *et al.* represented the chemical reaction of Zn_2SnO_4 nanowires through the equations below, which have been the most common in literature [35].

$$Zn^{2+} + Sn^{4+} + 6OH^- \rightarrow ZnSn(OH)_6\downarrow \quad (3)$$

$$Zn^{2+} + 4OH^- \rightarrow Zn(OH)_4^{2-} \quad (4)$$

$$ZnSn(OH)_6 + Zn(OH)_4^{2-} \rightarrow Zn_2SnO_4\downarrow + 4H_2O + 2OH^- \quad (5)$$

On the other hand, Fu *et al*. employed a different synthesis method to avoid the use of NaOH, using four different amines (surfactants) instead [53], represented as:

$$Sn^{4+} + 3H_2O \leftrightarrow H_2SnO_3 + 4H^+ \quad (6)$$

$$H_2SnO_3 \rightarrow SnO_2\downarrow + H_2O \quad (7)$$

$$Sn^{4+} + 6OH^- \rightarrow Sn(OH)_6^{2-} \quad (8)$$

$$Zn^{2+} + 4OH^- \rightarrow Zn(OH)_4^{2-} \quad (9)$$

$$Sn(OH)_6^{2-} + Zn(OH)_4^{2-} \rightarrow Zn_2SnO_4\downarrow + 4H_2O + 6OH^- \quad (10)$$

Several reports show that ZnO and SnO_2 crystals can co-exist with ZTO nanostructures when the synthesis parameters differ, even if slightly, from the ideal conditions for ZTO formation. Usually the formation of ZnO (Eq. 11) and SnO_2 (Eqs. 12–14) is associated with the two alkaline concentration extremes, higher and lower, respectively [54].

$$Zn(OH)_4^{2-} \rightarrow ZnO\downarrow + H_2O + 2OH^- \quad (11)$$

$$Sn^{4+} + 3H_2O \leftrightarrow H_2SnO_3 + 4H^+ \quad (12)$$

$$H_2SnO_3 \rightarrow SnO_2\downarrow + H_2O \quad (13)$$

$$H^+ + OH^- \rightarrow H_2O \quad (14)$$

In fact, in our previous work on seed-layer free synthesis of ZTO nanowires, using zinc and tin chloride precursors at a fixed concentration ratio it was shown that while for lower NaOH concentrations SnO_2 nanoparticles were obtained, for higher NaOH concentrations ZnO nanowires (mixed with fcc-$ZnSnO_3$ nanoparticles) were achieved, whereas intermediate NaOH concentrations yielded $ZnSnO_3$ nanowires [32]. As shown in **Figure 2**, a similar trend is seen even when increasing only the NaOH concentration (keeping the precursors' concentration fixed). This suggests that there is an optimal concentration of the mineralizer. These results agree with those reported by Zeng *et al.*, [54] however, while the authors suggest specific values of pH for obtaining the different structures (SnO_2, Zn_2SnO_4 and ZnO), in our case the pH is much higher due to the presence of ethylenediamine (EDA) which yields a pH of at least ≈ 12, showcasing the trend specifically with the variation of the NaOH concentration and not necessarily the overall pH [32].

As mentioned, the precursors' solubility is a key factor to achieve a well-controlled synthesis. Our previous work showed that for different zinc precursors (zinc chloride or zinc acetate), maintaining the same tin precursor (tin chloride), the reaction differs, being slower and less homogeneous when using zinc acetate, due its lower solubility in the EDA surfactant. The use of surfactants, such as EDA, cetrimonium bromide (CTAB) and sodium dodecyl sulfate (SDS), is very common specially when aiming to induce the growth of 1D nanostructures. Surfactants act as

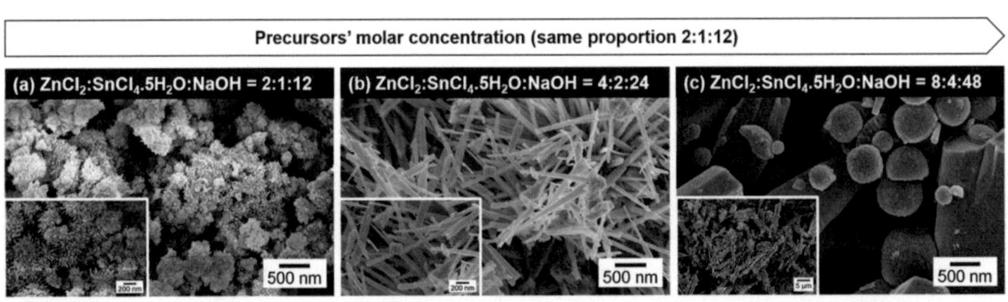

Figure 2.
SEM images of resultant nanostructures from synthesis with different precursors' molar concentrations, i.e., $ZnCl_2$:$SnCl_4$.$5H_2O$:NaOH of (a) 2:1:12 M, (b) 4:2:24 M (from reference [43]) and (c) 8:4:48 M, respectively, while maintaining the same proportion between them. Increasing the precursors' molar concentrations the materials obtained follow the common trend when increasing only the NaOH concentration: SnO_2 nanoparticles, orth-$ZnSnO_3$ nanowires and ZnO nanowires (mixed with fcc-$ZnSnO_3$ nanoparticles).

directing growth agents as their molecules aggregate to the surface of the metallic atoms inducing the growth of specific structures/shapes. The solubility of each precursor in the solvents is a key factor for achieving a better synthesis efficiency and homogeneity. This also influences the Zn to Sn precursor ratio required to optimize the achievement of the desired nanostructures.

The duration and temperature of the synthesis are also crucial to determine the achieved nanomaterials. Several reports showed that below 180°C no ZTO phases are obtained, with the intermediate phase $ZnSn(OH)_6$ being produced instead [43, 45]. Zeng *et al.* showed that to obtain Zn_2SnO_4 nanostructures a temperature of at least 200°C and 20 h of synthesis are necessary [54]. Meanwhile, Guo *et al.* observed that 12 h at 260°C are required to produce orth-$ZnSnO_3$ nanoplates for that specific solution process [45]. In our work on the synthesis of orth-$ZnSnO_3$ nanowires, it was observed that syntheses with 12 h at 200°C were necessary for a predominant growth of nanowires. However, for very long synthesis, or at higher temperatures (220°C), the decomposition of the $ZnSnO_3$ phase into the more stable phases starts to occur. It was also concluded that lower energy levels favor the growth of the more energetically stable phases (Zn_2SnO_4, ZnO and SnO_2), the metastable $ZnSnO_3$ is achievable for intermediate energy levels, and for higher energy levels the decomposition (into Zn_2SnO_4 and SnO_2) of $ZnSnO_3$ starts to occur. Higher solution volumes, corresponding to higher pressure in the synthesis, were found to be necessary for obtaining the $ZnSnO_3$ phase. A general growth mechanism for orth-$ZnSnO_3$ nanowires and Zn_2SnO_4 nanoparticles was proposed and is shown in **Figure 3** [43].

While tailoring the chemico-physical parameters is always necessary, the use of a seed-layer material, usually a thin film, can be very effective in strongly inducing the growth of a desired structure by means of an epitaxial growth mechanism. This approach is commonly used when 1D structures are aimed, as briefly mentioned in the previous section. The selection of the seed material depends on the desired material and structure (phase and shape) and while several reports for different structures exist, the relation between different seed materials and grown structures was not detailed yet in literature. This depends on a complex interrelation between preferential epitaxial growth and thermodynamical stability of the multiple phases and shapes within the Zn-Sn-O system. While the seed-layer route presents advantages for specific applications such as gate-all-around transistors or photocatalysis [55, 56], its absence also brings numerous benefits. For instance, one of the main

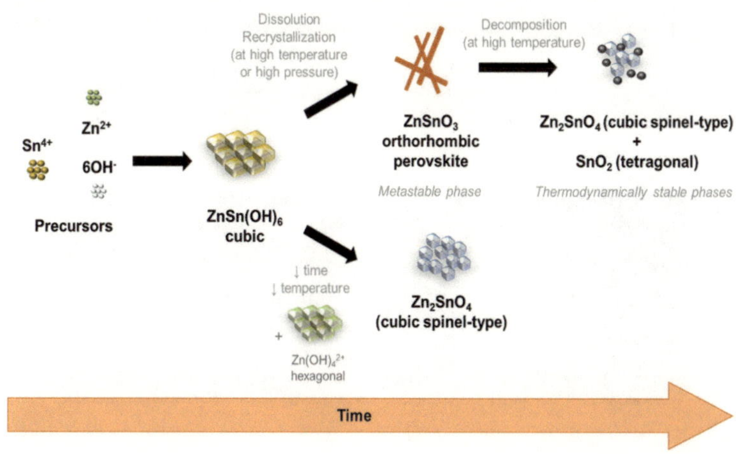

Figure 3.
Schematic of the growth mechanism of ZTO nanostructures ($ZnSnO_3$ nanowires and Zn_2SnO_4 nanoparticles) on a hydrothermal synthesis as a function of the energy available and the duration of the synthesis. From reference [43].

issues related with the use of seed-layers is the common residuals incorporated in the nanostructures, which are usually undesired for the applications. Also, without seed-layers the synthesis is less complex and this approach brings higher degree of freedom concerning the integration of nanostructures into devices [14, 32, 54].

3.1.2 ZTO phase transformations

Obtaining a single phase and shape of a multicomponent oxide as ZTO is highly desirable due to the different characteristic properties of each phase and shape, still by a hydrothermal synthesis is a challenging process as shown in the last section. In addition, the proper identification of the different possible phases obtained is a difficult task.

As previously presented, ZTO can grow in two different structural phases: Zn_2SnO_4 and $ZnSnO_3$ (fcc, orth and rhombohedral). Their identification by XRD analysis is challenging since both phases and intermediary compounds show very similar diffraction patterns. While the fcc-$ZnSnO_3$ (ICDD 00–011-0274) has a similar pattern to that of the intermediate phase $ZnSn(OH)_6$ (ICDD 01–073-2384), the orth-$ZnSnO_3$ (ICDD 00–028-1486) pattern can be confused with that of a mixture of Zn_2SnO_4 and SnO_2. In fact, the 00–028-1486 card was deleted from the ICDD database for this reason. **Figure 4** shows the XRD peaks of these phases. For clarification, the orth-$ZnSnO_3$ identification was performed by peak indexation, using both treor and dicvol methods, for which the determined crystalline structure was proven to be orthorhombic [32].

As previously stated, temperature conditions can induce different phase transformations. For instance, Bora *et al.* studied the phase transformation of fcc-$ZnSnO_3$ nanocubes into the inverse spinel Zn_2SnO_4 through Raman analysis during *in-situ* annealing treatment [57]. In this study the phase transformation occurred at 500°C.

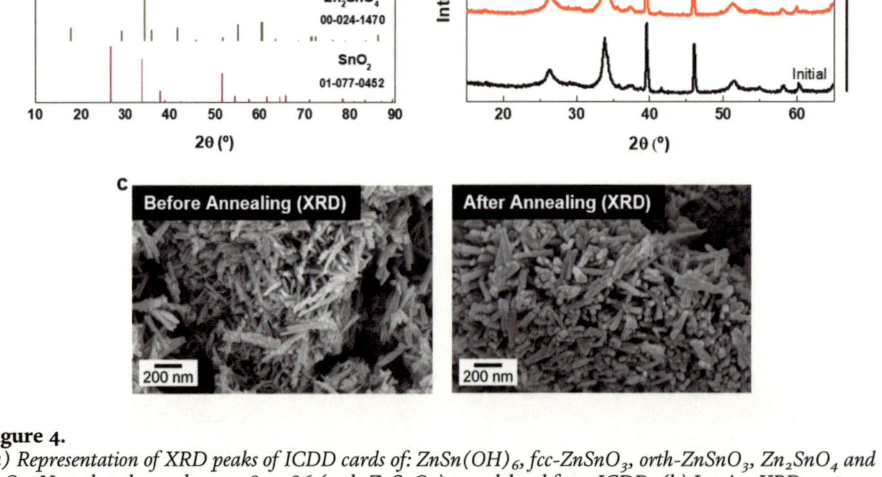

Figure 4.
(a) Representation of XRD peaks of ICDD cards of: $ZnSn(OH)_6$, fcc-$ZnSnO_3$, orth-$ZnSnO_3$, Zn_2SnO_4 and SnO_2. Note that the card 00–028-1486 (orth-$ZnSnO_3$) was deleted from ICDD. (b) In-situ *XRD patterns of $ZnsnO_3$ nanowires during annealing until 850°C. (c) SEM images of the $ZnsnO_3$ nanowires before and after the* in-situ *annealing experiments in XRD.*

Phase transformation in the $ZnSnO_3$ nanowires, synthesized by our group, was investigated by recording XRD patterns in the course of *in-situ* annealing treatment up to 850°C. **Figure 4b** shows the XRD patterns at different temperatures, where no phase transformation is observable bellow 750°C. At 850°C the characteristic peaks of Zn_2SnO_4 and SnO_2 start to be more pronounced, suggesting the phase transformation described in **Figure 3**. Nevertheless, a nanowire-like morphology is still obtained after this in-situ annealing experiment (**Figure 4c**), which was somehow unexpected from the experimental results used to propose the growth mechanism shown in **Figure 3**.

Thermogravimetry (TG) and differential scanning calorimetric (DSC) measurements up to 1350°C were also performed on $ZnSnO_3$ nanowires to shed light into this. A clear transformation occurred at ≈ 570°C with a mass loss of ≈ 4% (**Figure 5**), which can be attributed to the expected decomposition of $ZnSnO_3$ into Zn_2SnO_4 and SnO_2. Through XRD patterns (**Figure 5b**) the orth-$ZnSnO_3$ phase is identified before the annealing, while after the annealing a predominance of SnO_2 is noticeable (mixed with Zn_2SnO_4). SEM images presented in **Figure 5c** show the nanowires before and after the annealing. After annealing, larger and rounder structures are observed for which energy dispersive X-ray spectroscopy (EDS) analysis showed a predominance of Sn (Sn/Zn ratio of 14.5), in agreement with the XRD analysis.

The difference of the decomposition temperature observed between the DSC and the XRD annealing treatments can probably be attributed to the annealing process in both techniques. While the XRD annealing is performed through the heating of a platinum foil (where the nanostructures are placed), in DSC the nanostructures are placed in a melting pot, leading to a more efficient heating and faster decomposition of the $ZnSnO_3$ nanowires.

These results show that when annealing processes are demanded to improve the $ZnSnO_3$ crystallinity, it is important to consider phase transformations carefully. Furthermore, it is noticeable that the temperatures to achieve these phase

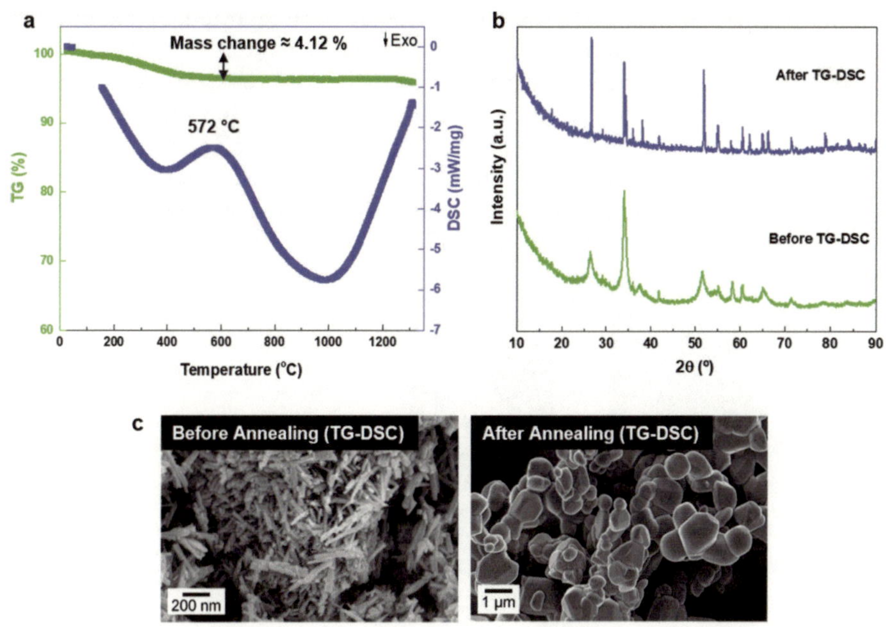

Figure 5.
(a) TG and DSC curves of as-prepared $ZnSnO_3$ nanowires at a heating rate of 10°C/min at N_2 atmosphere. (b) XRD patterns and (c) SEM images of $ZnSnO_3$ nanowires before and after annealing (TG-DSC measurements).

transformations as a post-synthesis treatment are significantly larger than those required during hydrothermal synthesis, owing to the higher energy provided during synthesis due to the combined effect of temperature and pressure.

3.2 Physico-chemical properties of ZTO nanostructures

The wide array of ZTO nanostructures present different physico-chemical properties which are imposed not only by the structures' shape but also by their phase (Zn_2SnO_4 or $ZnSnO_3$).

Concerning the optical properties, ZTO is a wide band gap semiconductor, with reported band gap values of 3.46–3.6 eV for Zn_2SnO_4 and 3.6–3.9 eV for $ZnSnO_3$ [18, 32, 35, 58]. Nevertheless, these values are not only dependent on the phase, but also on the shape and size of the nanostructures, with higher band gaps for smaller particles due to the quantum confinement effect [16].

While optical properties of nanostructures can be determined simply, their electrical properties are much more challenging to access, especially when considering the properties of a single nanostructure. For this reason, there are only a few reports on electrical characterization of single ZTO nanostructures. While most reports are focused on nanowires with lengths >10 μm (mostly produced by physical processes), smaller $ZnSnO_3$ nanowires (lengths <1 μm), produced in our group by hydrothermal synthesis, were probed individually by using nanomanipulators inside SEM, as shown in **Figure 6a**. For these, an average resistivity (in vacuum) of 7.80 ± 8.63 kΩ·cm was achieved [32, 59]. When compared to the ≈73 Ω·cm reported by Xue *et al.* for $ZnSnO_3$ nanowires produced by thermal evaporation (990°C), this resistivity is significantly higher [60], which can be attributed to the higher defect density expected for lower temperature (200°C) and solution-based processes. Concerning Zn_2SnO_4 nanowires, Karthik *et al.* reported a resistivity of 6 Ω·cm in vacuum for nanowires synthesized by vapor phase methods at 900°C [61]. Moreover, for Zn_2SnO_4 nanostructures, which are an n-type semiconductor, mobilities higher than 112 $cm^2V^{-1}s^{-1}$ have already been reported, highlighting the relevance of using this material for electronic applications [62].

The $ZnSnO_3$ phase is well-known for its piezoelectric properties. A piezoelectric polarization along the c-axis of ≈59 μC/cm^2 was reported by Inaguma *et al.* for $ZnSnO_3$, being much higher than the ≈5 μC/cm^2 reported for ZnO [64, 65]. Moreover, the piezoelectric constants of individual $ZnSnO_3$ and ZnO nanowires produced by hydrothermal synthesis were recently determined by piezoresponse force microscopy (PFM) measurements as 23 pm/V and 9 pm/V, respectively

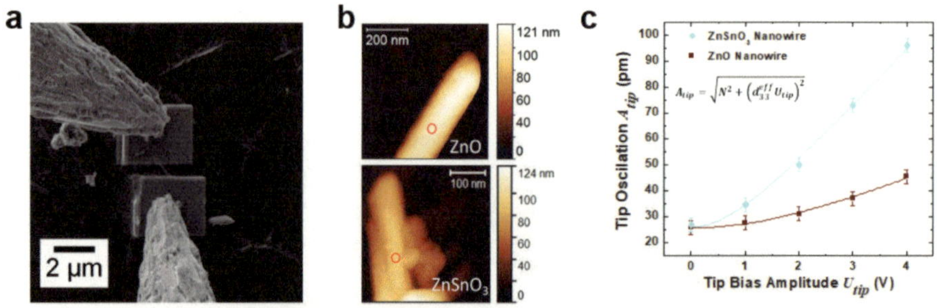

Figure 6.
(a) SEM image showing the tungsten tips of the nanomanipulators, which are contacting in-situ *deposited Pt electrodes for the electrical characterization of a single $ZnSnO_3$ nanowire. Atomic force microscopy characterization of individual ZnO and $ZnSnO_3$ nanowires: (b) topographies in noncontact mode, and (c) contact mode tip oscillation as a function of tip-bias ac-voltage. Reprinted with permission from [63]. Copyright 2020 American Chemical Society.*

(**Figure 6b** and **c**) [63]. The enhanced piezoelectric properties reported for $ZnSnO_3$ are related with the higher displacement of the Zn atom in the ZnO_6 octahedral cell when compared to the one of the Sn atom in the SnO_6 octahedral cell, leading to a higher polarization along the c-axis [66]. Even when compared with other 1D nanostructures produced by hydrothermal synthesis, only the piezoelectric constant of the well-known $BaTiO_3$ (31.1 pm/V) and $LiNbO_3$ (25 pm/V) exceeds the value reported for $ZnSnO_3$. Having sustainability in mind, $ZnSnO_3$ is then a very good alternative to both $BaTiO_3$ and $LiNbO_3$ as these contain critical raw materials [63].

Table 1 summarizes the optical, electrical, and piezoelectric properties of some of the most typical oxide semiconductor nanostructures. These properties show the potential of ZTO compared with other binary and ternary compounds to achieve the desired multifunctionality to meet the concepts of IoT and smart surfaces while avoiding the use of critical raw materials.

3.3 Application of ZTO nanostructures

The multicomponent nature, together with the wide range of different ZTO nanostructures provide this material system with truly impressive multifunctionality, which will be briefly covered next, mostly focusing on photocatalysis, energy harvesting and electronic applications.

3.3.1 Photocatalysis and piezo(photo)catalysis

Industrial actions and human activities play a negative environmental impact, raising water pollution [80]. Oxide nanostructured materials present great advantages for breakdown of water pollutants, as their band gaps are close to the visible light range and they have high surface-to-volume ratios [81]. Moreover, multicomponent oxides such as ZTO have a higher stability in aqueous environments when compared with binary compounds, which is significantly advantageous for photocatalytic applications [82].

The mechanism of photocatalytic activity of ZTO under UV light can be represented by the equations below [14, 54, 83]:

$$ZTO \rightarrow e^- + h^+ \tag{15}$$

$$e^- + h^+ \rightarrow energy \tag{16}$$

Properties / Material	Band gap (eV)	Resistivity (Ω·cm)	Piezoelectric constant or polarization	References
SnO_2	3.60	2.17	n/a	[67]
TiO_2	3.00–3.40	1.09	n/a	[68, 69]
ZnO	3.37	$1.4–2 \times 10^{-4}$ *	5 nC/cm^2, 9–26.7 pm/V	[63, 70–72]
IGZO	3.67	$>10^6–10^{-3}$ *	n/a	[73–75]
$BaTiO_3$	3.47	$6.25 \times 10^9–2.33 \times 10^8$	31.1 pm/V	[76–79]
$ZnSnO_3$	3.90	$73–1.4 \times 10^3$	59 nC/cm^2, 23 ± 4 pm/V	[18, 32, 60, 63]
Zn_2SnO_4	3.30–3.70	1.6	n/a	[18, 58]

*The properties marked with * are referent to the bulk materials. Abbreviations: n/a – not applicable.*

Table 1.
Optical, electrical and piezoelectric properties of some of the most typical oxide semiconductor nanostructures.

$$h^{+} + H_2O \rightarrow H^{+} + OH^{\cdot} \quad (17)$$

$$h^{+} + OH^{-} \rightarrow OH^{\cdot} \quad (18)$$

$$e^{-} + O_2 \rightarrow O_2^{\cdot -} \quad (19)$$

$$O^{\circ}{}_2^{-} + H^{+} \rightarrow HO_2^{\cdot} \quad (20)$$

$$\left(OH^{\circ}, O^{\circ}{}_2^{-}, HO^{\circ}{}_2\right) + dye \rightarrow degradation\ products \quad (21)$$

Considering photocatalytic activity under visible light, Jain *et al.* [84] proposes the following equations:

$$Zn_2SnO_4 + hv \rightarrow Zn_2SnO_4\left(e_{cb}^{-}\right) + Zn_2SnO_4\left(h^{+}\right) \quad (22)$$

$$Zn_2SnO_4\left(e_{cb}^{-}\right) + O_2 \rightarrow Zn_2SnO_4 + O_2^{-\cdot} \quad (23)$$

$$Zn_2SnO_4\left(h^{+}\right) + OH^{-} \rightarrow OH^{\cdot} \quad (24)$$

$$O_2^{-\cdot} + H_2O \rightarrow OH^{-} + HO_2^{\cdot} \quad (25)$$

$$HO_2^{\cdot} + H_2O \rightarrow OH^{\cdot} H_2O_2 \quad (26)$$

$$H_2O_2 \rightarrow 2OH^{\cdot} \quad (27)$$

$$Dye + OH^{\cdot} \rightarrow CO_2 + H_2O + degradation\ products \quad (28)$$

Zn_2SnO_4 nanocrystals were used for the degradation of 50% of reactive red 141 dye in 270 min under sunlight [85]. Different $ZnSnO_3$ structures such as nanowires and nanoplates were already used as photocatalysts for organic pollutants (for example, methylene blue and rhodamine B) [33, 40, 86]. Due to its high optical band gap (3.3–3.9 eV) UV light is usually required to photoactivate this material. Nevertheless, fcc-$ZnSnO_3$ nanoparticles were already reported with a very satisfactory photocatalytic behavior on methylene blue degradation under visible light (0.0156 min^{-1}) [81].

Alternatives to the conventional photocatalytic approach have also been explored, making use of the piezoelectric properties of materials such as $ZnSnO_3$ (nanowires and nanoplates) for piezocatalysis (in the dark) [87] or for piezophotocatalysis (under illumination) [42, 87, 88]. Indeed, piezoelectricity and ferroelectricity (associated with perovskite structures) have shown to play an important role in photocatalysis, since the photogeneration of electron–hole pairs is enhanced by the dipole moment formed by the polarization electric field across polar materials [89, 90]. A schematic representation of the piezocatalytic mechanism is presented in **Figure 7b** and shows the influence of the characteristic polarization of the piezoelectric materials, which contributes to the generation of hydroxyl radicals and consequently degradation of rhodamine B. The dye degradation was achieved in 2.5 h, with a degradation rate of 4.5×10^{-2} min^{-1} [87].

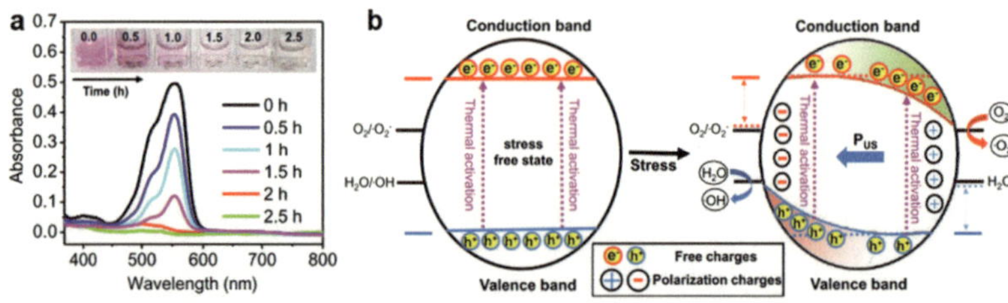

Figure 7.
(a) Piezocatalysis using $ZnSnO_3$ nanoparticles under ultrasound exposure. (b) Schematic of the piezocatalytic mechanism. Reprinted with permission from [87]. Copyright (2020) American Chemical Society.

Other interesting applications of photocatalytic properties have been reported, such as the photocatalytic inactivation of *Escherichia coli* using ZTO nanocubes under visible light. Only a 10% surviving rate was found for the bacteria, whereas the absorption of the visible light was attributed to the inherent surface defects enhancing the absorption edge in the visible region [82]. With this in mind, lower cost methods for nanostructure production (as hydrothermal methods), which typical result in more defective structures, might be advantageous for these applications as defect levels near the band edges may increase the absorption for lower energy levels.

3.3.2 Piezoelectric energy harvesting with ZTO nanostructures

Nanogenerators are devices that can convert external stimulus into electrical energy, being highly interesting for smart and self-sustainable surfaces, as they can be used for sustainable energy sources, biomedical systems and smart sensors [91]. Due to its excellent ferroelectric and piezoelectric properties, different $ZnSnO_3$ nanostructures (i.e., nanowires, nanoplates, nanocubes) have been widely explored for energy harvesting devices and sensitive human motion sensors, through their piezoelectric (induction of electrical charge by the applied mechanical strain) and piezoresistive (electrical resistivity change by the applied mechanical strain) effects, respectively [45, 66, 92–94]. The fcc-$ZnSnO_3$ nanocubes have been the most popular ZTO structures for these applications. For instance, Wang *et al*. reported the nanogenerators of fcc-$ZnSnO_3$ nanocubes mixed with polydimethylsiloxane (PDMS), reaching a maximum output of 400 V, 28 μA at a current density of 7 $\mu A{\cdot}cm^{-2}$ [95]. While, Paria *et al.* mixed fcc-$ZnSnO_3$ nanocubes with polyvinyl chloride (PVC), achieving a maximum output of ≈40 V and ≈1.4 μA, corresponding to a power density of 3.7 $\mu W{\cdot}cm^{-3}$ (**Figure 8a**) [94].

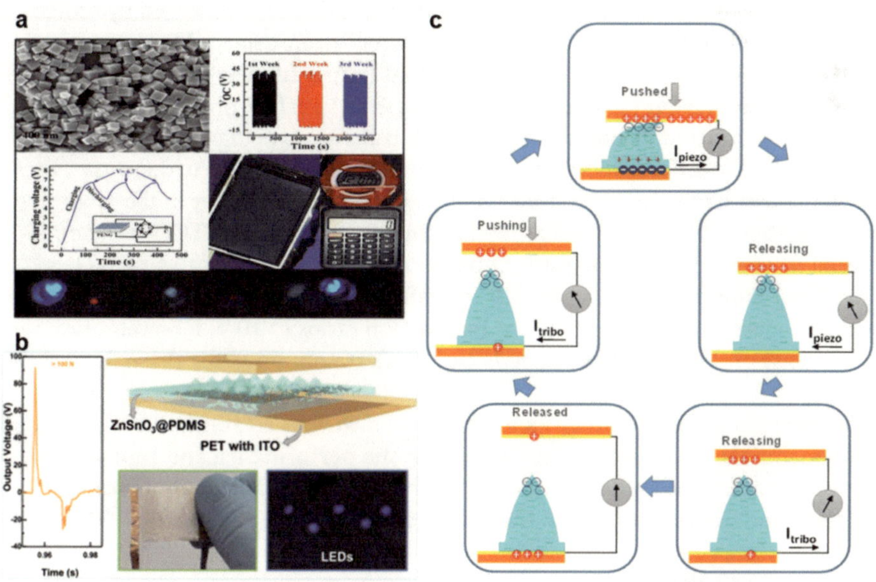

Figure 8.
Hybrid nanogenerators of: (a) a composite film based on fcc-$ZnSnO_3$ nanocubes and PVC. Reprinted with permission from [94], copyright 2020 American Chemical Society; and (b) a composite film based on orth-$ZnSnO_3$ nanowires and PDMS. (c) Schematic of the charge generation mechanism in the micro-structured devices of (b). Images (b) and (c) were reprinted with permission from [63], copyright 2020 American Chemical Society.

$ZnSnO_3$ nanoplates were also applied for nanogenerators. Guo *et al.* reported produced nanogenerators fabricated with orth-$ZnSnO_3$ nanoplates embedded in flat films of PDMS, reaching voltage and current outputs of 20 V and 0.6 μA, respectively, under bending stress [45]. More recently in our group, orth-$ZnSnO_3$ nanowires were mixed with PDMS to fabricate nanogenerators of micro-structured composites (**Figure 8b**) [63]. In the same work, a charge generation and displacement mechanism was proposed, as depicted in **Figure 8c**. Briefly, the micro-structures induced in PDMS are suggested to improve the force delivery to the nanowires, enhancing its piezoelectric signal, while bringing also a triboelectric contribution to the nanogenerator output. This results in an output voltage, current and instantaneous power of approximately 9 V, 1 μA and 3 $\mu W \cdot cm^{-2}$, respectively, when applying a force of only 10 N. For higher forces the devices were capable to reach outputs around 120 V and 13 μA, which was shown to be enough energy to light up LEDs and several small electronic devices [63].

3.3.3 Electronic applications

Electronic applications are always a relevant drive for materials. Multicomponent semiconductor nanostructures as ZTO are particularly interesting for these applications, with wide band gap semiconductors allowing for high-power and high-frequency operations [50]. Field-effect transistors (FETs) are the key elements enabling today's electronics, being 1D nanostructures particularly interesting in this regard, given the easiness of confining migratory direction of charge carriers through its length, i.e., between source and drain electrodes. Indeed, 1D nanostructures have already proven great usefulness for the upcoming generations of semiconductors in FETs [96]. While several reports already demonstrated ZTO as a candidate for replacement of IGZO in thin film technologies [13], similarly, ZTO is also one of the most promising multicomponent metal oxides for transistors with nanostructures [62]. Demonstrations of discrete Zn_2SnO_4 nanotransistors have already been made using nanotransfer molding of ZTO inks followed by annealing at 500°C, or by simple pick-and-place approach of drop-casted ZTO nanowires prepared by CVD above 700°C and by thermal evaporation at 1000°C [39, 97, 98]. While the achievement of on/off ratio $\approx 10^6$ and field-effect mobility $\approx 20\ cm^2/Vs$ is a good demonstration of the ZTO's potential, transistors using ZTO nanostructures synthesized by solution processes have not been reported yet. Furthermore, these nanostructures have also been used for the resistive switch layer in the emerging type of memory devices known as memristors. Reports show ZTO as the active material in memristors in the form of both Zn_2SnO_4 nanowires and $ZnSnO_3$ nanocubes, being the latter especially relevant for this application due to its ferroelectric properties. Properties such as high off/on ratios ($>10^5$), long retention times (>5 months) and fast response speeds (<20 ns) are obtained for these devices [99, 100].

Transforming ZTO or other nanostructures into well-established LAE semiconductor materials, while highly desirable from the performance and functionality point of view, will still require significant advances in reliable techniques for alignment and density control in transparent (and flexible) substrates [101].

3.3.4 Other applications

Besides the applications briefly presented above, ZTO nanostructures have also been widely used in sensing applications, with gas sensors being the most popular [102]. Their small crystallite size, high surface-to-volume ratios and surface reactivity result in enhanced sensitivities/selectivity, with multicomponent materials typically presenting smaller response times and superior stabilities compared to binary compounds [103]. Moreover, the implementation of these nanostructures in sensors allows miniaturization of the devices, as well as cost reduction. $ZnSnO_3$ has been reported as an excellent humidity sensor, in different nanostructure forms such as nanoparticles or even in composites of $ZnSnO_3$ nanocubes and Ag nanowires [29, 104]. Additionally, $ZnSnO_3$ nanoparticles were used as electrochemical biosensors for label free sub-femtomolar detection of cardiac biomarker troponin T and a composite of Zn_2SnO_4 nanoparticles and graphene was used for morphine and codeine detection [105, 106]. Recently, Durai *et al.*. reported ultra-selective sensors, based on $ZnSnO_3$ nanocubes modified glassy carbon electrode (GCE), for simultaneous detection of uric acid and dopamine through differential pulse voltammetry technique [107]. Zn_2SnO_4 and $ZnSnO_3$ nanostructures of different shapes such as nanoparticles, nanowires and nanocubes, have also been widely explored as photoconductors [23, 108–111]. While the optical band gap of these materials is typically in the UV energy levels (hence their transparency in visible range), quantum confinement effects or even defect levels near the band edges can be explored to increase the absorption for lower energy levels. Other applications that have been explored using ZTO nanostructures are related with energy storage and conversion. Zn_2SnO_4 has been widely used as photoanode for dye solar cells in different nanostructure morphologies such as nanoparticles and nanowires [21, 35]. Cherian *et al.* reported the performance of nanowires and compared with nanoplates of Zn_2SnO_4 for Li-batteries [34]. Supercapacitors (SC) have also started to be explored using ZTO nanostructures, with Bao *et al.* having reported the use of Zn_2SnO_4/MnO_2 core shell in carbon fibers showing a capacitance of 621.6 $F{\cdot}g^{-1}$ [112].

4. Conclusions

Expanding LAE to IoT and smart surface concepts requires an increasing number of objects to have embedded electronics, sensors and connectivity, driving a demand for compact, smart, multifunctional and self-sustainable technology with low associated costs. While nanomaterials are thought to be able to meet these requirements, playing an important role in the future technological world, low cost and sustainable technologies are demanded. For this, both low cost fabrication methods and sustainable materials must be considered. This chapter shows the versatility of the hydrothermal method to control the growth and morphology of zinc tin oxide (ZTO) nanostructures, and the variety of shapes that can be produced for each of the different ZTO phases. Compared to other preparation methods, especially vapor phase methods, hydrothermal synthesis reveals a large set of advantages from both research and industrial viewpoints. First, while the multitude of parameters to control requires an in-depth understanding of their role in the final products, it also brings enormous flexibility to tune the synthesis process for the desired results. Also, it can be performed at low temperature ($< 200°C$), which is compatible with a wide range of substrates for direct growth, while assuring lower costs. This links perfectly with the demonstrated upscaling capability of hydrothermal synthesis which is a crucial aspect for industrial implementation.

Furthermore, a summary of exciting results that have been reported regarding application in devices of these ZTO nanostructures over the past few years is presented. The multifunctionality of this material system is highlighted by its successful implementation in energy harvesters, photocatalysis, electronic devices, sensors, and others.

Acknowledgements

The authors would like to thank Ana Pimentel for the TG-DSC measurement.

This work is funded by FEDER funds through the COMPETE 2020 Programme and National Funds through the FCT – Fundação para a Ciência e a Tecnologia, I.P., under the scope of the project UIDB/50025/2020, and the doctoral grant research number SFRH/BD/131836/2017. This work also received funding from the European Community's H2020 program under grant agreement No. 716510 (ERC-2016-StG TREND), No. 787410 (ERC-2018-AdG DIGISMART) and No. 685758 (1D-Neon). This work is part of the PhD Thesis in Nanotechnologies and Nanosciences defended by Ana Rovisco at FCT-NOVA entitled "Solution-based Zinc Tin oxide nanostructures: from synthesis to applications" in December 2019.

Conflict of interest

The authors declare no conflict of interest.

Author details

Ana Isabel Bento Rovisco*, Rita Branquinho, Joana Vaz Pinto, Rodrigo Martins, Elvira Fortunato and Pedro Barquinha*
CENIMAT/i3N, Department of Materials Science, NOVA School of Science and Technology (FCT-NOVA) and CEMOP/UNINOVA, NOVA University Lisbon, Campus de Caparica, Caparica, Portugal

*Address all correspondence to: a.rovisco@fct.unl.pt and pmcb@fct.unl.pt

References

[1] Mancini L, Sala S, Recchioni M, Benini L, Goralczyk M, Pennington D. Potential of life cycle assessment for supporting the management of critical raw materials. Int J Life Cycle Assess. 2015;20: 100–16.

[2] Flexible, Printed and Organic Electronics 2019–2029: Forecasts, Players & Opportunities: IDTechEx. 2020.

[3] He Y, Wang X, Gao Y, Hou Y, Wan Q. Oxide-based thin film transistors for flexible electronics. J Semicond. 2018;39(1).

[4] Martins J, Bahubalindruni P, Rovisco A, Kiazadeh A, Martins R, Fortunato E, et al. Bias Stress and Temperature Impact on InGaZnO TFTs and Circuits. Materials (Basel). 2017;10 (12):680.

[5] Fortunato E, Barquinha P, Martins R. Oxide semiconductor thin-film transistors: a review of recent advances. Adv Mater. 2012;24(22):2945–86.

[6] Bahubalindruni PG, Kiazadeh A, Sacchetti A, , et al. Influence of Channel Length Scaling on InGaZnO TFTs Characteristics: Unity Current-Gain Cutoff Frequency, Intrinsic Voltage-Gain, and On-Resistance. J Disp Technol. 2016;12(6):515–8.

[7] Baraton M-I. The Future of TCO Materials: Stakes and Challenges. MRS Proc. 2009;1209:1209-P03–06.

[8] European Commission. COMMUNICATION FROM THE COMMISSION TO THE EUROPEAN PARLIAMENT, THE COUNCIL, THE EUROPEAN ECONOMIC AND SOCIAL COMMITTEE AND THE COMMITTEE OF THE REGIONS. 2020.

[9] Theerthagiri J, Salla S, Senthil RA, Nithyadharseni P, Madankumar A, Arunachalam P, Maiyalagan, T, Kim, H-S. A review on ZnO nanostructured materials: Energy, environmental and biological applications. Nanotechnology. 2019;30:392001.

[10] Baskoutas S. Special Issue: Zinc Oxide Nanostructures: Synthesis and Characterization. Materials (Basel). 2018;11(6):873.

[11] Shankar KS, Raychaudhuri AK. Fabrication of nanowires of multicomponent oxides: Review of recent advances. Mater Sci Eng C. 2005; 25(5–8):738–51.

[12] Facchetti A, Marks TJ, Wiley InterScience (Online service). Transparent electronics : from synthesis to applications. Wiley; 2010. 448 p.

[13] Fernandes C, Santa A, Santos Â, Bahubalindruni P, Deuermeier J, Martins R, et al. A Sustainable Approach to Flexible Electronics with Zinc-Tin Oxide Thin-Film Transistors. Adv Electron Mater. 2018;4(7):1800032.

[14] Baruah S, Dutta J. Zinc stannate nanostructures: hydrothermal synthesis. Sci Technol Adv Mater. 201;12:013004.

[15] Rabenau A. The Role of Hydrothermal Synthesis in Preparative Chemistry. Angew Chemie Int Ed English. 1985;24(12):1026–40.

[16] Annamalai A, Carvalho D, Wilson KC, Lee M-J. Properties of hydrothermally synthesized Zn2SnO4 nanoparticles using Na2CO3 as a novel mineralizer. Mater Charact. 2010;61: 873–81.

[17] Pimentel A, Ferreira S, Nunes D, Calmeiro T, Martins R, Fortunato E. Microwave Synthesized ZnO Nanorod Arrays for UV Sensors: A Seed Layer Annealing Temperature Study. Materials (Basel). 2016;9(4):299.

[18] Miyauchi M, Liu Z, Zhao Z-G, Anandan S, Hara K. Single crystalline zinc stannate nanoparticles for efficient photo-electrochemical devices. Chem Commun. 2010;46:1529–31.

[19] Sun S, Liang S. Morphological zinc stannate: synthesis, fundamental properties and applications. J Mater Chem A. 2017;5:20534–60.

[20] Lehnen T, Zopes D, Mathur S. Phase-selective microwave synthesis and inkjet printing applications of Zn2SnO4 (ZTO) quantum dots. J Mater Chem. 2012;22(34):17732.

[21] Tan B, Toman E, Li Y, Wu Y. Zinc Stannate (Zn2SnO4) Dye-Sensitized Solar Cells. J Am Chem Soc. 2007;129: 4162–3.

[22] Joseph LA, Jeronsia JE, Jaculine MM, Das SJ. Investigations on Structural and Optical Properties of Hydrothermally Synthesized Zn2SnO4 Nanoparticles. Phys Res Int. 2016;2016:1–6.

[23] Wang YF, Ding Y, Zhao JS, Wang X, Li DJ, Li XF. Optimized Zn2SnO4 nanoparticles with enhanced performance for photodetectors and photocatalysts. RSC Adv. 2016;6(73): 69191–5.

[24] Dimitrievska M, Ivetić TB, Litvinchuk AP, Fairbrother A, Miljević BB, Štrbac GR, et al. Eu3+–Doped Wide Band Gap Zn2SnO4 Semiconductor Nanoparticles: Structure and Luminescence. J Phys Chem C. 2016;120 (33):18887–94.

[25] Šepelák V, Becker SM, Bergmann I, Indris S, Scheuermann M, Feldhoff A, et al. Nonequilibrium structure of Zn2SnO4 spinel nanoparticles. J Mater Chem. 2012;22(7):3117.

[26] Wu YS, Chang WK, Jou M. Photocatalytic Analysis and Characterization of Zn2SnO4 Nanoparticles Synthesized via Hydrothermal Method with Na2CO3 Mineralizer. Adv Mater Res. 2010;97–101:19–22.

[27] Sun G, Zhang S, Li Y. Solvothermal Synthesis of Zn2SnO4 Nanocrystals and Their Photocatalytic Properties. Int J Photoenergy. 2014;2014:1–7.

[28] Jia T, Liu M, Yu D, Long F, Mo S, Deng Z, et al. A Facile Approach for the Synthesis of Zn2SnO4/BiOBr Hybrid Nanocomposites with Improved Visible-Light Photocatalytic Performance. Nanomaterials. 2018;8(5):313.

[29] Singh R, Yadav AK, Gautam C. Synthesis and Humidity Sensing Investigations of Nanostructured ZnSnO3. J Sens Technol. 2011;01(04): 116–24.

[30] Modeshia DR, Walton RI. Solvothermal synthesis of perovskites and pyrochlores: crystallisation of functional oxides under mild conditions. Chem Soc Rev. 2010;39(11):4303.

[31] Mageshwari K, Kim TG, Park J. Effect of alkaline concentration on the structural and luminescence properties of ZnSnO3 nanoparticles obtained by facile reflux method. J Mater Sci Mater Electron. 2016;27(4):4093–7.

[32] Rovisco A, Branquinho R, Martins J, Oliveira MJ, Nunes D, Fortunato E, et al. Seed-Layer Free Zinc Tin Oxide Tailored Nanostructures for Nanoelectronic Applications: Effect of Chemical Parameters. ACS Appl Nano Mater. 2018;1(8):3986–97.

[33] Beshkar F, Amiri O, Salehi Z. Synthesis of ZnSnO3 nanostructures by using novel gelling agents and their application in degradation of textile dye. Sep Purif Technol. 2017;184:66–71.

[34] Cherian CT, Zheng M, Reddy M V, Chowdari BVR, Sow CH. Zn2SnO4 Nanowires versus Nanoplates: Electrochemical Performance and

Morphological Evolution during Li-Cycling. ACS Appl Mater Interfaces. 2013;5(13):6054–60.

[35] Li Z, Zhou Y, Bao C, Xue G, Zhang J, Liu J, et al. Vertically building Zn2SnO4 nanowire arrays on stainless steel mesh toward fabrication of large-area, flexible dye-sensitized solar cells. Nanoscale. 2012;4:3490–4.

[36] Zhou T, Liu X, Zhang R, Wang Y, Zhang T. Shape control and selective decoration of Zn2SnO4 nanostructures on 1D nanowires: Boosting chemical–sensing performances. Sensors Actuators B Chem. 2019;290:210–6.

[37] Chen Z, Cao M, Hu C. Novel Zn2SnO4 hierarchical nanostructures and their gas sensing properties toward ethanol. J Phys Chem C. 2011;115:5522–9.

[38] Xue XY, Chen YJ, Wang YG, Wang TH. Synthesis and ethanol sensing properties of ZnSnO3 nanowires. Appl Phys Lett. 2005;86:1–3.

[39] Pang C, Yan B, Liao L, Liu B, Zheng Z, Wu T, et al. Synthesis, characterization and opto-electrical properties of ternary Zn2SnO4 nanowires. Nanotechnology. 2010;21: 465706.

[40] Fang C, Geng B, Liu J, Zhan F. d-fructose molecule template route to ultra-thin ZnSnO3 nanowire architectures and their application as efficient photocatalyst. Chem Commun. 2009;(17):2350.

[41] Men H, Gao P, Zhou B, Chen Y, Zhu C, Xiao G, et al. Fast synthesis of ultra-thin ZnSnO3 nanorods with high ethanol sensing properties. Chem Commun. 2010;46:7581.

[42] Lo M-K, Lee S-Y, Chang K-S. Study of ZnSnO3-Nanowire Piezophotocatalyst Using Two-Step Hydrothermal Synthesis. J Phys Chem C. 2015;119(9):5218–24.

[43] Rovisco A, Branquinho R, Martins J, Fortunato E, Martins R, Barquinha P. Growth Mechanism of Seed-Layer Free ZnSnO3 Nanowires: Effect of Physical Parameters. Nanomaterials. 2019;9(7): 1002.

[44] Joseph J, Saseendran SB, Achary SR, Sukumaran AA, Jayaraj MK. Zinc stannate flakes for optoelectronic and antibacterial applications. Dae Solid State Physics Symposium 2018. 2019. p. 030026.

[45] Guo R, Guo Y, Duan H, Li H, Liu H. Synthesis of Orthorhombic Perovskite-Type ZnSnO3 Single-Crystal Nanoplates and Their Application in Energy Harvesting. ACS Appl Mater Interfaces. 2017;9(9):8271–9.

[46] Chen Y, Yu L, Li Q, Wu Y, Li Q, Wang T. An evolution from 3D face-centered-cubic ZnSnO3 nanocubes to 2D orthorhombic ZnSnO3 nanosheets with excellent gas sensing performance. Nanotechnology. 2012;23(41):415501.

[47] Wang Y, Gao P, Bao D, Wang L, Chen Y, Zhou X, et al. One Pot, Two Phases: Individual Orthorhombic and Face-Centered Cubic ZnSnO3 Obtained Synchronously in One Solution. Inorg Chem. 2014;53(23):12289–96.

[48] Ji X, Huang X, Liu J, Jiang J, Li X, Ding R, et al. Hydrothermal synthesis of novel Zn2SnO4 octahedron microstructures assembled with hexagon nanoplates. J Alloys Compd. 2010;503:L21–5.

[49] Jie J, Wang G, Han X, Fang J, Yu Q, Liao Y, et al. Growth of Ternary Oxide Nanowires by Gold-Catalyzed Vapor-Phase Evaporation. J Phys Chem B. 2004;108:8249–53.

[50] Zhou Z, Lan C, Wei R, Ho JC. Transparent metal-oxide nanowires and their applications in harsh electronics. J Mater Chem C. 2019;7(2): 202–17.

[51] Einarsrud M-A, Grande T. 1D oxide nanostructures from chemical solutions. Chem Soc Rev. 2014;43(7):2187–99.

[52] Kumari V, Patra AK, Bhaumik A. Self-assembled ultra-small zinc stannate nanocrystals with mesoscopic voids via a salicylate templating pathway and their photocatalytic properties. RSC Adv. 2014;4:13626–34.

[53] Fu X, Wang X, Long J, Ding Z, Yan T, Zhang G, et al. Hydrothermal synthesis, characterization, and photocatalytic properties of Zn2SnO4. J Solid State Chem. 2009;182(3):517–24.

[54] Zeng J, Xin M, Li, Wang H, Yan H, Zhang W. Transformation Process and Photocatalytic Activities of Hydrothermally Synthesized Zn2SnO4 Nanocrystals. J Phys Chem C. 2008;112 (11):4159–67.

[55] Guerfi Y, Larrieu G. Vertical Silicon Nanowire Field Effect Transistors with Nanoscale Gate-All-Around. Nanoscale Res Lett. 2016;11(1):210.

[56] Barrocas B, Sério S, Rovisco A, Melo Jorge M. Visible-Light Photocatalysis in Ca0.6Ho0.4MnO3 Films Deposited by RF-Magnetron Sputtering Using Nanosized Powder Compacted Target. J Phys Chem C. 2014;118(1):590–7.

[57] Bora T, Al-Hinai MH, Al-Hinai AT, Dutta J. Phase Transformation of Metastable ZnSnO3 Upon Thermal Decomposition by In-Situ Temperature-Dependent Raman Spectroscopy. J Am Ceram Soc. 2015;98(12):4044–9.

[58] Lei M, Sheng Y, Wan L, Bi K, Huang K, Jia R, et al. A novel self-catalytic route to zinc stannate nanowires and cathodoluminescence and electrical transport properties of a single nanowire. J Alloys Compd. 2016; 657:394–9.

[59] Rovisco A. Solution-based Zinc-Tin Oxide nanostructures : from synthesis to applications [PhD Thesis]. Universidade NOVA de Lisboa; 2019.

[60] Xue XY, Chen YJ, Li QH, Wang C, Wang YG, Wang TH. Electronic transport characteristics through individual ZnSnO3 nanowires. Appl Phys Lett. 2006;88(18):182102.

[61] Karthik KRG, Andreasson BP, Sun C, Pramana SS, Varghese B, Sow CH, et al. Physical and Electrical Properties of Single Zn2SnO4 Nanowires. Electrochem Solid-State Lett. 2011;14:K5.

[62] Kim S, Kim H, Janes DB, Ju S. Interface studies of N2 plasma-treated ZnSnO nanowire transistors using low-frequency noise measurements. Nanotechnology. 2013;24:305201.

[63] Rovisco A, dos Santos A, Cramer T, Martins J, Branquinho R, Águas H, et al. Piezoelectricity Enhancement of Nanogenerators Based on PDMS and ZnSnO 3 Nanowires through Microstructuration. ACS Appl Mater Interfaces. 2020;12(16):18421–30.

[64] Inaguma Y, Yoshida M, Katsumata T. A Polar Oxide ZnSnO3 with a LiNbO3-Type Structure. J Am Chem Soc. 2008;130(21):6704–5.

[65] Dal Corso A, Posternak M, Resta R, Baldereschi A. Ab initio study of piezoelectricity and spontaneous polarization in ZnO. Phys Rev B. 1994; 50:10715–21.

[66] Wu JM, Xu C, Zhang Y, Yang Y, Zhou Y, Wang ZL. Flexible and transparent nanogenerators based on a composite of lead-free ZnSnO3 triangular-belts. Adv Mater. 2012;24 (45):6094–9.

[67] Hernández-Ramírez F, Tarancón A, Casals O, Rodríguez J, Romano-Rodríguez A, Morante JR, et al. Fabrication and electrical characterization of circuits based on

individual tin oxide nanowires. Nanotechnology. 2006;17:5577–83.

[68] Nunes D, Pimentel A, Santos L, Barquinha P, Fortunato E, Martins R. Photocatalytic TiO2 Nanorod Spheres and Arrays Compatible with Flexible Applications. Catalysts. 2017;7 (2):60.

[69] Fàbrega C, Hernández-Ramírez F, Daniel Prades J, J. On the photoconduction properties of low resistivity TiO2 nanotubes. Nanotechnology. 2010;21(44).

[70] Pimentel A, Samouco A, Nunes D, Araújo A, Martins R, Fortunato E. Ultra-Fast Microwave Synthesis of ZnO Nanorods on Cellulose Substrates for UV Sensor Applications. Materials (Basel). 2017;10(11):1308.

[71] Ellmer K. Resistivity of polycrystalline zinc oxide films: Current status and physical limit. J Phys D Appl Phys. 2001;34(21):3097–108.

[72] Zhao M, Wang Z, Mao SX. Piezoelectric Characterization of Individual Zinc Oxide Nanobelt Probed by Piezoresponse Force Microscope. Nano Lett. 2004;4(4):587–90.

[73] Santos L, Nunes D, Calmeiro T, Branquinho R, Salgueiro D, Barquinha P, et al. Solvothermal Synthesis of Gallium–Indium-Zinc-Oxide Nanoparticles for Electrolyte-Gated Transistors. ACS Appl Mater Interfaces. 2015;7(1):638–46.

[74] Barquinha P, Pereira L, Gonçalves G, Martins R, Fortunato E. Toward High-Performance Amorphous GIZO TFTs. J Electrochem Soc. 2009; 156(3):H161.

[75] Zan H-W, Tsai W-W, Chen C-H, Tsai C-C, Meng H-F. 4.3: High Performance a-IGZO TFT with Nano-Dots Doping. SID Symp Dig Tech Pap. 2011;42(1):28–31.

[76] Tang H, Zhou Z, Sodano HA. Large-scale synthesis of BaxSr1−xTiO3 nanowires with controlled stoichiometry. Appl Phys Lett. 2014;104 (14):142905.

[77] Huang X, Xie L, Jiang P, Wang G, Liu F. Electrical, thermophysical and micromechanical properties of ethylene-vinyl acetate elastomer composites with surface modified BaTiO3 nanoparticles. J Phys D Appl Phys. 2009;42(24).

[78] Salehi H, Shahtahmasebi N, Hosseini SM. Band structure of tetragonal BaTiO3. Eur Phys J B. 2003; 32(2):177–80.

[79] Suzuki K, Kijima K. Optical band gap of barium titanate nanoparticles prepared by RF-plasma chemical vapor deposition. Japanese J Appl Physics. 2005;44(4 A):2081–2.

[80] Kampouri S, Stylianou KC. Dual-Functional Photocatalysis for Simultaneous Hydrogen Production and Oxidation of Organic Substances. ACS Catal. 2019;9(5):4247–70.

[81] Najam Khan M, Al-Hinai M, Al-Hinai A, Dutta J. Visible light photocatalysis of mixed phase zinc stannate/zinc oxide nanostructures precipitated at room temperature in aqueous media. Ceram Int. 2014;40: 8743–52.

[82] Khan MN, Jaisai M, Dutta J. Photocatalytic Inactivation of *Escherichia coli* Using Zinc Stannate Nanostructures under Visible Light. Adv Mater Res. 2015;1131:203–9.

[83] Tatarchuk T, Peter A, Al-Najar B, Vijaya J, Bououdina M. Photocatalysis: Activity of Nanomaterials. Nanotechnology in Environmental Science. 2018. 209–92.

[84] Jain S, Shah AP, Shimpi NG. An efficient photocatalytic degradation of organic dyes under visible light using

zinc stannate (Zn2SnO4) nanorods prepared by microwave irradiation. Nano-Structures & Nano-Objects. 2020; 21:100410.

[85] Foletto EL, Simões JM, Mazutti M a., Jahn SL, Muller EI, Pereira LSF, et al. Application of Zn2SnO4 photocatalyst prepared by microwave-assisted hydrothermal route in the degradation of organic pollutant under sunlight. Ceram Int. 2013;39(4): 4569–74.

[86] Borhade A V., Baste YR. Study of photocatalytic asset of the ZnSnO 3 synthesized by green chemistry. Arab J Chem. 2017;10:S404–11.

[87] Biswas A, Saha S, Jana NR. ZnSnO3 Nanoparticle-Based Piezocatalysts for Ultrasound-Assisted Degradation of Organic Pollutants. ACS Appl Nano Mater. 2019;2(2):1120–8.

[88] Wang Y-T, Chang K-S. Piezopotential-Induced Schottky Behavior of Zn1−xSnO3 Nanowire Arrays and Piezophotocatalytic Applications. Xie R-J, editor. J Am Ceram Soc. 2016;99(8):2593–600.

[89] Chen F, Huang H, Guo L, Zhang Y, Ma T. The Role of Polarization in Photocatalysis. Angew Chemie Int Ed. 2019;58(30):10061–73.

[90] Liang Z, Yan CF, Rtimi S, Bandara J. Piezoelectric materials for catalytic/photocatalytic removal of pollutants: Recent advances and outlook. Appl Catal B Environ. 2019;241:256–69.

[91] Askari H, Khajepour A, Khamesee MB, Saadatnia Z, Wang ZL. Piezoelectric and triboelectric nanogenerators: Trends and impacts. Nano Today. 2018;22(1):10–3.

[92] Alam MM, Ghosh SK, Sultana A, Mandal D. Lead-free ZnSnO3/MWCNTs-based self-poled flexible hybrid nanogenerator for piezoelectric power generation. Nanotechnology. 2015;26(16):165403.

[93] Choi KH, Siddiqui GU, Yang B, Mustafa M. Synthesis of ZnSnO3 nanocubes and thin film fabrication of (ZnSnO3/PMMA) composite through electrospray deposition. J Mater Sci Mater Electron. 2015;26:5690–6.

[94] Paria S, Karan SK, Bera R, Das AK, Maitra A, Khatua BB. A Facile Approach To Develop a Highly Stretchable PVC/ZnSnO 3 Piezoelectric Nanogenerator with High Output Power Generation for Powering Portable Electronic Devices. Ind Eng Chem Res. 2016;55(40):10671–80.

[95] Wang G, Xi Y, Xuan H, Liu R, Chen X, Cheng L. Hybrid nanogenerators based on triboelectrification of a dielectric composite made of lead-free ZnSnO3 nanocubes. Nano Energy. 2015;18:28–36.

[96] Liu Z, Xu J, Chen D, Shen G. Flexible electronics based on inorganic nanowires. Chem Soc Rev. 2015;44(1): 161–92.

[97] Hwang JK, Cho S, Dang JM, Kwak EB, Song K, Moon J, et al. Direct nanoprinting by liquid-bridge-mediated nanotransfer moulding. Nat Nanotechnol. 2010;5:742–8.

[98] Lim T, Kim H, Meyyappan M, Ju S. Photostable Zn2SnO4 Nanowire Transistors for Transparent Displays. ACS Nano. 2012;6(6):4912–20.

[99] Dong H, Zhang X, Zhao D, Niu Z, Zeng Q, Li J, et al. High performance bipolar resistive switching memory devices based on Zn2SnO4 nanowires. Nanoscale. 2012;4(8):2571–4.

[100] Siddiqui GU, Rehman MM, Choi KH. Enhanced resistive switching in all-printed, hybrid and flexible memory device based on perovskite

ZnSnO3 via PVOH polymer. Polymer (Guildf). 2016;100:102–10.

[101] Nunes D, Pimentel A, Santos L, Barquinha P, Pereira L, Fortunato E, et al. Metal oxide nanostructures: Synthesis, properties and applications. Elsevier; 2018. 1–328 p.

[102] Chen Q, Wang Y, et al. Enhanced acetone sensor based on Au functionalized In-doped ZnSnO3 nanofibers synthesized by electrospinning method. J Colloid Interface Sci. 2019;543:285–99.

[103] Wan, Sun J, Liu H. Semiconducting Oxide Nanowires: Growth, Doping and Device applications. Nanowires - Implementations and Applications. InTech; 2011. p. 59–98.

[104] Sharma A, Kumar Y, Shirage PM. Structural, optical and excellent humidity sensing behaviour of ZnSnO3nanoparticles: effect of annealing. J Mater Sci Mater Electron. 2018;29(13):10769–83.

[105] Supraja P, Sudarshan V, Tripathy S, Agrawal A, Singh SG. Label free electrochemical detection of cardiac biomarker troponin T using ZnSnO3 perovskite nanomaterials. Anal Methods. 2019;11(6):744–51.

[106] Bagheri H, Khoshsafar H, Afkhami A, Amidi S. Sensitive and simple simultaneous determination of morphine and codeine using a Zn2SnO4 nanoparticle/graphene composite modified electrochemical sensor. New J Chem. 2016;40(8):7102–12.

[107] Durai L, Badhulika S. One pot hydrothermal synthesis of large area nano cube like ZnSnO3 perovskite for simultaneous sensing of Uric Acid and Dopamine using differential pulse voltammetry. IEEE Sens J. 2020;1748 (c):1–1.

[108] Dong Y, Wang S, Zou Y, Liu S, Zhu Z, Li J, et al. Zinc Stannate Nanocrystal-Based Ultrarapid-Response UV Photodetectors. Adv Mater Technol. 2018;3(6):1800085.

[109] Zhang Y, Wang J, et al. High performance ultraviolet photodetectors based on an individual Zn2SnO4 single crystalline nanowire. J Mater Chem. 2010;20(44):9858.

[110] Zhao Y, Hu L, Liu H, Liao M, Fang X, Wu L. Band Gap Tunable Zn2SnO4 Nanocubes through Thermal Effect and Their Outstanding Ultraviolet Light Photoresponse. Sci Rep. 2015;4(1):6847.

[111] Xue XY, Guo TL, Lin ZX, Wang TH. Individual core-shell structured ZnSnO3 nanowires as photoconductors. Mater Lett. 2008;62 (8–9):1356–8.

[112] Bao L, Zang J, Li X. Flexible Zn2SnO4/MnO2 Core/Shell Nanocable–Carbon Microfiber Hybrid Composites for High-Performance Supercapacitor Electrodes. Nano Lett. 2011;11(3):1215–20.

Chapter 2

Investigation of Alternative Techniques for Graphene Synthesis

Betül Gürünlü and Mahmut Bayramoğlu

Abstract

In recent years, a great deal of concentration has addressed the electronic and morphological characteristics of carbonaceous substances. Nowadays, particularly, graphene is one of the most popular materials in condensed-matter physics and materials science. It is used in different fields such as desalination of seawater, smartphones, computers, satellites, planes, cars, building materials, obtaining protective coatings and rust-free cars, nuclear clean up, transistors, sensors, electron microscopy, Li ion batteries, super capacitors, and bionics. Mechanical cleaving (exfoliation), chemical exfoliation, chemical synthesis, and thermal chemical vapor deposition (CVD) synthesis are the most commonly used methods today. Some other techniques are also reported such as unzipping nanotube and microwave synthesis. In graphene synthesis, starting material is usually graphite. On the other hand, different starting materials such as rice husks, fenugreek seeds, hibiscus flower petals, camphor, alfalfa plants, petroleum asphalt are used as a carbon source for graphene synthesis. In this study, alternative methods for graphene synthesis specially microwave irradiation and ultrasound energy were studied, and the performances of the final products were compared with the help of different characterization techniques. Advantages and drawbacks of these methods were clearly discussed for enhancing the understanding of the graphene synthesis phenomena.

Keywords: graphene synthesis, graphene characterization, microwave irradiation, ultrasound energy

1. Introduction

Graphene is a 2D material, which was firstly discovered by Geim and Novoselov in 2004. They won Nobel Prize in Physics by synthesizing graphene including of sp^2 carbon bonds via Scotch-tape method in 2010 [1, 2]. Graphene is a thin nanoplatelet, which can be produced by cleaving of graphite. Graphite can be downed into the single graphene sheet level [3]. Graphene is a one atomic layer having 0.34 nm thicknesses. Graphene is a single layer of carbon atoms organized in a honeycomb lattice [4]. It is the block of graphite that is used in pencil tips, but graphene is an extraordinary matter with a multitude of astounding specialties that named it as wonder material [5]. It is a hexagonal shaped plane consisting of sp^2-carbon atoms [6, 7]. Graphene can be seemed as either uncoiled single-walled carbon nanotubes or a wide atomic sheet of graphite. Graphene has superior mechanical strength, thermal conductivity, optical transparency, high mobility, room temperature quantum Hall effect and great electronic properties like Dirac-particles having a linear dispersion, transport energy gap and simply absorption coefficient of lights, thus it will become the favorable prospect after the silicon time [8, 9]. It is the thinnest substance at one atom thick, and also fabulously strong around 200 times stronger than steel [5]. Apart from that, graphene is a superb conductor of heat and electricity and has exciting light absorption capabilities. It is truthfully a material with wide potential for integrating in nearly any industry.

Graphene is a highly varied material and can be merged with other materials (involving gases and metals) to synthesize various materials with different exceptional qualities. Researchers proceed to examine its unexplored properties and possible applications such as touchscreens (for LCD or OLED displays), computer chips, transistors, batteries, supercapacitors, energy production, DNA sequencing, water filters, antennas, solar cells, and spintronics. This new 2D material has a prominent importance in present day. It is a quickly developing subject that flourishing novel concepts at incredible speed [10]. Graphene is extensively used substance in electronic industry such as field-effect transistor, transparent electrode, etc. The recent developments in surface area, optical, magnetic, and mechanical properties of functionalized graphene and the unique electronics have arisen new attitude of green technology and creative discovery for present complications such as photonic and electronic usages for ultrahigh-frequency graphene-based apparatus, anode for Li-ion battery, material science, ceramics, light natural gas tanks, medical science, sensors to identify sickness, supercapacitor, solar cell, desalination of seawater, smartphones, computers, satellites, planes, cars, building materials, obtaining protective coatings and rust free cars, nuclear clean up, transistors, sensors, electron microscopy, and bionics.

Graphene molecular structure includes of sp^2 hybrid carbon atoms that were presented in **Figure 1a**. Sp^2 hybrids supply σ bonds with adjacent carbon atoms. Each of σ bonds has the length of 1.42 A°. Excellent mechanical characteristics of graphene are obtained under favor of σ bonds.

Graphene gathers much interest particularly after Geim and Novoselov win the 2010 Nobel Prize in physics by obtaining it in 2004. To produce high-quality graphene in high amount is not easy and affordable. Most companies are using chemical vapor deposition (CVD) based processes. Also, mechanical and chemical exfoliation and chemical synthesis are the most preferred ways today. Other methods are unzipping of a nanotube and microwave irradiation [11].

In graphene synthesis, starting material is usually graphite. But different starting materials are also used in literature such as; rice husks [12], fenugreek seeds [13], hibiscus flower petals [14], camphor [15], alfalfa plants [16], petroleum asphalt [17]. Graphene synthesis ways are primarily separated under two main groups entitling as bottom-up and top-down methods as seen as in **Figure 2** [11].

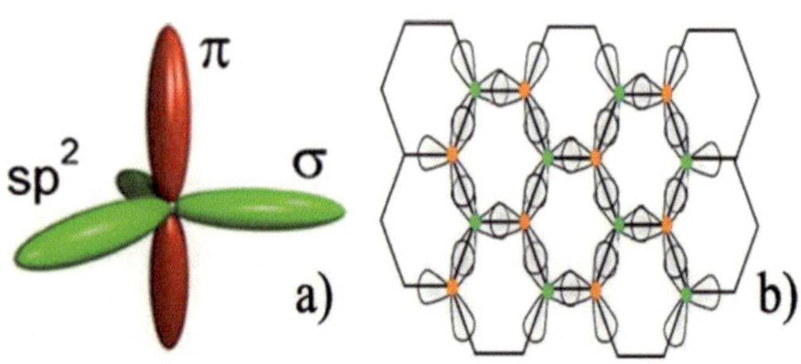

Figure 1.
(a) sp^2 hybrids carbon atoms in graphene (b) sp^2 hybrids of graphene carbon atoms connected to adjacent ones.

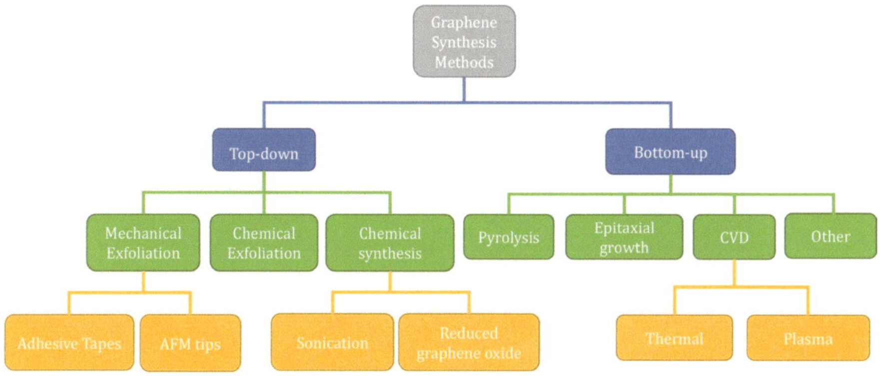

Figure 2.
Flow chart for available methods for synthesis of graphene sheets.

In top-down approach, graphene is synthesized by using graphite or graphite-oxide with the help of different methods. In this method, carbon materials such as graphite, carbon nanotubes are starting substances, and they are peeled by using chemical, electrochemical or physical ways [18]. Main top-down techniques are micromechanical exfoliation, cleavage of graphite intercalated compounds (GICs), unzipping of carbon nanotubes (CNTs), arc discharge, cleavage of graphene oxide, and liquid phase exfoliation.

2. Liquid phase exfoliation

Liquid phase exfoliation is an efficient and productive way for synthesizing of single and few layered graphene. It has been considered as one of the most feasible approach for industrial production of graphene due to its scalability and low cost. Solvent – carbon source suspension was first sonicated for preparation of exfoliation. Due to not having defects and oxide groups in the graphene products synthesized by LPE, they are more suitable for use in the electronics industry than that are produced by other techniques.

The LPE can form a stable dispersion of monolayer or few-layer defect-free graphene, which only involves the exfoliation of natural graphite via high-shear mixing or sonication [19]. Prepared graphene dispersion was stabilized by used solvent. Solvent type has also importance in productivity of the graphene dispersion [20]. Solvent ensures both the stability of synthesized graphene mixture and its productiveness. Tetrahydrofuran (THF) and N,N-dimethyl-formamide (DMF) are advantageous solvents to get high quality of graphene merely they are poisonous and show low efficiency. Dibasic ester (DBE) is an a nontoxic and environmental-friendly solvent and it was used for cleavage of graphite by Jiang et al. Its surface tension is 35.6 mJ/m^{-2} and solubility parameter is 9.7 [20].

Graphite can be exfoliated in liquid medium exploiting sound waves to form single layer, **Figure 3** [21]. Basically, exfoliation of carbon materials is a relatively economical and easy way to produce graphene [22].

The exfoliation step of the LPE can be conducted by the sonication of graphite in different solvents. There are two types of sonication: tip and bath sonication. In this study, tip sonication treatment was applied to the graphite-solvent dispersions. Epoxy/graphene composite shows better mechanical properties due to direct ultrasonication of tip sonication, that generates higher sound pressures and intensity compared to bath sonication which is indirect ultrasonication [23, 24]. The direct sonication of graphite in a solvent having similar surface energy to graphite enables a

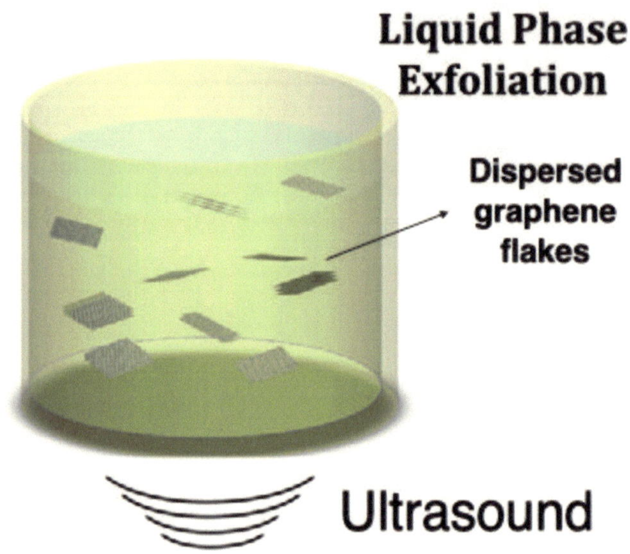

Figure 3.
Liquid phase exfoliation.

stable graphite dispersion [25]. Several studies have been performed in order to find the most appropriate solvent as well as the optimum operation conditions for the sonication process [26–29].

3. Experimental section

The experimental studies consist of two different methods; microwave (MW) energy-assisted method and ultrasound (US) energy-assisted method.

3.1 Materials used

In microwave energy method; graphite (natural flake graphite, grade 3061; purchased from Asbury Graphite Mills, Inc., New Jersey) was used as starting carbon source. Different solvents were used such as 25% ammonia solution (Merck KGaA), N,N-Dimethyl formamide (Merck KGaA), ethylene glycol (ZAG Chemicals) and ethylene diamine (Merck KGaA). Chemicals used in the second cycle of experiments were of analytical grade; n-Hexadecane (Merck, 99.5%), dimethyl sulfoxide (Merck, 99.9%), sodium hydroxide (J.T. Baker, 99%), 1-octanol (Merck, 99%), perchloric acid (Merck, 70–72%), N,N-Dimethyl formamide (Merck, 99.8%), ethylene glycol (ZAG Chemicals, 99.3%), and ethylene diamine (Merck, 99%).

Chemicals used in the ultrasound method are as follows: Graphite fine powder (Extra pure, Asbury Inc., New Jersey), graphene nanoplatelets (XG Sciences, Michigan, US) Dimethyl sulfoxide - DMSO (Merck), N,N-Dimethylformamide - DMF (Merck), Perchloric acid 70–72% - PA (Merck).

3.2 Preparation of graphene sheets via microwave method

The procedure of MW treatment was summarized as following: First, natural graphite is added to ammonia, then obtained suspension was sonicated by ultrasound energy device (BANDELIN ® HD 2200 SONOPULS), under conditions 200 W,

35 kHz, mode 5 and 50% power for 10 min. Secondly, reaction was performed in Milestone Start-S model microwave oven for half an hour at 120°C temperature and 1 bar pressure by applying 50, 100 or 200 Watt energy. Pressure controller was active, and thermocouple was adjusted carefully as shown in **Figure 4**.

3.3 Preparation of graphene sheets via ultrasound method

0.3 g graphite was dispersed in 50 ml solvent such as DMSO, DMF and PA. Obtained dispersions were sonicated by the means of BANDELIN ® HD 2200 SONOPULS (which is given in **Figure 5**) equipped with a VS 190 T sonotrode, 200 W, 50% amplitude for 3 hours.

Then, these dispersions were subjected to 60 minutes centrifugation (Elektromag, M 4812 P) at 3000 rpm to remove the unexfoliated part of graphite; after the heavier particles were settled down, supernatant parts were decanted and collected in separate vials.

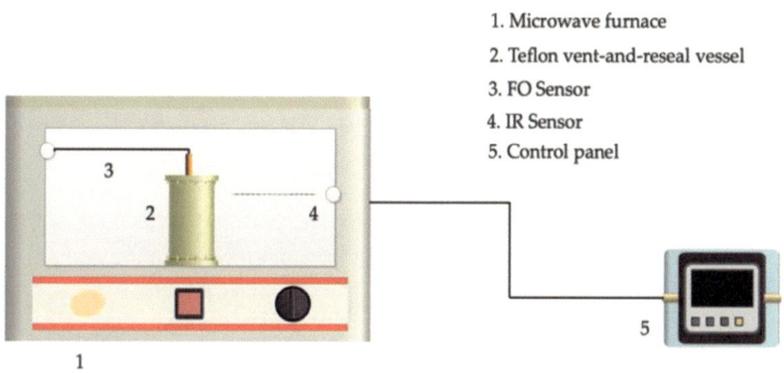

Figure 4.
The experimental system with a multimode microwave furnace: Reaction was performed inside a Teflon vent-and-reseal vessel.

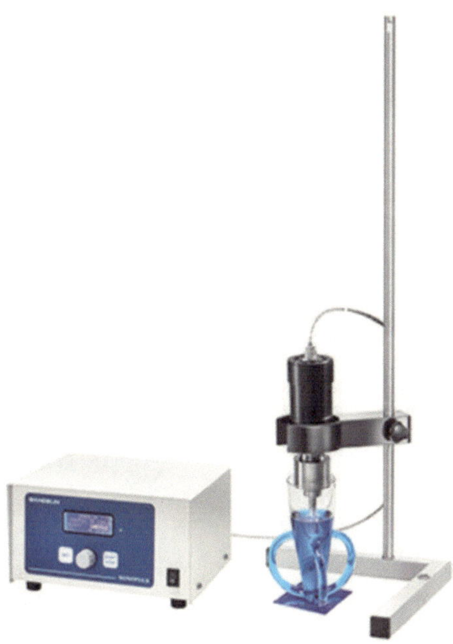

Figure 5.
Ultrasound device.

3.4 Characterization

Different characterization techniques were applied to the obtained final products via microwave energy method in order to determine their properties such as thickness, layer number, electrical conductivity. X-ray Diffraction (XRD) analysis was done via Rigaku D-Max 2200 Series equipped with Cu-Kα radiation (λ = 1.54 Å) at a scanning rate of 3° per minute. The tube voltage was 40 kV and the current were 40 mA. The intensity was determined over a 2θ° angular range of 2–90°. Electrical conductivities of synthesized products were measured by Keithley 2400 Sourcemeter which is seen in **Figure 6**.

Each sample was measured by applying following procedure; first, it was placed in a copper cylindrical container which has a copper cap and it was compressed by a hydraulic press under 50 bar for 30 min. The electrical resistivities of obtained products were determined by 4-point probe method. Synthesized powder sample were compressed in copper mold with the help of a joiner's clamp during the electrical conductivity measurement. The conductivity σ was then estimated according to $\sigma = l/AR$. The Fourier Transform Infrared (FTIR) spectra of synthesized products were measured by Perkin Elmer Spectrum Two equipped with a germanium (Ge) crystal (Pike Gladi ATR Ge-ATR) in the range of 650–4000 cm^{-1}. The obtained powder was characterized via ultraviolet–visible (UV–Vis) spectroscopy. For UV–vis analysis, the dried filtrate

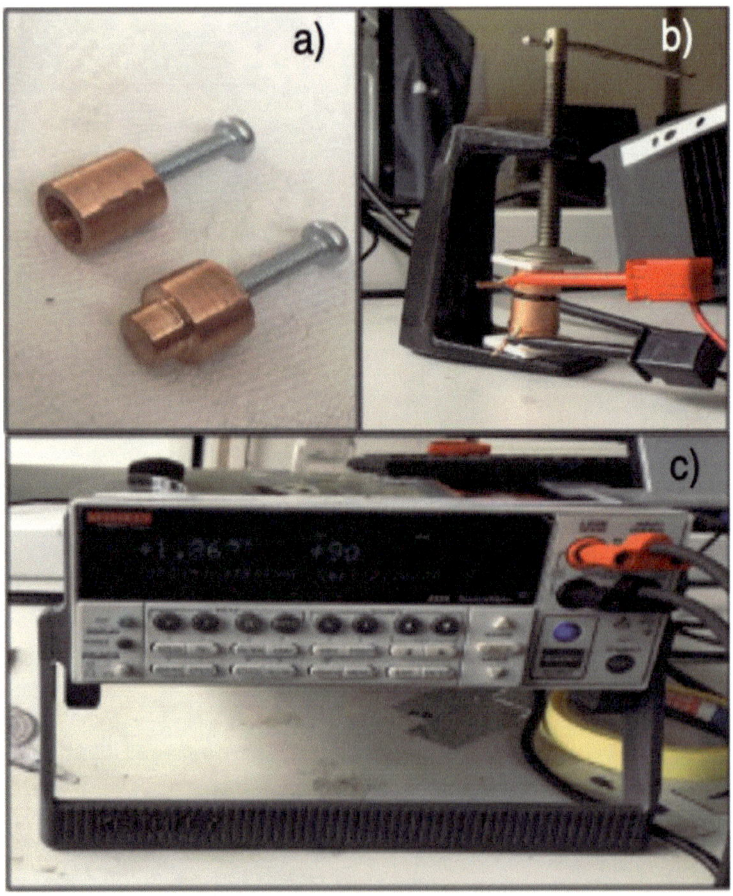

Figure 6.
Electrical resistivity measurement system: (a) copper cylindrical container and a copper cap. (b) Electrical resistivity measurement set-up (joiner'clamp and copper container). (c) Keithley 2400 Sourcemeter.

which is dried on drying oven at overnight was dispersed in distilled water by agitating via a magnetic stirrer. After that an amount of dispersion was taken into the 10x10 mm vial then it was analyzed by comparing with the water which is reference sample. The spectrum has an operation range (UV Perkin Elmer, Lambda 35) of 200 to 700 nm.

Also, the synthesized products via ultrasound energy method were analyzed via different characterization techniques such as UV–vis spectroscopy, Atomic Force Microscopy, X-ray Diffraction and dynamic light scattering analysis. UV–vis spectral measurements were acquired using a Perkin Elmer Precisely Lambda 35 UV/vis Spectrometer. UV–Visible spectra (Perkin Elmer, Lambda 35) were measured from 200 to 800 nm. Samples for AFM were prepared by dropping the graphene dispersions onto glass pieces (0.7 x 0.7 mm^2) and measurements were made in contact (tapping) mode, with 10.00 μm scan size, and 20.35 Hz scan rate by using Digital Instruments Nanoscope. Samples for XRD were prepared by depositing onto glass pieces (0.7 x 0.7 mm^2) and X-ray diffraction (XRD) patterns were obtained with a Rigaku D-Max 2200 Series equipped with Cu-Kα radiation (λ = 1.54 A°) at a scanning rate of 3° per minute. The tube voltage was 40 kV, and the current was 40 mA. Also, an extensive study of the particle size distribution was carried out by an analytical technique such as dynamic light scattering (DLS) method by using Malvern Zetasizer Nano ZS Laser Particle Size Distribution Meter.

4. Results & discussion

Microwave energy-assisted method and ultrasound energy-assisted method were studied, and the final products were obtained. Synthesized carbon products were analyzed by applying different characterization techniques such as XRD, AFM, TEM.

4.1 Microwave (MW) assisted method results

All the results of ammonia tests were summarized in **Table 1**. According to the results; sonication did not create a positive effect on electrical conductivity of final product. Lower temperature conditions give better yield and electrical conductivity results.

According to these results which were given in **Table 1**, low temperature showed better electrical conductivity results. Sonication step built a negative effect on electrical conductivity results. Also, after annealing step, electrical conductivity results slightly increased.

Another set of experiment were done in order to compare the effect of different solvents on graphene synthesis via microwave energy. The results of microwave tests that were conducted by using N,N-Dimethyl formamide (DMF), ethylene glycol (EG) and ethylene diamine (ED) were given in **Table 2**.

According to the results which were given in **Table 2**, the reaction yields of DMF, EG, and ED are 60, 88, and 75%, respectively. The electrical conductivity values of DMF, EG, and ED are 22.716, 6.0002, 7.0967 S/m, respectively. It can be concluded that; G-DMF shows better conductivity performance.

XRD spectra of natural graphite, MW assisted expanded graphite products which were obtained in different solvents such as ethylene glycol, ammonia, and DMF were given in **Figure** 7, respectively.

According to XRD results; all the spectrums show the 002 peak of graphite was predominant in all the four types of graphite, at 2θ° = 26.44° peak, which is characteristic for graphite. Natural graphite shows highest intensity peak at 2θ° = 26.44. The intensity of other two peaks 101, 004 was low at all the spectrums. Layer

Exp. No	Carbon source	Solvent	Sonication step	React. Cond.	Yield (%)	Elec. cond. (S/m)	E. cond. (After annealing) (S/m)
1	Natural graphite (0.5 g)	25% Ammonia	—	120°C, 1 bar, 50 watt	94	52.44	58.114
2	Natural graphite (0.5 g)	25% Ammonia	—	120°C, 1 bar, 50 watt	89	12.8	30.647
3	Natural graphite (0.5 g)	25% Ammonia	30 min mode 5 power 50%	200 °C, 1 bar, 50 watt	53.5	9.06	12.047

Table 1.
Results of experiments that were done by using ammonia.

Exp. No	Carbon source	Solvent	Sonication step	React. Cond.	Yield (%)	Elec. cond. (S/m)
4	Natural graphite (0.1 g)	DMF (50 ml)	10′, 200 W, 20kHz, mode 5, power 50%	30 min 180°C	60	22.7
5	Natural graphite (0.1 g)	EG (50 ml)	10′, 200 W, 20kHz, mode 5, power 50%	30 min 180°C	88	6
6	Natural graphite (0.1 g)	ED (50 ml)	10′, 200 W, 20 kHz, mode 5, power 50%	30 min 180°C	75	7.1

Table 2.
Microwave tests that were conducted by using DMF, EG and ED.

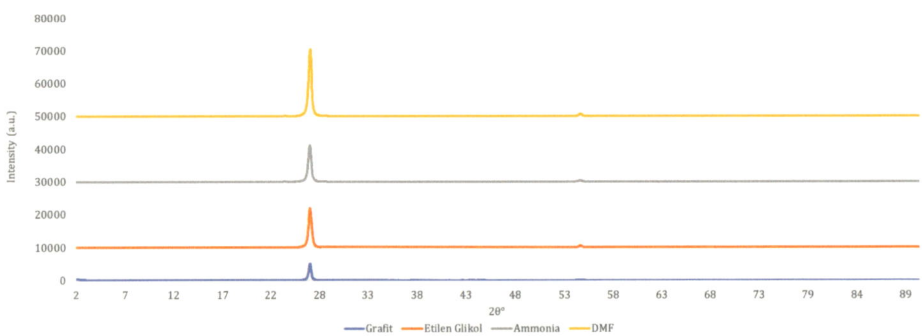

Figure 7.
XRD spectra of commercial graphite and the MW-assisted graphene products which were obtained in ethylene glycol, ammonia, and DMF.

numbers of final products calculating by using XRD data were presented at **Table 3**. Layer numbers of expanded graphite products, which were obtained in EG, ammonia, and DMF by using MW energy, were calculated as 1.5 for all solvents. Layer number of natural graphite was calculated as 1.75 by the help of XRD results.

The results of another experiment plan which covering the usage of wide scale of solvents including n-Hexadecane (n-Hexa), Dimethylsulfoxide (DMSO), Sodium

Code	Layer number
Ethylene glycol (EG)	1.5
Ammonia	1.5
N,N-Dimethyl formamide (DMF)	1.5
Natural graphite	1.75

Table 3.
Layer numbers of final products calculating from XRD results.

Hydroxide (50% aq.) (NaOH), 1-octanol (OCTA), Perchloric acid (PA), N,N-Dimethyl formamide (DMF), Ethylene glycol (EG), and Ethylene diamine (ED) were presented in **Table 4**.

According to the results, MW-G-DMF showed the highest electrical conductivity. Electrical conductivities of MW assisted graphene products were higher when the used chemicals have 2–4 Debye (D) dipole moments. These results are compatible with the dielectric constants and surface tensions of the used chemicals. Layer numbers were calculated by Scherrer equation and the half-width of the diffraction line $\beta(2\theta)$ (in rad) was taken as the experimental half-width (βexp) and was corrected for experimental broading ($\beta instr$) as described in Saberi et al.'s study [30]. Layer numbers show distribution between 10 and 16. MW-G-EG showed the thinnest layer number with the value of 5.5, which is seen at **Table 4**. Solvents that have surface tension bigger than 40 mN/m show better layer number results. Briefly, as the surface tensions increased, layer numbers decreased. These results are supported with Hernandez et al.'s study [29]. Electrical conductivities of MW assisted graphene products were higher when the used chemicals have 2–4 Debye (D) dipole moments as seen as in **Table 4**. When the dielectric constants (ε) get larger, electrical conductivity values of synthesized products increased.

MW-G-PA showed the optimum electrical conductivity and layer number values for the MW assisted graphene synthesis as seen in **Figure 8**.

All XRD spectrums showed peak at 26.5° which can be seen in **Figure 9**. XRD spectra of MW- G-PA also proved that graphite peak at 26.5° shows minimum intensity.

Solvent	Dipole moment (Debye)	Dielectric constant (ε)	Layer number	Surface Tension @20 °C (mN/m)	Elect. conductivity (S/m)
n-Hexadecane	0.06	2	15.81	27.47	8.174
Dimethylsulfoxide	3.96	46.7	12.36	43.54	7.581
Sodium Hydroxide (50% aq.)	6.832	57.5	10.33	74.35	10.664
1-octanol	1.76	3.4	14.02	27.6	1.784
Perchloric acid	2.146	115	10	69.69	20.619
N,N-Dimethyl formamide	3.86	36.7	15	37.1	22.716
Ethylene glycol	2.746	37	5.5	47.7	6.002
Ethylene diamine	1.83	16	10.61	42	7.097

Table 4.
Electrical conductivities, dipole moments, layer numbers and dielectric constants of MW supported graphene products.

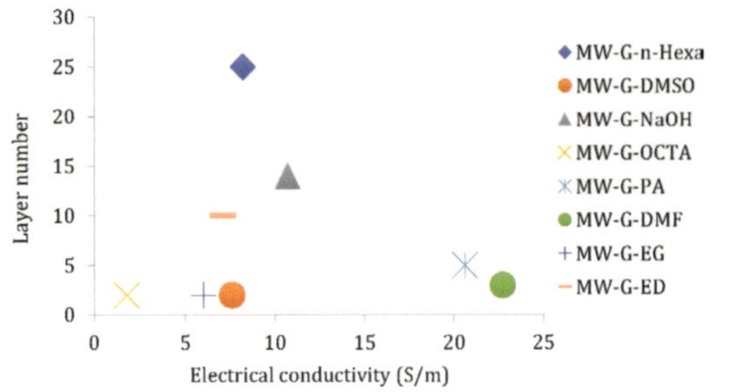

Figure 8.
Relation between layer numbers and electrical conductivity.

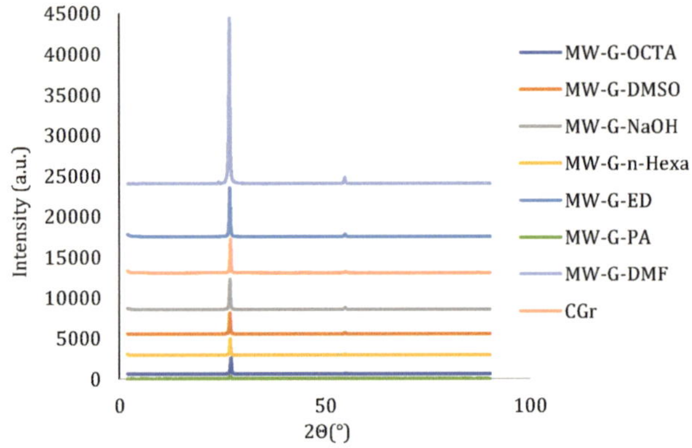

Figure 9.
XRD spectra of MW-assisted graphene products.

According to the UV–vis spectrums of MW-assisted graphene samples, which are presented in **Figure 10**, synthesized graphene samples, which were labeled as MW- G-PA, MW-G-NaOH, MW-G-n-Hexa, MW-G-ED, MW-G-DMSO, and MW-G-OCTA showed peak at 265 nm wavelength that referring sp^2 C=C bonds. This result is in line with the previous literature [31].

4.2 Ultrasound (US) assisted method results

The US-assisted synthesized graphene products were characterized by using UV–vis spectroscopy, AFM Spectroscopy, and DLS analysis. UV–vis spectrums of US-assisted graphene products are presented in **Figure 11**. Coleman's team calculated the absorption coefficient of graphene dispersion via UV/vis spectroscopy.

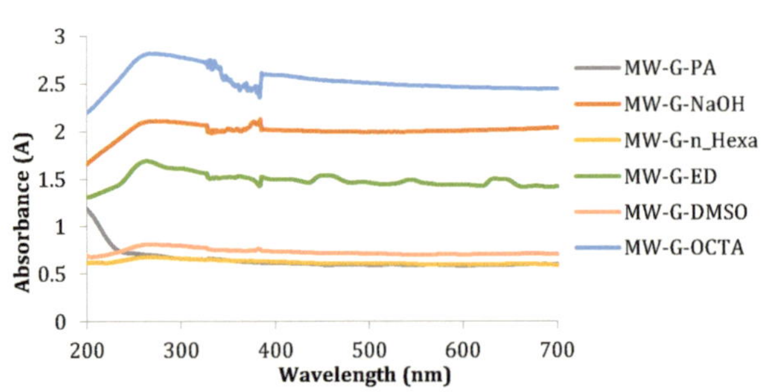

Figure 10.
UV spectrums of MW based synthesized graphene products.

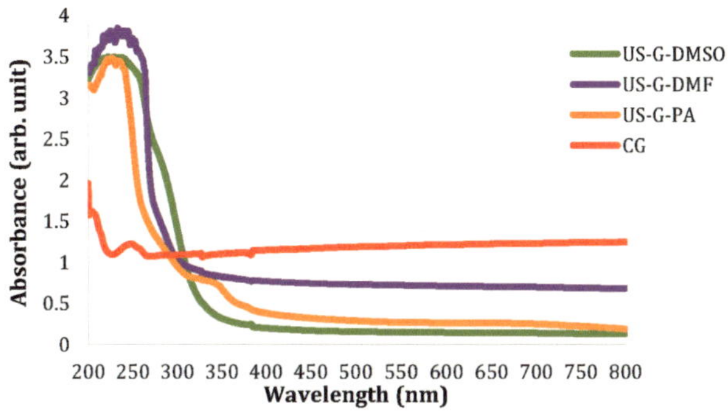

Figure 11.
UV–vis spectra of CG, US-G-DMSO, US-G-DMF, and US-G-PA products.

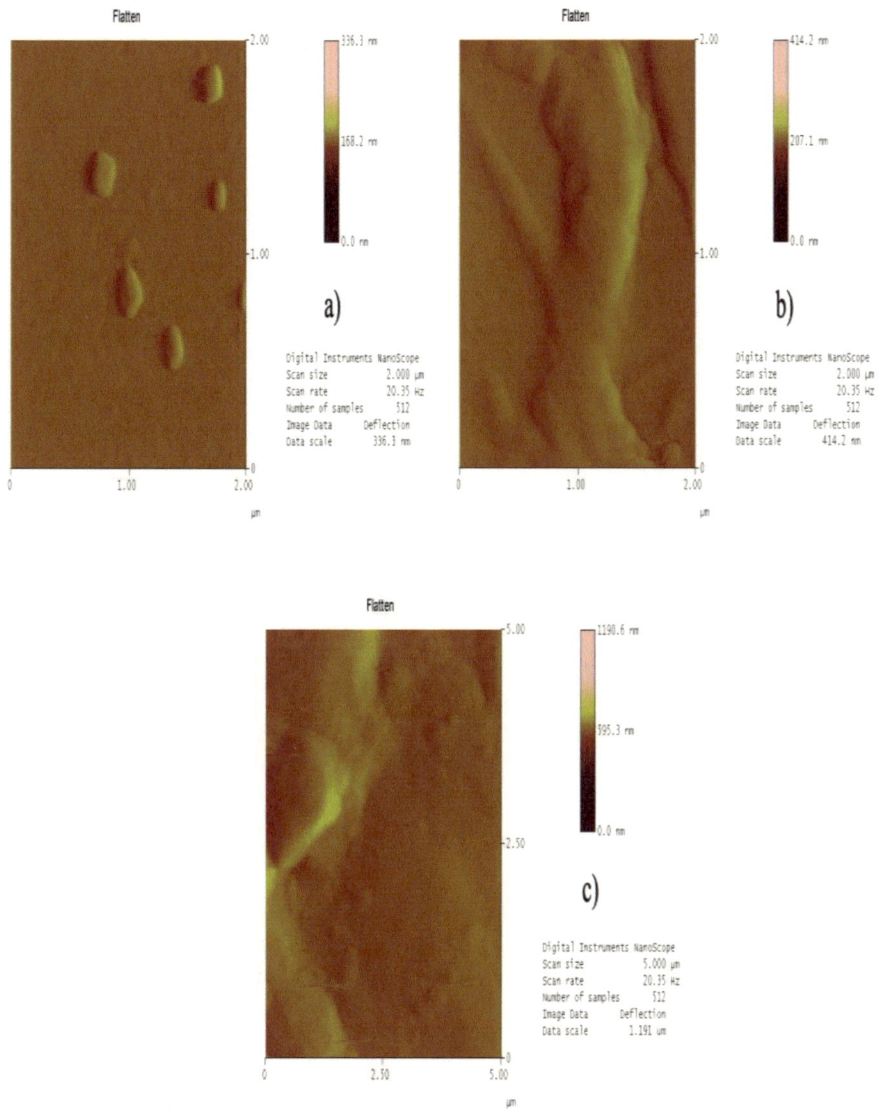

Figure 12.
The AFM images of (a) US-G-DMSO, (b) US-G-DMF, and (c) US-G-PA drop casted onto glass piece showing the homogeneous structure of the pristine graphene nanosheets.

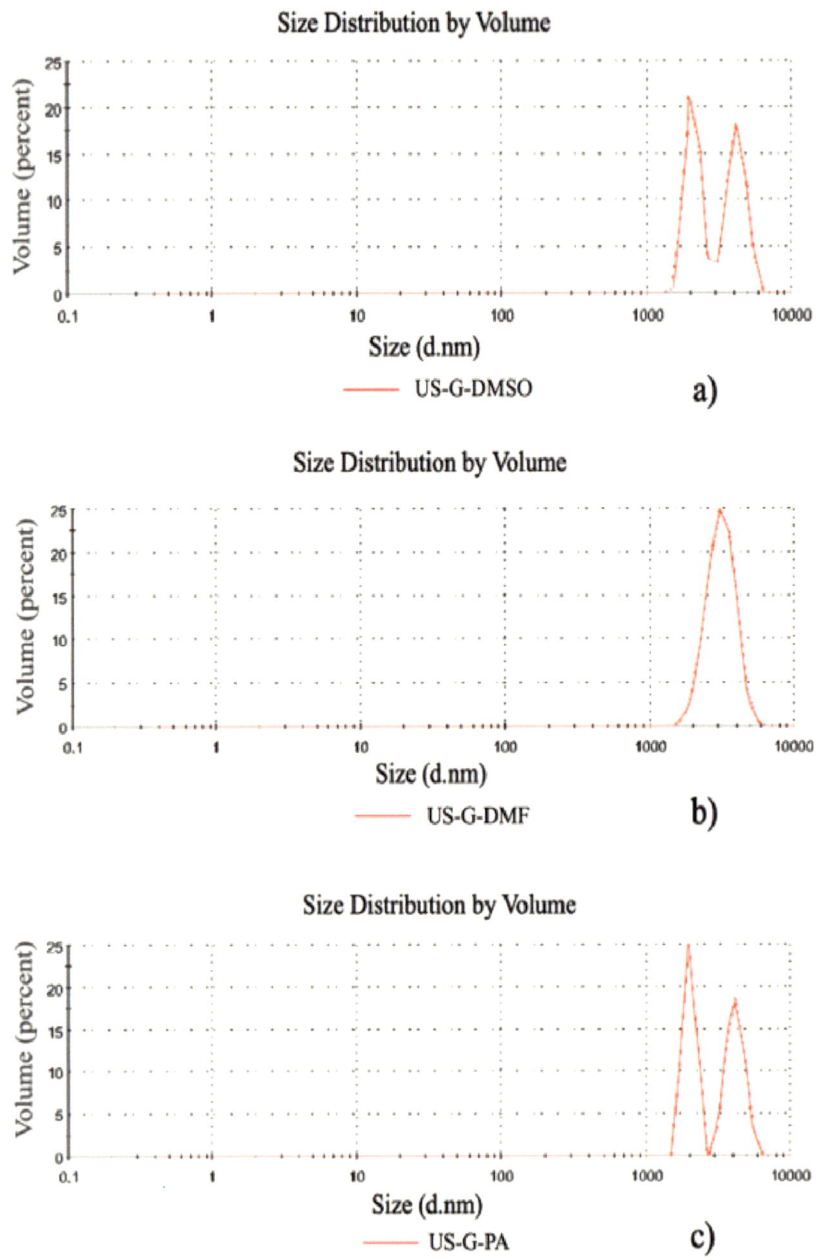

Figure 13.
Lateral size results of synthesized samples, (a) US-G-DMSO, (b) US-G-DMF, (c) US-G-PA.

Concisely, with the help of the Beer–Lambert law, absorption coefficient ($A = \alpha cl$) of graphene could be found by using dispersion at specific concentrations [29, 32–35]. UV–Vis absorbance spectroscopy was conducted at fixed wavenumbers of 253 nm for graphene. A piercing peak at 210 nm can be noticed and one more peak around 226 nm with a little bit less intensity of absorption peak is also observed due to Π-Π* bondings of the C-C aromatic rings.

The obtained graphene samples, which are labeled as US-G-DMSO, US-G-DMF and US-G-PA, show peak at 265 nm wavelength that referring sp^2 C=C bonds [31].

AFM characterization of final graphene products (US-G-DMF, US-G-DMSO, US-G-PA) were conducted to determine the optimal growth condition by measuring surface roughness and thickness. The AFM images of US-G-DMSO, US-G-DMF, and US-G-PA were presented in **Figure 12**. The Ra values of US-G-DMSO, US-G-DMF, and US-G-PA are 2.937, 6.343, and 10.103 nm, respectively. The Rq values of US-G-DMSO, US-G-DMF, and US-G-PA are 3.471, 8.046, and 11.748 nm, respectively. The RMS values of US-G-DMSO, US-G-DMF, and US-G-PA are 5.675, 8.842, and 11.910 nm, respectively. Vertical distance denotes the thickness of graphene and it is determined for US-G-DMSO, US-G-DMF, and US-G-PA as 1.638, 2.151, and 10.754 nm, respectively. The layer numbers were calculated via following equation: $N = (t_{measured} - 0.4)/0.335$.

The layer numbers of US-G-DMSO, US-G-DMF, and US-G-PA are calculated as 4, 5, and 31, respectively. According to AFM results, best result was obtained with DMSO. All these results confirmed that the US-G-DMSO materials had fewer layers and defects.

Although these techniques can determine the size of graphene products, dynamic light scattering (DLS) is also helpful to measure the lateral size. It is an easy and quick method for evaluating the size of graphene samples [36]. The size distribution of the synthesized graphene samples using DLS are shown in **Figure 13**. Z-average hydrodynamic radius (Rh) of US-G-DMF is 3846 nm, Rh of US-G-DMSO is 6930 nm, and Rh of US-G-PA is 7137 nm. According to these results, DMF provides graphene products with smallest lateral size.

5. Conclusion

Microwave (MW)-assisted method was developed. Although many solvents have been studied, carbon product, which was synthesized in DMF, showed the highest electrical conductivity. Electrical conductivities of MW-assisted graphene products were higher when the used solvents have 2–4 Debye (D) dipole moments. These results are compatible with the dielectric constants and surface tensions of the used chemicals. Layer numbers show distribution between 10 and 16. EG has minimum layer number with the value of 5.5. Solvents that have surface tension bigger than 40 mN/m show better layer number results. When the dielectric constants (ε) get larger, electrical conductivity values of synthesized products increased. As the surface tensions increased, layer numbers decreased. PA showed the optimum electrical conductivity and layer number values for the MW-assisted graphene synthesis. According to the UV–vis spectrums of MW assisted graphene samples. The obtained graphene samples, which were labeled as MW-G-PA, MW-G-NaOH, MW-G-n-Hexa, MW-G-ED, MW-G-DMSO, and MW-G-OCTA showed peak at 265 nm wavelength that referring sp^2 C=C bonds.

Ultrasound (US)-assisted method was studied. Graphene samples were easily synthesized via solution-based process. According to the UV–vis spectrums, all graphene products gave peak at 265 nm wavelengths, which may be caused by the ultrasonication required for proper suspension using the solution-based process. Also, as a result of AFM analyses, US-G-DMSO has four layers, US-G-DMF has five layers and US-G-PA has thirty-one layers. It can be understood that DMSO shows better solvent effect on graphite exfoliation by sonication process. Z-average hydrodynamic radius (Rh) of US-G-DMF is 3846 nm, Rh of US-G-DMSO is 6930 nm, and Rh of US-G-PA is 7137 nm. It can be concluded that, DMF provides graphene products with smallest lateral size.

Acknowledgements

This work has been partially supported by Research Fund of the Gebze Technical University (project no. 2018-A105-55). The authors also acknowledge the Materials Science and Engineering Department of Gebze Technical University for providing AFM and DLS measurements.

Conflict of interest

The authors declare no conflict of interest.

Author details

Betül Gürünlü[1*] and Mahmut Bayramoğlu[2]

1 Institute of Nanotechnology, Gebze Technical University, Kocaeli, Turkey

2 Chemical Engineering Department, Gebze Technical University, Kocaeli, Turkey

*Address all correspondence to: bgurunlu@gtu.edu.tr

References

[1] Eswaraiah V, Sankaranarayanan V, Ramaprabhu S. Graphene-Based Engine Oil Nanofluids for Tribological Applications. ACS Applied Materials & Interfaces. 2011; 3 (11):4221-4227. DOI: 10.1021/am200851z

[2] Akbar F, Kolahdouz M, Larimian S, Radfar B, Radamson HH. Graphene synthesis, characterization and its applications in nanophotonics, nanoelectronics, and nanosensing. J Mater Sci: Mater Electron. 2015;26(7):4347-4379. DOI: 10.1007/s10854-015-2725-9

[3] Lee S, Lim S, Lim E, Lee KK. Synthesis of aqueous dispersion of graphenes via reduction of graphite oxide in the solution of conductive polymer. Journal of Physics and Chemistry of Solids. 2010;71(4):483-486. DOI: 10.1016/j.jpcs.2009.12.017

[4] Tassin P, Koschny T, Soukoulis CM. Graphene for Terahertz Applications. Science. 2013;341(6146):620-621. DOI: 10.1126/science.1242253

[5] Lohar DV. Literature Review of Graphene Composites. In: International Conference on Recent Trends in Engineering and Science (ICRTES 2017); 20-21 January 2017; India. Tamilnadu: IJIRSET; 2017. p. 475-478.

[6] Baatar C. Promises of Graphene Nanoelectronics. In: 8th IEEE Conference on Nanotechnology; 18-21 August 2008; Arlington, TX. pp. 190-190, DOI: 10.1109/NANO.2008.62

[7] Bolotin KI, Sikes KJ, Hone J, Stormer HL, Kim P. Temperature-Dependent Transport in Suspended Graphene. Physical Review Letters. 2008;101(9):096802. DOI: 10.1103/PhysRevLett.101.096802

[8] Geim A, Novoselov K. The rise of graphene. Nature Mater. 2007;6:183-191. DOI: 10.1038/nmat1849

[9] Nair RR, Blake P, Grigorenko AN, Novoselov KS, Booth TJ, Stauber T, Peres NMR, Geim AK. Fine Structure Constant Defines Visual Transparency of Graphene. Science. 2008;320(5881):1308. DOI: 10.1126/science.1156965

[10] Gong JR. Graphene – Synthesis, Characterization, Properties and Applications. 1st ed. Croatia: InTech; 2011. 162 p. DOI: 10.5772/1742

[11] Bhuyan MSA, Uddin MN, Islam MM, Bipasha FA, Hossain SS. Synthesis of graphene. Int. Nano Lett. 2016; 6(2):65-83. DOI: 10.1007/s40089-015-0176-1

[12] Muramatsu H, Kim YA, Yang K-S, Cruz-Silva R, Toda I, Yamada T, Terrones M, Endo M, Hayashi T, Saitoh H. Rice Husk-Derived Graphene with Nano-Sized Domains and Clean Edges. Small. 2014;10:2766-2770. DOI:10.1002/smll.201400017

[13] Roy P, Periasamy AP, Chuang C, Liou Y-R, Chen Y-F, Joly J, Liang C-T, Chang H-T. Plant leaf-derived graphene quantum dots and applications for white LEDs. New Journal of Chemistry. 2014:38(10):4946-4951. DOI: 10.1039/C4NJ01185F

[14] Ray AK, Chatterjee S, Singh JK, Bapari H. Thermal Exfoliation of Natural Cellulosic Material for Graphene Synthesis. J. of Materi Eng and Perform. 2015; 24(1): 80-84. DOI: 10.1007/s11665-014-1224-0

[15] Ravani F, Papagelis K, Dracopoulos V, Parthenios J, Dassios KG, Siokou A, Galiotis C. Graphene production by dissociation of camphor molecules on nickel substrate. Thin Solid Films. 2013;527:31-37. DOI: 10.1016/j.tsf.2012.12.029

[16] Qu J, Luo C, Zhang Q, Cong Q, Yuan X. Easy synthesis of graphene

sheets from alfalfa plants by treatment of nitric acid. Materials Science and Engineering: B. (2013);178(6):380-382. DOI: 10.1016/j.mseb.2013.01.016

[17] Li Y, Chen Q, Xu K, Kaneko T, Hatakeyama VR. Synthesis of graphene nanosheets from petroleum asphalt by pulsed arc discharge in water. Chemical Engineering Journal. 2013;215-216:45-49. DOI: 10.1016/j.cej.2012.09.123

[18] Sridhar V, Jeon J-H, Oh I-K. Synthesis of graphene nano-sheets using eco-friendly chemicals and microwave radiation. Carbon. 2010;48(10):2953-2957. DOI: 10.1016/j.carbon.2010.04.034

[19] Randviir EP, Brownson DAC, Banks CE. A decade of graphene research: Production, applications and outlook. Mater. Today. 2014;17:426-432. DOI: 10.1016/j.mattod.2014.06.001

[20] Jiang F, Yu Y, Wang Y, Feng A, Song L. A novel synthesis route of graphene via microwave assisted intercalation-exfoliation of graphite. Materials Letters. 2017;200:39-42. DOI: 10.1016/j.matlet.2017.04.048

[21] Bonaccorso F, Lombardo A, Hasan T, Sun Z, Colombo L, Ferrari AC. Production and processing of graphene and 2D crystals. Materials Today. 2012;15(12):564-589. DOI: 10.1016/S1369-7021(13)70014-2

[22] Cai M, Thorpe D, Adamson DH, Schniepp HC. Methods of graphite exfoliation. Journal of Materials Chemistry. 2012;22(48):24992-25002. DOI: 10.1039/C2JM34517J

[23] Chun WW, Leng TP, Osman AF, Keat YC. Mechanical Properties and Morphology of Epoxy/Graphene Nanocomposite Using Bath Sonication and Tip Sonication. Solid State Phenomena. 2018;280:258-263. DOI: 10.4028/www.scientific.net/SSP.280.258

[24] Schnyder B, Alliata D, Kötz R, Siegenthaler H. Electrochemical intercalation of perchlorate ions in HOPG: an SFM/LFM and XPS study. Appl. Surf. Sci. 2001;173(3-4); 221-232. DOI: 10.1016/S0169-4332(00)00902-8

[25] Whitener KE Jr, Sheehan PE. Graphene synthesis. Diamond Relat. Mater. 2014:46;25-34. DOI: 10.1016/j.diamond.2014.04.006

[26] Yi M, Shen Z. A review on mechanical exfoliation for the scalable production of graphene. J. Mater. Chem. A. 2015:3;11700-11715. DOI: 10.1039/C5TA00252D

[27] Blake P, Brimicombe PD, Nair RR, Booth TJ, Jiang D, Schedin F, Ponomarenko LA, Morozov SV, Gleeson HF, Hill EW, Geim AK, Novoselov KS. Graphene-based liquid crystal device. Nano Lett. 2008:8(6);1704-1708. DOI: 10.1021/nl080649i

[28] Ciesielski A, Samori P. Graphene via sonication assisted liquid-phase exfoliation. Chem. Soc. Rev. 2014:43;381-398. DOI: 10.1039/C3CS60217F

[29] Hernandez Y, Nicolosi V, Lotya M, Blighe FM, Sun Z, De S, McGovern IT, Holland B, Byrne M, Gun'ko YK, Boland JJ, Niraj P, Duesberg G, Krishnamurthy S, Goodhue R, Hutchison J, Scardaci V, Ferrari AC, Coleman JN. High-yield production of graphene by liquid-phase exfoliation of graphite. Nat. Nanotechnol. 2008:3;563-568. DOI: 10.1038/nnano.2008.215

[30] Saberi A, Alinejad B, Negahdari Z, Kazemi F, Almasi A. A novel method to low temperature synthesis of nanocrystalline forsterite. Materials Research Bulletin. 2007:42(4);666-673. DOI: 10.1016/j.materresbull.2006.07.020

[31] Johra FT, Lee JW, Jung WG. Facile and safe graphene preparation on

solution based platform. Journal of Industrial and Engineering Chemistry. 2014:20(5);2883-2887. DOI: 10.1016/j.jiec.2013.11.022

[32] Khan U, O'Neill A, Lotya M, De S, Coleman JN. High Concentration Solvent Exfoliation of Graphene. Small. 2010:6;864-871. DOI: 10.1002/smll.200902066

[33] Khan U, Porwal H, O'Neill A, Nawaz K, May P, Coleman JN. Solvent-Exfoliated Graphene at Extremely High Concentration. Langmuir. 2011:27;9077-9082. DOI: 10.1021/la201797h

[34] Lotya M, Hernandez Y, King PJ, Smith RJ, Nicolosi V, Karlsson LS, Blighe FM, De S, Wang ZM, McGovern IT, Duesberg GS, Coleman JN. Liquid Phase Production of Graphene by Exfoliation of Graphite in Surfactant/Water Solutions. J. Am. Chem. Soc. 2009:131;3611-3620. DOI: 10.1021/ja807449u

[35] Lotya M, King PJ, Khan U, De S, Coleman JN. High-Concentration, Surfactant-Stabilized Graphene Dispersions. ACS Nano. 2010:4;3155-3162. DOI: 10.1021/nn1005304

[36] Lotya M, Rakovich A, Donegan JF, Coleman JN. Measuring the lateral size of liquid-exfoliated nanosheets with dynamic light scattering. Nanotechnology. 2013:24(26); 265703. DOI: 10.1088/0957-4484/24/26/265703

Chapter 3

Preparation of Hollow Nanostructures via Various Methods and Their Applications

Rudy Tahan Mangapul Situmeang

Abstract

The hollow nanomaterial is a unique material to be developed because of its characteristics, especially the surface area where it has more surfaces than other materials. In general, hollow nanomaterials could be synthesized using hard-templated, soft-templated, self-templated, template-free and simple methods. In this chance, the catalyst preparation focused on using a simple method to study its activity on the dyes photodegradation reaction, deNOx reaction, carbon dioxides utilization, and photoconversion of chemical compounds. The characterization is emphasized on Scanning electron and Transmission electron Microscopes were used to identify its structure and characteristics. Furthermore, the analysis of UV-Vis spectrophotometer and HPLC is done to point out its activity on the photodegradation of dyes, deNOx reaction, and photoconversion of cellulose and carbon dioxides utilization.

Keywords: simple method, hollow, nano material, spinel, perovskite

1. Introduction

Material that has space or cavity inside or not solid within is called a hollow material. The surface of hollow material has more area than regular materials. For example, a cube-shaped material (**Figure 1a**) has six surface areas, but if its shape changes to a hollow cubic structure (**Figure 1b**), so that it has eight surface areas. For instance, the surface area of the hollow cubic unit cell is 1.333 times the surface area of a regular cube per unit cell. The difference in the surface area depends on the geometric shape of the material if it is cylindrical or tubular, the difference in a surface area becomes much large.

In nature, some inorganic compounds have hollow structures such as zeolites even though the size of the hollow has not in the range of the nano category. However, the utilization of the hollow zeolite structures turned out to be quite a lot, for example, as function as molecular sieves [1], absorbents [1], and selective catalysts [2]. Although the application categories that can be covered come in microns.

In line with the development of nano and hollow materials, the manufacture of nano hollow single-crystal zeolites was carried out and shown in **Figure 2** below.

One of the applications that can be covered is the nano-sized material, such as zeolite, one of which is the molecular sieve where the application of purification

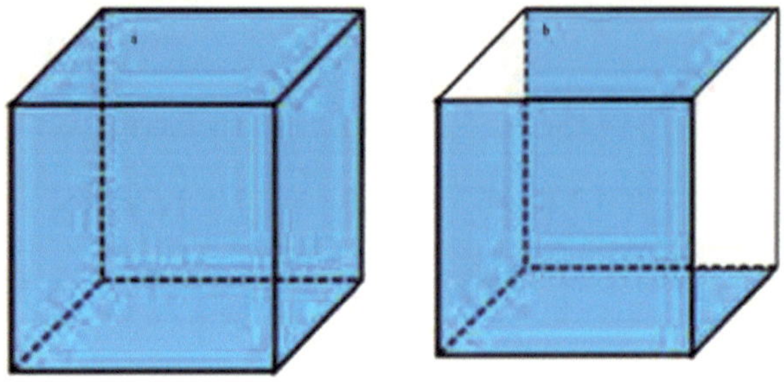

Figure 1.
The structure area of regular and hollow cubic shapes.

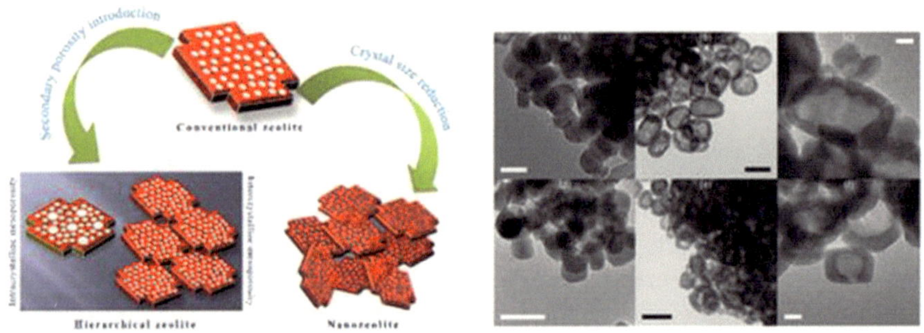

Figure 2.
A flowchart and the example of Zeolite nano hollow formation [3, 4].

or separation of pollutant particles from plastic contaminated water with nano-microns or microbes was able to be done [5].

Based on the study of specific surface area, load capacity, material transfer as well as storage, the size of the cavity makes hollow materials have extraordinary advantages in their characteristics. Having driven by these unique characteristics, the research groups eager to explore the more possible applications such as catalysis, photocatalysis, drug delivery, solar cells, supercapacitors, lithium-ion batteries, electromagnetic wave absorption, and sensors. The challenge faced in producing hollow materials at this time is to synthesize nano hollow materials which have a series of controlled structures in terms of composition and geometric configuration so that their applicative development is still constrained. However, the progress regarding the ability to manipulate both structure and morphology of nano hollow scale solid materials will have greater control over the local chemical environment [6–9].

Furthermore, the simple method used in the manufacture of nano hollow materials emphasizes the preparation process, economic review, and environmental friendliness for each of the chemicals used. This simple method is possible to produce nano hallow materials of various shapes such as nano hollow spheres (NHS), nano hollow cubes (NHC), nano hollow squared tubes (NHST), and related fibers. The applications described are the catalytic utilization of carbon dioxide into alcohol compounds, degradation of dyes, and the conversion of nano-cellulose to alcoholic sugars by photocatalysis.

2. Preparation methods

Hollow materials, in general, can be prepared using the Kirkendall effect and Ostwald ripening based on events, as well as the templating method (hard, soft, or one-pot/self-templating and free) based on the use of templates. In more detail, it described below:

2.1 Kirkendall effect

Kirkendall effect, a vacuum ordering occurs due to a change in the rate of diffusion between two or more components diffusing simultaneously. The process of different diffusion movements was proven experimentally by Smigelkas and Kirkendall [10] in 1947 that atomic diffusion occurs through the exchange of vacancies rather than by the direct replace of atoms. One example of this method is the preparation of metal oxides that can change the morphology of nanowires to nanotubes [11]. The example of nanowire formation based on Kirkendall effect is shown in **Figure 3**.

The mechanism explaining the formation of a cavity or hollow material in the inner direction could be described as follows: cations will flow rapidly outward through the oxide layer and flow inward from the void as a counterweight to the metal oxide interfacial void. Then, the direction of flow of the material is equalized by the direction of flow of the void through condensation into the pore or eliminating the crystalline defects. The direction of material flow can also result from the phenomena of diffusion and reaction pairs at the gas/solid or liquid/solid interfaces, the formation of deformations and vacancies, or both during the growth of metal oxide or sulfide layers [13, 14]. It should be remembered that the hollows produced in the metal-metal diffusion pair or near the metal oxide interfaces of an oxide growth do not produce mono-spheres in regular directions but form a very heterogeneous molecular collection.

2.2 Ostwald ripening

Ostwald Ripening is a phenomenon that is observed in solid solutions or liquid soles and explains changes in the structure of inhomogeneity with time, for example, small crystals or sol particles dissolving and being deposited back into crystals or larger sol particles. This phenomenon was first described by Wilhelm Ostwald in 1896 [15, 16] and is commonly found in oil-in-water [17] emulsions

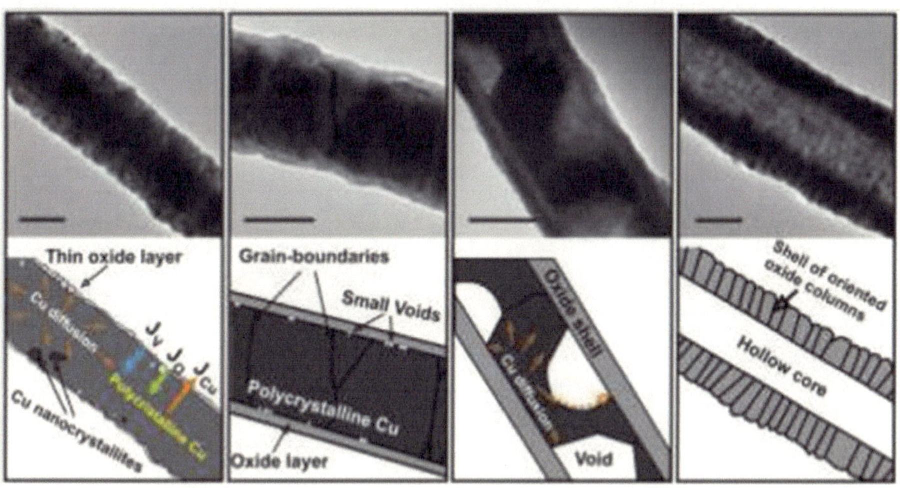

Figure 3.
The schematic formation of Hollow Cu nanowires based on Kirkendall effect during the thermal oxidation process in air at 300°C [12].

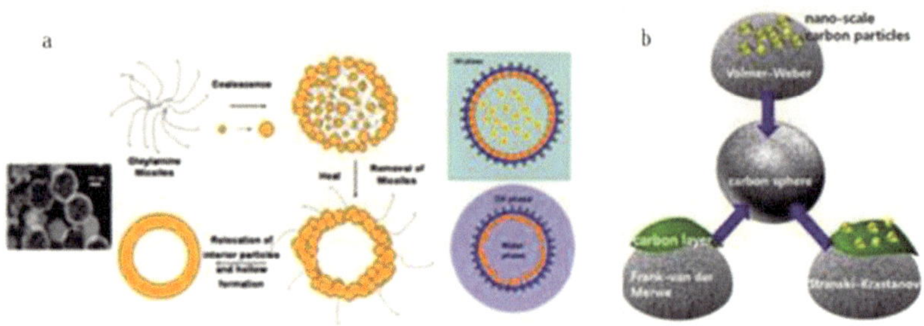

Figure 4.
Schematic of both w/o and o/w emulsion and hollow particles formation (a) using oleyamine micelles [19], and the growth of solid carbon sphere (b) based on Ostwald repining mechanism [23].

when flocculation is found in water-in-oil [18] emulsions. Schematically the w/o and o/w emulsions are presented below in **Figure 4a**.

Ostwald ripening mechanism is well-known through several growth methods, such as island formation [20], layer by layer formation [21], and the mixed layers and islands formation [22] as illustrated in a solidified growth of carbon sphere in **Figure 4b**.

The emulsion produced in the w/o or o/w system is affected by various factors such as pressure (Laplace and osmotic), the concentration of the dispersed phase, the concentration of surfactants, and the additives used. Furthermore, the emulsifiers or surfactants used are generally biopolymers such as various proteins (whey protein isolate (WPI), β-lactoglobulin, casein, soy protein isolate (SPI), and pea protein [24], polysaccharides such as xanthan, Arabic gum, modified starch, carrageenan, pectin, and modified celluloses frequently utilized to stabilize emulsions, especially O / W and W/O/W double emulsions [25].

2.3 The Smoluchowski process

The Smoluchowski process is a process to produce nano hollow complex materials in an "integrative" nature from colloidal particles. An example of this preparation was the manufacture of titanium oxide, TiO_2, and the yield observed by a high-resolution TEM [26]. The HRTEM TiO_2 micrograph showed that the tiny nanocrystallites stuck to each other in the aggregated end product while keeping the overall orientation unchanged. An example of the formation of particles based on the Smoluchowski mechanism is presented in **Figure 5** below.

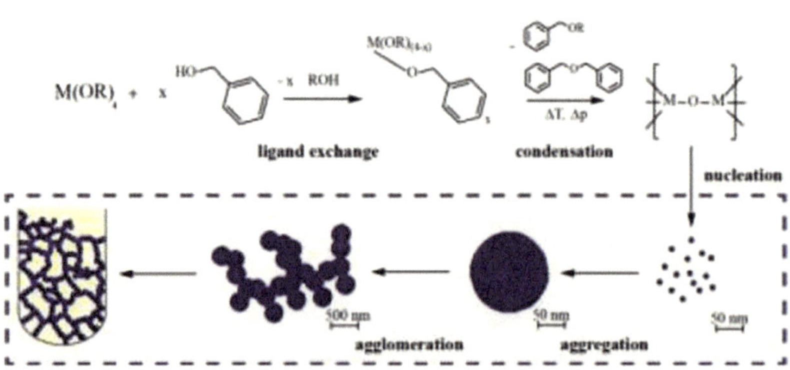

Figure 5.
An example of a particle formation mechanism based on the Smoluchowski process with an emphasis on agglomeration and aggregation [27].

2.4 Template methods

These methods can effectively control the morphology, particle size, and structure during the nanomaterial manufacturing process. In general, these methods consist of two types/categories, namely: hard methods and soft (or one-pot or self) templates according to different structures. The methods of templates in their preparation are insensitive, easy to operate, and practice.

2.4.1 Hard-template method

In principle, this method is for the preparation of one-dimensional hollow materials. Materials used as hard-templates are polymer microspheres, porous membranes, plastic foam, ion exchange resins, carbon fiber, and anodic aluminum oxide (AAO) [28, 29]. Because the templates and the resulting target products have a unique structure and influence the particle size range, they play an important role in many areas of application. Furthermore, after the desired target is obtained so that a template used is moved/separated or modified.

One example of using the hard template method is making the ordered mesoporous CeO_2 prepared via a hard-template method using SBA-15 as a structure-directing agent. Leaching with NaOH and thermal treatment at 500°C enabled the removal of the inorganic template, thus resulting in the formation of long-range ordered CeO_2. Nevertheless, small amounts of silica were present in the final oxides. The resulting CeO_2 samples were used as supports for Au nanoparticles as shown in **Figure 6** below.

2.4.2 The soft templating or the endotemplate method

The soft templating or the endotemplate method refers to supramolecular entities like self-assembled arrangements of structure-directing molecules such as surfactants, leading to mesopores up to 30 nm [31, 32].

In the soft template method as shown schematically in **Figure 7**, compounds that function as templates are organic compounds whose molecules form aggregates through inter-molecular or intra-molecular interactions such as hydrogen bonds, chemical bonds, and electrostatic forces. The metal cations as the target as the hollow material are deposited on the surface or in the inside of the aggregate. The process of placing metal cations in the aggregate carried out using electrochemical methods, precipitation, and other synthesis/preparation methods to form metal oxide or composite materials of various shapes and sizes. Organic compounds that commonly function as templates are surfactants, polymers, biopolymers, supramolecules,

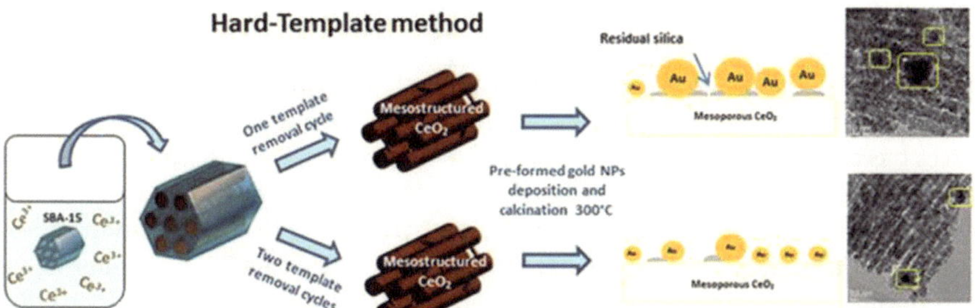

Figure 6.
Schematic pathways of Au doped CeO using hard template method [30].

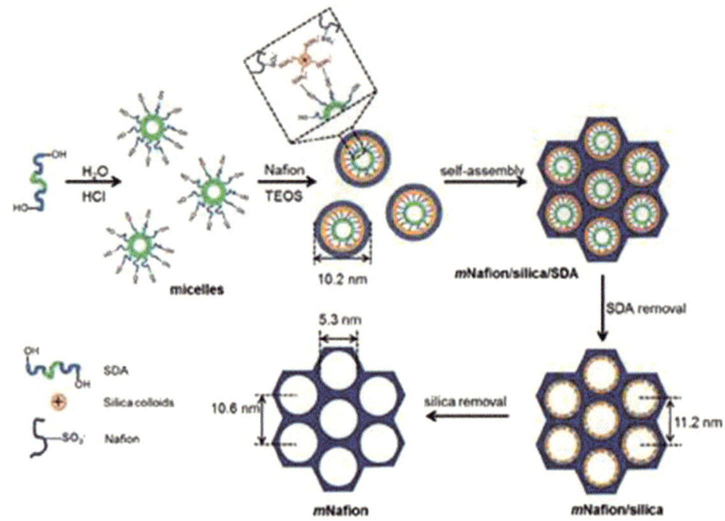

Figure 7.
Soft template pathways to produce hollow material [33].

and inorganic compounds. Based on the type of compound that can act as a soft template, it is possible to develop nanomaterial synthesis because this method has advantages such as simplicity of the process, repetition of the process with good results, and does not require removal of targets from the aggregates [34–38].

One example of a soft template method to generate ABO_3/AB_2O_4 nano hollow is spinel compounds of both Fe_3O_4 and $CoFe_2O_4$, respectively [39, 40]. Magnetite hollow spheres, Fe_3O_4 were prepared using a soft/free template with the solvothermal method described by Chen et al. [39] as follows: 13 g $FeCl_3.6H_2O$ was dissolved in 350 mL of ethylene glycol and diethylene glycol. Subsequently, 2 g NaAc, 2 g polyvinyl pyrrolidone (PVP), and sodium citrate (Na_3Cit) were added to the solution's ultrasonic processing. After an hour, the solution was sealed in a 400 mL Teflon-lined stainless-steel autoclave. The autoclave was heated to 210°C for 12 h and then cooled to room temperature naturally. The black products were collected by magnetic decantation and centrifugation, followed by repeated washing with deionized water and ethanol. The final products were dried in a vacuum oven at 50 C for 12 h. Another procedure with the same steps and only differs in the number of materials used and the washing process of the solution which turned black was washed with alcohol several times and dried at 60°C overnight. The diameter size of the product magnetite hollow spheres can be adjusted by changing the concentration of the added PVP [41]. Preparation of Fe_3O_4 using urea and PVP as a binder for Fe^{n+} cations gives nano hollow spheres as shown in the following figure.

Mandal et al. [41] have synthesized of hollow Fe_3O_4 particles via a one-step solvothermal approach for microwave absorption materials: effect of reactant concentration, reaction temperature, and reaction time as shown in **Figure 8j** below.

Then, another method of a template-free preparation of Fe_3O_4 nano hollow spheres has prepared by researcher Shi et al. [42] using the following procedure, hydrated ferrous chlorine salt ($FeCl_3.6H_2O$, 1.084 g) was dissolved in 80 mL of deionized water under rigorous and constant stirring for 10 minutes. Then added Na-citrate salt (2.352 g), PAM (0.8 g), and urea (0.72 g) while stirring vigorously for 30 minutes. The mixture was then transferred to Teflon and tightly closed before being placed into the autoclave and heated at 200°C and held at the temperature for 24 hours. Then cooled naturally with air. The result of a black precipitate Fe_3O_4 was washed with water and ethanol, separated by magnetic attraction, and finally dried

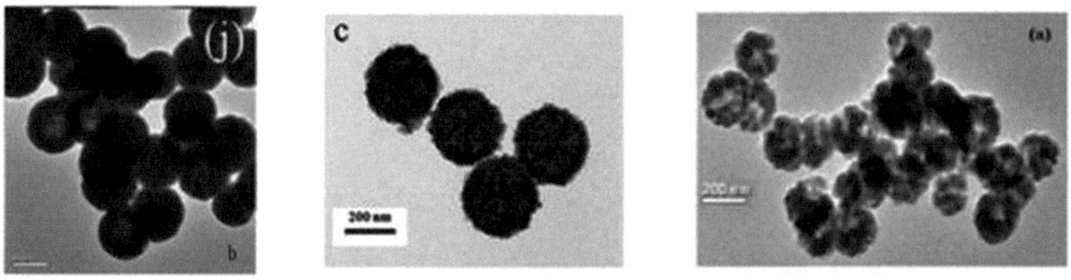

Figure 8.
The TEM results of NHS Fe_3O_4 (j), NHS Fe_3O_4 (c), and NHS $NiFe_2O_4$ spinel (a) using the solvothermal method.

at 50°C for 12 hours in an oven. An example of the results obtained by the research group of Shi et al. [42] is shown in **Figure 8c** below. Furthermore, $NiFe_2O_4$ nano hollow spinel preparation used a template-free method, namely the solvothermal process was carried out using oleyl amine capping agent. Hydrated chlorine salts of nickel ($NiCl_2.6H_2O$) and iron ($FeCl_3.6H_2O$) respectively mixed with urea with a 1:2 molar ratio. The solvent uses a mixture of ethylene glycol and ethanol with a ratio of 2:1. After all these substances put into a glass chemical 100 mL, added as much as 1 mL while stirring. After 30 minutes stir, the solution becomes transparent and homogeneous, then put the Teflon which is tightly closed and put into the autoclave steel and heated at 200°C for 24 hours. The product was then passed with ethanol and collected by separation and heated at 60°C for 30 minutes. Product samples were analyzed by TEM with a result in the following **Figure 8a** below [43].

2.5 Simple method

The simple method for producing hollow nanomaterials in question is in terms of the use of chemicals to produce nano hollow materials and environmentally friendly products. In the nano hollow material preparation, water and pectin or egg white solution is used as media. The procedure to obtain the nano hollow material is explained in brief here. The procedure to obtain the nano hollow material is explained in brief here. A stoichiometric amount of Ni (II) nitrate hydrates, ammonium vanadates, and Fe (III) nitrate hydrates were dissolved in distilled water, having compositions of $Ni_{1-x}V_xFe_2O_4$ under magnetic stirring for 1 h, respectively, followed by mixing each solution to make the final solution weight ratio between nitrates to pectin is 3:2. Adjust the pH = 11 in the above solution by an addition of ammonia, and heat it at 80 °C with continuous stirring to form a viscous gel. Then, dried the gel using the freeze dryer for 7 h to form the precursors' networks and calcined at 600 °C for 3 h. The results are shown in **Figure 9** below.

Figure 9.
TEM results of hollow material $Ni_{1-x}Fe_2O_4$ (where x = 0.1 – 0.5) were prepared using sol-gel method [44].

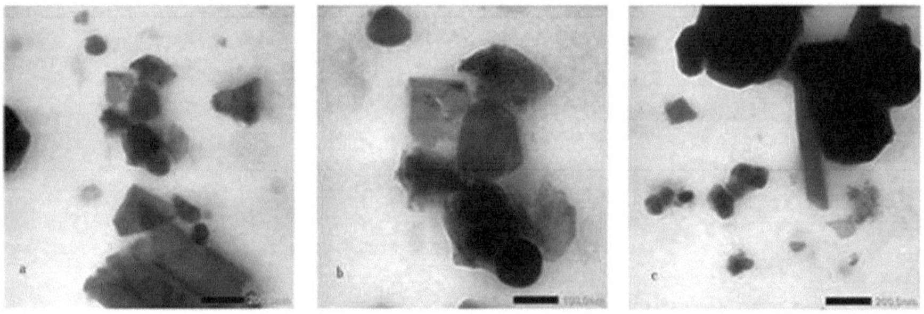

Figure 10.
TEM results of hollow nanomaterial $LaCr_{1-x}Mo_xO_3$ (x = 0.01-0.05) were prepared using sol-gel method [45].

Figure 11.
TEM and SEM results of $LaCrO_3$ and $LaCr_{1-x}V_xO_3$ materials prepared using pectin and egg-white solution [46, 47].

Figure 9b and **c** clearly show the formation of nano hollow cube (NHC) from $Ni_{1-x}V_xFe_2O_4$ (x= 0.1 – 0.5) spinel. Furthermore, in **Figure 9a**, if you notice there are the cubic hollow aggregate and also a squared nano hollow tube (SNHT).

Then, in **Figure 10a** the micrograph shows that squared hollow pipes, hollow cube, and hollow tubes formed. In **Figure 10b**, you can see the nano hollow cubes (NHC) and micron sizes and nano spherical tubes (NST). Whereas in **Figure 10c**, you can see the interconnected pillars of micron and nano hollow cube sizes.

In the preparation of both pure $LaCrO_3$ and modified $LaCrO_3$ by the sol-gel method [46] gave SEM micrograph results shown in **Figure 11a** and **b**. It seemed that the shapes of material are varied that are nano hollow cubes (NHC), nano hollow tubes (NHT), and the blended shapes presented in **Figure 11a**. In **Figure 11b**, the interconnected microfiber structure and the hollow micro material formed. Meanwhile, **Figure 11c** shows the homogeneous nanoscale grains of hollow $NiFe_{2-x}Co_xO_4$ spinels prepared using the egg white solution.

3. Applications

After the preparation of all the catalysts is done, it is used respectively for both thermic catalytic reactions and photocatalysis. The compounds that are the research targets are CO_2, NOx, dyestuffs, and cellulose. The selection of the four targets intensely focused on the impact factor and the benefits that can gain.

Carbon dioxide (CO_2) and NOx gas emitted from the use of fossil energy sources containing the main elements H, C, and O as well as other minor elements N, and S. The overall reaction can be described below:

$$\text{Substrate}(C, H, O, N, S) + O_2 \rightarrow CO_2 + NO_x + SO_x + H_2O + \text{Energy}$$

The greater use of energy sources for activities, causing the emission of CO_2, NOx, and SOx gases to increase [48]. Continuous emissions without treatment will cause acid rain and the greenhouse effect. This emission will stimulate global warming and even higher. One way to participate in the handling of COx and NOx wastes is through its utilization. One of the handling methods is using the nanocatalysts to handle thermally and photonically by converting the organic wastes (solid, liquid, and gaseous) such as cellulose, dyes, and COx and NOx pollutant into products that are economically valuable and environmentally friendly as described below.

3.1 Catalytic reaction: thermis

Catalytic reaction - thermic is a catalytic reaction that takes place with the help of thermal energy. These catalytic reactions control more than 90% of processes in the chemical industry [49]. In thermic catalytic research, the study is the hydrogenation reaction of CO_2 and the decomposition of NOx exhaust gases. The research results of this reaction are briefly presented below.

3.1.1 CO_2 hydrogenation reaction

The CO_2 hydrogenation reaction was carried out using the perovskite $LaCrO_3$, and spinel $Ni_{1-x}Fe_2M_xO_4$ catalysts (M = Cu, Co, and Zn) with the reactor scheme shown in **Figure 12a** below.

The catalytic reaction takes place at a temperature of 100 to 400°C with a composition of CO_2/H_2 = 1/3 in the gas flow. Examples of reaction results using rapid tests and several quantitative analyzes are shown in **Figure 12b** and **c**, respectively.

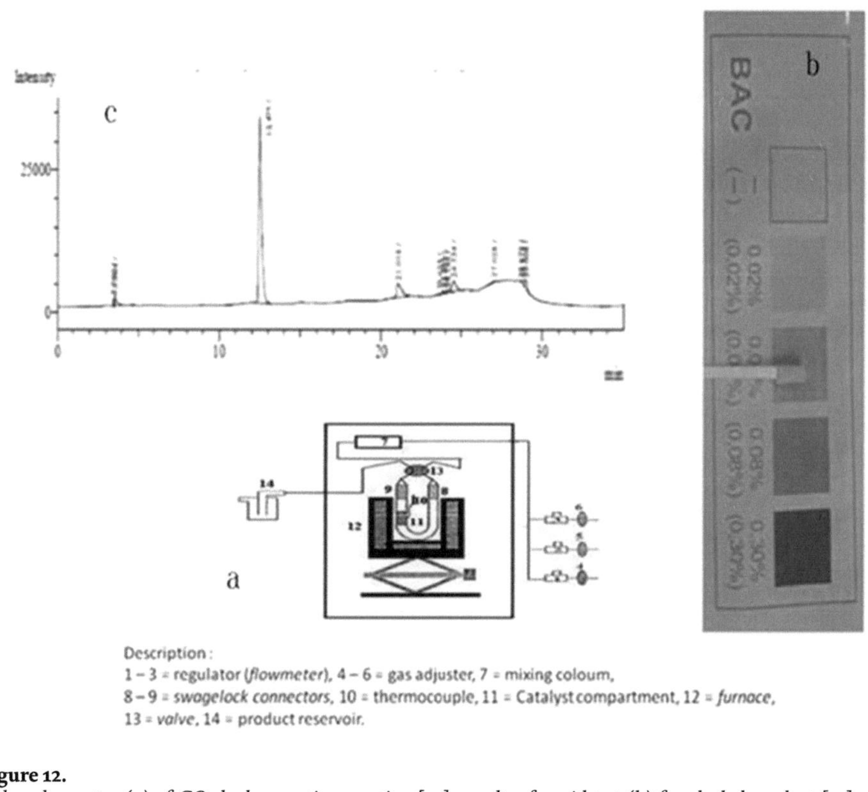

Figure 12.
Lab scale reactor (a) of CO_2 hydrogenation reaction [50], results of rapid test (b) for alcohol product [51], and chromatogram results (c) of the CO_2 hydrogenation reaction [52].

3.1.2 deNO$_x$ reaction

The decomposition reaction of NO_2 and NO or NOx is a type of reaction that uses a selective catalyst reduction (Selective Catalyst Reduction). In general, the catalyst (SCR) is used to reduce NOx, COx, and SOx emissions with the ability to reduce more than 90% of emission gases from boilers [53], power stations [54], and motorized vehicles [55] to be applicable. The results of the deNOx reaction research conducted by our team are presented in **Figure 13** below.

The NO_2 conversion results obtained using NiO/$LaCrO_3$ nanocatalyst (**Figure 13a**) is relatively better than those obtained using Fe/Zeolite Catalyst (**Figure 13b**) at the same reactant conditions and reaction temperature ranges.

3.2 Photocatalysis

Photocatalytic reactions are catalytic reactions that take place with the help of photon energy, so they are often called catalytic reactions - photonics. This reaction has been going on for a long time while the development is taking quite a while. It was a German chemist, Dr. Alexander Eibner who is firstly doing research in photocatalysis by irradiated ZnO in a concentrated Prussian blue solution and the solution became clear [58, 59]. Then, it has grown rapidly from 1964 until now, for various chemical reactions such as the production of hydrogen gas [60], and to photosynthetic-mimic reactions [61, 62]. Furthermore, our research related to photocatalysis is described below.

3.2.1 Dyes decomposition reaction

The textile and other industries usually use dyes in their products to make them look attractive. However, the remaining dyes have gone through a waste treatment process, especially in large factories but not necessarily in medium and small factories. As usual, the dye waste is thrown away into water bodies such as rivers and seas. Since the dye waste is very toxic and difficult to degrade naturally, so it can disturb the aquatic biota. One of the dyes that difficult to degrade and widely used in the small batik textile industry (home industry) is methylene golden yellow. Our research team also studied the decomposition of these dye compounds using $NiFe_2O_4$ nanocatalysts stimulated by sunlight and UV rays. An example of the result of the decomposition reaction is shown in **Figure 14** below [44].

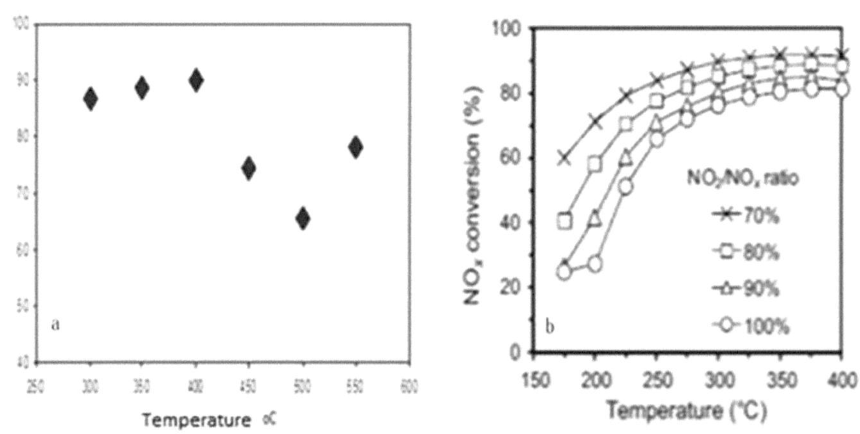

Figure 13.
Decomposition of NO$_x$ using catalysts (a) NiO/LaCrO$_3$ [56], and (b) Fe/Zeolite [57].

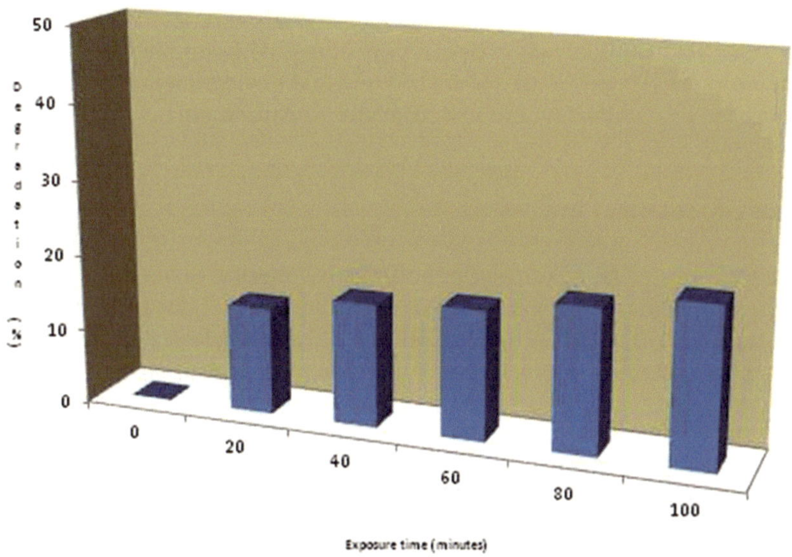

Figure 14.
RGY decomposition using $NiFe_2O_4$ nano hollow catalyst under the irradiated light of: (a) Sun, and (b) UV.

In the decomposition reaction of the remazol golden yellow dye under solar and UV irradiation, as shown in **Figure 14**, the difference in activity occurs because of sunlight contains UV rays and the nanocatalysts are active for both rays [63].

3.2.2 Cellulose conversion reaction

This type of reaction was studied considering the abundant availability of residual raw materials for agri-industrial products in Lampung Province and various conversion results such as glucose, xylitol, mannitol, sorbitol to fuel alcohol. The research team's target in the conversion of cellulose is a sugar alcohol, and the reaction takes place at room temperature and is environmentally friendly. The results achieved are shown in **Figure 15** below.

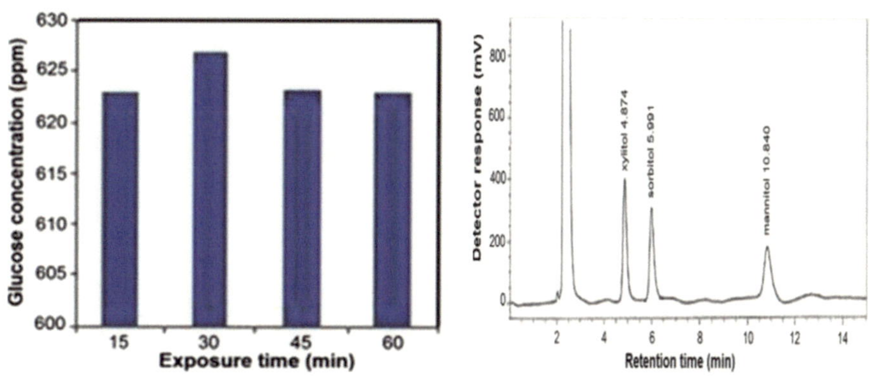

Figure 15.
Results of nano cellulose conversion (a) and the chromatogram of alcohol sugar (b) using HPLC [64].

4. Conclusion

The brief description of nano hollow materials presented in this paper is basically to provide an overview of the potential for nano hollow materials in managing reactions with results that are environmentally friendly and have economic value. Furthermore, nano hollow materials can be resulted using simple methods in terms of the chemicals used, economics point of view, and environmental considerations such as pectin, egg white, and monosaccharides in water media.

Acknowledgements

The author gratefully acknowledge both the Indonesian Government through the Directorate Research, Ministry of Research and Higher Education on the contract number 179/SP2H/LT/ADM/DRPM/2020 and the Research Institution and Community services of the University of Lampung for supporting this book chapter.

Author details

Rudy Tahan Mangapul Situmeang
Department of Chemistry, Faculty of Mathematics and Natural Sciences, University of Lampung, Indonesia

*Address all correspondence to: rudy.tahan@fmipa.unila.ac.id

References

[1] Gleichmann K, Unger B, Brandt A. Industrial Zeolite Molecular Sieves. IntechOpen Book Chapter. 2016: DOI:10.5772/63442.

[2] Zhang L, Wu Q, Meng X, Müller U, Feyen M, Dai D, Maurer S, McGuire R, Moini A, Parvulescu A-N, Zhang W, Shi C, Yokoi T, Pan X, Bao X, Gies H, Marler B, De Vos DE, Kolb U, and Xiao F-S. Recent advances in the preparation of zeolites for the selective catalytic reduction of NOx in diesel engines. RSC Reaction Chemistry & Engineering. 2019;4:975-985

[3] Koohsaryan E, Anbia M. Nanosized and hierarchical zeolites: A short review. Chinese Journal of Catalysis. 2016; 37(4):447-467

[4] Fodor D, Pacosová L, Krumeich F, and van Bokhoven JA. Facile synthesis of nano-sized hollow single crystal zeolites under mild conditions. Chemical Communication. 2014; 50(1): 76 – 78

[5] Wang D, Yang M, Zhang W, Liu Z. Hollow nanocrystals of silicoaluminophosphate molecular sieve synthesized by an aminothermal co-templating strategy. RSC CrystEngComm. 2016; 18 : 1000 – 1008. DOI: 10.1039/C5CE01798J

[6] Mao, D., Wan, J., Wang, J., and Wang, D. Sequential Templating Approach: A Groundbreaking Strategy to Create Hollow Multishelled Structures. Advanced Materials. 2018: 1802874

[7] Trong On D, Desplantier-Giscard D, Danumah C, Kaliaguine S. Perspectives in catalytic application of mesostructured materials. Applied Catalysis A: General. 2001;222: 299 - 357

[8] Davis ME. Ordered porous materials for emerging applications. Nature 2002;417: 813 - 821

[9] Gu W, Warrier M, Ramamurthy V, Weiss RG. Photo-Fries Reactions of 1-Napththyl Esters in Cation-Exchanged Zeolite Y and Polyethylene Media. Journal of the American Chemical Society. 1999;121: 9467 – 9468

[10] Smigelskas AD, and Kirkendall EO. Zinc diffusion in alpha brass. Trans. AIME 1947; **171:** 130-142

[11] Nakamura R, and Nakajima H. 2011. Book Chapter : Application of the Kirkendall Effect to Morphology Control of Nanowires: Morphology Change from Metal Nanowires to Oxide Nanotubes, The Institute of Scientific and Industrial Research, Osaka University, Japan. www.intechopen.com, Book Title : **Nanowires - Implementations and Applications** Edited by Dr. Abbass Hashim)

[12] El Mel A-A, Nakamura R, and Bittencourt C. The Kirkendall Effect and Nanoscience: hollow nanospheres and nanotubes. Beilstein J. Nanotechnology, 2015;6:1348 – 1361 doi: 10.3762/bjnano.6.139

[13] Birchenall CE. Closure to "Discussion of Kinetics of Formation of Porous or Partially Detached Scales". Journal of Electrochemistry Society. 1957; 103: 619 - 624

[14] Colson JC, Lambertin M, Barret P. in Proceeding 7th Int. Symp. Reactivity of Solids, J. S. Anderson, F. S. Stone., M. W. Robert Eds. (Chapman and Hall, London). 1972; 283 – 293.

[15] Ostwald, W. Z. Phys. Chem. 1900;34:495 - 503

[16] Ostwald, W. (1896). *Lehrbuch der Allgemeinen Chemie*, vol. 2, part 1. Leipzig, Germany.

[17] Soares SF, Fernandes T, Daniel-da-Silva AL, Trindade T. 2019. The

controlled synthesis of complex hollow nanostructures and prospective applications. Proc. R. Soc. A475: 20180677. http://dx.doi.org/10.1098/rspa.2018.0677

[18] Sheth T, Shesadri S, Prileszky T, Helgeson ME. Multiple nanoemulsions Nature Reviews Materials. 2020; 5:214-228

[19] Goswami MM. Synthesis of Micelles guided Magnetite (Fe_3O_4)hollow spheres and their application for AC Magetic Field Responsive Drug Release. Scientific Reports 2016; 6:35721/ DOI:10.1038/srep35721

[20] Roghabadi FA, Ahmadi V, and Aghmiuni KO. High coverage solution-processed planar perovskite solar cell grown based on the Stranski–Krastanov mechanism at low temperature and short time. RSC Advances 2016; **6**: 112677-112685

[21] Andersson TG. The Initial Growth of Vapour Deposited Gold Films. Gold Bulletin. 1982; 15: 7 - 11

[22] Schuck CF, Roy SK, Garrett T, Yuan Q, Wang Y, Cabrera CI, Grossklaus KA, Vandervelde TE, Liang B, and Simmonds PJ. Anomalous Stranski-Krastanov growth of (111)-oriented quantum dots with tunable wetting layer thickness. Scientific Reports. 2019; 9: 18179

[23] Ham H, Park N-H, Kim SS, Kim H W. Evidence of Ostwald Ripening during evolution micro-scale solid carbon spheres, Scientific Reports, 4: 3579 / DOI: 10.1038/srep03579

[24] He, W., Tan, Y., Tian, Z., Chen, L., Hu, F., & Wu, W. (2011). Food protein-stabilized nanoemulsions as potential delivery systems for poorly water-soluble drugs: Preparation, in vitro characterization, and pharmacokinetics in rats. International Journal of Nanomedicine, 6, 521-533

[25] Dickinson, E. (2011). Double emulsions stabilized by food biopolymers. Food Biophysics, 6 (1), 1-11).

[26] Penn RL, and Banfield JF. Imperfect oriented attachment: dislocation generation in defect-free nanocrystals. Science 1998;281: 969 - 971

[27] Stolzenberg P, Hamisch B, Richter S, Huber K, Garnweitner G. Secondary Particle Formation during the nonaqueous Synthesis of Metal Oxides NanoCrystals. Langmuir 2018; 34 (43): 12834 – 12844

[28] Ren N, and Tang Y. "Template-induced assembly of hierarchically ordered zeolite materials," *Petrochemical Technology*, 2005;34(5): 405-411.

[29] Masuda H, and Fukuda K. "Ordered metal nanohole arrays made by a two-step replication of honeycomb structures of anodic alumina," *Science*, 1995;268(5216):1466-1468

[30] Lolli A, Amadori R, Lucarelli C, Cutrufello MG, Rombi E, Cavani F, Albonetti S. Hard-template preparation of Au/CeO_2 mesostructured catalysts and their activity for the selective oxidation of 5-hydroxymethylfurfural to 2,5-furandicarboxylic acid. Microporous and Mesoporous Materials. 2016;226(15): 466-475

[31] Serrà A, and Vallés E. Microemulsion-Based One-Step Electrochemical Fabrication of Mesoporous Catalysts. Catalysts 2018; 8: 395

[32] Bradshaw D, El-Hankaria S, and Lupica-Spagnoloa L. Supramolecular templating of hierarchically porous metal-organic frameworks. RSC Chemical Society Reviews 2014; 000127

[33] Zhang J, Li J, Tang H, Jiang SP. Comprehensive strategy to design highly ordered mesoporous Nafion membranes

for fuel cells under low humidity conditions. Journal of Materials Chemistry A 2014;2(48): 20578-20587 DOI: 10.1039/C4TA02722A

[34] Lee J, Kim J, and Hyeon T, "Recent progress in the synthesis of porous carbon materials," *Advanced Materials*, 2006;18(16): 2073-2094

[35] Wan Y, Shi Y, and Zhao D, "Designed synthesis of mesoporous solids via nonionic-surfactant-templating approach," *Chemical Communications*, 2007;38(28):897-926

[36] Lu A-H, and Schüth F, "Nanocasting: a versatile strategy for creating nanostructured porous materials," *Advanced Materials*, 2006;18(14):1793-1805

[37] Y. Lu, "Surfactant-templated mesoporous materials: from inorganic to hybrid to organic," *Angewandte Chemie—International Edition*, vol. 45, no. 46, pp. 7664-7667, 2006.

[38] Wan Y, Yang H, and Zhao D. "'Host-guest' chemistry in the synthesis of ordered nonsiliceous mesoporous materials," *Accounts of Chemical Research*, 2006; 39(7): 423-432

[39] Chen Y, Zhang J, Wang Z, and Zhou Z. Solvothermal Synthesis of Size-Controlled Monodispersed Superparamagnetic Iron Oxide Nanoparticles. MDPI Applied Sciences 2019; 9: 5157; doi:10.3390/app9235157

[40] Borade RM, Shinde PR, Kale SB, and Pawar RP. Preparation, characterization, and catalytic application of $CoFe_2O_4$ nanoparticles in the synthesis of benzimidazoles. AIP Conference Proceedings 2018; 1953: 030194 doi: 10.1063/1.5032529

[41] Sarkar D, Mandal A, and Mandal K. Magnetic properties of Fe_3O_4 nano-hollow spheres. Journal of Applied Physics. 2012;112: 064318

[42] Shi L, He Y, Hu Y, Wang X, Jiang B, Huang Y. Synthesis of Size controlled hollow Fe_3O_4 nanospheres and their growth mechanism. Paticuology 2020; 49: 16 - 23

[43] Mandal D, Alam M, Mandal K. $NiFe_2O_4$ nano-hollow spheres with improved magnetic and dielectric properties. Physica B : Condensed Matter 2019; 554: 51 - 56

[44] Situmeang R, Romiyati R, Saputra AA, Sembiring S. $Ni_{0.5}V_{0.5}Fe_2O_4$ nano photocatalyst: Preparation, Characterization, and its activity on Remazol Golden Yellow degradationunder sunlight irradiation. Key Engineering Materials 2020; 840: 71 - 78

[45] Yuliarni T, Situmeang R, Simanjuntak W, Ratri CR. Synthesis and Characterization of $LaCr_{1-x}Mo_xO_3$. IOP Journal of Physics: Conference Series 2020; 1572: 012064

[46] Situmeang R, Supriyanto R, Kahar LAN, Simanjuntak W, Sembiring S. Characteristics of nano-size $LaCrO_3$ prepared through sol-gel route using pectin as emulsifying agent. Oriental Journal of Chemistry 2017; 33(4): 1705-1713

[47] Situmeang R, Manurung P, Sulistio ST, Hadi S, Simanjuntak W, Sembiring S. Sol-gel method for preparation of nanosize $NiFe_{2-x}Co_xO_4$ using egg-white. Asian Journal of Chemistry 2015; 27 (3): 1138 – 1142

[48] Situmeang R. Pectins as Emulsifying Agent on the Preparation, Characterization, and Photocatalysis of Nano-$LaCrO_3$ IntechOpen Publishing, London UK 2019:

[49] Herrmann J-M. Heterogeneous photocatalysis: fundamentals and applications to the removal of various types of aqueous pollutants. CatalysisToday 1999; 53: 115-129

[50] Situmeang R. unpublished reactor

[51] Djayasinga R. Pembuatan dan Karakterisasi Nanokatalis $Ni_{1-x}Cu_xFe_2O_4$ serta Uji aktivitas pada konversi (CO_2 + H_2). Thesis, Fakultas MIPA Universitas Lampung 2015.

[52] Situmeang R. *Unpublished chromatogram data*

[53] Sorrels JL, Chapter 2 Selective Catalytic Reduction Editor : David D. Randall, Karen S. Schaffner, Carrie Richardson Fry. RTI International Research Triangle Park, NC 27709 2019: 1 - 93

[54] Suárez-Ruiz I, and Ward C R. Chapter 4: *Coal Combustion*. Applied Coal Petrology 2008: 85 - 117

[55] Miller BG. Chapter 6: Emissions control Strategies for Power Plants. Academic Press, Coal Energy Systems 2005: 283 – 392

[56] Situmeang R. NiO/$LaCrO_3$ prepared by citric acid – freeze drying method for NO_2 conversion. Satek II 2007

[57] *Lambert CK. Perspective on SCR Noxfor diesel Vehicles. Reaction Chemistry and Engineering, 2019; 4: 969 – 974.*

[58] Iwasaki M, Shinjoh H. A comparative study of "standard", "fast" and "NO_2" SCR reactions over Fe/zeolite catalyst. Applied Catalysis A: General 2010; 390: 71-77.

[59] Eibner A. "Action of Light on Pigments I". Chem-ZTG. 1911; **35**: 753-755.

[60] Wang X, Maeda K, Thomas A, Takanabe K, Xin G, Carlsson JM, Domen K, and Antonietti M. A metal-free polymeric photocatalyst for hydrogen production from water under visible light. Nature Materials 2009; 8: 76 - 80

[61] *Coronado JM, Fresno F, Hernández-Alonso MD, Portela R. Design of Advanced Photocatalytic Materials for Energy and Environmental Applications. London: Springer. 2013: 1-5.*

[62] *Chu S, Yan Y, Hamann T, Shih I, Wang D, Mi Z. "Roadmap on Solar Water Splitting: Current Status and Future Prospects". Nano Futures. IOP Publishing Ltd. 2017;* **1***(2): 022001.*

[63] Situmeang R, Romiyati R, Simanjuntak W, Yuwono SD, Saputra AA, Sembiring S. $Ni_{1-x}V_xFe_2O_4$ Nano photocatalysts: The effect of Vanadium addition into its activity on Remazol Golden Yellow Degradation under visible light Irradiation. Hindawi Journal of Nano Materials 2020 under reviewing process.

[64] Situmeang R, Tamba M, Simarmata E, Yuliarni T, Simanjuntak W, Sembiring Z, and Sembiring S. $LaCrO_3$ nano photocatalyst: the effect of calcination temperature on its cellulose conversion activity under UV-ray irradiation. Advances in Natural Sciences: Nanoscience and Nanotechnology 2019; 10: 015009

Chapter 4

Preparation, Properties and Use of Nanocellulose from Non-Wood Plant Materials

Valerii Barbash and Olga Yaschenko

Abstract

The chapter describes the chemical structure and hierarchical organization of cellulose fibers, characteristics of non-wood plant raw materials (NWPM), and methods for preparing pulp and nanocellulose (NC). NWPM have the necessary reserves and properties to make up for a possible shortage of wood fiber for pulp production. The methodology for evaluating the efficiency of the delignification processes of plant raw materials is presented. A two-stage technology for producing pulp for the preparation of NC by environmentally friendly organosolvent methods of NWPM delignification is proposed. Methods for preparing nanocellulose are described. The technological parameters of the extraction of NC from pulp are discussed. The influence of NC on the properties of composite materials is analyzed. Areas of use for NC from NWPM are shown.

Keywords: non-wood plant, wheat straw, flax, kenaf, miscanthus, pulp, nanocellulose, paper, cardboard, thermoelectric material, composite

1. Introduction

In recent years, there has been a growing interest in the development of new biodegradable materials from environmentally friendly renewable plants. They are able to replace materials made from exhaustible natural resources—oil, gas, coal. Polymers from these fossils take hundreds of years to decompose, causing irreparable damage to the environment. Plastic accounts for 85 percent of all waste in the world's oceans, half of which are disposable plastic products [1, 2].

The European Parliament in March 2019 approved a new law banning single-use plastic products such as plates, cutlery, straws, plastics and food containers and expanded polystyrene cups [3]. Scientists and civil society organizations are working together to create new consumption patterns that meet the needs of all people, while eliminating waste and overconsumption, where the production of consumer goods is less dependent on the use of natural resources and makes the most of recycled materials [4, 5]. The use of natural polymers from cellulosic plant materials is being seen as an alternative to plastics and could be a viable approach to reducing deforestation, increasing the use of agricultural surplus and developing biodegradable materials. The development of environmentally friendly technologies of processing renewable plant sources contribute to the sustainable development of society, solving economic and environmental problems in the production of consumer goods [6–8]. The processing products of such renewable plant materials are widely used in the chemical, pharmaceutical, paper, medicine, textile and electronic industries [9, 10].

The main component of all plants is cellulose, which is the most abundant renewable biopolymer in nature with an estimated annual production of 1.5×10^{12} ton [11]. Cellulose is a structural component of the cell walls of softwood and deciduous wood, stalks and leaves of non-wood plants. A source of cellulose can be also bacteria, algae, and fungi [12, 13]. Cellulose $(C_6H_{10}O_5)_n$ is a stereoregular, semicrystalline polysaccharide consisting of a linear chain from several hundred to several tens of thousands of repeating units of β-D-glucopyranose (n), covalently linked by 1–4 glycosidic bonds (**Figure 1**).

The number of repeating units n is highly dependent on the source of the original cellulose (e.g. 10,000 in natural wood, 15,000 in cotton and 44,000 in the genus Valonia). To a lesser extent, the number of repeating unit n depends on the methods of preparation and purification (e.g., n = 250–500 in regenerated cellulose and n = 1000 in bleached kraft pulp) [11, 14]. The β-D-glucopyranose ring of the middle units of cellulose macromolecules contains three hydroxyl groups, which determine the chemical reactions, the ability to form intramolecular and intermolecular between different chains hydrogen bonds and properties of cellulose - solubility, thermal stability and mechanical properties [13, 15]. In the process of biosynthesis due to enzymatic polymerization of glucose monomers, glucan chains are formed, which independently form chains through van der Waals forces and are held together through hydrogen bonds, forming elementary fibrils. It was found that 36 cellulose chains lead to the production of elementary cellulose fibrils, the cross section of which has a size of 3–5 nm [16]. These elementary fibrils or nanofibrils have highly ordered regions (crystallites) that alternate with less organized (amorphous regions) [16, 17]. Then multiple elementary fibrils are brought together into larger units called microfibrils with a diameter of ~20–30 nm and length of several micrometers [18, 19]. The hierarchical organization of cellulose macromolecules in elementary fibrils and microfibrils of the plant cell wall is shown in **Figure 2**.

Cellulose is in the form of microfibrils, consisting of amorphous and crystalline domains in combination with other substances such as lignin, hemicelluloses, proteins, extractives and minerals, which constitute the main structural unit of plant cell walls [11], as shown schematically in the **Figure 2**. The proportion of plant fiber constituents depends on parameters like botanical origin, maturation time, climatic conditions, age of the plant etc. [20].

In world practice, the main consumers of cellulose are the pulp and paper industry for the production of paper and cardboard, and the chemical industry for the production of cellulose derivatives. Recently, cellulose has also attracted considerable interest as a source of raw materials for the production of nanocellulose (NC). Nanocellulose belongs to a group of nanomaterials consisting of the nanosized cellulose particles. The NC exhibit unique properties, such as high elastic modulus, high specific surface area, optical transparency, low thermal expansion

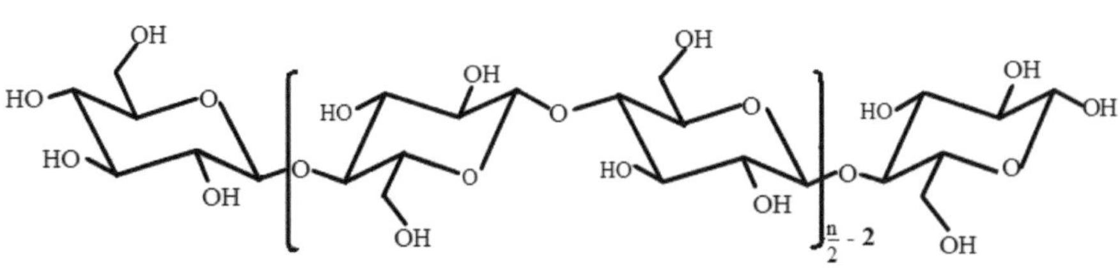

Figure 1.
Chemical structure of cellulose.

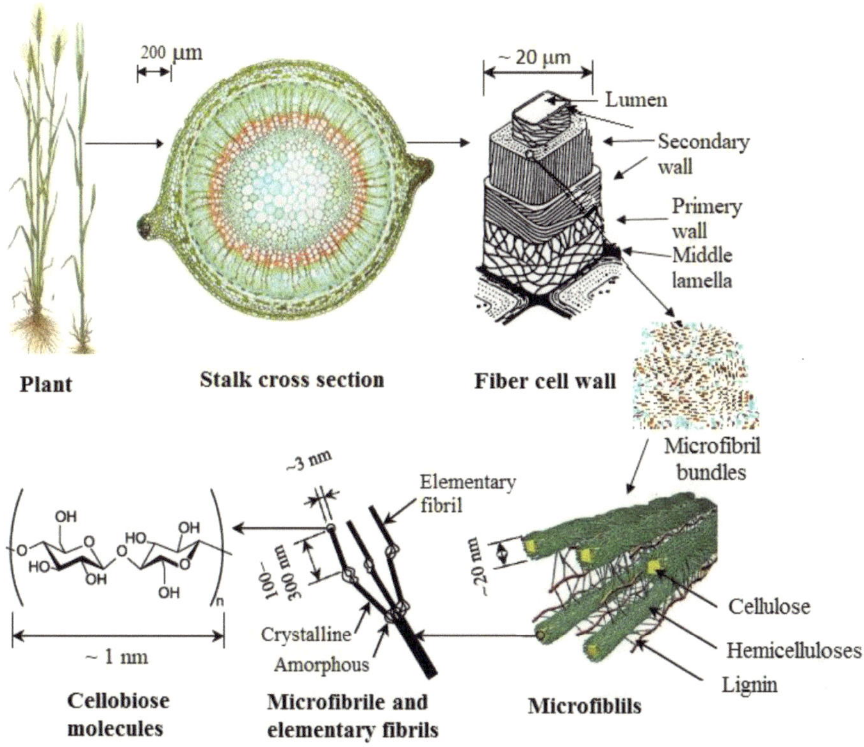

Figure 2.
The hierarchical organization of cellulose macromolecules in elementary fibrils and microfibrils of the plant cell wall (adapted from [15, 19]).

coefficient, and chemical reactivity [21–23]. NC has high transparency, biodegradability and biocompatibility, a low lightweight and production cost in comparison with synthetic polymers [24, 25].

The main raw material for cellulose production in the world pulp and paper industry is wood. For countries that do not have large reserves of free wood, alternative sources of fibrous raw materials may be non-wood plant raw materials (NWPM) - annual and perennial plants and fibrous waste from agricultural production. For example, in 2014, 172.6 million tons of pulp were produced from wood and only 13 million tons from non-wood fibers [26]. At the same time, in the world, forests occupy 3937 million hectares and agricultural plants 4932 million hectares [27]. World reserves of NWPM are estimated at 2.527 billion tons [28]. Almost half of all NWPM stocks are cereal stalks (1250 million tons), of which about half are wheat straw [29]. Non-wood fibers have a wide range of properties that are used for the production of cellulose-containing products [30].

In general, NWPM can be divided into two broad categories [31]:

a. common non-forest plants, which are considered as an alternative to deciduous wood, which include: straw of cereals, corn stalks, sorghum, bagasse, reeds, bamboo, esparto, natural herbs, etc.;

b. special types of plants that are considered as an alternative to coniferous wood, which include: cotton linter, flax, hemp, kenaf, etc.

The first category includes the predominant absolute reserves of non-wood plants. It contains 35 - 62% cellulose, 10 - 25% lignin and 18 - 36% pentosanes.

The fibers in it are shorter than the fibers of the second category and softwood, the length of the fibers of which is 0.3 - 2 mm. Fibers of the second category of plants contain 55 - 85% of cellulose, 1 - 10% of lignin and have stronger and longer (larger than 5 mm) fibers [32, 33]. For most annuals, the average fiber length is close to the length of wood fibers of some deciduous species, but less than the average length of coniferous wood fibers and some industrial crops. The fiber width of annual plants is 2 - 3 times thinner than the fibers of coniferous wood, but the ratio of the length of the fibers to their width in annual plants and in wood have the same order [34].

The chemical composition of the main components of NWPM differs from coniferous and deciduous wood. The content of cellulose, as the main component of raw materials, varies in wide ranges of values from 26% (bamboo) to 98% (cotton). Non-wood plant materials are distinguished by a high content of hemicellulose, especially pentosans to 30% [35]. Most NWPM have a lower lignin content from 6% (hemp) to 24% (bagasse) compared to wood (to 34%), which indicates the possibility of their use for pulp extraction [36]. Lignin of NWPM consists of guaiacylpropane, syringylpropane and oxyparaphenylpropane structural units, connected by simple ether and carbon–carbon bonds [37]. NWPM contain more minerals, but less lignin than wood, which a priori gives reason to expect a lower consumption of reagents for their delignification in comparison with the production of pulp from wood.

2. Preparation of nanocellulose

Properties of NC particles depend on the properties of plant raw materials and methods used for their production [38]. In world practice, for the production of NC, pulp with a minimum content of lignin, mineral and extractive substances, is usually used as a feedstock [39–42]. In this case, such pulp is obtained either by traditional methods of cooking with subsequent bleaching [43], or by environmentally friendly organosolvent methods of delignification [44].

In the global practice of pulp and paper industry, the dominating technologies to obtain pulp are sulphate and sulfite methods, which lead to environmental pollution [45]. Cooking pulp from NWPM in alkaline liquor is predominant because lignin from NWPM has a lower molecular weight than softwood lignin [37]. During cooking in an alkaline solution, the main ingredient of hemicellulose, xylan, is easily dissolved, which also opens additional channels for the cooking solution to penetrate into lignin, thereby facilitating the removal of lignin from the cell wall [46]. Alkaline methods for producing pulp from non-wood plants also include the NACO method based on alkaline oxygen cooking and the SAICA method. Experiments have shown that NACO-derived straw pulp has a lower yield and is inferior to the physico-mechanical properties of soda pulp and pulp obtained by the SAICA method [47, 48].

Alkaline methods include the neutral-sulfite method, which is used to obtain high yield pulp from hardwood and annual plants. Neutral sulfite pulp in comparison with sulphate pulp with the same degree of delignification has a 3–5% higher yield from plant raw materials due to less destruction of hemicelluloses and, therefore, is easier to grind [49]. It should be noted that among the main problems of organizing the process of obtaining pulp from NWPM by alkaline methods is the high content of silicates in them, which turn into black liquor during cooking and require additional technological solutions [45, 46].

Increased environmental requirements to the quality of wastewater and gas emissions of industrial enterprises requires the development of new technologies for processing of plant raw materials with the use of different organic solvents [50–52]. Organic solvents used in organosolvent methods of delignification differ in the chemistry of interaction with the components of plant raw materials and technological parameters of the pulp cooking process. The most developed organosolvent methods of plant delignification include the following methods: ASAE [53], ALCELL [54], Acetosolv [55], MILOX [56], Chempolis [57] and CIMV [58]. Each of them has its own advantages and disadvantages, but they are all united by relative environmental safety.

Methods for producing NC include mechanical, chemical, oxidative and enzymatic treatment of cellulose fibers [59–61]. The essence of the mechanical methods is an application of different forces to reduce the size of the natural cellulose fibers to nanoscale. For this, various mechanical processing is used: homogenization, grinding, microfluidization, ultrasonic treatments, ball milling, and cryocrushing [42, 62]. The use of mechanical methods for obtaining nanocellulose is characterized by significant energy consumption, for example, with multiple passages of the cellulose fibers through a high-pressure homogenizer, it is above 25 kW/kg [38]. In [63] have shown that the homogenization process is the most expensive method for nanomaterial isolation. To reduce energy consumption and fiber damage during mechanical processes, various pretreatments of cellulose are used: enzymatic treatment, alkaline treatment and chemical oxidation. As a result of the rupture of strong interfibrillar hydrogen bonding, the power required for the production of NC is significantly reduced, for example, from 20 to 30 kW/kg to 0.5 kW/kg of sulfite pulp [38].

Chemical methods are based on the cleavage of 1–4 glycosidic bonds of cellulose chains and isolation of cellulose nanocrystals with the removal of a part of the amorphous cellulose under the action of acids [64, 65]. For these purposes, the different acids are used: sulfuric, hydrochloric, phosphoric, maleic, hydrobromic, nitric, formic, p-toluenesulfonic [66–68]. Sulfuric acid is the most widely used acid for making NC. It reacts with the surface hydroxyl groups of cellulose to form negatively charged sulfonic groups and a stable gel. Otherwise, upon hydrolysis with hydrochloric acid, uncharged nanocellulose particles tend to flocculate in aqueous dispersions [69].

Recently, oxidizing agents such as 2,2,6,6-tetramethylpiperidine-1-oxyl radical (TEMPO) and phthalimide-N-oxyl (PINO) have been used to obtain NC. They improve the environmental friendliness and shorten the duration of the nanocellulose production process compared to hydrolysis, but have a higher cost than the above acids [70–72].

Enzymatic methods are based on the biosynthesis from monosaccharides or decreasing the size of the cellulose fibers by the fermentation. The enzymatic methods are time-consuming and require reagents that are more expensive. However, preliminary treatment of cellulose by enzymes before the mechanical grinding can decrease the energy consumption required for preparation of NC [73, 74]. For these reasons, a pre-treatment of the fibrous material is usually performed in order to decrease the size of the cellulose fibers and to ease the fibrillation and the process of nanocellulose preparation. The method of NC production by combining mechanical, chemical or biological pretreatment with homogenization treatment can not only reduce energy consumption, but also obtain NC with controllable size [75].

The various types of NC can be classified into different subcategories based on their shape, dimension, function, and preparation method, which in turn primarily depend on the cellulosic source and processing conditions [69].

Different terminologies have been used for the various types of NC. The Technical Association of the Pulp and Paper Industry (TAPPI) proposed standard terms and their definitions for cellulose nanomaterial WI 3021, based on the NC size [76]. NC is categorized into following kinds, such as cellulose nanofibrils (CNF), cellulose nanocrystals (CNC), amorphous nanocellulose, and cellulose nanoyarn [78]. CNF consist of a network of intertwined elementary nanofibrils, consisting of alternating crystalline and amorphous areas. CNF particles are

10–60 nm in width, 500–2000 nm in length, and high aspect ratio L/D > 50 [77]. CNF is usually obtained by some kind of mechanical treatment of softwood pulp without any pretreatment or after chemical or enzymatic pretreatment. CNC particles are extracted from pulp, usually by hydrolysis, and have a diameter from 4 to 20 nm, a length 100–300 nm, and low aspect ratio L/D > 5 [78].

The hydrolysis of cellulosic materials remains the most common commercial-scale CNC production method. The CNC yield after acid hydrolysis of pulp is 30–50% [62], and CNC films have brittle and rigid characteristics, which limits its use, for example, in flexible electronics. CNF has higher yield, good strength and good elasticity [79]. The combination of the intrinsic strength of CNF particles with the strong interaction between nanoparticles during drying makes it possible to obtain a more rigid and flexible film from CNF than from CNC [80].

Typically, higher acid concentrations, longer reaction times, and higher temperatures lead to higher surface charge and narrow sizes, but to lower yield and decreased crystallinity and thermal stability of cellulose nanocrystals [81].

3. Results and discussion

3.1 Methods of obtaining pulp for preparation of nanocellulose

Many factors determine the efficiency of the plant-based pulp production process, to which there has been a lot of research. These include technological, economic and environmental factors [31, 45, 57]. We are proposing to estimate efficiency of processes of delignification of plant raw materials by the diagram of dependence of pulp yield on the maintenance in it of residual lignin. For example, the dependence of pulp yield on the content of residual lignin for different methods of delignification of wheat straw is shown in **Figure 3**.

The proposed methodology for constructing diagram differs from the known lignin-carbohydrate diagrams of Ross, Geertz and Schmidt in the simplicity of construction, the essence of which is consists of the following [82]. On the y-axis the pulp yield

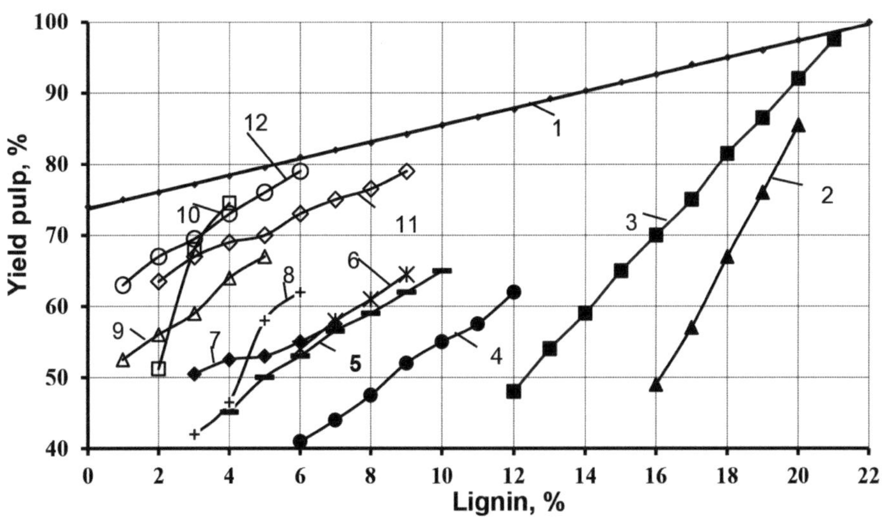

Figure 3.
The dependence of the pulp yield on the residual lignin content for different methods of delignification of wheat straw: 1- line of "ideal" delignification; 2 - Acetic; 3 - Ester; 4 - Soda; 5 - two-stage alkali-alcohol; 6 - two-stage alkali-alcohol + AQ; 7 - Neutral-sulfite; 8 - Bisulfite; 9 - Alkaline- sulfite-alcohol + AQ; 10 - Peracetic; 11 - Ammonium-sulfite-alcohol; 12 - Ammonium-sulfite-alcohol + AQ.

is indicated from 30% (for better visualization on the few percent lesser than cellulose content is in the plant raw material) to 100%. On the y-axis the point corresponding to holocellulose content is also indicated. On the x-axis, the percentage value of the lignin content in pulp is indicated from zero to maximum value in plant raw material. The intersection of horizontal axis at 100% yield and vertical axis of lignin content creates the point corresponding to initial composition of all plant components. The line, which links this point with the point of holocellulose content in plant raw material, can be considered as the line of ideal delignification‖. It characterizes maximal polysaccharide content for certain residual lignin content in pulp. Further, on the lignin-carbohydrate diagram, the dependencies of the yield on the residual content of lignin in the pulps obtained by different methods are plotted. So the closer the line of certain delignification method is to the line of ideal delignification, the higher is polysaccharide yield in the obtained pulp and thus delignification method is more efficient.

The dependencies presented in the diagram (**Figure 3**) allowed concluding that investigated delignification methods with approaching to the line of ideal delignification, i. e. with the increased efficiency of obtaining pulp from wheat straw, can be located in following sequence: Acetic – Ester – Soda – Neutral- sulfite – Bisulfite – two-stage alkali-alcohol – two-stage alkali-alcohol + AQ - Alkaline- sulfite-alcohol + AQ – Peracetic – Ammonium-sulfite-alcohol – Ammonium-sulfite-alcohol + AQ. This methodology is applicable to assess the efficiency of the processes of obtaining pulp from one type of raw material using different methods [83], and for a comparative assessment of the delignification of various types of plant raw materials by one method [84].

To obtain cellulose with a minimum residual content of lignin and minerals, we used a two-stage method of delignification of NWPM [85–89]. Peracetic acid (PAA) was used as a reagent for the delignification of NWPM, which has bleaching properties. PAA leads to minimal fiber damage and is environmentally friendly [90]. We have already demonstrated the possibility of obtaining straw pulp by means of organosolv delignification in the system of isobutyl alcohol–H_2O–KOH–hydrazine, which makes it possible to reuse the organic component and waste cooking liquor without regeneration [85]. At the same time, the waste liquor is divided into two layers: the upper organic solvent layer and the lower aqueous layer to which has moved the bulk of soluble minerals and organic substances from plant raw material (lignin, hemicelluloses, and extractives). The use of potassium and nitrogen compounds in the cooking liquor allows the use of waste liquor in the manufacture of fertilizers. **Table 1** shows the stages of preparation and indicators of the pulps from various plants, which were used to obtain nanocellulose.

The data in **Table 1** show that carrying out the two-stage thermochemical treatment of NWPM makes it possible to almost completely remove lignin and minerals and obtain a pulp with a content of non-cellulose components of no more than 1%. The obtained organosolvent pulps are not inferior in quality, if not superior, to bleached sulphate pulp from softwood, and therefore were used for the production of nanocellulose.

3.2 Properties of nanocellulose

We obtained nanofibrillated cellulose from air-dry bleached sulphate pulp of softwood using mechanochemical treatment [91]. It was carried out on grinding equipment, and hydrolysis with sulfuric acid solutions of various concentrations at temperatures from 20 to 60° C for 5–60 minutes. An increase of the acid concentration from 18–43% has led to an increase of the mechanical properties of the nanocellulose films. Further increase of the acid concentration above 50% leads to a sharp decrease of all strength properties and led to the formation of films with the

Pulp from a plant and a method of obtaining	Stages of preparation of pulp	Yield, %	Lignin, %	Ash, %
Wheat straw, isobutanol* [85]	I – isobutanol II – PAA**	49.0 41.5	1.1 0.2	1.63 0.2
Wheat straw, PAA** [86]	I – NaOH*** II - PAA	54.9 51.1	9.8 0.4	0.98 0.09
Flax fiber, PAA [87]	I – PAA II - NaOH	68.2 52.8	1.7 0.02	0.5 0.04
Kenaf, PAA [88]	I – PAA II - NaOH	61.2 51.2	0.37 0.29	0.24 0.18
Miscanthus, PAA [89]	I – PAA II - NaOH	50.7 57.0	0.25 0.08	0.96 0.04
Bleached sulphate softwood pulp	I – grinding to 93 °SR	—	0.23	0.21

**isobutanol: mixture of isobutanol-KOH–hydrazine, 120 min at 95 ± 2°C.*
***PAA: mixture of acetic acid and hydrogen peroxide in a volume ratio of 70:30%, 120 min, at 95 ± 2°C.*
****- solution 5% NaOH, 120 min, at 95 ± 2°C.*

Table 1.
Indicators of pulps from various plant materials for the production of nanocellulose.

brownish color. We recommended a reduced sulfuric acid concentration of 43% at 60°C during 60 minutes as the main process parameters for the production of NC by hydrolysis of organosolvent pulp from NWPM [87–89]. Such conditions agree well with data in [92] and are economically more favorable than traditional conditions for hydrolysis of cellulose with 60–65% sulfuric acid at 40–50°C for 1–2 h [93]. We used never-dried organosolvent pulps from NWPM to prepare NC. Never dried pulp is better than once dried sample, as the drying process leads to cornification of the fibers, which reduces the impregnation of the fibers with chemicals during their hydrolysis. Using of never-dried pulp does not require the consumption of energy for drying and grinding since dried cellulose fibers lose the ability to swell and percolate due to irreversible cornification.

Hydrolyzed NC was washed three times with distilled water by centrifugation at 8000 rpm, followed by dialysis to achieve a neutral pH and ultrasonic treatment during 30–60 min. As a result, the suspension took the form of a homogeneous gel-like dispersion and was stored in sealed containers for further research in order to determine the physical and mechanical characteristics of the NC. The prepared suspensions were poured into Petri dishes and dried in the air at a room temperature to obtain NC films. The structural change and crystallinity index of organosolvent pulps and NC were studied by means of SEM and XRD techniques. TEM and AFM methods were used to determine the particle size of nanocellulose (**Figure 4**). Transparency of the NC films was determined by electron absorption spectra, and tensile strength - according to ISO 527-1. The indicators of the obtained NC from NWPM is shown in the **Table 2**.

As can be seen from **Figure 4** and the data in **Table 2**, the process of hydrolysis and ultrasonic treatment of pulps leads to the formation of nanosized particles. NC had homogeneous and stable nanocellulose suspension. The nature of stabilization of the colloidal suspension is explained by the presence of charged groups on the surface of nanocellulose, which are formed by the interaction of cellulose with sulfate acid due to the esterification reaction. The structure of the NC films, according to SEM, TEM and AFM data, was similar to the structure of the film obtained from nanofibrillar cellulose [95]. The obtained NC had high tensile strength from 42.3 to 70 MPa and Young's modulus from 8.9 to 11.45 GPa. The obtained samples

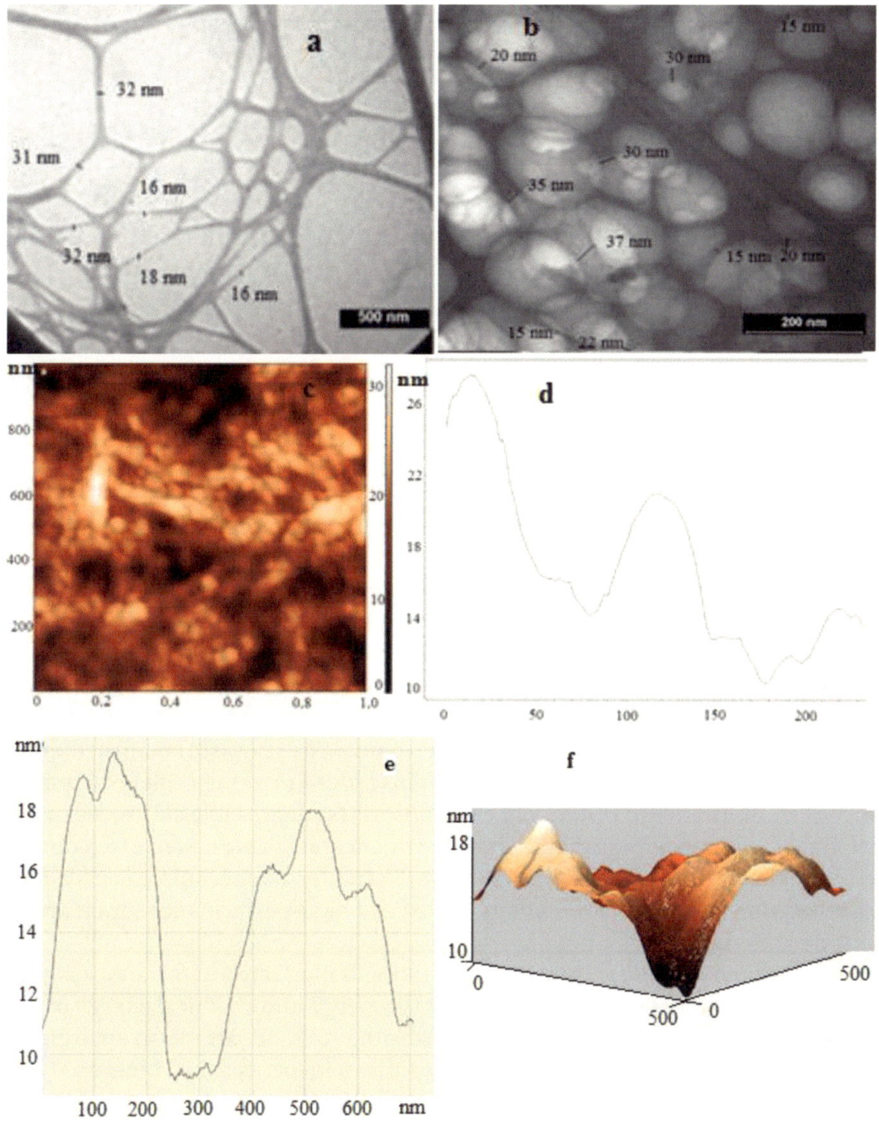

Figure 4.
TEM images of nanocellulose prepared by hydrolysis from bleached sulphate pulp (a) and organosolvent wheat straw pulp (b); AFM images of nanocellulose from kenaf (c) and lateral size of its nanocellulose surface (d); and AFM images of nanocellulose from miscanthus: The lateral size of the nanocellulose surface (e) *and 3D projection (f) with definition of sample height tapping mode.*

of NC from NWPM physical and mechanical parameters, comparable to the values obtained by other researchers. For instance, Young's modulus values of films generated from nanofibrillated bleached pulp, wheat straw, and recycled newspaper were between 6,0 and 9,0 GPa [96]. The positive results of obtaining NC from NWPM are given in other sources [97–99]. The properties of NC from NWPM exhibit great potential for their application to new nanocomposite materials and consumer goods.

3.3 Application of nanocellulose

Due to its unique properties, nanocellulose is widely used in various fields: in the production of electronic devices and composites, as a natural material for replacing synthetic reinforcing substances in the paper, chemical, pharmaceutical, cement

Nanocellulose from plant	Density, g/cm^3	Particle diameter, nm	Tensile strength, MPa	Crystallinity index, %	Transparency, %
Wheat straw isobutanol [94]	1.3	10–40	42	72.2	70
Wheat straw PAA [86]	1.27	16–20	123	71.3	78
Flax [87]	1.37	20–60	70	62.0	60
Kenaf [88]	1.39	10–28	65	80.0	72
Miscanthus [89]	1.32	10–18	62	76.7	74
Bleached sulphate softwood pulp [91]	1.38	15–30	88	79.8	78

Table 2.
The indicators of the obtained nanocellulose from NWPM.

industries [100–104]. In the last few years, the cellulosic biopolymer-based green electronics is supplemented by lightweight, portable, flexible NC-based power generators to provide energy to wearable electronics through harvesting mechanical energy in triboelectric and piezoelectric appliances [105, 106]. Huge ecological advantage of the NC-based electronic and thermoelectric devices is their inherent biodegradability. As these devices have become ubiquitous in modern society, and are prevalent in every facet of human activities, and the lifetime of electronics get shorter and shorter, the pressure on electronic waste (e-waste) management systems is mounting with no abate insight. This poses a growing ecological problem, and an alternative to traditional electronics is biodegradable electronics as the most viable replacement to address the issue of uncontrollable e-waste to reduce the environmental footprint of devices [107, 108]. Recently, nanocellulose is used for the preparation of porous carbon that be used as a high-performance supercapacitor electrode [109–111].

The use of nanocellulose in energy harvesting is illustrated in the following articles [112–114]. We used conversion of solar energy into chemical energy of biomass of fast-growing perennial herb Miscanthus to fabricate via an environmentally friendly method of organosolvent delignification low-cost NC films used as substrates for the creation of new biodegradable thin film thermoelectric material [115]. To do this, we applied a 0.72-μm CuI film onto a 12-μm NC substrate using a low-temperature, low-cost and scalable sequential ion layer adsorption method and thus obtained a lightweight and flexible biodegradable CuI thermoelectric material CuI/NC. We found out that nanostructured p-type semiconductor CuI film in the CuI/NC thermoelectric material is quite dense and completely covers the NC surface. The determined value of the Seebeck coefficient (*S*) is about 228 μVK^{-1} more than an order of magnitude higher than the Seebeck coefficients that were measured earlier for the nanocellulose-derived organic and composite materials contained poly(3,4-ethylenedioxythiophene)–poly(styrenesulfonate) (PEDOT:PSS), silver nanoparticles and carbon nanotubes [116]. The coefficient *S* obtained by us even exceeds the *S* values for the different thin-film composites of such well-known inorganic thermoelectric material as Bi_2Te_3 [117]. At that, *S* is constant in the temperature range 290–335 K, which is favorable for the use of CuI/NC as new thermoelectric material for an in-plane design of biodegradable flexible thin film thermoelectric generator. The thermoelectric power factor of CuI/NC is about 36 $\mu W{\cdot}m^{-1}{\cdot}K^{-2}$ is higher than that of the best examples bacterial nanocellulose films with embedded highly dispersed carbon nanotubes networks (20 $\mu W{\cdot}m^{-1}{\cdot}K^{-2}$) in article [118]. At temperature gradient of 50 K, the single *p*-CuI thermoelectric leg made from CuI/NC generates Voc = 8.4 mV. It corresponds to the power density 10 μW/m^2. Due to the complex of these properties, the developed environmentally friendly biodegradable flexible thin-film thermoelectric material can be effectively used to convert low-grade waste heat into electricity at temperatures close to room temperature [115].

The use of NC during paper formation or even coating of dry formed paper can improve interfiber bonding, softness and printability and, consequently, physical-mechanical properties [119–121]. **Figures 5–7** show the results of using NC from NWPM for the production of mass grades of paper and cardboard [122, 123].

As can be seen from the data in **Figures 5–7**, the application of NC to the surface of paper and cardboard has a positive effect on their physical and mechanical properties. Low consumption of NC allows production of the paper and cardboard with properties that meet the requirements to appropriate standards and replacement of synthetic reinforcing materials [123]. The increase in the values of the indicators of paper and cardboard occurs due to the creation of new hydrogen bonds between the fibers of cellulose and NC, which is confirmed by other authors [124, 125]. NC often replaces such well-known material, as glass and certain polymers, which are not biodegradable at ambient conditions. Modern technologies allow use NC

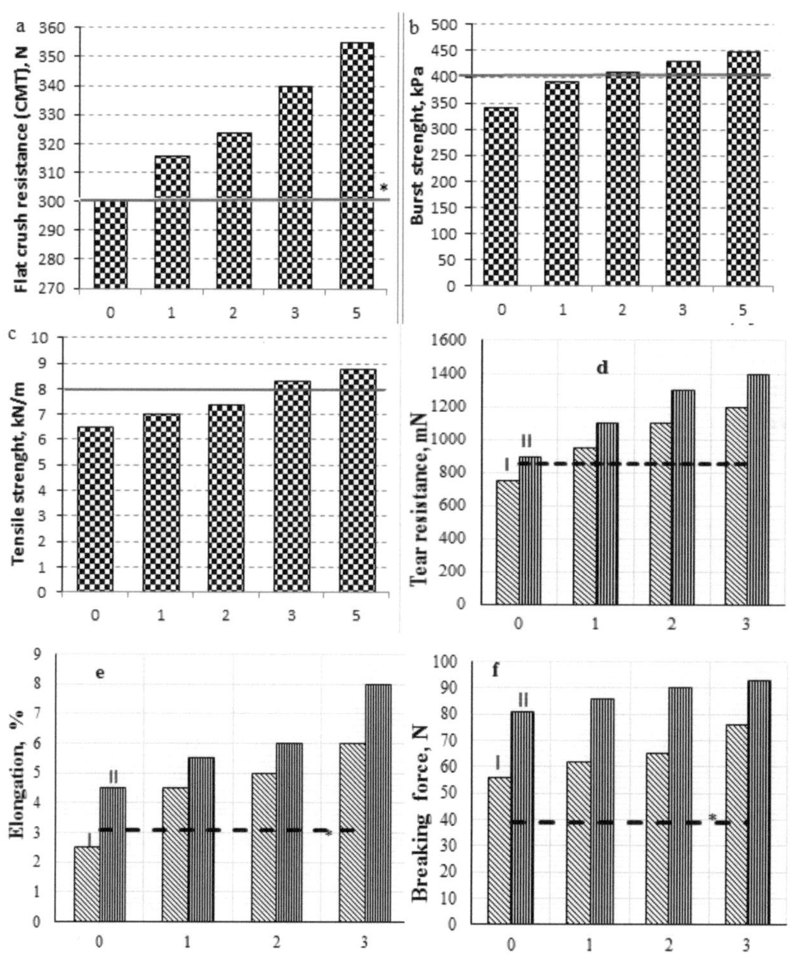

Figure 5.
Properties of paper for corrugating (a, b, c) and paper for bags (d, e, f) from waste paper (I) and unbleached pulp (II) with different consumption (g/m²) of nanocellulose from NWPM per paper surface; asterisk or dotted line - line of standards requirements.

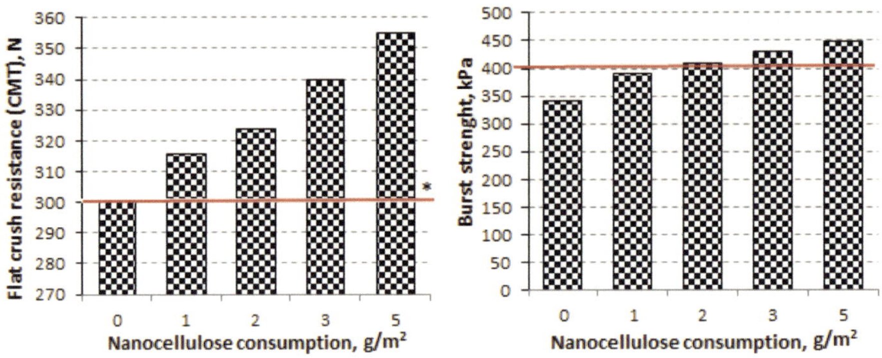

Figure 6.
Properties of cardboard for flat layers of corrugating cardboard with different consumption of sizing agents on 1 m²: Without application (1); 7 g of glue (2); 3.5 g of nanocellulose (3); 3.5 g of glue and 3.5 g of nanocellulose (4); 7 g of nanocellulose (5); asterisk: Line of standard requirements.

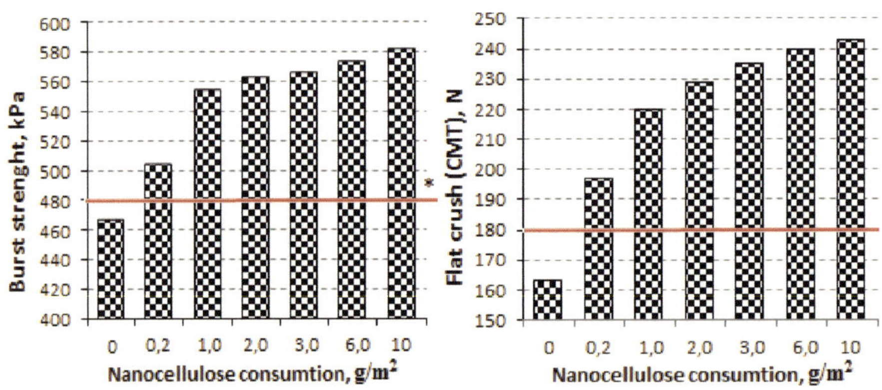

Figure 7.
Properties of recycled cardboard with different nanocellulose consumption; asterisk: Line of standard requirements.

in energy storage systems [126], biosensors [127], as well as in various electronic and optoelectronic devices [128, 129]. Among them, transparent transistors, light emitting diodes, solar cells, antennas and radiofrequency identification devices, high-performance loudspeakers, and lightweight actuators [130, 131].

Nano-sized cellulose fibers are considered as promising candidates for the production of nanocomposites. NC was added to polymer matrices to obtain reinforced composites with mechanical strengths from ten to one hundred times and to improve barrier properties [132–134].

Loading of structural materials with NC particles makes it possible to reduce their weight while maintaining the strength of the composites [135, 136]. For instance, addition of 6.5% nanofibrils improved the tensile strength and elongation at the break of the nanocomposite from cassava starch and polyvinyl alcohol by 24% and 51%, respectively. At the same time, the water vapor permeability and water solubility of the nanocomposite containing high contents of nanofibrils decreased up to 20% and 30%, respectively, in relation to the control blend [137]. A high effect of reinforcement was observed even at a low content of CNC when used to obtain nanocomposites with a matrix of natural rubber. With the addition of only 2.5 wt % CNC, which were isolated from soybean husks by acid sulfur hydrolysis, the elastic modulus of the composite was about 21 times higher than that of a pure rubber matrix [138]. In [139] it was shown that the addition of 10% NC from miscanthus to a composite based on epoxy resin Eposir-7120 with a polyethylene polyamine hardener increases the elastic modulus of the composite by 12.2% with respect to the control mixture. Otherwise, adding 5% NC from Colombian fique to acrylic hydrogels made it possible to obtain a reinforced hydrogel with 2.5 times higher compression resistance values than the resistance of the original hydrogel [140].

4. Conclusions

With the increasing requirements for environmental protection, there is a need to replace exhaustible sources - oil, gas, coal, and existing forest resources with biodegradable and renewable, including non-wood plant raw materials (NWPM). NWPM have the necessary reserves and properties to make up for a possible shortage of wood fiber for pulp production. To obtain pulp suitable for the production of nanocellulose (NC), a two-stage technology for delignification of NWPM with reagents that does not contain sulfur and chlorine has been proposed. NC has unique physical and mechanical properties and can replace well-known materials such as glass and some polymers, which are not biodegradable under ambient conditions. Methods for preparing nanocellulose are described. The influence of the main technological parameters of the cellulose hydrolysis process on the properties of nanocellulose is discussed. It is proposed to carry out the hydrolysis of cellulose using 43% concentration of sulfuric acid. Examples of the use of nanocellulose in various industries are given.

Acknowledgements

The authors are grateful to the co-authors of the previous articles for carrying out our joint research and to the Ministry of Education and Science of Ukraine for financial support.

Conflict of interest

The authors declare that they have no known competing financial interests or personal relationships that could have appeared to influence the work reported in this paper.

Author details

Valerii Barbash* and Olga Yaschenko
National Technical University of Ukraine "Igor Sikorsky Kyiv Polytechnic Institute", Kyiv, Ukraine

*Address all correspondence to: v.barbash@kpi.ua

References

[1] Zarfl C, Flet D, Fries E, et al. Microplastics in oceans. Marine Pollution Bulletin.2011;62:1589-1591. DOI: 10.1016/j.marpolbul.2011.02.040

[2] Jambeck JR, Andrady A, Geyer, R., Narayan, R., Perryman, M., Siegler, T., Wilcox, C., Lavender Law, K. Plastic waste inputs from land into the ocean, Science. 2015;347:768-771. https://jambeck.engr.uga.edu/landplasticinput

[3] Schlanger Z. The problem with turning to paper after the EU's plastic ban. 2019. [Internet]. Available from: https://qz.com/1582520/the-problem-with-turning-to-paper-after-the-eus-plastic-ban.[Accessed: 2020-07-23]

[4] Haggith M, Kinsella S, Baffoni S, Anderson P, Ford J, Leithe R, Neyroumande E, Murtha N, Tinhout B. The state of the global paper industry. Environmental Paper Network. 2018. 89 p. http://www.environmentalpaper.org

[5] Huang C, Wang X, Liang C, Jiang X, Yang G, Xu J, Yong Q. A sustainable process for procuring biologically active fractions of high-purity xylooligosaccha-rides and water-soluble lignin from Moso bamboo prehydrolyzate. Biotechnology for Biofuels. 2019;12(1):189. DOI: 10.1186/s13068-019-1527-3

[6] Baghban MH, Mahjoub R. Natural Kenaf Fiber and LC3 Binder for Sustainable Fiber-Reinforced Cementitious Composite: A Review. Appl. Sci. 2020;357:1-15. DOI: 10.3390/app10010357.

[7] Dong H, Zeng L, Yu P, Jiang Q, Wu Y, Huang C, Yin B. Characterization and application of lignin-carbohydrate complexes from lignocellulosic materials as antioxidant for scavenging in vitro and in vivo reactive oxygen species. ACS Sustainable Chemistry & Engineering. 2019;8(1):256-266. DOI: 10.1021/acssuschemeng.9b05290

[8] Jasmani L, Thielemans W. Preparation of nanocellulose and its potential application. Forest Res. 2018;7:1-8. DOI: 10.4172/2168-9776.10002 22

[9] Sagar NA, Pareek S, Sharma S, Yahia EM, Lobo MG. Fruit and vegetable waste: bioactive compounds, their extraction, and possible. utilization. Compr Rev Food Sci Food Saf. 2018;17:512-531. DOI: 10.1111/1541-4337.12330

[10] Thomas B, Raj MC, Athira KB, Rubiyah MH, Joy J, Moores A, Drisko GL, Sanchez C. Nanocellulose, a versatile green platform: from biosources to materials and their applications. Chem Rev. 2018;118(24):11575-11625. DOI: 10.1021/acs.chemrev.7b00627

[11] Klemm D, Heublein B, Fink H-P, Boh A. Cellulose: Fascinating biopolymer and sustainable raw material. Angew. Chem. Int. Ed. 2005;44:3358-3393. DOI: 10.1002/anie.200460587

[12] Henriksson M, Berglund LA. Structure and properties of cellulose nanocomposite films containing melamine formaldehyde. J Appl Polym Sci. 2007;106:2817-2824. DOI: 10.1002/app.26946

[13] Iwamoto S, Nakagaito AN, Yano H. Nano-fibrillation of pulp fibers for the processing of transparent nanocomposites. Appl Phys A Mater. 2007;89:461-466

[14] Calvini P. The Influence of Levelling-off Degree of Polymerisation on the Kinetics of Cellulose Degradation. Cellulose. 2005;12(4):445-447. DOI: 10.1007/s10570-005-2206-z

[15] Dufresne A. Nanocellulose: from nature to high performance tailored

materials. De Gruyter; 2012. 460 p. DOI: 10.1515/9783110254600

[16] Moon RJ, Martini A, Nairn J, Simonsen J, Yougblood J. Cellulose nanomaterials review structure, properties and nanocomposites. Chemical Society Reviews. 2011;40:3941-3994. DOI: 10.1039/c0cs00108b

[17] Chen H. Biotechnology of lignocellulose: theory and practice. Beijing: Chemical Industry Press; 2014. 211 p.

[18] Kumar R, Sharma RK, Singh AP. Cellulose based grafted biosorbents-Journey from lignocellulose biomass to toxic metal ions sorption applications-a review. J.Mol. Liq. 2017;232:62-93. DOI: 10.1016/j.molliq.2017.02.050

[19] Ritter SK. Sustainability. Chemical & Engineering News. 2008;86:59-68.

[20] Célino A, Fréour S, Jacquemin F, Casari P. The hygroscopic behavior of plant fibers: a review. Frontiers in Chemistry. 2013;1:43. DOI: 10.3389/fchem.2013.00043

[21] Neto WP, Silvério HA, Dantas NO, Pasquini D. Extraction and characterization of celulose nanocrystals from agroindustrial residue-soy hulls. Ind Crop Prod. 2013;42:480-488. DOI: 10.1016/j.indcrop.2012.06.041

[22] Nie S, Hao N, Zhang K. et al. Cellulose nanofibrils-based thermally conductive composites for flexible electronics: a mini review. Cellulose. 2020;27: 4173-4187. DOI: 10.1007/s10570-020-03103-y

[23] Clemons C, Sedlmair J, Illman B, Ibach R, Hirschmug C. Chemically imaging the effects of the addition of nanofibrillated cellulose on the distribution of poly(acrylic acid) in poly(vinyl alcohol). Polymer. 2013;54(8):2058-2061. DOI: 10.1016/j.polymer.2013.02.016

[24] Mondal S. Preparation, properties and applications of nanocellulosic materials. Carbohydr Polym. 2017;163:301-316. DOI: 10.1016/j.carbpol.2016.12.050

[25] Phanthong P, Reubroycharoen P, Hao X, Xu G, Abudula A, Guan G. Nanocellulose: Extraction and application. Carbon Resources Conversion. 2018;1:32-43. DOI: 10.1016/j.crcon.2018.05.004

[26] FAOSTAT Resources. ResourceSTAT data archives, Land, Updated 30 April 2009. Available from: http://faostat.fao.org/site/377/default.aspx#ancor

[27] FAO, Yearbook 2015 Forest products, [Internet]. Available from: http://www.fao.org/3/a-i5542m.pdf

[28] Fahmy Y, Fahmy TY, Mobarak F, El-Sakhawy M, Fadl M. Agricultural Residues (Wastes) for Manufacture of Paper, Board, and Miscellaneous Products: Background Overview and Future Prospects. Int. J. ChemTech Res. 2017;2(10):424-448.

[29] Hunter RW. Non-wood Fiber – 2010 and belong. Prospects for non-wood paper production in Asia Pacific. APPITA. 2010;56.

[30] Alireza A. Nonwood fibers a potential source of raw material in papermaking. Journal of Macromolecular Science: Part D–Reviews in Polymer Processing. 2006;45(10):1133-1136. DOI: 10.1080/03602550600728976

[31] El-Sayed ES, El-Sakhawy M, El-Sakhawy MA-M. Non-wood fibers as raw material for pulp and paper industry. Nordic Pulp & Paper Research Journal. 2020;35(2):215-230. DOI: 10.1515/npprj-2019-0064

[32] Zeng Y, Himmel ME, Ding S. Visualizing chemical functionality in plant cell walls. Biotechnol Biofuels. 2017;10:263. DOI: 10.1186/s13068-017-0953-3

[33] Sorieul M, Dickson A, Hill SJ, Pearson H. Plant Fibre: Molecular Structure and Biomechanical Properties, of a Complex Living Material, Influencing Its Deconstruction towards a Biobased Composite. Materials. 2016;9(8):618. DOI: 10.3390/ma9080618

[34] Jawaid M, Abdul Khalil HPS. Cellulosic/synthetic fibre reinforced polymer hybrid composites: A review. Carbohyd Poym. 2011;86:1-18. DOI: 10.1177/0731684413516393

[35] Atchison JE. Twenty five years of global progress in non-wood plant fibre pulping - historical highlights, present status and future prospects. In: Proceedings of TAPPI pulping conference (Book 1); 1995; TAPPI Press, Atlanta, GA, USA; 1995. p. 91-101.

[36] Pande H. Non-wood fiber and global fiber supply. Uunasylva. 1998;193: 44-50

[37] Tarasov D, Leitch M, Fatehi P. Lignin–carbohydrate complexes: properties, applications, analyses, and methods of extraction: a review. Biotechnol Biofuels. 2018;11:269. DOI: 10.1186/s13068-018-1262-1

[38] Klemm D, Kramer F, Moritz S, Lindström T, Ankerfors M, Gray D, Dorris A. Nanocelluloses: a new family of nature-based materials. Angew. Chem. Int. Edn. 2011;50:5438-66. DOI: 10.1002/anie.201001273

[39] Chunhong W, Suyue B, Xinmin Y, Long B. Relationship between chemical composition, crystallinity, orientation and tensile strength of kenaf fiber. Fibers Polymers. 2016;17(11):1757-1764. DOI: 10.1007/s12221-016-6703-5

[40] Płażek A, Dubert F, Kopeć P, et al. In vitro-propagated Miscanthus x giganteus plants can be a source of diversity in terms of their chemical composition. Biomass and Bioenergy, 2015;75:142-149. DOI: 10.1007/s42452-020-2529-2

[41] Silva LE, Claro PI, Sanfelice RC, Junior MG, Oliveira JE, et al. Cellulose nanofibrils modification with polyaniline aiming at enhancing electrical properties for application in flexible electronics. Cellulose Chem Technol. 2019;53(7-8):775-785

[42] Rol F, Belgacem MN, Gandini A, Bras J. Recent advances in surface-modified cellulose nanofibrils, Progress in Polymer Science. 2019;88:241-264. DOI: 10.1016/j.progpolymsci.2018.09.002

[43] Li W, Wang R, Liu S. Nano-crystalline cellulose prepared from softwood kraft pulp via ultrasonic-assisted acid hydrolysis. Bioresources. 2011;6(4):4271-4281. DOI: 10.15376/biores.6.4.4271-4281

[44] Hernández J, Romero V, Escalante A, Toriz G, Rojas O, Sulbarán B. Agave tequilana bagasse as source of cellulose nanocrystals via organosolv treatment. BioRes. 2018;13(2):3603-3614. DOI: 10.15376/biores.13.2.3603-3614

[45] Smook G.A. Handbook for Pulp & Paper Technologists. 3rd edition, Angus Wilde Publications, Inc.; 2003, 425 p.

[46] Liu Z, Wang H, Hui L. Pulping and Papermaking of Non-Wood Fibers. Chapter 1: Intechopen; 2018. p. 1-31. DOI: 10.5772/intechopen.79017

[47] Fiala W, Nardi F. The NACO Process, Pap. Technol. Ind. 1985;26(2):75-79

[48] Leponiemi A. Non-wood pulping possibilities – a challenge for the chemical pulping industry. Appita J. 2008;61(3):234-243

[49] Nassar MM. High yield acid and neutral sulfite cooking of rice straw. Cellul. Chem. Technol. 2004;37(5-6):487-495

[50] Saberikhan E, Rovsseh JM, Rezayati-Charani P. Organosolv pulping

of wheat straw by glycerol. Cellul Chem Technol. 2011;45(1-2):67-75

[51] Correia VC, dos Santos V, Sain M, Santos SF, Leão AL, Savastano JH. Grinding process for the production of nanofibrillated cellulose based on unbleached and bleached bamboo organosolv pulp. Cellulose. 2016;23:2971-2987

[52] Sarkanen KV. Chemistry of solvent pulping. Tappi J. 1990;73(10):215-219

[53] Usta M, Eroglu H, Karaoglu C. ASAE pulping of wheat straw (*Triticum aestivum* L.). Cellul. Chem. Technol. 1999;33(1-2):91-102

[54] Winner SR, Minogue LA, Lora JH. ALCELL pulping of annual fibers. In: Proceedings of 9th International symposium on wood and pulping chemistry; December 1997; Montreal. Poster presentation; 1997. p. 120-1-120-4 1997:120-1 –120-4

[55] Nimz HH. Pulping and bleaching by the Acetosolv process, Papier. 1989;43(10A):102-108

[56] Sundquist J, Poppius-Levlin K. MILOX pulping and bleaching. In: Young RA, Akhtar M, editors. Environmentally Friendly Technologies for the Pulp and Paper Industry. New York: Wiley; 1998. p. 157-190

[57] Rousu PP, Rousu P, Rousu E. Process for producing pulp with a mixture of formic acid and acetic acid as cooking chemical. Pat. US 6562191. 2003

[58] Lam HQ, Le Bigot Y, Denis G, Thao VH, Delmas M. Location and composition of silicon derivatives in rice straw pulp obtained by organic acid pulping. Appita J. 2005;58(3):214-217

[59] Cruz J, Fangueiro R. Surface modification of natural fibers: a review. Procedia Eng. 2016;155:285-288. DOI: 10.1016/j.proeng.2016.08.030

[60] Sacui IA, Nieuwendaal RC, Burnett DJ, Stranick SJ, et al. Comparison of the properties of cellulose nanocrystals and cellulose nanofibrils isolated from bacteria, tunicate, and wood processed using acid, enzymatic, mechanical, and oxidative methods. ACS Appl Mater Interfaces. 2014;6(9):6127-6138. DOI: 10.1021/am500359f. Epub 2014 Apr 18

[61] Zhou Y, Shuji S, Saito T, Isogai A. Characterization of Concentration-Dependent Gelation Behavior of Aqueous TEMPO-Cellulose Nanocrystal Dispersions Using Dynamic Light Scattering. Biomacromolecules. 2018; DOI: 10.1021/acs.biomac.8b01689

[62] Fang Z, Zhu H, Preston C, Hu L. Development, application and commercialization of transparent paper Transl. Mater. Res. 2014;1:015004. DOI: 10.1088/2053-1613/1/1/015004

[63] Spence KL, Venditti RA, Rojas OJ, Habibi Y and Pawlak J J. A comparative study of energy consumption and physical properties of microfibrillated cellulose produced by different processing methods. Cellulose. 2011;18(4):1097-1111. DOI: 10.1007/s10570-011-9533-z

[64] Siro I, Plackett D. Microfibrillated cellulose and new nanocomposite materials: review. Cellulose. 2010;17:291-294. DOI: 10.1007/s10570-010-9405-y

[65] Um BH, Karim MN, Henk LL. Effect of Sulfuric and Phosphoric Acid Pretreatments on Enzymatic Hydrolysis of Corn Stover. Applied Biochemistry and Biotechnology. 2003;105-108(1-3):115-25. DOI: 10.1385/ABAB:105:1-3:115

[66] Mahmud MM, Asma M, Asma P, Perveen A, Jahan RA, Arafat MT. Preparation of different polymorphs of cellulose from different acid hydrolysis medium. International Journal of

Biological Macromolecules. 2019;130. DOI: 10.1016/j.ijbiomac.2019.03.027

[67] Yan CF, Yu HY, Yao JM. One-step extraction and functionalization of cellulose nanospheres from lyocell fibers with cellulose II crystal structure. Cellulose. 2015;22:3773-3788. DOI: 10.1007/s10570-015-0761-5

[68] Biana H, Gaoa Y, Yanga Y, Fang G, Daia H. Improving cellulose nanofibrillation of waste wheat straw using the combined methods of prewashing, p-toluenesulfonic acid hydrolysis, disk grinding, and endoglucanase post-treatment. Bioresource Technology. 2018;256:321-327. DOI: 10.1016/j.biortech.2018.02.038.

[69] Kargarzadeh H, Ioelovich M, Ahmad I, Thomas S, Dufresne A. Methods for Extraction of Nanocellulose from Various Sources. In: Kargarzadeh H, Ahmad I, Thomas S, Dufresne A, editors. Handbook of Nanocellulose and Cellulose Nanocomposites, 1st ed. Weinheim: Wiley-VCH Verlag GmbH & Co. KGaA; 2017. p. 1-50

[70] Isogai A. Development of completely dispersed cellulose nanofibers. Proc. Jpn. Acad., Ser. B. 2018;94:161-178. DOI: 10.2183/pjab.94.012

[71] Madivoli ES, Kareru PG, Gachanja AN, Mugo SM, Sujee DM, Fromm KM. Isolation of Cellulose Nanofibers from Oryzasativa Residues via TEMPO Mediated Oxidation. Journal of Natural Fibers, 2020. DOI: 10.1080/15440478.2020.1764454

[72] Coseri S. Phthalimide-N-oxyl (PINO) Radical, a Powerful Catalytic Agent: Its Generation and Versatility Towards Various Organic Substrates', Catalysis Reviews, 2009;51(2):218 — 292. DOI: 10.1080/01614940902743841

[73] Amezcua-Allieri MA, Durán TS, Aburto J. Study of chemical and enzymatic hydrolysis of cellulosic material to obtain fermentable sugars. J Chem 2017:1-9. DOI: 10.1155/2017/5680105

[74] Long L, Tian D, Hu J, Wang F, Saddler J. A xylanase-aided enzymatic pretreatment facilitates cellulose nanofibrillation. Bioresour Technol 2017;243:898-904. DOI: 10.1016/j.biortech.2017.07.037

[75] Zhanga Y, Haoa N, Lina X, Niea S. Emerging challenges in the thermal management of cellulose nanofibrilbased supercapacitors, lithium-ion batteries and solar cells: A review. Carbohydrate Polymers. 2020;234:115888. DOI: 10.1016/j.carbpol.2020.115888

[76] Mariano, M., El Kissi, N., and Dufresne, A. Cellulose nanocrystals and related nanocomposites: review of some properties and challenges. J.Polym. Sci., Part B: Polym. Phys. 2014;52:791-806. DOI: 10.1002/polb.23490

[77] Davoudpour Y, Hossain S, Abdul Khalil HPS, Mohamad Haafiz MK, Mohd Ishak ZA, Hassan A, Sarker ZI. Optimization of high pressure homogenization parameters for the isolation of cellulosic nanofibers using response surface methodology. Ind. Crops Prod. 2015;74:381-387. DOI: 10.1016/j.indcrop.2015.05.029

[78] Peng BL, Dhar N, Liu HL, Tam KC. Chemistry and applications of nanocrystalline cellulose and its derivatives: a nanotechnology perspective. Can. J. Chem. Eng., 2011;89(5):1191-1206. DOI: 10.1002/cjce.20554

[79] Moon, R.J., Schueneman, G.T. & Simonsen, J. Overview of cellulose nanomaterials, their capabilities and applications. JOM. 2016;68:2383-2394. DOI:10.1007/s11837-016-2018-7

[80] Henriksson M, Berglund LA, Isaksson P, Lindström T, Nishino T.

Cellulose nanopaper structures of high toughness. Biomacromolecules. 2008;9(6):1579-85. DOI: 10.1021/bm800038n

[81] Ioelovich M. Peculiarities of Cellulose Nanoparticles. Tappi Journal. 2014;13(5). DOI: 10.32964/TJ13.5.45

[82] Barbash V, Trembus I, Sokolovska N. Performic pulp from wheat straw. Cellulose Chem. Technol. 2018;52(7-8):673-680

[83] Barbash V, Trembus I, Alushkin S, Yashchenko O. Comparative pulping of sunflower stalks, Scientific Journal Science Rise. 2016;3/2(20):71-78

[84] Barbash VA, Trembus IV, Oksentyuk NN, Primakov SF. Alternative raw materials for the production of organosolvent pulp products. Ecotechnology and resource conservation. 2012;1:27-32

[85] Barbash V, Yashchenko O. Obtaining a straw pulp in the isobutanol medium. Research Bull NTUU "KPI" 2015;6(104):80-86. Available from: http://bulletin.kpi.ua/article/view/51145/67710

[86] Barbash VA, Yaschenko OV, Kovalchuk VO. Optically transparent nanocellulose films from wheat straw. In: International research and practice conference "Nanotechnology and Nanomaterials" (NANO-2017); 23-26 August 2017; Chernivtsi, 2017; p. 543

[87] Barbash VA, Yashchenko O, Kedrovska A. Preparation and properties of nanocellulose from peracetic flax pulp. Journal of Scientific Research and Reports. 2017;16(1):1-10. DOI: 10.9734/JSRR/2017/36571

[88] Barbash VA, Yashchenko OV, Opolsky VO. Effect of Hydrolysis Conditions of Organosolv Pulp from Kenaf Fibers on the Physicochemical Properties of the Obtained Nanocellulose. Theor Exp Chem. 2018;54(3):193-198. DOI: 10.1007/s11237-018-9561-y

[89] Barbash VA, Yashchenko OV, Vasylieva OV. Preparation and properties of nanocellulose from Miscanthus x giganteus. Hindawi Journal of Nanomaterials. 2019, Article ID 3241968, 8 p. DOI: 10.1155/2019/3241968

[90] Paschoal G, Muller CM, Carvalho GM, Tischer CA, Mali S. Isolation and characterization of nanofibrillated cellulose from oat hulls. Quim Nova. 2015;38(4):478-482. DOI: 10.5935/0100-4042.20150029

[91] Barbash VA, Yaschenko OV, AlushkinSV, KondratyukAS, Posudievsky OYu, Koshechko VG. The Effect of Mechanochemical treatment of cellulose on characteristics of nanocellulose films. Nanoscale Research Letters. 2016;11:410. DOI: 10.1186/s11671-016-1632-1

[92] Ioelovich M. Superposition of Acid Concentration and Temperature at Production of Nanocrystalline Cellulose Particles. J. Chem. Edu. Res. Prac. 2017;1:1-7

[93] Lin N and Dufresne A. Surface chemistry, morphological analysis and properties of cellulose nanocrystals with gradiented sulfation degrees. Nanoscale, 2014;6:5384-5393. DOI: 10.1039/C3NR06761K

[94] Barbash VA, Yaschenko OV, Shniruk OM. Preparation and Properties of Nanocellulose from Organosolv Straw Pulp. Nanoscale Research Letters, 2017;12:241. DOI: 10.1186/s11671-017-2001-4

[95] Nogi M, Iwamoto S, Nakagaito AN, Yono H.Optically transparent nanofiber paper. Adv Mater. 2009;20:1-4. DOI: 10.1002/adma.200803174

[96] Josset S, Orsolini P, Siqueira G, Tejado A, Tingaut P, Zimmermann T.

Energy consumption of the nanofibrillation of bleached pulp, wheat straw and recycled newspaper through a grinding process. Nordic Pulp Paper Res J. 2014;29(1):167-75. DOI: 10.5281/zenodo.32932

[97] Correia VC, dos Santos V, Sain M, Santos SF, Leão AL, Savastano JH. Grinding process for the production of nanofibrillated cellulose based on unbleached and bleached bamboo organosolv pulp. Cellulose. 2016;23:2971-2987. DOI: 10.1186/s11671-017-2001-4

[98] Paschoal G, Muller CM, Carvalho GM, Tischer CA, Mali S. Isolation and characterization of nanofibrillated cellulose from oat hulls. Quim Nova. 2015;38(4):478-482. DOI: 10.5935/0100-4042.20150029

[99] Sánchez R, Espinosaa E, Domínguez-Roblesa J, Mauricio Loaiza J, Rodrígueza A. Isolation and characterization of lignocellulose nanofibers from different wheat straw pulps. Int J of Biological Macromolecules. 2016;92:1025-1033. DOI: 10.1016/j.ijbiomac.2016.08.019

[100] Kang YJ, Chun SJ, Lee SS, Kim BY, Kim JH, Chung H, Kim W. All-solid-state flexible supercapacitors fabricated with bacterial nanocellulose papers, carbon nanotubes, and triblock-copolymer ion gels. ACS Nano. 2012;6(7): 6400-6406. DOI: 10.1021/nn301971r. Epub 2012 Jun 25

[101] Bacakova L, Pajorova J, Bacakova M, et al. Versatile Application of Nanocellulose: From Industry to Skin Tissue Engineering and Wound Healing. Nanomaterials. 2019;9(2):164. DOI: 10.3390/nano9020164

[102] Johnson DA, Paradis MA, Bilodeau M, Crossley B, Foulger M, Gélinas P. Effects of cellulosic nanofibrils on papermaking properties of fine papers. Tappi J. 2016;15:395-402

[103] Lin N, Dufresne A. Nanocellulose in biomedicine: Current status and future prospect. European Polymer Journal. 2014;59:302-325. DOI: 10.1016/j.eurpolymj.2014.07.025

[104] Ahmat HO, Wilson W, Sorelli L, et al. Nanocellulose for improved concrete performance: A macro-to-micro investigation for disclosing the effects of cellulose filaments on strength of cement systems. Construction and Building Materials. 2019;206:84-96. DOI: 10.1016/j.conbuildmat.2019.02.042

[105] Zhu H, Luo W, Ciesielski PN, Fang Z, Zhu JY, Henriksson G, Himmel ME, Hu L. Wood-derived materials for green electronics, biological devices, and energy applications. Chem. Rev. 2016;116(16): 9305-9374. DOI: 10.1021/acs.chemrev.6b00225

[106] Zhang Y, Park S-J. Flexible organic thermoelectric materials and devices for wearable green energy harvesting. 909-1-909-18. Polymers. 2019;11(5). DOI: 10.3390/polym11050909

[107] Irimia-Vladu M, Głowacki ED, Voss G, Bauer S, Sariciftci NS. Green and biodegradable electronics. Mater. Today. 2012;15(7-8):340-346. DOI: 10.1016/s1369-7021(12)70139-6

[108] Tan MJ, Owh C, Chee PL, Kyaw AKK, Kai D, Loh XJ. Biodegradable electronics: cornerstone for sustainable electronics and transient applications. J. Mater. Chem. C. 2016;4(24):5531-5558. DOI: 10.1039/c6tc00678g

[109] Cai J, Niu H, Li Z, Du Y, Cizek P, Xie Z, Xiong H, Lin T. High-Performance Supercapacitor Electrode Materials from Cellulose-Derived Carbon Nanofibers. ACS Appl. Mater. Interfaces 2015; 7(27): 14946-14953. DOI: 10.1021/acsami.5b03757

[110] Zhao Z, Hao S, Hao P, Sang Yu, Manivannan A, Wu N, Liu H.

Lignosulphonate-cellulose derived porous activated carbon for supercapacitor electrode. J. Mater. Chem. A, 2015;3:15049-15056. DOI:10.1039/C5TA02770E

[111] Liua C, Xiaochen W, Haolin Z, Yiwei L, Lin S, Minghua T, Shia J, Shia Z. Cellulose-derived carbon-based electrodes with high capacitance for advanced asymmetric supercapacitors. J. of Power sources 2020;457:228056. DOI: 10.1016/j.jpowsour.2020.228056

[112] Kim I, Jeon H, Kim D, You J, Kim D. All-in-one cellulose based triboelectric nanogenerator for electronic paper using simple filtration process. Nano Energy 2018;53:975-981. DOI:10.1016/j.nanoen.2018.09.060

[113] Kim HJ, Yim EC, Kim JH, Kim SJ, Park JY, Oh IK. Bacterial Nano-Cellulose Triboelectric Nanogenerator. Nano Energy 2017;33:130-137. DOI: 10.1016/j.nanoen.2017.01.035

[114] Zhang L, Liao Y, Wang YC, Zhang S, Yang W, Pan X, Wang ZL. Cellulose II Aerogel-Based Triboelectric Nanogenerator. Advanced Functional Materials 2020;2001763. DOI:10.1002/adfm.202001763

[115] Klochko NP, Barbash VA, Klepikova KS, Kopach VR, Tyukhov II, Yashchenko OV, Zhadan DO, Petrushenko SI, Dukarov SV, Lyubov VM, Khrypunova AL. Use of biomass for a development of nanocellulose-based biodegradable flexible thin film thermoelectric material. Solar Energy. 2020;201:21-27. DOI: 10.1016/j.solener.2020.02.091

[116] Jung M, Kim K, Kim B, Lee K-J, Kang J-W, Jeon S. Vertically stacked nanocellulose tactile sensor. Nanoscale. 2017;9(44):17212-17219. DOI: 10.1039/c7nr03685j

[117] Zhao X, Han W, Jiang Y, Zhao C, Ji X, Kong F, Xu W, Zhang X. Honeycomb-like paper-based thermoelectric generator based on Bi2Te3/bacterial cellulose nanofiber coating. Nanoscale. 2019;11:17725-17735. DOI: 10.1039/c9nr06197e

[118] Abol-Fotouh D, Dörling B, Zapata-ArteagaO, Rodríguez-MartínezX, Gómez A, Reparaz JS, Laromaine A, Roig A, Campoy-Quiles M. Farming thermoelectric paper. Energy Environ. Sci. 2019;12:716-725. DOI: 10.1039/c8ee03112f

[119] Luzi F, Torre L, Kenny LM, Puglia D. Bio- and fossil-based polymeric blends and nanocomposites for packaging: structure-property relationship. Materials. 2019;12(3):471. DOI: 10.3390/ma120 30471

[120] Balea A, Merayo N, Fuente E, et al. Cellulose nanofibers from residues to improve linting and mechanical properties of recycled paper. Cellulose. 2018;25:1339-1351. DOI: 10.1007/s10570-017-1618-x

[121] Kim KM, Lee JY, Jo HM, Kim SH. Cellulose nanofibril grades' effect on the strength and drainability of security paper, BioRes, 2019;14(4):8364-8375

[122] Barbash VA, Yashchenko OV. Preparation and application of nanocellulose from non-wood plants to improve the quality of paper and cardboard. Applied Nanoscience. 2020. DOI: 10.1007/s13204-019-01242-8

[123] Barbash VA, Yashchenko OV, VasylievaOA. Preparationandapplication of nanocellulose from Miscanthus x giganteus to improve the quality of paper for bags. SN Applied Sciences 2020;2:727. DOI:10.1007/s42452-020-2529-2

[124] Miao C, Du H, Parit M, et al. Superior crack initiation and growth characteristics of cellulose nanopapers. Cellulose. 2020;27:3181-3195. DOI: 10.1007/s10570-020-03015-x

[125] Hubbe MA, Ferrer A, Tyagi P, Yin Y, Salas C, Pal L, Rojas OJ. Nanocellulose in Thin Films, Coatings, and Plies for Packaging Applications: A Review. BioResources. 2017;12(1):2143-2233

[126] Zheng, G., Cui, Y., Karabulut, E., Wågberg, L., Zhu, H., Hu, L. Nanostructured paper for flexible energy and electronic devices. MRS Bull. 2013;38(04):320-325. DOI: 10.1557/mrs.2013.59

[127] Burrs SL, Bhargava M, Sidhu R, Kiernan-Lewis J, Gomes C, Claussen JC, McLamore ES. A paper based graphene-nanocauliflower hybrid composite for point of care biosensing. Biosens Bioelectron. 2016;85:479-487. DOI: 10.1016/j.bios.2016.05.037. Epub 2016 May 11

[128] Feig VR, Tran H, Bao Z. Biodegradable polymeric materials in degradable electronic devices. ACS Cent. Sci. 2018;4(3):337-348. DOI: 10.1021/acscentsci.7b00595

[129] Hoeng F, Denneulin A, Bras J. Use of nanocellulose in printed electronics: a review. Nanoscale. 2016;8 (27):13131-13154. DOI: 10.1039/c6nr03054h

[130] Huang J, Zhu H, Chen Y, Preston C, Rohrbach K, Cumings J, Hu L, Highly transparent and flexible nanopaper transistors. ACS Nano. 2013;7(3): 2106-2113. DOI: 10.1021/nn3044071

[131] Fang Z, Hou G, Chen C, Hu L. Nanocellulose-based films and their emerging applications. 100764-1-100764-11. Curr. Opin. Solid State Mater. Sci. 2019;23. DOI: 10.1016/j.cossms.2019.07.003

[132] Lee K-Y, Aitomaki Y, Berglund LA On the use of nanocellulose as reinforcement in polymer matrix composites. Compos Sci Technol. 2014;105:15-27. DOI: 10.1016/j.compscitech.2014.08.032

[133] Chattopadhyay DP, Patel BH. Synthesis, characterization and application of nanocellulose for enhanced performance of textiles. J Text Sci Eng. 2016;6:248. DOI: 10.4172/2165-8064.1000248

[134] Dufresne A. Cellulose nanomaterial reinforced polymer nanocomposites. Current Opinion in Colloid & Interface Science. 2017;29:1-8. DOI: 10.1016/j.cocis.2017.01.004

[135] Al-Turaif HA. Relationship between tensile properties and film formation kinetics of epoxy resin reinforced with cellulose. Prog Org Coat. 2013 nanofibrillated;76:477-481. DOI: 10.1016/j.porgcoat.2012.11.001

[136] Chen J, Huang X, Zhu Y, Jiang P. Cellulose nanofiber supported 3D interconnected BN nanosheets for epoxy nanocomposites with ultrahigh thermal management capability. Advanced Functional Materials. 2017;27(5):1604754. DOI: 10.1002/adfm.201604754

[137] Guimarães JrM, Botarob VR, Novackc KM, Teixeirad FG, Tonolie GH. Starch/PVA-based nanocomposites reinforced with bamboo nanofibrils. Industrial Crops and Products. 2015;70:72-83. DOI: 10.1016/j.indcrop.2015.03.014

[138] Flauzino Neto WP, Marianoc M, Vieira da Silvaa IS, Silvério HA, Putauxe JL, Otaguroa H, Pasquini D, Dufresnec A. Mechanical properties of natural rubber nanocomposites reinforced with high aspect ratio cellulose nanocrystals isolated from soy hulls. Carbohydrate Polymers. 2016;153:143-152. DOI:10.1016/j.carbpol.2016.07.073

[139] Sigareva NV, Barbash VA, Yashchenko OV, Shulga SV, Starokadomsky DL, Gorelov BM. Influence of cellulose particles on chemical resistance, mechanical and

thermal properties of epoxy composites. Biophysical bulletin iss. 2020;43:57-70

[140] Guancha-Chalapud MA, Gálvez J, Serna-Cock L, Aguilar CN. Valorization of Colombian fique (Furcraea bedinghausii) for production of cellulose nanofibers and its application in hydrogels. Scientific Reports. 2020;10:11637. DOI: 10.1038/s41598-020-68368-6

Chapter 5

Synthesis and Applications of Organic-Based Fluorescent Carbon Dots: Technical Review

Musa Yahaya Pudza and Zurina Z. Abidin

Abstract

New ways of synthesizing organic-based fluorescent carbon dots (CDs) are required in environmental application. This is crucial for mitigation and control of pollutants without increasing the risk of releasing byproduct pollutants as the case with non-organic (metallic) quantum based substrate. Notably, this study provides current research on sustainable synthesis of CDs and their applications through analytical concept of recent and advance techniques for efficient and optimized processes. New scrutinized methods of synthesis and applications of CDs are beneficial and essential to optimize the state-of-art practices. The value distilled in this study adds to the field of sustainable production and application of CDs.

Keywords: carbon dots, fluorescent, organic-based, optimized, sustainable

1. Introduction

1.1 Origin of carbon dots

Carbon dots (CDs) are nanoparticles generated from organic/inorganic sources, was first discovered in 2004 when single-wall carbon nanotubes were electrophoretically purified [1]. CDs can be classified as carbon nanomaterials that are less than 10 nm in size, they are the latest class of fluorescent nanoparticles [2]. CDs have attracted the interest of researchers in diverse fields of science and technology such as; optoelectronics [3], environmental pollution and remediation [4], biosensor [5], bio-imaging and biomedical applications [6, 7].

CDs possess properties such as being dimensionless, durable, large surface area, enhanced porosity and stability, ease of being functionalized, fluorescence emission, biocompatibility and low toxicity [5, 8]. These properties of CDs can be applied to improve the environment and human health [4, 9, 10].

A toxic rival to the CDs is the popular semiconductor nanocrystals popularly known as quantum dots (QDs). The QDs are a type of semiconductor nanoparticles with diameter range from 1 to 10 nm [11]. More so, QDs normally are made from semiconducting materials, especially iron and cadmium, which are highly toxic and expensive to acquire [12]. Compared to QDs, CDs are considered best option with a high degree of biocompatibility, cost-effectiveness and non-toxic. It also serves as a suitable substitute to QDs in numerous areas of research such as bio-imaging, bio-sensing, pharmaceutical and fuel cells [13, 14].

Carbon dots (CDs) are suitable for the modification of electrode sensors. It combines fundamental aspects of biology, chemistry, and physical sciences, computer science and electrical engineering to meet various needs in a wide application field. Therefore, carbon as a sensor portrays several meanings, conditional upon what field the user subscribes [15, 16].

Over the years, various bulk materials and several processes and techniques have been developed and adopted by a wide range of researchers in the synthesis of CDs. These processes include the hydrothermal and microwave-assisted routes, heating, biogenic synthesis, thermal oxidation, ultra-sonification, subcritical water process (use of oil bath and salt bath), refluxing and chemical oxidation [17–26].

Three important factors must be considered in synthesizing CDs which are control of size, uniformity of CDs in solvents, and mitigated aggregation [27]. Wang and Hu [28] confirmed that CDs carbonaceous aggregation tends to form during carbonization but this can be prevented when synthesized by methods such as electrochemical synthesis, hydrothermal, or by pyrolysis method.

The application of biological and agro-waste to synthesize CDs have been advocated in numerous research such as; cooking oil waste [29], egg-white and egg-yolk [30], orange juice [6] as well as eggshells [31]. Though it is advantageous to use waste biomaterials in the synthesis of CDs to avoid competition with essential food production [32], however, the downside of the application of biomass in the synthesis of CDs is lacking of essential purity and structural homogeneity to obtain homogenous fluorescent CDs for purposes of sensing minute concentrations of analytes [21, 33]. These had caused the application of clean materials to be used in the synthesis of homogeneous fluorescent CDs [2].

A competent carbon source for soluble CDs synthesis is needed to comply with the goals of green chemistry and not be in direct competition with essential food production and should be cheap to synthesize [34–36]. Research in the synthesis of CDs must consider low price of additives and less purification steps in case of using biomass as a precursor material.

Thus, the emphasis is necessary on the cost of producing typical CDs, not to be a replica of the currently observed situation with semiconductor QDs, with the high cost and potential environmental negative impact and yet to achieve its full potential in commercial applications [37–39].

2. Green and sustainable carbon dots

Carbon dots (CDs) have emerged to be attractive materials due to their excellent photoluminescence (PL) properties and wide surface areas, which are needed for sensitive and selective sensing of analytes [40]. These qualities are owed to the characteristics of the carbon element at nano-dimension and five valence electrons to bind carbon atoms [31, 41]. The green and sustainable carbons dots refer to CDs that are synthesized from agro and biomaterials that can be readily available without depleting their sources [42].

CDs can be obtained from various source [3]. These sources include plants and animal origins such as bamboo leaves, woods, green algae, sugar cane, mangosteen, carica papaya, saffron, gringko, neem gum, prawn shells, orange, cucumber and pineapple [32, 43–45]. Further interesting applications of CDs have been reported in diverse sectors of the environment and health fields of science and technology [3, 32].

For instance, Pattanayak and Nayak, in 2013 [43] presented an eco-friendly synthesis of iron nanoparticles from various plants and spices extract. The synthesis of nanoparticles from plant parts (leaf) is essential since this will not require expensive processes that are involved mostly in biomaterial processing. Iravani et al., [46] demonstrated a green synthesis of metal nanoparticle using plants (*emblica officinalis* fruit extract) as a mean of mitigating the synthesis process of metal nanoparticles that are efficient and able to enhance green chemistry procedure for nanoparticles synthesis.

Liu [44], reported a research work on one-step green synthesized fluorescent carbon nano-dots from bamboo leaves for copper (ll) ion detection and demonstrated the exploration of bamboo leaves as a carbon source. Carbon nano-dots were synthesized hydrothermally and a resultant high quantum yield quantum dots, with sensitive Cu^{2+} detection at a limit of detection as low as 115 nM on a dynamic range from 0.333 to 66.6 μM. The zeta potential of the pristine carbon quantum dots was measured at −4.78 mV which changes to +13.8 mV after treatment with positively charged polyethyleneimine (a water-soluble cationic polymer).

Wembo et al., [47], researched on the economical and green synthesis of fluorescent carbon nanoparticles and their use as probes for sensitive and selective detection of mercury (II) ions. The adopted process by Wembo and colleagues was based upon the economy and green preparative strategy toward water-soluble fluorescent carbon nanoparticles with a quantum yield of 6.9% by a hydrothermal process using a low-cost waste from pomelo peel as a carbon source.

Piyushi et al., [45] cultivated *chlorella* (a genus of single-cell green algae belonging to the phylum *Chlorophyta*) on brewery wastewater for nanoparticle biosynthesis. The method of bio-nanoparticle synthesis using *chlorella* algal biomass grown in single water sample were harvested from the culture medium by centrifugation at 4000 rpm for 5 min followed by washing with ultrapure water to eliminate impurities. Iron nanoparticles were synthesized by mixing 0.5 g (dry weight) *Chlorella* sp. MM3 with 5 mL of 0.1 M $FeCl_3$ solution followed by incubation at 37 C for 48 h which entails long and tedious process.

Till et al., [48] synthesized CDs by microwave-assisted hydrothermal treatment of starch and Tris-acetate-EDTA. The process confirmed that nitrogen-doped CDs have emerged to be complementary to starch-derived CDs. Addition of nitrogen to CDs improved the yield of photoluminescence from 19% to 28%, making them promising luminescent materials for improving fluorescence of CDs. However, there is no added value in incurring additional chemicals during synthesis process of CDs. Starch is a better alternative to the use of nitrogen for synthesizing CDs. Till and colleagues observed the effect of nitrogen (N) additives, through the use of ethylenediaminetetraacetic acid (EDTA); tris (hydroxymethyl) aminomethane (Tris) and a combination of both (TAE-buffer) on the photophysical properties of CDs. Temperature (45 min at 230°C) plays an important role in the improved nitrogen-doped carbon structures [48].

Some researchers have adopted nitrogen for fluorescence and photoluminescence enhancement, but this approach has shown indistinct composition which required extensive purification steps. This, however, is environmentally not suitable and contravenes the concept of green chemistry since it involves many chemicals in the synthesis process [49].

2.1 Methods of carbon dots (CDs) synthesis

Synthesis methods of CDs can be divided into two major parts; top-down and bottom-up as in **Figure 1**. Top-down starts from cutting the carbon materials into carbon particles or cleavage of larger carbonaceous materials such as carbon nanotube by laser ablation, arc discharge, electrochemical and candle/natural gas burner soot, and recently the hydrothermal route.

The bottom-up route involves the use of molecules as support for localizing the growth of CDs by blocking aggregation during high-temperature treatment.

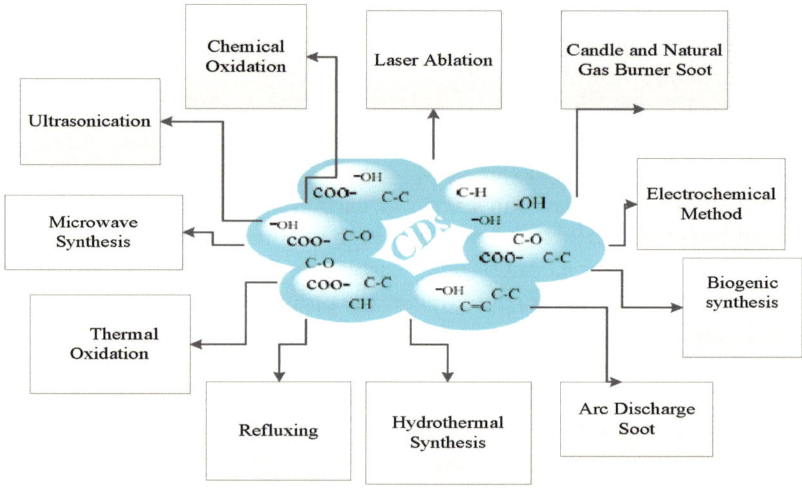

Figure 1.
Synthesis methods of carbon dots (CDs).

However, this study explores the top-down process of CDs synthesis. As earlier stated the top-down approach concentrates on precursor carbonization that include microwave-assisted method, chemical oxidation, heating, and hydrothermal process [50].

2.1.1 Laser ablation

It is a process removal of material from solid or liquid by irradiating it with a laser beam [51–54]. Material evaporates or sublimates when the laser flux is low and converted to plasma at high laser flux. Goncalves and colleagues [51] reported the synthesis of CDs from carbon targets immersed in deionized water by direct laser ablation (UV pulsed laser irradiation). CDs were optimized and synthesized after being functionalized with NH_2– polyethylene-glycol (PEG200) and N-acetyl-l-cysteine (NAC). To produce particles in tens of nanometer range by laser ablation, the energy is controlled within the incidence area of the precursor [54, 55].

Yu et al., [53] demonstrated the possibilities of relying on irradiating a toluene sample with a non-focused pulsed laser that is very different from the high powered laser irradiation employed in conventional ablation. This process by Yu and colleagues revealed an induced transformation of toluene into graphene sheaths, which subsequently produced fluorescent CDs. These nanoparticles can simply be functionalized using more than one molecule and stayed stable in an aqueous solution. It can also be applied to optical fiber devices through immobilization due to its stability in a specific optical nano-analytical sensor [56]. However, the equipment to conduct laser ablation is quite expensive and it needs technically skilled personnel to operate.

2.1.2 Arc discharge soot

CDs were first discovered through this method accidentally when the separation of single-walled carbon nanotubes (SWNTs) were made using gel electrophoresis from carbon soot by arc discharge method. Carbon is formed when direct current arc voltage is applied in an inert gas across two graphite electrodes. The biggest challenge of this method is that it generates impurities that are difficult to purify [1].

2.1.3 Electrochemical method

Electrochemistry is another top-down approach used in synthesizing CDs, this process is facile and the product yield is normally high [57]. CDs with a size of 6 to 8 nm and 2.8% to 52% can be obtained through exfoliation that utilizes graphite rods and Pt wire in ionic liquid or water solution.

The mechanism of the exfoliation was due to complex interplay of anodic oxidative cleavage of water and anionic intercalation from the ionic liquid using titanium cathode and spectrum pure graphite in the center of electrolyzer to yield pure blue fluorescent CDs without the urgency of complex purification [28, 58–62].

2.1.4 Candle and natural gas burner soot

Application of carbon soot in the synthesis of CDs have been reported by Tian et al., [63], the carbon source was from a carbon-processing reaction. Due to the simplicity to obtain the starting material, this method has been used widely by researchers. It also provides a new use for a complicated by-product. At the same time, it possesses disadvantages such as uncontrolled chemical surface, production of many byproducts that can harm human health with a broad dispersion [64].

2.1.5 Microwave synthesis

Wang et al., [13] prepared CDs by microwave method. It proved reaction time can be shortened to 30–45 minutes with microwave-assisted technique. Similarly, Choi et al., [10] made effective use of lysine as a precursor to synthesize CDs within 5 minutes in a home type of microwave oven and the CDs were soluble in water with deep blue photoluminescence at a high mass yield of 23.3%.

Compared with other methods, the microwave route is more convenient since the heating of the carbon precursor is rapidly achieved within few minutes. It also exhibits high quantum yield and provides a long fluorescence lifetime. The procedure of microwave synthesis is much easier compared to others as it only utilizes heating via irradiation technique [65].

2.1.6 Chemical oxidation

This method is mostly applied to produce CDs on an industrial scale. CDs can be obtained through oxidation treatment of carbon precursors by a strong oxidant. CDs from natural products have been researched and developed, by the synthesis of large scale CDs from human hair, coffee, and biomass by adding it into concentrated sulfuric acid and then heating at different temperatures. The time range is from hours to days [66]. By varying the temperature of synthesizing CDs, the quality of CDs such as diameter and quantum yields can be controlled [67–69].

2.1.7 Hydrothermal synthesis

The hydrothermal route of synthesizing CDs is considered as environmental-friendly, low cost and involves few synthesis steps that are non-toxic [2, 70–74]. Musa et al explored the hydrothermal method at a temperature range between 75°C to 175°C where the researchers reacted the precursor in a sealed hydrothermal reactor that resulted into a high yield photoluminescent quantum yield at 34.9% [2].

As illustrated in **Figure 2**, tapioca was added to an aldehyde solvent (acetone + sodium hydroxide) to improve the mobility of glucose molecules in starch [2]. The mixture underwent stages of reactions such as hydrolysis, adsorption, and gelatinization to particle disintegration simultaneously [2]. The carbonization temperature breaks the bond between the starch, making it available for the reactive solvent which leads to hydrolysis to form disaccharide and gelatinized glucose. The disaccharides polymerized into polysaccharides and the gelatinized glucose yielded CDs for functional group characterizations [9].

Like all-natural products, starch undergoes seasonal changes and particularly the amylose/amylopectin ratio is influenced by plant species and area of plant cultivation, which could influence the CDs formation. Independently of seasonal changes and origin, a starch will provide CDs with highly reproducible photoluminescent properties [2].

Substances such as glucose, citric acid, banana juice, and protein are examples of many precursors used to prepare CDs by adopting the hydrothermal route of synthesis [75]. Success has been reported in the synthesis of CDs through one-step hydrothermal carbonization using chitosan applied directly as a bioimaging agent [76]. The hydrothermal method is promising in producing CDs and is suitable for industrial or large scale production [75]. However, it is notable in 2010 where Zhang et al. [77] first reported a one-pot hydrothermal method to synthesize CDs from ascorbic acid in the presence of ethanol as solvent. Quantum yield and average particle sizes of their synthesized CDs were 6.79% and ~2 nm, respectively [77].

Several methods of synthesizing CDs has been explored in this section, to prevent the use of expensive precursor and energetic systems like in laser ablation. The hydrothermal synthesis route is being recommended as foremost for the sake of ecological sustainability [25]. Chemical oxidation and exfoliations provide an inexpensive alternative although it employs large amounts of strong acid which is hazardous and undesirable [77].

The other methods of synthesizing CDs need multi-step experimental operations and some of them require post-treatments to improve their water solubility, stability, and luminescent. Besides, several other methods suffer from drawbacks

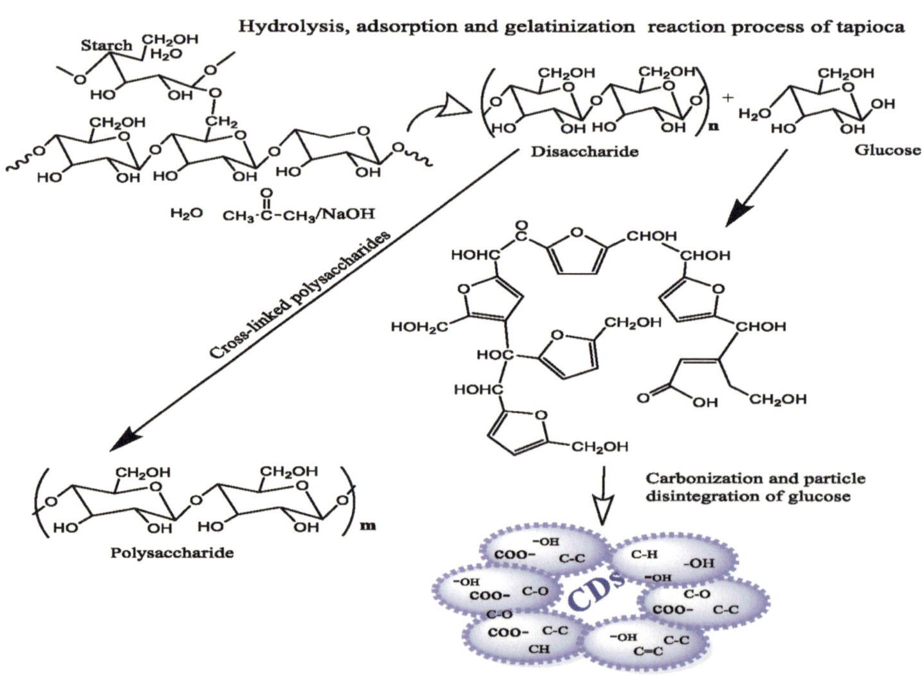

Figure 2.
Mechanism for synthesis of carbon dots [2].

such as they require complex process and high temperature, time-consuming, harsh synthetic materials, and are expensive. This causes their applicability to be limited [78]. Several research successes proved that hydrothermal route to be a green method for the synthesis of CDs since the procedure produce soluble fluorescence CDs at reduce time and cost [2, 46, 79].

CDs derived from organic sources are excellent for the researcher and environment. Because the adoption of such material presents the choice to eliminate the need for metallic quantum dots, and any doping requirements, either through the use of sulfur (S) or nitrogen (N) agents. The use of metallic quantum dots and possible inclusion of S and N in enhancing their functionality contravenes the purpose of sustainable applications of nanomaterials in the modern field of nanotechnology [48].

Table 1 is a list of different synthesis techniques that have been attractive to researchers in recent years. The table provides a list of interesting techniques such as hydrothermal, microwave assisted, biogenic synthesis, thermal oxidation, ultrasonication, refluxing and chemical oxidation with excellent particle sizes [80–91]. The hydrothermal synthesis of CDs proves to be efficient and effective since it provides relatively smaller sizes of the nanoparticles as synthesized by Du et al., [81] at 1.8 nm, when compared to other methods such as chemical oxidation by Thambiraj and Shankaran [85] at 4.1 nm, biogenic synthesis by Phadke et al., [20] at 5–8 nm, and refluxing by Himaja et al., [84] at ~50 nm.

2.2 Properties and characterization of fluorescent carbon dots (CDs)

One of the CDs properties is that it shows strong optical absorption in the UV region (200–800 nm) with a tail extending to the visible range, see **Figure 3**. CDs possess low toxicity with excellent photostability as compared to semi-conductor quantum dots [7, 23, 50, 71].

Absorption shoulders in the spectrum are due to the π-π* (pi to pi star transition) of C=C bonds or n-π* (n to pi star transition) of C=O and other fringe functional elements present [69, 92].

Method	Size (nm)	Reference
Microwave-assisted	2.7	[7]
Microwave	5–10	[10]
Biogenic synthesis	5.0–8.0	[20]
Thermal oxidation	5.0–10	[22]
Heating	3.0	[78]
Hydrothermal	2.3	[80]
Hydrothermal	1.8	[81]
Ultrasonication	5.0	[82]
"Oil bath"	2.59	[83]
Refluxing	~50	[84]
Chemical-oxidation	4.1	[85]
Chemical oxidation	2.5	[86]

Note: n/a = not available.

Table 1.
Carbon dots sizes and synthesis techniques.

The uniqueness of CDs is the availability of wide surface area for trace detection of analytes and provision of adsorptive sites through the availability of heteroatomic carbon in nano-dimension along with photoluminescence emission. Based on past study, CDs is dependent on intensity and wavelength emission towards its excitation wavelength [93]. This is due to the different sizes of particles and surface chemistry and/or different emissive traps on CDs' surface. The wavelength dependence behaviour makes CDs possible to be applied in multi-colour imaging and adsorptive purposes. Vinci et al., [93] suggest that CDs' core, surface states, and size are responsible for their emission and adsorptive properties [93].

Table 2 shows the excitation wavelengths of CDs through the UV-lamp excitation process to obtain fluorescent characteristics [2].

The colour of CDs most of the time is related to the surface groups which corresponds to particle sizes [93]. Normally CDs show strong photoluminescence from blue to green wavelength. To enhance the quantum yield (QY) of CDs or change photoluminescence (PL) emission to meet desired applications, surface passivation and functionality play a vital role. Besides, CDs show great photostability as there are no reductions in PL intensity with continuous exposure to excitation. In terms

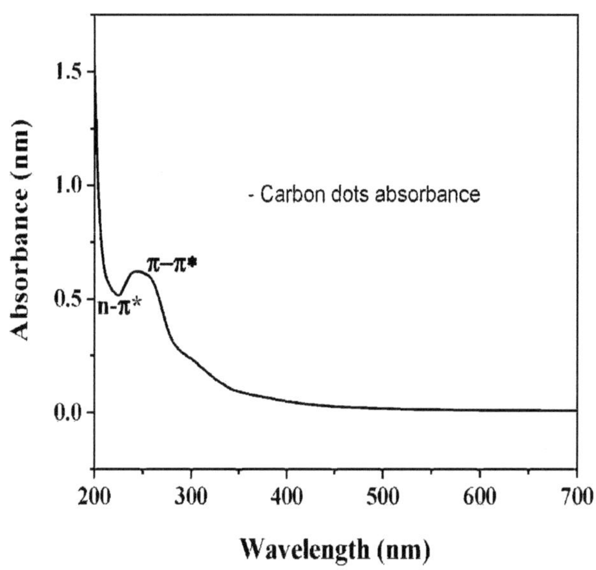

Figure 3.
Optical properties of carbon dots at UV-visible absorption and emission spectra.

Colour	Interval of wavelength (nm)
Red	700–635
Orange	635–590
Yellow	590–560
Green	560–520
Cyan	520–490
Blue	490–450
Violet	450–400

Table 2.
The color range of visible light spectrum.

of chemical properties, different synthesis methods of CDs lead to different chemical structure and abundance of surface sites. They are usually connected or modified by polymer chains, oxygen-based, amino based groups, and others [93].

Characterization of CDs by high resolution transmission electron microscopy (HRTEM), Xray photoelectron spectroscopy (XPS), Fourier transform infrared (FTIR), Atomic force microscopy (AFM) and Zeta Potential provide deep insights into the attributes of CDs such as hybridization and coefficient between functional groups and carbon core that take parts in the provision of the abundance of surface sites and the photoluminescence behaviour [94]. In comparison to graphene and metallic quantum dots, the CDs serves as the way out of toxicity concerns in environmental monitoring and medical applications [95].

2.2.1 High-resolution transmission electron microscopy (HRTEM)

The sizes and texture of CDs are important for fundamental applications in the field of environmental science and nanotechnology. **Figure 4(A–C)**, shows the HRTEM images of CDs at different resolutions between 1 nm to 10 nm. Synthesized CDs revealed amorphous quasi-spherical morphology with a lattice spacing of ca 0.24 nm (**Figure 4A**), CDs characteristics are suitable absorbent of pollutants that are larger than 0.24 nm [2, 71, 89].

High-Resolution Transmission Electron Microscopic images of the CDs characterized in magnifications of 5 nm and 10 nm (**Figure 4A** and **B** respectively). **Figure 4A** is the lattice spacing for carbon dots at 5 nm magnification. While **Figure 4B** is the size distribution within 10 nm magnification. **Figure 4C** is the histogram chart, demonstrating the nanoparticle sizes of CDs. The synthesis of nanoparticle with low lattice space is needed for research applications of CDs in environmental chemistry, pollutant entrapment in aqueous media and water purification [74]. The interplanar distance (lattice spacing) of 0.24 nm (**Figure 4**) is lower than the lattice spacing planes of graphitic materials (0.34 nm), the larger interlayer spacing could be attributed to the abundant oxygen-containing groups. In other words, the oxygen-containing groups could expand the layer spacing. The synthesized CDs is in consonance with recent reports by Arumugam and colleagues, CDs was hydrothermally synthesized from broccoli [79], ginkgo fruits [87], and cabbage [8].

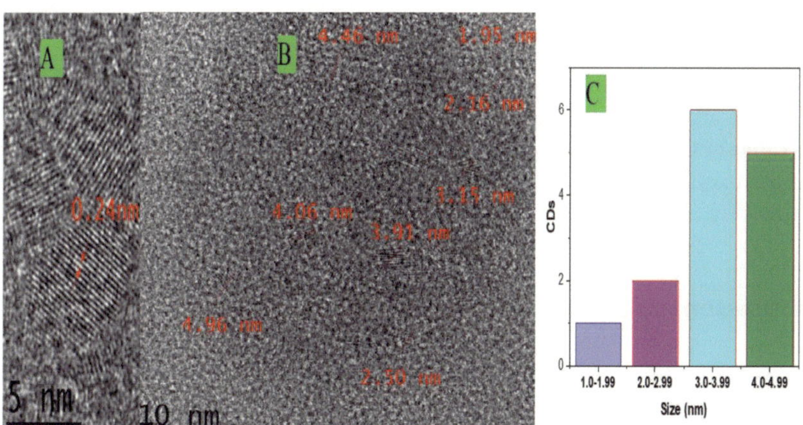

Figure 4.
High resolution transmission electron microscopic (HRTEM) (A) Lattice space of carbon dots (CDs) characterized in magnifications of 5 nm. (B) Images of CDs at 10 nm. (C) Size distribution of CDs within 10 nm magnification [2].

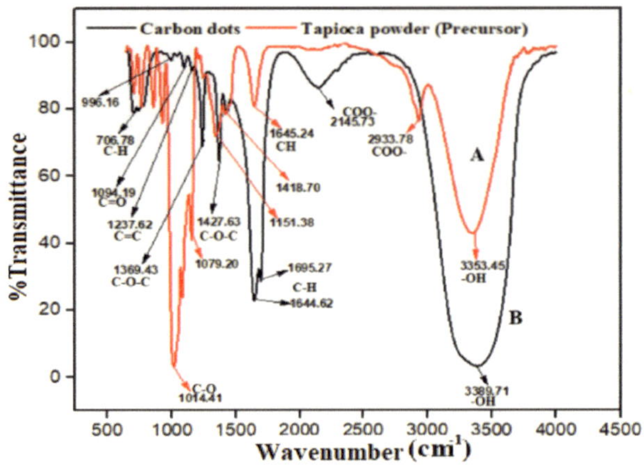

Figure 5.
FT-IR spectrum of the carbon dots and tapioca [2].

2.2.2 Fourier-transform infrared spectroscopy (FTIR)

Fourier-transform Infrared spectroscopy (FTIR) portrays the functional structure of CDs. It reveals the intrinsic functional groups and other useful compounds present in CDs. **Figure 5** provides functional groups that exist before and after the hydrothermal treatment of tapioca as a precursor for CDs.

As shown in **Figure 5(A)** representing tapioca. Peaks associated with the stretching vibrations of hydroxyl (–OH) and carboxylic (COO–) groups are at 3353.45 and 2933.78 cm^{-1} [75]. Further stretching vibration of C–H occurred from 1645.24 to 1341.82. The peaks at 1151.38, 1079.20, 1014.41 cm^{-1} can be due to the C–O stretching vibrations and out-of-plane bending modes of sp^2 and sp^3 –CH group [75].

There were substantial changes observed in the spectra of CDs (**Figure 5B**). The hydroxyl (–OH) group of 3389.71 cm^{-1} increased on the carbon structure as a result of hydrolysis. While the carboxylic (COO–) group 2145.73 cm^{-1} reduced by thermal destruction of saccharides structure [34]. The peaks at 1695.27 cm^{-1} and 1644.62 cm^{-1} showed the increase in the C–H stretching vibrations of the bending modes of the sp^2 and sp^3 –CH group. The peaks around 1427.63 cm^{-1} until 1369.43 cm^{-1} are due to C–O–C [34]. The peak at 1237.62 cm^{-1} corresponds to the C=C stretching vibration while 1094.19 cm^{-1} and 996.19 cm^{-1} represents the C=O stretching vibration and the last group at 706.78 cm^{-1} denotes the C=C bond of the unsaturated glucose structure in the starch. These attributes were responsible for the water-soluble nature of CDs [34]. The FTIR graph shows the formation of unsaturated carbon. Along with oxygen-rich groups such as hydroxyl, carboxyl, and carbonyl situated on the CDs surface, which agree with the hydrothermal synthesized CDs from the organic origin [23, 25, 26, 81, 90].

3. Applications of carbon nanoparticles

There are numerous applications of carbon nanoparticles due to the abundant properties they possess [3]. These applications are being discussed in the subsequent sections of the report.

3.1 Application in bioimaging and biomedicals

Carbon dots have shown great potential to act as a sensor and can be used for environmental monitoring and control of pollutants, more so in the medical field for biosensor applications. It can donate or accept electrons that make it suitable for detection of ions, vitamins, nucleic acid, protein, enzyme and biological pH value [7, 11, 96–98]. Even though different materials are used to detect specific ions, the detection mechanisms are identical [99].

The functional groups on the surface of CDs specify distinctive affinities to different target ions, through an electron or energy transfer process and high selectivity to other ions [100]. CDs has been involved in the detection of 2,4,6-tri-trotoluene (TNT) and also applied as a dual-sensing platform for fluorescent and electrochemical detection of TNT [101]. Other reports utilized CDs as pH sensors for in-vitro and in-vivo investigations [102].

Research showed CDs able to detect intracellular pH inside a living pathogenic fungal cell and has been developed to sense nucleic acid in the DNA [103]. In other cases, CDs have been used in bioimaging because of their low toxicity and excellent photostability compared to semi-conductor quantum dots that posed health problems and environmental concerns [8]. Its visible excitation, emission wavelengths, and high brightness confirm CDs as a suitable candidate in this area. Several studies have been conducted using CDs in cell imaging, including pig kidney cell line [104], *Escherichia coli* [105], Hela Cells [106], liver diseases [95], see **Figure 6**.

Chengkun et al., [98] discovered photoluminescence in CDs synthesized from Nescafe original instant coffee and applied it in the field of bioimaging. From their investigation, CDs from Nescafe are found to be amorphous and the cytotoxicity study revealed that the CDs did not cause any toxicity to human hepatocellular carcinoma cells at a concentration as high as 20 mg/ml. Yang et al., [107] also worked on novel green synthesis of high-fluorescent CDs from honey for sensing and imaging. It was an innovative and green approach towards a CDs of high fluorescent quantum yield and excellent photostability, employed for HeLa cells imaging and coding. Rui-jun et al., [108] produced photoluminescent CDs from polyethylene glycol (PEG) for cellular imaging. The PEG is a biocompatible non-conjugated polymer, used as both carbon source and passivating agent [108].

3.2 Application in heavy metal ions detection

The application of CDs in the selective detection of heavy metals have been reported in several scientific and experimental research [34, 40, 79]. However, there are gaps and lapses needing redress, such applications are predominantly in

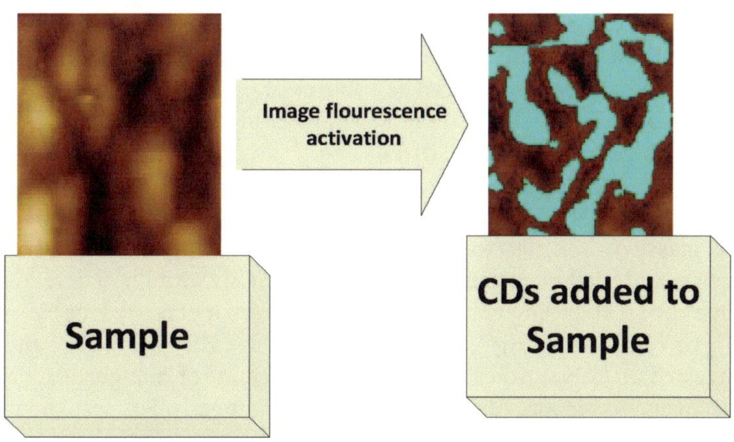

Figure 6.
Graphical description of fluorescence images of carbon dots.

photoluminescent quenching of heavy metals. whereas, current section looks into reliable and robust CDs for applications in electrochemical sensing of multiple ranges of heavy metal ions.

The development of a convenient and sustainable technique for detecting and identifying human and environmentally toxic metal ions is of great interest. The following are reports concerning CDs application in heavy metal detection.

Zhang and Chen [109] worked on nitrogen-doped carbon quantum dots application as a turn-off fluorescent probe for the detection of Hg^{2+} ions at a detection limit of 0.23 μM. The fluorescent quenching mechanism is attributed to the surface-state triggered by the mercury-induced conversion of special functional group (–CONH–) from spirolactam structure to an opened-ring amide [109].

Sandhya et al., [110] applied nanostructures for heavy metal ion sensing in water using surface plasmon resonance of metallic nanostructures. They reviewed on techniques to improve selectivity and sensitivity of surface plasmon response sensors with attention to homogeneity. Effects of particle size, shape, material type, and surrounding environment were found to be effectual in the surface plasmon surface frequency.

Similarly, Qu et al., [111] developed CDs to detect Fe^{3+} ions by using dopamine as a starting material with a detection limit of 0.32 μM. Quenching of photoluminescence intensity occurred when there was an interaction between CDs and ions. Meanwhile, Liu [44] reported a research work on one-step green synthesized fluorescent carbon nanodots from bamboo leaves for copper (ll) ion detection and demonstrated the exploration of bamboo leaves as a carbon source with high carbon constituent. Carbon quantum dots were synthesized hydrothermally with sensitive Cu^{2+} detection at limit of detection as low as 115 nM and a dynamic range from 0.333 to 66.6 μM. The zeta potential of the pristine carbon quantum dots was measured at −4.78 mV which improved to +13.8 mV after treatment with positively charged polyethyleneimine (a water-soluble cationic polymer). More so, Rao et al., [112] reported on the ability of CDs generated from citrus acid anhydrous to detect heavy metal such as Fe^{3+}, with a detection limit of 0.239 μM.

Methionine has been used as a material for the synthesis of CDs [113]. These CDs were co-doped with nitrogen and sulfur to enhance surface functionalization for the detection and environmental monitoring of heavy metal pollutants [113]. Similarly, Shen et al., [4] applied fresh pomelo in the synthesis of CDs co-doped with nitrogen and sulfur for the detection of chromium (Cr (VI)).

A fluorescent probe for selective detection of metal ions such as mercury (Hg^{2+}, $1.00 \times 10^{-8} - 1.50 \times 10^{-3}$ M, 1.00×10^{-7} M) with wide linear range and satisfactory detection limits was discovered when citric acid monohydrate was used for the synthesis of fluorescent CDs [114]. More essentially and effective is the burning of ash from waste paper and further utilized as a source of CDs by Lin et al., [115]. They succeeded in synthesizing CDs without any surface modification and subsequently, the fluorescent CDs were quenched by Fe^{3+}.

Simpson et al., [21] synthesized carbon nanoparticle from glycerol and phosphoric acid mixed in a Berghof high-pressure reactor at 250°C for 4 hours. Afterward, glassy carbon electrodes were fabricated by drop-casting the carbon nanoparticles, and further applied for heavy metal (Cu^{2+} and Pb^{2+}) detection by square wave anodic stripping voltammetry [21]. Heavy metals such as Na^+, K^+, Mg^{2+}, Ca^{2+}, Cr^{3+}, Co^{2+}, Ag^+, Hg^{2+}, Cd^{2+}, Pb^{2+}, Ni^{2+}, Cu^{2+}, Zn^{2+}, Al^{3+}, Fe^{2+}, and Fe^{3+} have been tested on CDs synthesized from carbon source of mangosteen pulp and a ground discovery was made. Among the listed heavy metals, Fe^{3+} was the favourite in detection with a detection limit of 52 nM. Further application was found for cell imaging, which reveals their diverse potential applications [89].

Abhishek et al., [14] made a paper strip based live cell ultrasensitive lead sensor using CDs synthesized from biological media. They reported a formulation of a sensor through microwave heating of potato-dextrose agar (PDA) for the detection of lead (pb^{2+}) in solution but again involved a long and laborious process.

Pajewska et al. [116] explored the fluorescence of synthesized CDs from citric acid with glutathione for the sensing of mercury (Hg^{2+}) ion. A high recovery of Hg^{2+} was achieved at 115.1%. The method of synthesizing CDs with low toxicity is embedded in the green chemistry principles. Thus, it fulfills the criteria of being eco-friendly. **Table 2** provides a list of applications of CDs in the detection of heavy metals ions.

As seen in **Table 3**, the mechanism of action for the application of CDs largely depends on the analyte of concern. In the case of CDs from citric acid monohydrate for application in fluorescence quenching of Hg^{2+}, it relies on Förster resonance energy transfer (FRET) [114]. This is similar to CDs synthesized from biomass [117], polyacrylamide [118], lotus root [119], degreased cotton [120], gold nanoclusters [111], and Petroleum coke [127].

Fluorescent carbon nanoparticle sensing is largely dependent on changes or disturbances that are caused by an analyte that interacts with a fluorescent probe. This shift mostly will lead to a measurable change in the emission characteristics of the probe (emission wavelength, intensity, lifetime, or anisotropy), which can be directly linked to analytes (e.g heavy metal) concentration. More so, fluorescence probe strategies are based on quenching (turn-off) or enhancing (turn-on) emission, and surface-enhanced Raman scattering (SERS) techniques [33, 125].

Source of carbon nanoparticles	Sensing mechanism	Type of metal ions and linear range	Sensing (LOD)	Reference
Biomass	Fluorescence	Hg^{2+}/Fe^{3+} 0.002 mol L^{-1}	10.3 and 60.9 nM	[117]
Polyacrylamide	Fluorescence	Hg^{2+} 0.25–50 μM	13.48 nM	[118]
Lotus plant	Fluorescence	Hg^{2+} 0.1 to 60.0 μM	18.7 nM	[119]
Degrease cotton	Fluorescence	Cr(VI) 1.00–6.00 mmol/L	0.12 μg/mL	[120]
Gold nanoparticles	Luminescence	Pb^{2+} 1×10^{-5} M	n/a	[121]
Biomass from peanut shells	Fluorescence	Cu^{+2} 0–5 mM	4.8 mM	[122]
Metal oxides	Electrochemical oxidation.	Cu^{2+} 0.1 to 1.3 μM	0.04 μM	[123]
Metal nitrates	Isotherm	Cd^{2+} 10 mg/L	12.60 mg/g	[124]
Mushroom	Fluorescent	Hg^{2+} 0 to 100 nM	4.13 nM	[125]
Penaeus merguiensis enzyme	Isotherm	Cu^{+2} 1–5 mM	2 mM	[126]
Coke	Fluorescent	Cu^{+2} 0.25–10 μM	0.0295 μM	[127]

Table 3.
Carbon nanoparticles for heavy metal sensing.

Other notable techniques for the detection and quantification of heavy metals ions include, inductively coupled plasma mass spectrometry (ICP-MS). This instrumentation is efficient among several other methods, but it is expensive. It was developed since the 1980s [128–130], used mostly by multivariate analysis along with the ICP-MS technique to unravel heavy metal elements present samples. However, inductively coupled plasma atomic emission spectroscopy (ICP-AES) have also been used to identify heavy metal pollutants. But, the method is expensive and requires sophisticated instrumentations and a highly trained technician [131].

Nowadays, marine pollution is becoming a global phenomenon and seafood safety has played a crucial role in human health [129]. Fatema et al., [132] applied atomic absorption spectroscopy (AAS) to measure the absorbed quantity of Pb^{+2}, $Cd^{+2,}$ and Hg^{+3} in shrimps. Heavy metals have been detected by other means such as energy dispersive x-ray fluorescence (EDXRF), electrothermal atomic absorption method (ETAAS), and flame atomic absorption spectroscopy (FAAS) [133, 134]. But, all of the aforementioned techniques have disadvantages in the detection of heavy metals, such that they are expensive and require strenuous experimental steps [135]. Therefore, environmental researchers have continued to strive to develop a cheap, simple, sensitive, specific, accurate, user-friendly, and eco-friendly means of detection for heavy metal pollutants.

3.3 Application in non-metal detection

CDs are very useful in detecting non-metallic elements. Several types of research have been reported, CDs synthesized from potato are well able to detect phosphate [106]. Zhaoxia et al., [136] utilized CDs with turnable emission and controlled size for sensing hypochlorous acid. As a class of carbohydrate that is widely distributed in a living organism, sucrose was chosen as a carbon source with assistance of microwave irradiation. A strongly fluorescent CDs without post-passivation was produced. By increasing the concentration of phosphoric acid as fluorescence enhancer under UV lamp, various fluorescent emissions of CDs of variable sizes were obtained. It was found that green CDs have excellent sensitivity for the detection of hypochlorous acid.

Kuo et al., [85], experimented with percutaneous fiber-optic nanosensors for instant evaluation of chemotherapy efficacy for in-vivo strategy of assay design aimed at monitoring non-homogeneously distributed biomarkers. They identified optimal exogenous fluorophores for the cell distribution indicators that are independent of the treatment of the apoptotic initiator and without interfering with the optical characteristics of fluorophores.

Huilin et al., [137], investigated on CDs as a fluorescent probe for off-on detection of sodium dodecyl-benzenesulfonate (SDBS) in aqueous solution. The pristine CDs were synthesized from sodium citrate through a simple, convenient, and one-step hydrothermal method. Fluorescent recovery was achieved with the application of SDBS. Detection of SDBS in real water samples was proportional to the concentration in the range of 0.10 to 7.50 ug/mL. Furthermore, fluorescence sensing probe has been used to detect kaempferol (flavonoid that is present in a variety of plants and plant-derived foods) using fluorescent CDs synthesized from chiefly acetic acid with a detection limit of 38.4 nM in the concentration range of 3.5–49 μM. Finally, organophosphorus as pesticides have been detected through the use of CDs as a detector for pollutants without surface modification [115].

3.4 Application in adsorption studies

CDs and carbon structured nanoparticles have attracted researchers to explore their effectiveness and optimization ability in the fields of pollution research [3].

DOI: http://dx.doi.org/10.5772/intechopen.94511

Adsorbent material	Adsorbate/analyte	Reference
Gold nanoparticles (AuNPs)	4-nitrophenol	[145]
Carbon dots (sodium citrate)	Mercury (II) ions.	[146]
Fluorescent carbon dots from o- phenylenediamine	Cell imaging and sensitive detection of Fe^{3+} and H_2O_2	[18]
Silica gel	Aromatic volatile organic compounds (VOCs)	[147]
Graphene oxide	Nitrobenzene in sulfide	[148]
TiO_2, SiO_2, and ZnO nanoparticles	Neptunium (V)	[149]
Polystyrene latex nanoparticles	Alumina	[150]
Graphene oxide	Radionuclide removal	[151]
Polyaniline modified graphene oxide	Uranium(VI)	[152]
Carbon nanotubes	Mingle-ringed N- and S-heterocyclic aromatics	[153]
Graphene oxide	Minerals such as montmorillonite, kaolinite, and goethite, in aqueous phase	[154]

Table 4.
Nanostructured materials in adsorption processes.

Because of their excellent properties; carbon material performs concurrently as adsorbent and a transducing-agent [138–141].

Due to abundant surface sites provided by CDs, it is a suitable candidate for studies in the detection and adsorption of heavy metals [142, 143]. For instance, Ghiloufi et al., [144] used gallium doped zinc oxide (ZnO) nanoparticle in the adsorption of heavy metals (Cd^{2+} and Cr^{6+}) in aqueous solution. The adsorption of heavy metals was analyzed through the effect of pH and it revealed favourable adsorption at a low pH level, less than pH-3 and temperature of 298 K [144].

Table 4 provides harmonized presentation of nanomaterials applied for the purpose of absorbing environmental pollutants and contaminants in aqueous systems [145–154].

So far the concept of applying nanoparticles for environmental objectives have been successful. Meanwhile it is recommended that comparisons be made with bulk counterparts of the same substance to measure efficiency. On this note a study on the application of bulk agro material from *jatropha curcas* demonstrated efficiency in adsorption of pollutants and is recommended for comparison with its nano-dimension counterparts [155]. Similarly, a report on the application of sesame straw biochar in adsorption of heavy metal analyte concluded that further adsorption studies for nano-range agro-based materials are necessary for accurate estimation of adsorption in natural environments [156].

4. Conclusion

A suitable carbon source for CDs synthesis should be soluble in water (green chemistry), accessible worldwide (i.e. geographical abundance) with defined and well-known properties (i.e. functional attributes), should not be in direct competition with essential food production (i.e. sustainable), and it should be cost-effective (i.e. cheaply accessible). While the price of additives or carbon source plays a minor role in fundamental research, it may play a major role when large quantities are considered.

Acknowledgements

The authors would like to thank Universiti Putra Malaysia (UPM), Malaysia for funding this article.

Author contributions

M.Y.P., as the first author; made the study conception and design acquisition of reports and drafting of manuscript. Z.Z.A., contributed in the study conception and design, critical revision of major scientific ideas through clinical experience.

Funding

This research was funded by Universiti Putra Malaysia, grant number GP-IPS/2017/9556800.

Conflicts of interest

The authors hereby declare that there is no conflict of interest.

Author details

Musa Yahaya Pudza* and Zurina Z. Abidin*
Department of Chemical and Environmental Engineering, Faculty of Engineering, Universiti Putra Malaysia, Serdang, Selangor, Malaysia

*Address all correspondence to: pudzamusa@gmail.com and zurina@upm.edu.my

References

[1] Xu X, Robert R, Yunlong G, Harry JP, Latha G, Kyle R, Walter A. Electrophoretic Analysis and Purification of Fluorescent Single-Walled Carbon Nanotube Fragments. Journal of American Chemical Society. 126: 2004: 12736-12737.

[2] Musa YP, Zurina ZA, Suraya AR, Faizah MY, Noor ASM, Mohammed A. Synthesis and Characterization of Fluorescent Carbon Dots from Tapioca. ChemistrySelect, 4: 2019: 1– 8.

[3] Das R, Rajib B, Panchanan, P. Carbon quantum dots from natural resource: A review. Materials Today Chemistry. 8: 2018: 96-109.

[4] Shen J, Shaoming S, Xiuying C, Dan W, Yan, C. Highly fluorescent N, S-co-doped carbon dots and their potential applications as antioxidants and sensitive probes for Cr (VI) detection. Sensors and Actuators B. 248: 2017: 92-100.

[5] Da Silva-Souza DR., Larissa DC., Joao PM., Fabiano VP. Luminescent carbon dots obtained from cellulose. Materials Chemistry and Physics. 203: 2018: 148-155.

[6] Sahu S, Behera B, Maiti TK, Mohapatra S. Simple one-step synthesis of highly luminescent carbon dots from orange juice: application as excellent bio-imaging agents. Chemical Communication. 2012: 48: 8835.

[7] Bhattacharyya D, Prashant, KS, Michael LF. Quantum dots and carbon dots based fluorescent sensors for TB biomarkers detection. Vacuum. 2017: 146: 606-613.

[8] Alam AM, Park BY, Ghouri ZK, Park M, Kim,HY. Synthesis of carbon quantum dots from cabbage with down- and up-conversion photoluminescence properties: excellent imaging agent for biomedical applications. Green Chemistry. 17: 2015: 3791-3797.

[9] Bin W, Feng L, Yuanya W, Yanfen C, Chang ML. Synthesis of catalytically active multielement-doped carbon dots and application for colorimetric detection of glucose. Sensors and Actuators B: Chemical. 255: 2018: 2601-2607.

[10] Choi Y, Nichaphat T, Ari C, Seongho J, Insik I. Microwave-assisted synthesis of luminescent and biocompatible lysine-based carbon quantum dots. Journal of Industrial and Engineering Chemistry. 47: 2017: 329-335.

[11] Ajay KG, Mona G. Synthesis and surface engineering of iron oxide nanoparticles for biomedical applications. Biomaterials. 26: 2005: 3995-4021.

[12] Derfus AM, Chan WCW, Bhatia SN. Probing the cytotoxicity of semiconductor quantum dots. Nano Letters. 4: 2004: 11-19.

[13] Wang F, Chen YH, Liu CY, Ma DG. White light-emitting devices based on carbon dots' electroluminescence. Chemical Communication. 47: 2011: 3502-3504.

[14] Abhishek G, Navneen CV, Syamantak K, Shalini T, Abhishek C, Chayan KN. Paper strip based and live cell ultrasensitive lead sensor using carbon dots synthesized from biological media. Sensors and Actuators B: Chemical, 47. 232: 2016: 107-114.

[15] Rasheed T, Muhammad B, Faran N, Hafiz MNI, Chuanlong L, Yongfeng Z. Fluorescent sensor based models for the detection of environmentally-related toxic heavy metals. Science of the Total Environment. 615: 2018: 476-485.

[16] Hayat A, Marty JL. Disposable screen printed electrochemical sensors: tools for environmental monitoring. Sensors. 14: 2014: 10432-10453.

[17] Monoj KB, Bikash J, Santanu B, Amitava P. Photophysical Properties of Doped Carbon Dots (N, P, and B) and Their Influence on Electron/Hole Transfer in Carbon Dots–Nickel (II) Phthalocyanine Conjugates. Journal of Physics and Chemistry C, 118: 2014: 20034-20041.

[18] Song Y, Hao L, Fang L, Huibo W, Mengling Z, Jinjing Y, Jian H. Fluorescent carbon dots with highly negative charge as sensitive probe for real-time monitoring of bacterial viability. Journal of Material Chemistry B. 5: 2017: 6008-6015.

[19] Edison TNJI, Raji A, Mathur GS, Jae-Jin S, Yong RL. Microwave assisted green synthesis of fluorescent N-doped carbon dots: Cytotoxicity and bio-imaging applications. Journal of Photochemistry and Photobiology B: Biology. 161: 2016: 154-161.

[20] Phadke C, Mewada A., Dharmatti R., Thakur M, Pandey S, Sharon M. Biogenic synthesis of fluorescent carbon dots at ambient temperature using azadirachta indica (Neem Gum). Journal of Fluorescent. 8: 2015: 1103-1107.

[21] Simpson A, Pandey RR, Charles CC, Kartik G, Adam KW. Fabrication characterization and potential applications of carbon nanoparticles in the detection of heavy metal ions in aqueous media. Carbon. 127: 2018: 122-130.

[22] Mewada A, Pandey S, Shinde M, Mishra N, Oza G, Thakur M, Sharon M. Green synthesis of biocompatible carbon dots using aqueous extract of Thrapabispinosa peel. Material Science and Engineering, 56: 2013: 2914-2917.

[23] Gedda G, Chun-Yi L, Yu-Chih L, Hui-fen W. Green synthesis of carbon dots from prawn shells for highly selective and sensitive detection of copper ions. Sensors and Actuators B: Chemical. 224: 2016: 396-403.

[24] Mehta,VN, Jha S, Singhal RK, Kailasa SK. Preparation of multicolor emitting carbon dots for HeLa cell imaging. New Journal of Chemistry, 38: 2014: 6152-6160.

[25] Mehta VN, Jha S, Basu H, Singhal RK, Kailasa SK. One step hydrothermal approach to fabricate carbon dots from apple juice for imaging mycobacterium and fungal cells. Sensors and Actuators B. 213: 2015: 434-443.

[26] Kumar D, Singh K., Verma V, Bhatti HS. Synthesis and characterization of carbon quantum dots from orange juice. Bionanoscience. 4: 2014: 274-279.

[27] Fernandez M, Dyna M, Suneesh KS, Sobhi D. Green synthesis of nitrogen and sulphur doped carbon dot composites for the sensing of glucose. Materials Today: Proceedings. 9: 2019: 54-60.

[28] Wang Y, Hu A. Carbon quantum dots: synthesis, properties and applications. Journal of Materials Chemistry C. 34: 2014: 6921-6939.

[29] Mahardika PA, Pradita AW, Jotti K, Annisa LW. Removal of Heavy Metal Nickel-Ions from Wastewaters Using Carbon Nanodots from Frying Oil. Procedia Engineering. 170: 2017: 36 – 40.

[30] Wang J, Wang CF, Chen S. Amphiphilic Egg-Derived Carbon Dots: Rapid Plasma Fabrication, Pyrolysis Process, and Multicolor Printing Patterns. Angewandte Chemie International Edition. 51: 2012: 9297-9301.

[31] Ke Y, Garg B, Ling YC. Waste chicken eggshell as low-cost precursor

for efficient synthesis of nitrogen-doped fluorescent carbon nanodots and their multi-functional applications. Royal Socity of Chemistry Advances. 4: 2014: 58329-58336.

[32] Titirici MM, Robin JW, Nicolas B, Vitaliy LB, Dang SS, Francisco DM, James HC, Mark JM. Sustainable carbon materials. Chemical Society Review. 44: 2015: 250-290.

[33] Wanekaya AK. Applications of nanoscale carbon-based materials in heavy metal sensing and detection. Analyst. 136: 2011: 4383-4391.

[34] Lim SY, Shen W, Gao Z. Carbon quantum dots and their applications. Chemical Society Review. 44: 2015: 362-371.

[35] Prat D, Hayler J, Wells A. A survey of solvent selection guides. Green Chemistry. 16: 2014: 4546-4551.

[36] Anastas PT, Warner JC. Green chemistry: Theory and practice, Oxford (England) and New York: Oxford University Press. 1998.

[37] Wang Y, Hu R, Lin G, Roy I, Yong KT. Functionalized quantum dots for biosensing and bioimaging and concerns on toxicity. American Chemical Society Applied Matererial Interfaces 5: 2013: 2786-2799.

[38] Jiang L, Nelson GW, Kim H, Sim I, Han SO, Foord JS. Cellulose-Derived Supercapacitors from the Carbonisation of Filter Paper. Chemistry Open. 4: 2015: 586-589.

[39] Derfus AM, Chan AC Warren CW, Bhatia SN. Probing the Cytotoxicity of Semiconductor Quantum Dots. Nano Letter 4: 2004: 11-18.

[40] Yuqian P, Hui G, Shaohui W, Xiaolong L. Facile synthesis the nitrogen and sulfur co-doped carbon dots for selective fluorescence detection of heavy metal ions. Materials Letters. 193: 2017: 236-239.

[41] Sayan G, Poushali D, Madhuparna B, Subhadip M, Amit, KD, Das NC. Strongly blue-luminescent N-doped carbogenic dots as a tracer metalsensing probe in aqueous medium and its potential activity towardsin situ Ag-nanoparticle synthesis. Sensors and Actuators B. 252: 2017: 735-746.

[42] Huang CC, Yi-Shan H, Yih-Ming W, Wenlung C, Yen-Shi L. Sustainable development of carbon nanodots technology: Natural products as a carbon source and applications to food safety. Trends in Food Science & Technology. 86: 2019: 144-152.

[43] Pattanayak M, Nayak PL. Ecofriendly green synthesis of iron nanoparticle from various plants. International Journal of Plant. Animal and Environmental Sciences. 78: 2013: 68-78.

[44] Liu Y, Xiao N, Gong N, Wang H, Shi X, Gu W, Ye L. One-step microwave-assisted polyol synthesis of green luminescent carbon dots as optical nanoprobes. Carbon. 68: 2014: 258-264.

[45] Piyushi N, Subramanian KA, Dastidar MG. Adsorptive removal of dye using biochar derived from residual algae after in-situ transesterification: Alternate use of waste of biodiesel industry. Journal of Environmental Management. 32: 2016: 187-197.

[46] Iravani S. Green synthesis of metal nanoparticles using plants. Green Chemistry. 34: 2011: 2638-2650.

[47] Wembo L, Xiaoyun Q, Sen L, Guohui C, Yingwei Z, Yonglan L, Abdullah MA, Abdulrahaman OA, Xuping S. Economical, Green synthesis of fluorescent carbon nanoparticles and their use as probes for sensitive and selective detection of mercury (II) ions. Analytical chemistry. 78: 2012: 5351-5357.

[48] Till TM, Piotr JC, Ilko B. White carbon: Fluorescent carbon nanoparticles with tunable quantum yield in a reproducible green synthesis. Scientific Reports. 6: 2016: 28557-28569.

[49] Han S, He, Z, Jian Z, Yujie X, Liangliang L, Hangxing W, Xiangkai L, Weisheng L, Yu T. Fabrication, gradient extraction and surface polarity-dependent photoluminescence of cow milk-derived carbon dots. Royal Society of Chemistry Advances. 63: 2014: 58084-58089.

[50] Biswajit G, Soubantika P, Joydeep C. Carbon Dots: A Mystic Star in the World of Nanoscience. Journal of Nanomaterials. 10: 2019: 1-19.

[51] Gonçalves H, Pedro AS, Jorge JRA, Joaquim CGES. Hg(II) sensing based on functionalized carbon dots obtained by direct laser ablation. Sensors and Actuators B. 145: 2010: 70-82.

[52] Thongpoola V, Asanithia P, Limsuwana P. Synthesis of Carbon Particles using Laser Ablation in Ethanol. Procedia Engineering. 32: 2012: 1054 – 1060.

[53] Yu H, Li X, Zeng X, Lu Y. Preparation of carbon dots by non-focusing pulsed laser irradiation in toluene. Chemical Communications. 52: 2016: 819-822.

[54] David A, Mouhamed D, Julien L, Gilles L, Christophe D. Origin of the nano-carbon allotropes in pulsed laser ablation in liquids synthesis. Journal of Colloid and Interface Science. 489: 2017: 114-125.

[55] Myungjoon K, Saho O, Taesung K, Hidenori H, Takafumi S. Synthesis of Nanoparticles by Laser Ablation: A Review. KONA Powder and Particle Journal. 34: 2017: 80-90.

[56] Mendonça LT, Azevedo WM. A fast bottom-up route for preparing CdS quantum dots using laser ablation in a liquid environment. Journal of Luminescence.23: 2016: 79-84.

[57] Li H, Xiaodie H, Zhenhui K, Hui H, Yang L, Jinglin L, Suoyuan L. Water-soluble fluorescent carbon quantum dots and photocatalyst design. Angewandt Chemie International Edition, 49: 2010: 4430-4434.

[58] Shinde DB, Pillai VK. Electrochemical resolution of multiple redox events for graphene quantum dots. Angewandte Chemie International Edition. 52: 2013: 2482-2485.

[59] Deng J, Lu Q, Mi N, Li H, Liu M, Xu M, Tan L, Xie Q, Zhang Y, Yao S. Electrochemical synthesis of carbon nanodots directly from alcohols. Chemistry European Journal. 20: 2014: 4993-4999.

[60] Yao S, Hu Y, Li G. A one-step sonoelectrochemical preparation method of pure blue fluorescent carbon nanoparticles under a high intensity electric field. Carbon. 65: 2014: 77-83.

[61] Zhang P, Xue Z, Luo D, Yu W, Guo Z, Wang T. Dual-peak electrogenerated chemiluminescence of carbon dots for iron ions detection. Analytical Chemistry. 86: 2014: 5620-5623.

[62] Yang Y, Zhao B, Gao Y, Liu H, Tian Y, Qin D, Wu H, Huang W, Hou L. Novel hybrid ligands for passivating PbS colloidal quantum dots to enhance the performance of solar cells. Nano-Micro Letters. 7: 2015: 325-331.

[63] Tian L, Ghosh D, Chen W. Nanosized Carbon Particles From Natural Gas Soot. Chemistry of Materials, 21: 2009: 2803-2809.

[64] Liu H, Ye T, Mao C. Fluorescent Carbon Nanoparticles Derived from Candle Soot. Angewandte Chemie

International edition. 63: 2007: 6473-6475.

[65] Zhai X, Zhang P, Liu C, Bai T, Li W, Dai L, Liu W. Highly luminescent carbon nanodots by microwave-assisted pyrolysis. Chemical Communications. 48: 2012: 7955-7957.

[66] Xinyue Z, Mingyue J, Na N, Zhijun C, Shujun L, Shouxin L, Jian L. Natural-Product-Derived Carbon Dots: From Natural Products to Functional Materials. ChemSusChem. 11: 2018: 11-24.

[67] Sun D, Ban R, Zhang P, Wu G, Zhang J, Zhu J. Hair fiber as a precursor for synthesizing of sulfur- and nitrogen-co-doped carbon dots with tunable luminescence properties. Carbon. 56: 2013: 424-434.

[68] Zhou P, Lu X, Sun Z, Guo Y, He H. A review on syntheses, properties, characterization and bioanalytical applications of fluorescent carbon dots. Microchimica Acta. 54: 2016: 519-542.

[69] Fang Y, Guo S, LI D, ZHu C, Ren W, Dong S, Wang E. Easy synthesis and imaging applications of cross-linked green fluorescent hollow carbon nanoparticles. American Chemical Society Nano: 6: 2012: 400-409.

[70] Tang L, Ji R, Cao X, Lin J, Jiang H, Li X, Teng KS, Luk CM, Zeng S, Hao J, Lau SP. Deep ultraviolet photoluminescence of water-soluble self-passivated graphene quantum dots. American Chemical Society Nano. 6: 2012: 5102-5110.

[71] Ali AE, Hghighat S, Kazemifard N, Rezaei B, Moradi F. A novel one-step and green synthesis of highly fluorescent carbon dots from saffron for cell imaging and sensing of prilocaine. Sensors and Actuators B: Chemical. 253: 2017: 451-460.

[72] Luyao L, Xingxian W, Zheng F, Fengling C. One-step hydrothermal synthesis of nitrogen- and sulfur-co-doped carbon dots from ginkgo leaves and application in biology. Materials Letters. 196: 2017: 300-303.

[73] Yongli L, Qingxiang Z, Yongyong Y, Yalin W. Hydrothermal synthesis of fluorescent carbon dots from sodium citrate and polyacrylamide and their highly selective detection of lead and pyrophosphate. Carbon. 115: 2017: 550-560.

[74] Jie W, Yuhao C, Fengxian Q, Xin L, Dongya Y. One-pot simple green synthesis of water-soluble cleaner fluorescent carbon dots from cellulose and its sensitive detection of iron ion. Journal of Cleaner Production. 10: 2017: 23-30.

[75] Sharma S, Ahmad U, Swati S, SurinderKM,SushilKK.Photoluminescent C-dots: An overview on the recent development in the synthesis, physiochemical properties and potential applications. Journal of Alloys and Compounds. 748: 2018: 818-853.

[76] Yang Y, Cui J, Zheng C. One step synthesized of amino-functionalised fluorescent carbon naoparticles by hydrothermal carbonization of chitosan. Chemical Communications. 39: 2012: 380-382.

[77] Zhang B, Liu CY, Liu Y. A Novel One-Step Approach to Synthesize Fluorescent Carbon Nanoparticles. European Journal of Inorganic Chemistry. 28: 2010: 4411-4414.

[78] De B, Karak N. A green and facile approach for the synthesis of water soluble fluorescent carbon dots from banana juice. Royal Society of Chemistry Advances. 36: 2013: 1-9.

[79] Arumugam N, Jongsung K. Synthesis of carbon quantum dots from

Broccoli and their ability to detect silver ions. Materials Letters. 219: 2018: 37-40.

[80] Chunxi Z, Yang J, Feng H, Yaling Y. Green synthesis of carbon dots from pork and application as nanosensors for uric acid detection. Spectrochimica Acta Part A: Molecular and Biomolecular Spectroscopy. 190: 2018: 360-367.

[81] Du F, Zhang M, Li X, Li J, Jiang X, Li Z, Gong A. Economical and green synthesis of bagasse derived fluorescent carbon dots for biomedical applications. Nanotechnology. 53: 2014: 315702-315718.

[82] Li HT, He XD, Liu Y, Huang H, Lian SY, Lee ST, Kang ZH. One step ultrasonic treatment synthesis of water soluble carbon nanoparticles with excellent photoluminescent properties. Carbon. 49: 2011: 605-609.

[83] Liu X, Chunlan Y, Baozhan Z, Jianyuan D, Dan X. Green anhydrous synthesis of hydrophilic carbon dots on large-scale and their application for broad fluorescent pH sensing. Sensors and Actuators B: Chemical. 255: 2018: 572-579.

[84] Himaja A, Karthik P, Sreedhar B, Singh SP. Synthesis of Carbon dots from kitchen waste: Conversion of waste to value added product. Fluorescence. 24: 2014: 1767-1773.

[85] Thambiraj S, Shankaran DR. Green synthesis of highly fluorescent carbon quantum dots from sugarcane bagasse pulp. Applied Surface Science. 390: 2016: 435-443.

[86] Gaddam RR, Sudip M, Neelambaram P, Vasudevan D, Raju VSNK. Facile synthesis of carbon dot and residual carbon nanobeads: Implications for ion sensing, medicinal and biological applications. Materials Science and Engineering: C. 73: 2017: 643-652.

[87] Lingling L, Luyao L, Chang-Po C, Fengling C. Green synthesis of nitrogen-doped carbon dots from ginkgo fruits and the application in cell imaging. Inorganic Chemistry Communications. 86: 2017: 227-231.

[88] Yaling W, Jingxia Z, Junli W, Yongzhen Y, Xuguang L. Rapid microwave-assisted synthesis of highly luminescent nitrogen-doped carbon dots for white light-emitting diodes. Optical Materials. 73: 2017: 319-329.

[89] Yang R, Xiangfeng G, Lihua J, Yu Z, Fedor L. Green preparation of carbon dots with mangosteen pulp for the selective detection of Fe^{3+} ions and cell imaging. Applied Surface Science. 423: 2017: 426-432.

[90] Kasibabu BSB, D'souza SL, Jha S, Singhal RK, Basu H, Kailasa SK. One-step synthesis of fluorescent carbon dots for imaging bacterial and fungal cells. Analytical Methods. 7: 2015: 2373-2382.

[91] Xu Y, Wu M, Liu Y, Feng XZ, Yin XB, He XW, Zhang YK. Nitrogen-doped carbon dots: a facile and general preparation method, photoluminescence investigation, and imaging applications. Chemistry: A European Journal. 19: 2013: 2276-2283.

[92] Cao L, Sheng-Tao Y, Xin W, Pengju GL, Jia-Hui L, Sushant S, Yamin L, Ya-Ping S. Competitive Performance of Carbon "Quantum" Dots in Optical Bioimaging. Theranostics. 2: 2012: 295-301.

[93] Vinci JC, Ferrer IM, Seedhouse SJ, Bourdon AK, Reynard JM, Foster BA, Colon LA. Hidden properties of carbon dots revealed after HPLC fractionation. Physical Chemistry Letters. 65: 2012: 239-253.

[94] Zhu GM, Zeng Y, Zhang L, Tang J, Chen M, Cheng LH, Zhang L,

He Y, Guo XX, He MY. Highly sensitive electrochemical sensor using a MWCNTs/GNPs-modified electrode for lead (II) detection based on lead(II)-induced G-rich DNA conformation. Analyst. 139: 2014: 5014-5020.

[95] Gho EJ, Ki SK, Yi RK, Ho SJ, Songeun B, Won HK, Giuliano S, Seok HY, Sei KH. Bioimaging of Hyaluronic Acid Derivatives Using Nanosized Carbon Dots. Biomacromolecules. 13: 2012: 2554-2561.

[96] Vadivel R, Senthil KT, Kaviyarasan R, Ragupathy S, Rajkumar S, Perumal R. Outright Green Synthesis of Fluorescent Carbon Dots from Eutrophic Algal Blooms for In Vitro Imaging. American Chemical Society Sustainable Chemical Engineering. 9: 2016: 4724-4731.

[97] Cheng IW, Huan TC, Chia HL, Yu WS, Binesh U, Yu J, Chih CH. One-step synthesis of biofunctional carbon quantum dots for bacterial labeling. Biosensors and Bioelectronics. 68: 2015: 1-6.

[98] Chengkun J, Hoa W, xiaojie S, xiaojun M, Jihui W, Mingquian T. Presence of photoluminescent carbon dots in Nescafe original instant coffee: Application to bioimaging. Talanta. 10: 2014: 68-74.

[99] Kuo-Chih L, Han-sheng C, Shu-yu F, You-Di T, Pei-Hsuan L. Percutaneous fiber-optic biosensors for immediate evaluation of chemotherapy efficacy in vivo (part 1): Strategy of assay design for monitoring non-homogeneously distributed biomarkers. Sensors and Actuators B: Chemical. 45: 2016: 544-550.

[100] Musa YP, Zurina ZA, Suraya AR, Faizah MY, Noor ASM, Jaafar A. Selective and simultaneous detection of cadmium, lead and copper by tapioca-derived carbon dot–modified electrode. Environmental Science and Pollution Research. 27: 2020: 13315-13324.

[101] Lingling Z, Yujie H, Zhu Y, Shaojun D. Simple and Sensitive Fluorescent and Electrochemical Trinitrotoluene Sensors Based on Aqueous Carbon Dots. Analalytical Chemistry. 874: 2015: 2033-2036.

[102] Miaoran Z, Rigu S, Jian Z, Ling F, Wei C, Qingwen G, Weijun L, Neng L, Yusheng C, Lulu, C. Red/orange dual-emissive carbon dots for pH sensing and cell imaging. Nano Research. 12: 2019: 815-832.

[103] Li Y, Hu Y, Zhao Y, Shi G, Deng L, Hou Y, Qu L. An electrochemical avenue to green-luminescent graphene quantum dots as potential electron-acceptors for photovoltaics. Advance Materials. 23: 2011: 776-780.

[104] Hsu PC, Shih ZY, Lee CH, Chang HT. Synthesis and analytical applications of photoluminescent carbon nanodots. Green Chemistry. 14: 2012: 917-921.

[105] Sun Y, Zhou B, Lin Y. Quantum-sized carbon dots for bright and colorful photoluminescence. Journal of American Chemical Society. 53:: 2006: 7756-7757.

[106] Xu J, Zhou Y, Cheng G, Dong M, Liu S, Huang C. Carbon dots as a luminescence sensor for ultrasensitive detection of phosphate and their bioimaging properties. Luminescence. 42: 2015: 411-415.

[107] Yang X, Zhuo Y, Zhu S, Luo Y, Feng Y, Dou Y. Novel and green synthesis of high-fluorescent carbon dots originated from honey for sensing and imaging. Biosensors Bioelectronic. 60: 2014: 292-298.

[108] Rui-Jun F, Qiang S, Ling Z, Yan Z, An-Hui L. Photoluminescent carbon

dots directly derived from polyethylene glycol and their application for cellular imaging. Carbon. 71: 2014: 87-93.

[109] Zhang R, Chen W. Nitrogen-doped carbon quantum dots: facile synthesis and application as a "turn-off" fluorescent probe for detection of Hg^{2+} ions. Biosensors and Bioelectronic. 55: 2014: 83-90.

[110] Sandhya BDB, Tanujjal B, Sunandan B, Joydeep D. Heavy Metal ion Sensing in water using surface plasmon resonance of metallic nanostructures. Groundwater for sustainable development. 1:2015: 1-11.

[111] Qu K, Wang J, Ren J, Qu X. Carbon dots prepared by hydrothermal treatment of dopamine as an effective fluorescent sensing platform for the label-free detection of iron(III) ions and dopamine. Chemistry – A European Journal. 22: 2013: 7243-7252.

[112] Rao L, Yong T, Zongtao L, Xinrui D, Binhai Y. Efficient synthesis of highly fluorescent carbon dots by microreactor method and their application in Fe^{3+} ion detection. Materials Science and Engineering: C. 81: 2017: 213-223.

[113] Pang Y, Hui G, Shaohui W, Xiaolong L. Facile synthesis the nitrogen and sulfur co-doped carbon dots for selective fluorescence detection of heavy metal ions. Materials Letters. 193: 2017: 226-239.

[114] Han M, Liping W, Siheng L, Liang B, Zhenhui K. High-bright fluorescent carbon dot as versatile sensing platform. Talanta. 174: 2017: 265-273.

[115] Lin B, Yun Y, Manli G, Yujuan C, Duo W. Modification-free carbon dots as turn-on fluorescence probe for detection of organophosphorus pesticides. Food Chemistry. 77: 2017: 6-11.

[116] Pajewska-Szmyt M, Bogusław B, Renata GK. Sulphur and nitrogen doped carbon dots synthesis by microwave assisted method as quantitative analytical nano-tool for mercury ion sensing. Materials Chemistry and Physics: 242: 2020: 1-9.

[117] Ye Q, Fanyong Y, Yunmei L, Yinyin W, Li C. Formation of N, S-codoped fluorescent carbon dots from biomass and their application for the selective detection of mercury and iron ion. Spectrochimica Acta Part A: Molecular and Biomolecular Spectroscopy. 173: 2017: 854-862.

[118] Hua J, Jian Y, Yan Z, Chunxi Z, Yaling Y. Highly fluorescent carbon quantum dots as nanoprobes for sensitive and selective determination of mercury (II) in surface waters. Spectrochimica Acta Part A: Molecular and Biomolecular Spectroscopy. 187: 2017: 149-155.

[119] Gu D, Shaoming S, Qin Y, Jie S. Green synthesis of nitrogen-doped carbon dots from lotus root for Hg(II) ions detection and cell imaging. Applied Surface Science. 390: 2016: 38-42.

[120] Wang J, Fengxian Q, Xin L, Haiyan W, Dongya Y. A facile one-pot synthesis of fluorescent carbon dots from degrease cotton for the selective determination of chromium ions in water and soil samples. Journal of Luminescence 188: 2017: 230-237.

[121] Fang Y, Song J, Li J, Wang Y, Yang H, Sun J, Chen G. Electrogenerated chemiluminescence from Au nanoclusters. Chemical Communications. 47: 2011: 2369-2371.

[122] Ma X, Yuanhua D, Hanyuan S, Ningsheng C. Highly fluorescent carbon dots from peanut shells as potential probes for copper ion: The optimization and analysis of the synthetic process. Materials Today Chemistry. 5: 2017: 1-10.

[123] Liu W. Preparation of a Zinc Oxide-Reduced Graphene Oxide Nanocomposite for the Determination of Cadmium(II), Lead(II), Copper(II), and Mercury(II) in Water. International Journal of Electrochemical Science. 12: 2017: 5392 – 5403.

[124] Rahmanian O, Mohammad D, Mahmood KA. Carbon quantum dots/ layered double hydroxide hybrid for fast and efficient decontamination of Cd(II): The adsorption kinetics and isotherms. Applied Surface Science. 428: 2018: 272-279.

[125] Venkateswarlu S, Buddolla V, Ankireddy SR, Minyoung Y. Fungus-derived photoluminescent carbon nanodots for ultrasensitive detection of Hg^{2+} ions and photoinduced bactericidal activity. Sensors and Actuators B: Chemical. 258: 2018: 172-183.

[126] Homaei A. Immobilization of Penaeus merguiensis alkaline phosphatase on gold nanorods for heavy metal detection. Ecotoxicology and Environmental Safety. 136: 2017: 1-7.

[127] Wang Y, Wen-ting W, Ming-bo W, Hong-di S, Jie-shan Q. Yellow-visual fluorescent carbon quantum dots from petroleum coke for the efficient detection of Cu^{2+} ions. New Carbon Materials. 30: 2015: 550-559.

[128] Simona T, Michaela VG, Kristýna Š, Zuzana Č, Viktor K. Study of metal accumulation in tapeworm section using laser ablation-inductively coupled plasma-mass spectrometry (LA-ICP-MS). Microchemical Journal. 133: 2017: 380-390.

[129] Beauchemin D, Inductively Coupled Plasma Mass Spectrometry Methods. Encyclopedia of Spectroscopy and Spectrometry. 3: 2017: 236-245.

[130] Tokalıoğlu S. Determination of trace elements in commonly consumed medicinal herbs by ICP-MS and multivariate analysis. Food Chemistry. 134: 2012: 2504-1512.

[131] Isai KA, Shrivastava VS. Detection and Identification of Organics and Metals from Industrial Wastewater by ICP-AES. Journal of Advanced Chemical Sciences. 1: 2015: 164-166.

[132] Fatema K, Naher K, Choudhury TR, Islam MA, Tamim U. Determination of Toxic Metal Accumulation in Shrimps by Atomic Absorption Spectrometry (AAS). Journal of Environmental Analytical Chemistry. 2: 2015: 140-153.

[133] Teerawat P, Natnicha S, Atitaya S. Use of electrothermal atomic absorption spectrometry for size profiling of gold and silver nanoparticles. Analytica Chimica Acta. 1000: 2018: 75-84.

[134] Hisham KF, Randa ME, Mohamed DH. The Application of Flame Atomic Absorption Spectrometry for Gold Determination in Some of Its Bearing Rocks. American Journal of Analytical Chemistry. 6: 2015: 411-421.

[135] Steve JH, Andy SF. Atomic Absorption, Methods and Instrumentation. Encyclopedia of Spectroscopy and Spectrometry (Third Edition). 2017: 37-43.

[136] Zhaoxia H, Feng L, Ming H, Chunxiang L, Ting X, Chuan C, Xiangqun G. Carbon dots with turnable emission, Controlled size and their application for sensing hypochlorous acid. Journal of Luminescence. 51: 2014: 100-105.

[137] Huilin T, Xiufen L, Qingyi W, Xiangli X, Fuxin Z, Zhongsheng Y, Mei Q, Zhenli W. Carbon dots as fluorescent probe for "off-on" Detecting sodium dodecyl-benzenesulfonoate in aqueous solution. Spectrochimica Acta part A: Molecular and Biomolecular Spectroscopy. 153: 2016: 268-272.

[138] Pumera M. The electrochemistry of carbon nanotubes: fundamentals and

applications. Chemistry- A European Journal. 20: 2009: 4970-4978.

[139] Musa YP, Zurina ZA, Suraya AR, Faizah MY, Noor ASM, Mohammed A. Eco-Friendly Sustainable Fluorescent Carbon Dots for the Adsorption of Heavy Metal Ions in Aqueous Environment. Nanomaterials. 10: 2020: 315-331.

[140] Trojanowicz M. Analytical applications of carbon nanotubes: a review. Trends in Analytical Chemistry. 25: 2006: 480-489.

[141] Musameh MM, Hickey M, Kyratzis IL. Carbon nanotube-based extraction and electrochemical detection of heavy metals. Research on Chemical Intermediates. 37: 2011: 675-687.

[142] Burakova AE, Evgeny VG, Irina VB, Anastassia EK, Shilpi A, Alexey GT, Vinod KG. Adsorption of heavy metals on conventional and nanostructured materials for wastewater treatment purposes: A review. Ecotoxicology and Environmental Safety. 148: 2018: 702-712.

[143] Musa YP, Zainal ZA, Abdul RS, Md Yasin F, Noor ASM, Issa MA. Sustainable Synthesis Processes for Carbon Dots through Response Surface Methodology and Artificial Neural Network. Processes. 10: 2019: 704-723.

[144] Ghiloufi I, ElGhoul J, Modwi L. Ga-doped ZnO for adsorption of heavy metals from aqueous solution. Materials Science in Semiconductor Processing. 42: 2016: 102-106.

[145] Ruihua L, Juan L, Weiqian K, Hui H, Xiao H, Xing Z, Yang L, Zhenhui K. Adsorption dominant catalytic activity of a carbon dots stabilized gold nanoparticles system. Dalton Transactions. 43: 2014: 10920-10929.

[146] Guo Y, Wang Z, Shao H, Jiang X. Hydrothermal synthesis of highly fluorescent carbon nanoparticles from sodium citrate and their use for the detection of mercury ions. Carbon. 52: 2014: 583-589.

[147] Susmita D, Susanta KB, Raz J. Carbon-dot-aerogel sensor for aromatic volatile organic compounds. Sensors and Actuators B. 241: 2017: 607-613.

[148] Fu H, Zhu D. Graphene oxide-facilitated reduction of nitrobenzene in sulfide-containing aqueous solutions. Environmental Science and Technology. 47: 2013: 4204-4210.

[149] Müller K, Foerstendorf H, Brendler V, Bernhard G, Sorption of Np(V) onto TiO2, SiO2, and ZnO: An in situ ATR FT-IR spectroscopic study. Environmental Science and Technology. 43: 2009: 7665-7670.

[150] Quevedo IR, Olsson AL, Tufenkji N. Deposition kinetics of quantum dots and polystyrene latex nanoparticles onto alumina: role of water chemistry and particle coating. Environmental Science and Technology. 47: 2013: 2212-2220.

[151] Romanchuk AY, Slesarev AS, Kalmykov SN, Kosynkin DV, Tour JM. Graphene oxide for effective radionuclide removal. Physical Chemistry Chemical Physics. 15: 2013: 2321-2327.

[152] Shao D, Hou G, Li J, Wen T, Ren X, Wang X. PANI/GO as a super adsorbent for the selective adsorption of uranium (VI). Chemical Engineering Journal. 255: 2014: 604-612.

[153] Wang L, Zhu D, Duan L, Chen W. Adsorption of single-ringed N- and S-heterocyclic aromatics on carbon nanotubes. Carbon. 48: 2010: 3906-3915.

[154] Zhao J, Liu F, Wang Z, Cao X, Xing B. Heteroaggregation of graphene

oxide with minerals in aqueous phase. Environmental Science and Technology. 49: 2015: 2849-2857.

[155] Musa YP, Zurina Z A, A sustainable and eco-friendly technique for dye adsorption from aqueous media using waste from Jatropha curcas (isotherm and kinetic model). Desalination and Water Treatment. 182: 2020: 365-374.

[156] Park JH, Yong SO, Seong-Heon K, Ju-Sik C, Jong-Soo H, Ronald D, Dong-Cheol S, Competitive adsorption of heavy metals onto sesame straw biochar in aqueous solutions. Chemosphere. 142: 2016: 77-83.

Section 2

Energy Storage and Harvesting Applications of Nanomaterials

Chapter 6

Carbon-Based Nanocomposite Materials for High-Performance Supercapacitors

Prasanta Kumar Sahoo, Chi-Ang Tseng, Yi-June Huang and Chuan-Pei Lee

Abstract

Lightweight, flexible, wearable, and portable electronic gadgets have drawn significant attention in modern electronics industry. To power these gadgets, great efforts have been made to develop highly efficient energy-storage equipment. Among various power sources, a supercapacitor, acting as a bridge between the conventional battery and electrolytic capacitor, has been considered a promising portable energy storage device because of its high power density, fast charge/discharge rate, adequate operational safety, and excellent working lifetime. Hybrid supercapacitors, which combine redox materials with carbon-based materials, exhibit tremendous potential to fulfill the requirement of practical applications. In this chapter, we will review recent reports focusing on composite materials (*i.e.* metal oxide, metal hydroxide, and metal dichalcogenide composited with carbon materials) for the application in supercapacitors. The conclusion and futuristic prospects and challenges of highly efficient supercapacitors are briefly discussed.

Keywords: energy storage, composites, metal oxides, metal hydroxides, transition metal dichalcogenides, supercapacitor

1. Introduction

There is sharply increasing demand for energy with the rapid growth of the global economy. The energy generation from sustainable sources, such as wind and solar, plays an important role in power supply. However, the intermittent nature and imbalanced regional distribution of the sustainable energy make them unable to stably supply the power [1]. The development of energy storage systems is an urgent requirement to meet the sufficient and stable power supply for industrial and residential usage. Although rechargeable lithium-ion batteries, dominant energy sources in each field, as high energy density providers have filled their position [2], lithium-ion batteries still have the limitations of poor cycle life and low power performance [3]. Supercapacitors (SCs), also known as ultracapacitor and electrochemical capacitors, are an emerging class of energy storage device, which possess high power density and tens of thousands of charge/discharge cycles [4, 5]. **Figure 1** shows the Ragone plot of different energy conversion and storage devices. SCs have a unique position to bridge the gap between conventional capacitors and batteries. Compared with conventional capacitor, SCs possess higher specific energy density

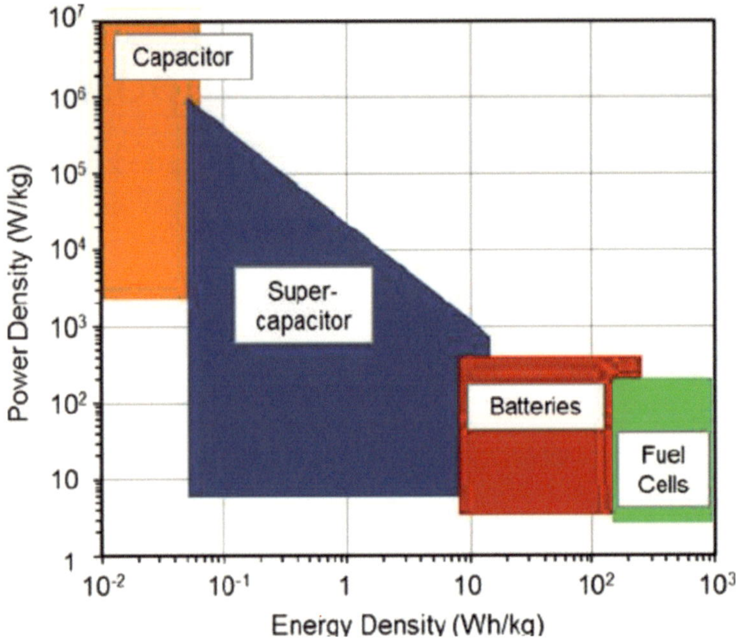

Figure 1.
Ragone plot for various energy storage and conversion devices [6].

in several orders of magnitude. Moreover, SCs provide higher specific power density than batteries due to its unique charge storage mechanism.

2. Theoretical background for supercapacitors

2.1 Principle and mechanism of supercapacitors

Based on different charge storage mechanisms, SCs are mainly divided into two categories, electrical double layer capacitors (EDLCs) and pseudocapacitors, as shown in **Figure 2**. EDLCs store the electrical charge by electrostatic force at the electrode-electrolyte interface, which is a physical process without involving electrochemical reactions on the electrode surface. In order to increase the capacitance and energy density of SCs, some electrochemically active materials, such as transition metal oxide and conducting polymers, have been explored as electrode materials for pseudocapacitors. The energy storage in pseudocapacitors originates from reversible surface faradaic redox reactions at the interface of electrolyte and electroactive materials.

2.2 Factors affecting the performance of Supercapacitors

The capacitance of EDLCs is strongly dependent on effective surface area and the pore size distribution of the electrode [7, 8]. Typically, the carbon-based materials and their derivatives, including activated carbon, carbon nanotubes (CNTs) and graphene, with high conductivity, chemically-stability, and large surface area are widely utilized in EDLCs. Although the EDLCs possess high power density and excellent charge/discharge cycling stability, they suffer from low energy density owing to the relatively low capacitance of carbon-based materials. Pseudocapacitors can achieve significantly higher energy density, as compared to EDLCs, because they have a variety of oxidation states for redox charge transfer reactions. However, relatively low electrical conductivity and poor rate capability and cycle stability of pseudocapacitive materials limit their widespread commercial applications [9]. Therefore, carbon-based materials with high conductivity and distinct structures can be combined with pesudocapacitive materials to exhibit synergistic effects for supercapacitive performance, known as hybrid SCs.

3. Carbon based composite electrode materials

Carbon material is EDLCs type for supercapacitor. In section 2.1, EDLCs has introduced their property, which store the electrical charge by electrostatic force at the electrode-electrolyte interface, as shown in **Figure 2**. It is not involving electrochemical reactions on the electrode surface. There are different types of carbon nanostructured materials, which can be used as single electrode materials due to their unique structural, mechanical, and electrical properties.

3.1 Zero-dimensional (0-D) carbon nanoparticles

They are round-shaped particles such as ultrafine activated carbon (AC), mesoporous carbon, carbon nanosphere, and carbon quantum dot, with a high specific area (AC: ~3000 $m^2\ g^{-1}$) and an aspect ratio of nearly [10]. In addition, by tuning the pore size distribution and pore content, they can use as suitable supporting materials for composite electrodes.

3.2 One-dimensional (1-D) carbon nanostructures

These are the high aspect ratio materials with fiber shaped and good electronic properties e.g. carbon nanotubes (CNT), carbon nanocoils, and carbon nanofibers (CNF), which facilitates the electrochemical reaction kinetics by 1-D charge transfer pathway.

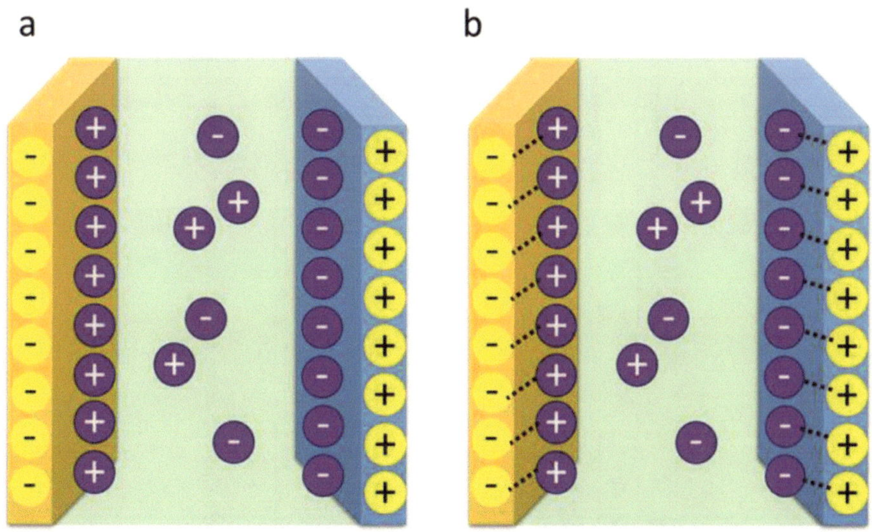

Figure 2.
Schematic diagram of (a) an electrical double layer capacitor and (b) a pseudocapacitor.

Materials	Carbon onions	Carbon nanotubes	Graphene	Templated carbon
Dimensionality	0D	1D	2D	3D
Conductivity	High	High	High	Low
Volumetric Capacitance	Low	Low	Moderate	Low
Cost	High	High	Moderate	High
Structure		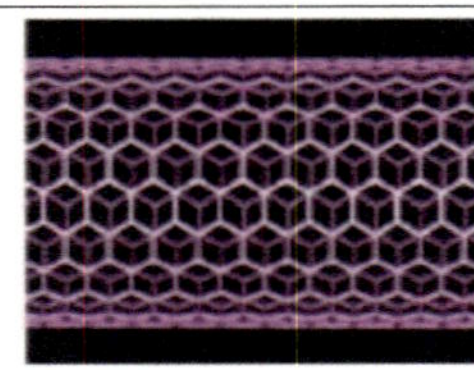		

Table 1.
Different carbon nanostructures used as electrode materials for EDLCs with onion-like carbon, carbon nanotubes, graphene, activated carbon, carbide-derived carbon, and templated carbon [16].

3.3 Two-dimensional (2-D) nanosheets

They are sheet like structures with high aspect ratio such as graphene, graphene oxide (GO) or reduced graphene oxide (rGO). In addition, they have high specific surface area, good mechanical strength, and excellent electrotonic conductivity, which helps them as promising electrode materials for SCs. For an example, single layered graphene has theoretical surface area of 2756 $m^2 g^{-1}$ and charge mobility of 200000 $cm^2 V^{-1} s^{-1}$ [11].

3.4 Three-dimensional (3-D) porous nanostructures

These are the low dimensional building blocks such as carbon nanofoams or sponges with hierarchical porous channels, rich pore structures, higher electrical conductivity and better structural mechanical stability, which are extensively used in composite electrode materials for SCs. For an example, foam has high specific surface area with continuous electron transport path and large area of electrolyte-electrode interface.

Table 1 shows some examples of different carbon nanostructured materials such as carbon onions, carbon nanotubes, graphene, and templated carbon, which are used as electrode materials for EDLCs. Each carbon nanostructured materials have its advantages and disadvantages. For example, carbon onions have high power performance due to excellent conductivity with high accessible ion adsorption capacity but low capacitance of ~30 F g^{-1} [12]. On the other hand, CNTs have high energy density due to superior electrical properties and unique tubular structures for fast charge transportation but due to the high cost, their widespread applications are limited [13]. Recently, graphene has been attracted much attention as electrode materials for EDLC applications due to unique properties, like as ultrahigh specific surface area, unique conductivity, and exceptionally high mechanical strength [14]. However, the aggregation of sheets during electrode preparation limits the aspect of application. More recently, 3D porous carbon nanostructured materials are widely used for EDLCs because of rich pore structures and high surface areas but due to relative low conductivity and presence of micropores specific capacitance is insufficient at a high current density [15]. Therefore, it is necessary to construct composite materials by coupling the advantages of different types of carbon nanostructured materials and high energy electrode materials such as transition metal oxides, metal hydroxides and metal dichalcogenides (TMDs) to enhance the energy density without the compromise of power density and also meet the requirement for fabrication of high energy storage devices. In the composite electrode material, different types of carbon nanostructured materials not only contribute to high capacitance but also provide an easy conductive path for charge transportation due to conductive nature.

4. Carbon-metal oxide composite electrode materials

Many metal oxide such as RuO_2, MnO_2, Fe_3O_4, V_2O_5, NiO, Co_3O_4, and TiO_2, has been received significant attention and extensive studied as SC electrode materials due to Pseudo capacitance nature, which depends on the fast reversible redox reaction of electroactive species directly as well as in the vicinity of electrode surface [17–20]. The redox behavior is due to the multivalent property of the above oxides which changes their oxidation states by interaction with protons or hydroxide ions reversibly. In spite of their excellent specific capacitance, they still suffer from low conductivity, low rate capability, poor stability and durability during the process of charge/discharge. In contrast carbon materials shows excellent performance in these regards but suffer from comparatively limited specific capacitance. Hence, the synergic integration of metal oxides with conducting carbon supports may form high potential carbon-metal oxide composite electrodes materials for SCs and hybrid devices because of their enhanced electrochemical performance through the combined effect of pseudocapacitive/faradaic charge storage and electrical double layer capacitance mechanisms [21–23].

4.1 Carbon-ruthenium oxide (RuO_2)-based composite electrode materials

Among the metal oxides, ruthenium oxide (RuO_2) has been considered as very common electrode materials for SCs in acidic medium due to their excellent pseudocapacity which is arising from high conductivity, good thermal stability, highly reversible redox reactions, three different oxidation states within 1.2 V, and high specific capacitance natures. The pseudocapacitance mechanism of RuO_2 for SC electrodes can be described as equation [24]:

$$RuO_2 + xH^+ + xe^- = RuO_{2-x}(OH)_x \ (0 \le x \le 2) \tag{1}$$

Or

$$RuO_2 + H^+ + e^- = RuOOH \tag{2}$$

However, its scarcity and high cost limits the fabrication of RuO_2 based electrodes for potential applications. But, smartly use of composite materials by synergic integration of pseudocapacitive RuO_2 materials with conductive carbonaceous substrates not only improves the capacitance but also reduces the cost of the electrode. Recent studies are more focus about the selecting the best carbonaceous substrate and the synthesis procedures to fabricate ruthenium oxide (RuO_2)-coated on the porous carbonaceous substrates.

RuO_2-CNT composite has been prepared by uniformly coating of RuO_2 on the vertically aligned porous carbon nanotubes porous through atomic layer deposition (ALD) technique and further activation by voltammetry potential coulometry (**Figure 3(a-c)**) [25]. This ALD technique has many advantages such as deposition on large surface area, accurate thickness and exceptional uniformity for electrode designing in energy storage devises. The as-prepared RuO_2-CNT composite shows excellent electrochemical performance as an electrode material for SC in respect of capacitance, power density and stability. Several publications have been reported the specific capacitance and power density of RuO_2-CNT composite, which are around 650 F g^{-1} and 17 kW kg^{-1}, respectively. Kaner *et.al* recently demonstrated the synthesis and processing of 3D porous RuO_2/laser-scribed graphene (LSG) composite electrode for miniaturized and interdigitated SC that exhibit ultrahigh energy and power density (**Figure 3(d)**) [26]. The high-resolution TEM (HRTEM) image of 3D porous RuO_2/LSG composite in **Figure 3(e)** shows that multiple layers of the graphene sheets wrap around each RuO_2 nanoparticle. 3D porous RuO_2/LSG composite electrode showed an ultrahigh specific capacitance of 1139 Fg^{-1} with outstanding rate capability and the asymmetric supercapacitor (ASC) made of 3D porous RuO_2/LSG composite electrode as positive electrode exhibited an extremely high energy density of 55 W h kg^{-1} at a power density of 12 kW kg^{-1} (**Figure 3(f)**). Other interesting composite of RuO_2 made of RuO_2 decorated nitrogen-doped reduced graphene oxide aerogel (NGA) are used as high-performance transparent

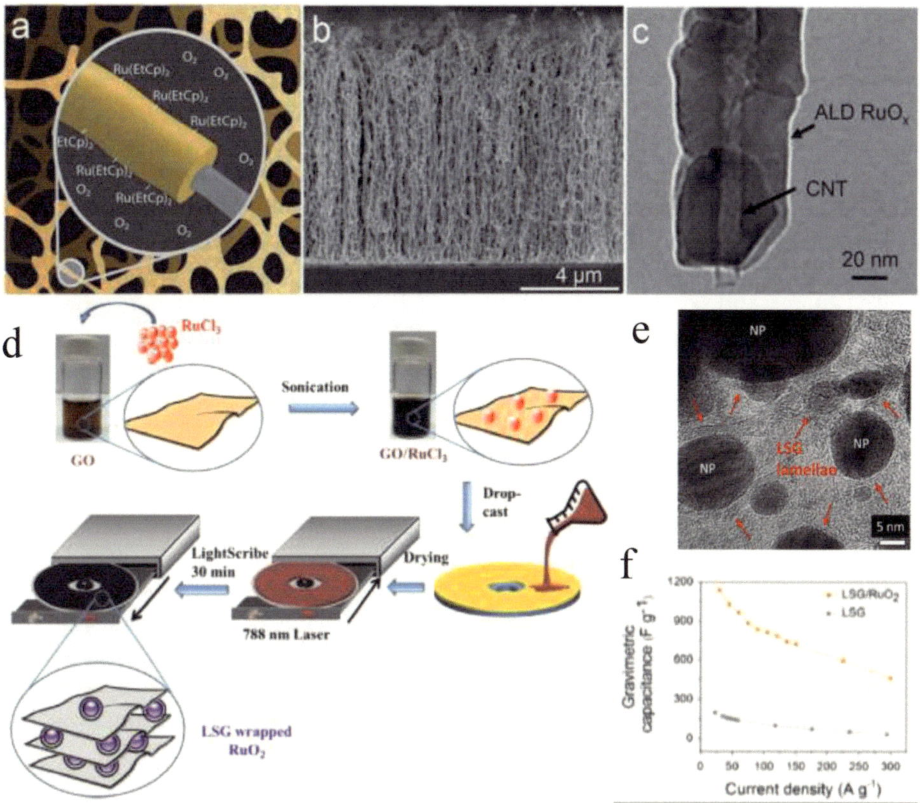

Figure 3
(a) Schematic presentation of RuO_x deposited on the vertically aligned porous carbon nanotubes porous through ALD by sequential pulsing of Ru $(EtCp)_2$ and oxygen. (b) and (c) SEM and TEM images of vertically aligned CNTs coated with ALD RuO_x [25]. (d) Microfabrication process of 3D porous RuO_2/LSG interdigitated micro-supercapacitors through direct laser writing on a DVD disc using a LightScribe DVD burner. (e) A high-magnification TEM image of 3D porous RuO_2/LSG composite showing complete wrapping of the RuO_2 nanoparticles (NP) by multiple layers of the graphene sheets. (f) The gravimetric capacitance retention of laser scribed graphene (LSG) and RuO_2/LSG electrodes as a function of the applied current density [26].

solid-state supercapacitors. RuO_2/NGA composite with finely tuned mass loading of 16.3 μg cm^{-2} and transmittance of 34.1% (λ = 550 nm) demonstrated maximum areal energy of 0.074 μW h cm^{-2} and power of 64 μW cm^{-2} with cyclic stability of 100% over 2000 cycles [27]. This RuO_2/NGA based high transparent SC can be practically used in many advanced transparent electrical devices.

4.2 Carbon-manganese oxides (MnO_2)-based composite electrode materials

MnO_2 has been considered as a promising pseudocapacitive electrode materials for energy storage applications due to low price, abundant reserve, high specific capacitance, and environmental environment benign nature and low toxicity in comparison to other transition-metal oxides. In general, the charge storage mechanism of MnO_2 involves change in manganese oxidation state from +3 to +4 and the contribution of protons or alkali cations, which can be shown in the following equation [28].

$$MnO_2 + C^+ + e^- \leftrightarrow MnOOC \qquad (3)$$

Where C⁺ represents protons or alkali cations (Li^+, Na^+, K^+).

However, MnO_2 based electrodes limits the capacity and power density due to their low surface area and poor electronic/ionic conductivity. Therefore, the composite of MnO_2 with high-surface area and conducting carbonaceous materials may improve the electrochemical performance in terms of specific capacity, energy and power densities by providing the larger interfacial area between the MnO_2 particles and the electrolyte solution [29].

Gao *et al.* fabricated a MnO_2/activated carbon (AC) based hybrid SC, where AC not only acted as a conducting support but also increase the capacitance as well as energy and power densities [30]. In addition, engineering the morphology of MnO_2 into different nanostructures is considered to be a practical approach to increase its electrochemical performance. It is reported that the pore sizes of the mesoporous-MnO_2/AC are greatly affected the specific capacitance and the rate capability of the SCs. Huang *et al.* demonstrated the influence of CNT on the electrochemical properties of MnO_2-CNT composite electrode by controlling the growth of MnO_2 nanostructures on CNTs through a facile redox approach (**Figure 4(a-c)**) [31]. The as-prepared MnO_2-CNT composite electrode showed a maximum specific capacitance of 247.9 F g^{-1} with outstanding cyclic stability of 92.8% after 5000 cycles. In addition, it has been noticed that the aligned CNTs are more favoured as SC electrodes over nonaligned CNTs due to their large specific surface area, low contact resistance, and fast electron-transfer kinetics. Graphene is being used as a supporting material for MnO_2 nanostructures due to its large surface area, high conductivity, and high stability nature. For example, microwave

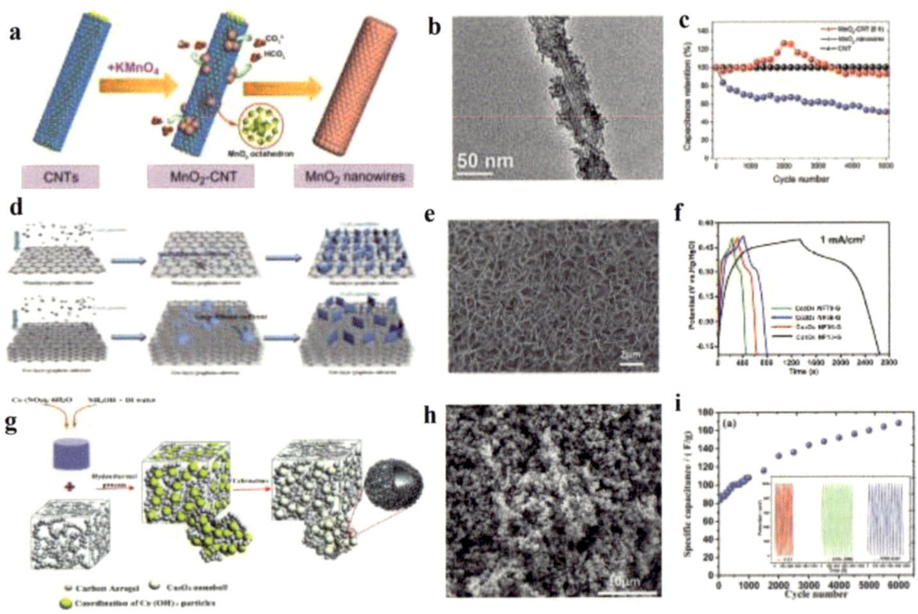

Figure 4.
(a) Controlled growth of MnO_2 nanostructured on CNT surface through facile redox method. (b) TEM images displaying coverage of MnO_2 on the surface of CNT. (c) The cyclic curve of a MnO_2-CNT nanowire composite at current density of 2 A g^{-1} [31]. (d) Schematic illustration of fabrication of Co_3O_4 nanoflake/graphene@ Ni hybrid electrode materials by in situ synthesis method. (e) Top-view SEM images of the Co_3O_4 nanoflake/ graphene/Ni hybrid electrode. (f) GCD curves of Co_3O_4 nanoflake/graphene/Ni hybrid electrode at current density of 1 mA cm^{-2} [39]. (g) A schematic of the synthesis of the porous Co_3O_4 nanoball/CA hybrid. (h) FE-SEM images of the porous Co_3O_4 nanoball/CA hybrid. (i) The specific capacitance test of the porous Co_3O_4 nanoball/CA hybrid electrode at a current density of 1 A g^{-1} as a function of cycle number (inset: 11 cycles continuous GCD curves obtained for porous Co_3O_4 nanoball/CA hybrid electrode for the different cycle numbers) [41].

irradiation synthesised MnO_2-graphene composites exhibited the maximum capacitance of 310 F g^{-1}, which is much higher than the bare graphene and MnO_2 (110 F g^{-1}) [32]. Beside their high capacitance, MnO_2-graphene composites have better cyclic stability of 95% over 15000 cycles. The excellent electrochemical performance of MnO_2-graphene composites is due to large surface area and high conductivity of graphene network. Recently, Zhang *et al.* reported highly flexible ASCs based on graphene hydrogel (GH)/copper wire (CW) as the negative electrode and hierarchical MnO_2/graphene/carbon fiber (CF) as the positive electrode, which exhibited excellent areal energy density of 18.1 μW h cm^{-2} and operated reversibly at potential window of 0-1.6 V [33]. 3D porous carbon nanostructures can also be used as MnO_2 support for supercapacitor (SC) electrodes as they provided large surface area, well-defined pathways to electrolyte access, and better mechanical stability. Fang *et al.* demonstrated a novel solid-state symmetric supercapacitor (SSC) based on 3D rGO@MnO_2 foam electrode and Polyacrylic Acid (PAA)-Portland cement-KOH electrolyte, which showed a very high areal capacity of 1.84 F cm^{-2} at current density of 0.5 mA cm^{-2} and excellent capacitance retention of 61% at a current density of 40 mA cm^{-2} [34].

4.3 Carbon-cobalt oxides (CoO/Co_3O_4)-based composite electrode materials

Cobalt oxides has been received considerable attention as highly promising SC electrode materials due to their non-toxic, low cost, easy synthesis, environmentally friendly, and more importantly high theoretical capacitance (CoO: 4292 F g^{-1}, Co_3O_4: 3560 F g^{-1}) [35]. In addition, cobalt oxides exhibits outstanding electrochemical behaviour in alkaline as well as organic electrolyte, which is possible due to their ability to interact with the ions at the electrolyte surface as well as through the bulk of the material. The pseudocapacitance of cobalt oxides (CoO/Co_3O_4) are originates from the following redox reaction: [36]

CoO:

$$\begin{gathered} \mathrm{Co}O + \mathrm{OH}^- \leftrightarrow 4\,\mathrm{CoOOH} + \mathrm{e}^- \\ \mathrm{CoOOH} + \mathrm{OH}^- \leftrightarrow 4\,\mathrm{CoO_2} + \mathrm{H_2O} + \mathrm{e}^- \end{gathered} \tag{4}$$

Co_3O_4:

$$\begin{gathered} \mathrm{Co_3O_4} + \mathrm{OH}^- + \mathrm{H_2O} \leftrightarrow 3\mathrm{CoOOH} + \mathrm{e}^- \\ \mathrm{CoOOH} + \mathrm{OH}^- \leftrightarrow \mathrm{CoO_2} + \mathrm{H_2O} + \mathrm{e}^- \end{gathered} \tag{5}$$

However, the low electrical/ionic conductivity of cobalt oxides hinders their practical performance as SC electrodes. Most efficient way to improve their electrochemical performance is to form composites of cobalt oxides by incorporation into a carbon-based conducting supports. A Co_3O_4/AC composite SC electrode was reported by Iqbal *et al.* [37]. The electrode exhibited maximum achievable specific capacitance 567 F g^{-1} and maximum energy density of 63 W h kg^{-1} at 0.7 A g^{-1}. In addition to the high specific capacitance, Co_3O_4/AC composite of capacitive retentivity is 82% after 6000 charge/discharge cycles and safe to handle due to no leakage. The specific capacitance of the cobalt oxide strongly depends on the microstructure and morphology of the materials, which facilitate the electrolyte ion transport through the material more effectively. Sun *et al.* demonstrated a simple and effective approach to grow well-aligned 3D cobalt oxide nanowire arrays (Co_3O_4 NWAs) directly on carbon nanotube fibers (CNTFs) through CVD process [38]. The Co_3O_4 NWAs/CNFs showed a specific capacitance of 734.25 F cm^{-3} (2210 mF cm^{-2}) at 1.0 A cm^{-3} and a high energy density of 13.2 mW h cm^{-3} at a current density of 1.0 A cm^{-3}. Graphene along with cobalt oxides can be used as a composite material for SCs because of its high conductivity, high surface area, high carrier mobility, and excellent mechanical strength. For example, an in situ synthesised Co_3O_4/graphene@NF hybrid composite electrode with a thickness of 13 nm exhibited a high specific capacitance of 1.75 F cm^{-2} at 1 mA cm^{-2} and a capacitance increase of 12.2% after 5000 cycles at 10 mA cm^{-2} (**Figure 4(d-f)**) [39]. Tseng *et al.* demonstrate a binder-free and flexible SC based on CoO/graphene hollow nanoballs (GHBs) composite electrode [40]. The as fabricated CoO/GHBs composite electrode exhibits high specific capacitance of 2238 F g^{-1} at a current density of 1 A g^{-1} and good rate capability of 1170 F g^{-1} at a current density of 15 A g^{-1}. The excellent capacitive performance and high rate capability were accomplished by the synergistic combination of conductive GHBs with large surface areas and highly pseudocapacitive CoO. In addition, as fabricated SSC demonstrated a very high power density (6000 W kg^{-1} at 8.2 W h kg^{-1}), high energy density (16 W h kg^{-1} at 800 W kg^{-1}), good cycling stability (~100% capacitance retention after 5000 cycles), and excellent mechanical flexibility at various bending positions. Recently, 3D-carbon aerogels (3D-CA) with appropriate electrical conductivity, high specific surface area and rich dielectric electrochemical stability when combined with the porous cobalt oxides can enabled the fabrication of an composite electrode with outstanding electrochemical performance.Co_3O_4/CA composite electrode which was synthesized through in situ growth method showed a specific capacitance of 350 F g^{-1} at 1 A g^{-1} and Energy density of 23.82 kW kg^{-1} at a power density of 95.96 W kg^{-1} (**Figure 4(g-i)**) [41]. The as-prepared ASC device could be cycled reversibly in a potential range of 0.0 to 1 V at 1 A g^{-1} and showed a capacity retention of 210% over 6000 cycles. Zhu *et al.* adopted a facile hydrothermal method to synthesize self-assembled cobalt oxide (CoO) nanorod cluster on 3D-graphene foam (CoO-3DGF) which exhibits a very high performance compared with CoO nanorod clusters grown on Ni foam (680 F g^{-1}) in terms of specific capacitance 980 F g^{-1} at 1 A g^{-1} and cycling stability of 103% over 10,000 cycles [42].

4.4 Carbon-binary metal oxide based composite electrode materials

Recently, binary metal oxides such as $NiCo_2O_4$, $NiFe_2O_4$, $CoFe_2O_4$, $ZnMnO_4$, and $ZnCo_2O_4$ have attracted much attention due to higher electrical conductivity than individual metal oxide and provide higher capacitance due to more affluent redox reaction than individual components [43]. Even though binary meal oxides possess better electrochemical performance than individual metal oxide extremely, they still suffer from inferior rate performance, low utilization rate and poor cycle stability. However, by incorporating carbon based materials improve their conductivity as well as power density due to high surface area, high conductivity and stable chemical properties of carbon based materials [44]. Kumar *et al.* fabricated Carbon black (CB) decorated Ni/Co oxide composite electrode through by using the successive ionic layer adsorption and reaction (SILAR) method [45]. Carbon black (CB) decorated Ni/Co oxide composite electrode with 7% weight percentage of CB exhibited a high specific capacitance of 1811 F g^{-1} at 0.5 mA cm^{-2} with excellent cyclic retention of 92% over 8000 cycles and delivered an impressive high energy density of 91 W h Kg^{-1} at a power density of 151 W Kg^{-1}, which is significantly higher than pure Ni/Co oxide composite electrode as well as other carbon embedded composites. Veerasubramani *et al.* have adopted a novel approach to fabricate

CNT-deposited $CoMoO_4$/Ni foam through a hydrothermal method followed by dry reforming reaction (DRR) of propane and CO_2 [46, 47]. The as fabricated CNT-deposited $CoMoO_4$/Ni foam electrode achieved a maximum areal capacity of 160 μAh cm^{-2} at 1 mA cm^{-2} with excellent cyclic stability of ~105% over 3000 cycles and showed 22-fold higher performance than the heat-treated $CoMoO_4$/Ni foam. The high electrochemical performance is due to the presence of CNTs on the surface of $CoMoO_4$/Ni foam electrode, which increases the conductivity of the electrode and enhances the ion transport kinetics. Further as fabricated ASC device, consists of CNT-deposited $CoMoO_4$/Ni foam as the positive electrode and reduced graphene oxide (rGO)-coated carbon cloth (CC) as the negative electrode stored a maximum areal energy density of 122 μWh cm^{-2} (29.04 Wh kg^{-1}) at 2 mA cm^{-2} and delivered a high power density of 7,727 μWcm^{-2} (1835 W kg^{-1}) 10 mA cm^{-2} with excellent capacitance retention of more than 95% of its initial capacitance over 1500 cycles. Soam *et al.* synthesized porous type of $NiFe_2O_4$/graphene nanocomposite electrode by a solution based process for supercapacitor application [48]. The as-prepared $NiFe_2O_4$/graphene nanocomposite electrode exhibited a maximum specific capacitance of 207 Fg^{-1} at a scan rate of 5 mV/sec, which is almost 4 times larger than pure $NiFe_2O_4$ (60 Fg^{-1}) and showed the capacitance retention of 95% over 1000 cycles. The significantly enhanced specific capacitance of the $NiFe_2O_4$/graphene nanocomposite electrode material is due to the synergic effect of high porous graphene sheets and $NiFe_2O_4$ particles, which are strongly interconnected together leading to a good electric/ionic conduction on the electrode and better contact of ions with the electrode materials. Zhou *et al.* reported a novel and green Cu_2O template-assisted route based on "coordinating etching and precipitating" process for the synthesis of 3D porous reduced graphene (rGN)/$NiCo_2O_4$ film [49]. The as-synthesized 3D rGN/$NiCo_2O_4$ film exhibited high specific capacitance of 708.36 F g^{-1} at a current density of 1 A g^{-1} with a rate retention of 82.2% as current density ranges from 1 to 16 Ag^{-1}, and remarkable capacitance retention of 94.3% after 6000 cycles at a high current density of 10 A g^{-1}.

5. Carbon-metal hydroxide composites electrode materials

Among the active materials, metal hydroxides have also been considered promising electrode materials for electrochemical SCs because of extremely high specific capacitance. Metal hydroxide in several forms such as $Ni(OH)_2$, $Co(OH)_2$, $NiCo(OH)_2$, $Cu(OH)_2$, FeOOH have been investigated as electrodes for SC [50–52]. These materials have large internal spaces for fast insertion and desertion of electrolyte ions. Moreover, these metal hydroxides can be synthesized using simple synthetic approaches. Metal hydroxide consists of stacked layers intercalated having interlayer space to occupy more ions hence larger capacitance.

5.1 Carbon-nickel hydroxide ($Ni(OH)_2$) composite electrode materials

$Ni(OH)_2$ is being considered as an attractive candidate as electrode in SCs because of its high theoretical capacitance (2358 F g^{-1}). It can be prepared by a simple and low cost process. It has demonstrated good stability in alkaline electrolytes. Its low electrical conductivity is a barrier to achieve higher capacitance. Therefore, a thin region near the surface of nickel hydroxide contributes to the charge storage process due to diffusion-limited redox reactions. To obtain larger capacitance, it has to be utilized completely in the charge storage process. In this regard, researchers have generally adopted conductive additives to effectively improve utilization of active materials and result in larger capacitance. Kang *et al.* have used the same concept

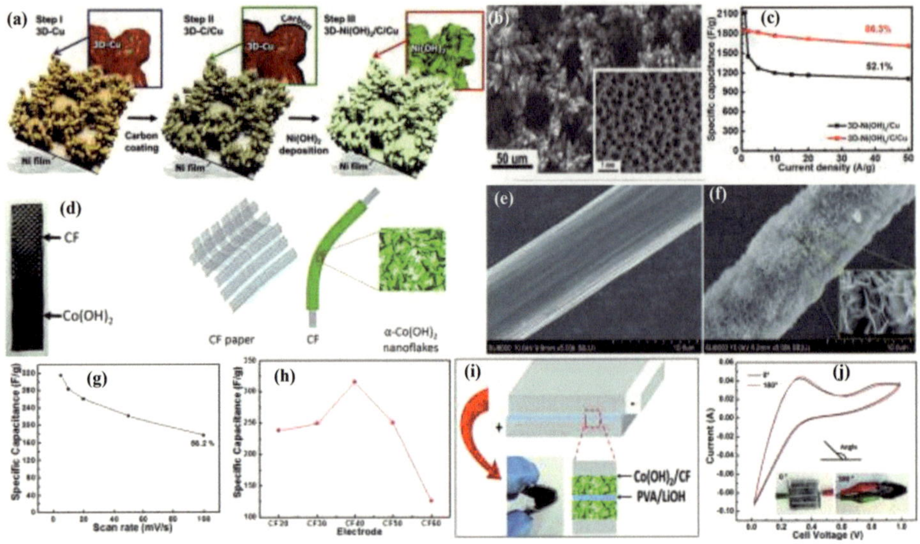

Figure 5.
(a) A schematic of the growth process of 3D-Ni(OH)$_2$/C/Cu, (b) Morphology of the as-synthesized 3D-Ni(OH)$_2$/C/Cu electrode (inset: large-area uniform porous morphology of the 3D-Ni(OH)$_2$/C/Cu), (c) Specific capacitance of 3DNi(OH)$_2$/C and 3D-Ni(OH)$_2$/C/Cu as a function of the current density based on the galvanostatic charge/discharge measurement [53], (d) Photograph of CF paper coated with cobalt hydroxide nanoflakes and schematic diagram illustrating the loading procedure of cobalt hydroxide on CF, (e) SEM image of bare CF, (f) SEM image of cobalt hydroxide nanoflakes coated on CF (Inset: magnified SEM image of the nanoflakes), (g) variation of specific capacitance with mass loading of each electrode, (h) specific capacitances of CF electrode at scan rates of 5, 10, 20, 50 and 100 mV/s, (i) schematic of flexible SC fabrication, (j) CV curves at bending conditions of 0° and 180° at scan rate of 20 mV/s [56].

and deposited an ultrathin nickel hydroxide film on carbon-coated 3D porous copper structure in order to prepare binder-free conductive electrode (**Figure 5(a-b)**) [53]. This electrode has short electron path distances and large electrochemical active sites, which improved structural stability for high performance SCs. A carbon coating was used to improve the electron transport behavior and to prevent the oxidation of Cu. Nickel hydroxide supported on mesoporous hollow dendritic three-dimensional-nickel exhibited a specific capacitance of 1860 F g^{-1} at a current density of 1 A g^{-1} (**Figure 5(c)**). It could retain 86.5% capacitance over 10,000 cycles. Tang *et al.* have prepared an additive-free, nano-architectured nickel hydroxide/carbon nanotube ($Ni(OH)_2$/CNT) electrode for high performance SCs [54]. This $Ni(OH)_2$/CNT electrode was fabricated by depositing $Ni(OH)_2$ nano-flakes on CNT bundles which were directly grown on Ni foams. The above electrode exhibited the specific capacitance of 3300 F g^{-1} and an aerial capacitance of 16 Fcm^{-2}. Ma *et al.* have synthesized electrode of $Ni(OH)_2$ nanosheet/3D GF framework using two methods, CVD and hydrothermal [55]. They have compared the capacitive properties of Ni(OH)2 electrode/graphene fiber with $Ni(OH)_2$/Ni foam and $Ni(OH)_2$ nanosheet/carbon fiber cloth electrodes. $Ni(OH)_2$ electrode with graphene fiber exhibited better performance in terms of specific capacitance and rate capability. The $Ni(OH)_2$ nanosheet/graphene fiber electrode exhibited electrochemical capacitance as high as 2860 F g^{-1} at a current density of 2 A g^{-1}, and maintains 1791 F g^{-1} at 30 A g^{-1}.

5.2 Carbon-cobalt hydroxide ($Co(OH)_2$) composite electrode materials

$Co(OH)_2$ has recently received increasing attention as electrode for SC application because of its low cost and high capacitance. Jagadale *et al.* have used cobalt hydroxide nanoflakes which were uniformly loaded on flexible carbon fiber (CF) paper as electrode for SC (**Figure 5(d)**) [56]. The carbon fiber was basically used to provide unique porous nanostructure offering low ion diffusion and charge transfer resistance to the electrode (**Figure 5(e, f)**). The electrode exhibited maximum specific capacitance of 386.5 F g^{-1} at a current density of 1 mA cm^{-2} with a mass loading of 2.5 mg cm^{-2} (**Figure 5(g, h)**). An energy density of 133.5 W h kg^{-1} has been obtained with power density of and 1769 W kg^{-1}. The carbon fiber has improved the cyclic stability of 92% over 2000 cycles. To check applicability of electrodes, these electrodes further employed to fabricate flexible solid state supercapacitor. CV curves of SC at bending conditions of 0° and 180° at scan rate of 20 mV/s. It is clearly seen that the area under curve doesn't change significantly after bending which proves that SC is highly flexible and does not lose its structural integrity under bending conditions (**Figure 5(i, j)**).

Two possible reactions are suggested for the electrochemical reactions of $Co(OH)_2$ in KOH electrolyte [57]:

$$\mathrm{Co}(OH)_2 + \mathrm{OH}^- = \mathrm{CoOOH} + \mathrm{H_2O} + \mathrm{e}^- \tag{6}$$

$$\mathrm{CoOOH} + \mathrm{OH}^- = \mathrm{Co}O_2 + \mathrm{H_2O} + \mathrm{e}^- \tag{7}$$

$Co(OH)_2$ nano-sheet-decorated graphene-CNT composite structure has been designed for SC application [58]. Suspensions method was used to prepare graphene-CNT composite by sonication and vacuum filtration. The graphene-CNT composite may offer high porosity with high conductivity, chemical stability and a three-dimensional structure. The vertically aligned $Co(OH)_2$ nano-sheets were then deposited on 3D graphene-CNT composite by solution based process. The ASC of $Co(OH)_2$ with graphene-CNT has shown a specific capacitance of 310 F g^{-1}. The electrode exhibited an energy density of 172 W h kg^{-1} and maximum power density of 198 kW kg^{-1} in ionic liquid electrolyte 1-ethyl-3-methylimidazoliumbis (trifluoromethanesulfone)imide (EMI-TFSI). Zhang *et al.* have deposited $Co(OH)_2$ on multi-walled CNT which were grown on the carbon paper substrate [59]. The composite electrode showed the specific capacitance of 1083 F g^{-1} determined at a current density of 0.83 A g^{-1} in aqueous electrolyte. CNTs were added to $Co(OH)_2$ in order to improve the electrical conductivity of the electrode. The interconnected nanosheets of the $Co(OH)_2$ would help to facilitate the contact of the electrolyte with active materials, exhibiting good cycling stability and lifetime.

5.3 Carbon-iron oxy hydroxide (FeOOH) composite electrode materials

FeOOH has been recognized is an attractive electrode material for SC due to low cost, high theoretical specific capacitance, and broad potential window. In addition, the unique tunnel structure of FeOOH with open permeable channels are beneficial for ion transportation and shorten the diffusion path for electrolyte ion diffusion [60]. However, the poor electrical conductivity and low specific surface area limited the use of FeOOH as a potential electrode for SC, which limited specific capacitance and rate capability [61]. Alternatively, composite system by assembling FeOOH on the carbon based supporting materials (AC, carbon black, graphene, etc.) can be enhance the capacitive performance. Shen *et al.* synthesized radiating γ-FeOOH Nanosheets on CC substrate (γ-FeOOH NSs/CC) by a simple one-step electrodeposition method and investigated its pseudocapacitive behaviour in a typical ionic liquid [1-ethyl-3-methylimidazolium bis imide (EMIM-NTF2)] through electrochemical quartz crystal microbalance (EQCM). The charge storage is mainly due

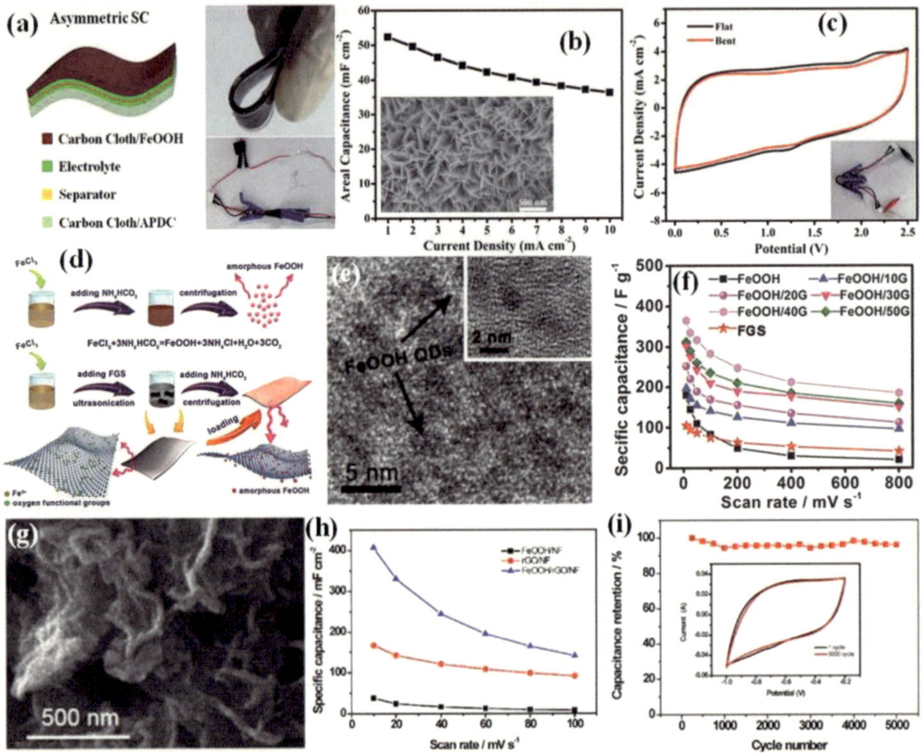

Figure 6.
(a) Schematic illustrations of the fabrication procedure for the FeOOH//APDC f-SSC electrodes and flexibility and operating status as supercapacitor device, (b) The areal capacitance as a function of the discharge current density (Inset: SEM images of as-prepared γ-FeOOH nanosheets on a carbon cloth substrate), (c) CV curves of the FeOOH//APDC f-SSC at bent and flat statuses [62]. (d) Schematic illustration of the synthesis of amorphous FeOOH QDs and amorphous FeOOH/FGS hybrid nanosheets, (e) HRTEM images of the FeOOH QDs (Inset: enlarged HRTEM for FeOOH QDs), (f) The specific capacitances of the FeOOH, functionalized graphene sheet (FGS), and FeOOH/FGS composite electrodes as a function of the scan rate [64]. (g) High-magnification SEM images of as-prepared 3D FeOOH/rGO/NF, (h) Areal capacitance of FeOOH/NF, rGO/NF and FeOOH/rGO/NF electrodes calculated from CV curves as a function of scan rate, and (i) Cycling performance of MnO_2//FeOOH-ASC collected at a scan rate of 100 mV s^{-1} for 5000 cycles [66].

to the insertion and extraction of $[EMIM]^+$ cations through the transport pathways offered by the crystalline network of γ-FeOOH during charging-discharging process. γ-FeOOH NSs/CC exhibited a good areal capacitance of 210 mF cm^{-2} at a current density of 1 mA cm^{-2} and the ASC device made of γ-FeOOH||APDC (activated polyaniline-derived carbon nanorods) solid-state flexible SCs acquired a high energy density of 1.44 mW h cm^{-3} at a current density of 3 A g^{-1} with a cycling stability of 80.5% retention over 2000 cycles (**Figure 6(a-c)**) [62]. An amorphous FeOOH nanoflowers@multi-walled CNT (FeOOH NFs@ MWCNTs) composite was prepared by Sun *et al.* [63]. The as-prepared composite electrode displays a high specific capacitance of 345 F g^{-1} at 1 A g^{-1} current density and outstanding rate performance (167 F g^{-1} at 11.4 A g^{-1}) with good cycling stability of 76.4% over 5000 cycles. The outstanding electrochemical performance of the composite electrode is due to the mesoporous structure and high surface area of the electrode materials as well as fast ion/electronic transport and easy accessibility of the active materials to electrolytes. Liu *et al.* demonstrated FeOOH quantum dots (QDs)/graphene hybrid nanosheets, which exhibited a high specific capacitance of 365 F g^{-1} at a current density of 1 A g^{-1} with excellent capacitance retention of 89.7% of initial capacitance over 20000 cycles as well as a great rate capability (189 F g^{-1} at a high current density of 128 A g^{-1}) (**Figure 6(d-f)**) [64]. In addition, specific capacitance of the SC increased to 1243 F g^{-1} at 5 mV s^{-1} while the voltage window was extended from -0.8 to 0 V to -1.25 to 0 V but the cycling performance declined sharply. Wei *et al.* synthesized ultrathin α-FeOOH nanorods/graphene oxide (GO) composite by hydrothermal method, which exhibited high specific capacitance of 127 F g^{-1} at a current density of 10 A g^{-1}, good cyclic performance of 85% capacitance retention over 2000 cycles, and excellent rate capability (100 F g^{-1} at 20 A g^{-1}) as compared to than bare α-FeOOH nanorods [65]. The outstanding electrochemical performance of α-FeOOH nanorods/GO composite is due to its unique structure, which provides fast electron/ions transport and high charging/discharging rate. 3D FeOOH/reduced graphene oxide/Ni foam (FeOOH/rGO/NF) based hybrid electrodes fabricated by the electrodeposition of FeOOH nanosheets on the rGO/Ni foam surface exhibited an exception high areal capacitance of 406.5 mF cm^{-2} at a scan rate of 10 mV s^{-1}, which is 10-fold higher than the bare FeOOH/NF electrode (**Figure 6(g-i)**) [66]. This high areal capacitance of FeOOH/rGO/NF is due to the improved conductivity and increased surface area, which not only provide a superior pathway for electron transfer, but also offer more active sites for energy storage. In addition, an ASC device made of 3D FeOOH/rGO/NF electrode as anode and MnO_2@TiN electrode as cathode attained a remarkable maximum power density of 0.19 W cm^{-3} with maximum energy density of 0.48 mW h cm^{-3}.

6. Carbon-transition metal dichalcogenides (TMDs) composite electrode materials

TMDs are layered inorganic materials with a chemical configuration of MX_2, in which M is a transition metal element (M: Ti, Mo, V, W, Re, Ta), and X can be any chalcogenide element (X: S, Se, Te) (**Figure 7(a)**). Each MX_2 unit cell is stacked

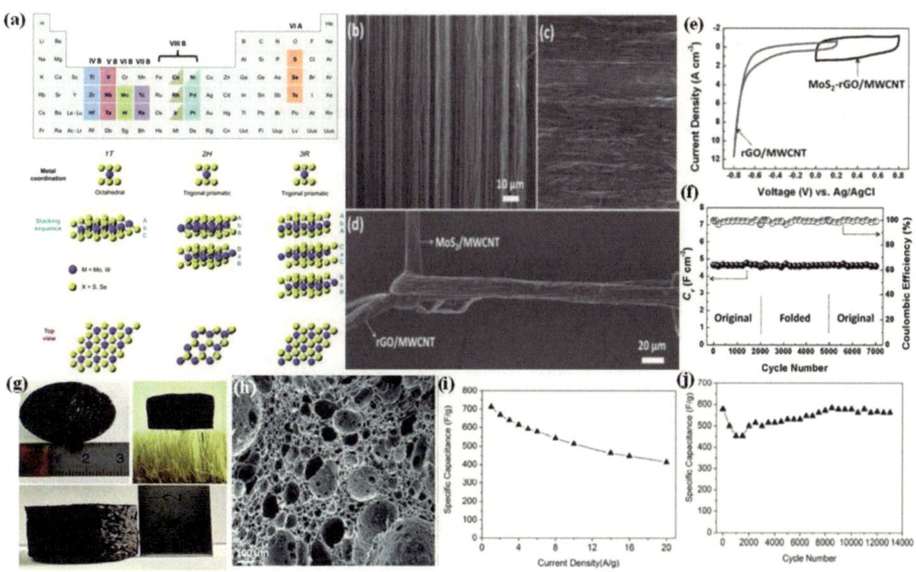

Figure 7.
(a) Different metal coordination and stacking sequence in TMD unit cells [67]. (b-d) SEM images of aligned MWCNT sheets, MWCNT/MoS_2 hybrids, and tightly knotted MoS_2/MWCNT and rGO/MWCNT fibers, respectively. (e) CV curves of rGO/MWCNT (cathode) and MoS_2-rGO/MWCNT (anode) at different potential windows. (f) Cycle stability test of the fiber-based asymmetric device at 0.55 A cm^{-3} current density [74]. (g) Optical photographs and (h) SEM images of the MoS_2/C composite aerogel. (i) specific capacitances at different current densities and (j) long-term cycle stability at a current density of 6 A g^{-1} of the MoS_2/C composite aerogel electrode material [75].

together through Vander Waals force in such a way that transition metal layer is present in between the two chalcogen sheets [67]. On the basis of crystal structure, there are two types of phases of TMDs, which are metallic 1T phase with an octahedral structure and semiconducting 2H phase with a trigonal structure. Recently, TMDs have been attracted great attention as SC electrode materials due to their large surface area, low cost, variable oxidation states, high mechanical properties, high chemical stability and easy synthesis [68]. The variable oxidation states, large surface area, and active edges of TMDs allow electrical double layer and fast/reversible redox charge storage mechanisms and offer high energy storage capability in SCs. However, due to the inherently low conductivity, poor cycle life, large volume change during cycling and restacking limits their electrochemical performance as SC electrodes [69]. For example, Soon *et al.* has synthesized sheet-like morphology of MoS_2 by chemical vapor deposition method, which has a very large surface area favorable for double layer storage. But due to its poor electrical conductivity, it showed low specific capacitance of ~100 F g^{-1} at a scan rate of 1 mV s^{-1} [70]. Therefore, in order to improve the electrochemical performance of TMDs, they have been compositing with highly conducting/electroactive carbon based supporting materials by various top-down/bottom-up and both synthetic approaches. The synergic effect of carbon-TMDs based composite materials such as carbon offers conductive channels and increasing the interfacial contact, whereas TMDs provide a short ion diffusion path and followed by short electron transport path enhances the overall electrochemical performance of the SC.

6.1 Carbon-MoS_2 composite electrode materials

MoS_2/MWCNT nanocomposite synthesized by a hydrothermal method exhibited a large surface area and fast ionic transport properties and showed a high specific capacitance of 452.7 F g^{-1} with good cycling stability (95.8% retention after 1000 cycles), which is almost three times larger than the bare MoS_2 (149.6 to 452.7 F g^{-1}) [71]. Ali *et al.* fabricated MoS_2/graphene composite from bulk MoS_2 and graphite rod through facile electrochemical exfoliation method and exhibited high specific capacitance of 227 F g^{-1} as compared with the exfoliated MoS_2 (70 F g^{-1}) and exfoliated graphene (85 F g^{-1}) at a current density of 0.1 A g^{-1} [72]. The high specific capacitance of MoS_2/graphene composite is due to the synergistic effect between MoS_2 and graphene. Ali *et al.* demonstrated the electrochemical performance of MoS_2/CNT/GNF composite and compared the performance with MoS_2/CNTs, MoS_2/graphene nanoflakes [73]. It has been noticed that the electrochemical charge storage performance has been improved by incorporation of the carbon materials into the composite and the composite showed a maximum specific capacitance of 104 F g^{-1} at a current density of 0.5 A g^{-1} with capacitance retention of 75% after the 1000 cycle at a scan rate of 10 mV/s. Another interesting MoS_2-rGO/MWCNT fiber electrode was fabricated by incorporating rGO nanosheets and MoS_2 into aligned MWCNT, which operated at a stable potential window of 1.4 V and exhibited high coulombic efficiency of 100% over 7000 cycles in the bending state (**Figure 7(b-f)**) [74]. Zhang *et al.* reported an agarose induced technique to synthesize MoS_2/carbon composite aerogel, which showed a high specific capacitance of 712.6 F g^{-1} at a current density of 1 A g^{-1} with cyclic stability of 97.3% over 13000 charge-discharge cycles (**Figure 7(g-j)**) [75]. The high specific capacitance of MoS_2/carbon composite aerogel is because of 3D intercalated network with hierarchal porous and interlayer MoS_2 expanded structures, which were beneficial for easy ion transportation. 3D graphene/MoS_2 composite electrode material has been synthesized by Sun et al and co-workers through a simple and facile one-step hydrothermal process [76]. The as-synthesized composite electrode exhibited gravimetric capacitance of 410 F g^{-1} at a current density of 1 A g^{-1} and an excellent cycling stability of 80.3% over 10,000 continuous charge-discharge cycles at 2 A g^{-1} current density. The outstanding electrochemical performance of 3D graphene/MoS_2 composite electrode is due to the 3D architecture of conducting network graphene and flower-like structure of MoS_2, which enhances the electrolyte ions diffusion process.

6.2 Carbon-WS_2 composite electrode materials

WS_2 nanoplates supported on carbon fiber cloth (WS_2/CFC) have been synthesized by a facile solvothermal process and used as electrode material for SC [77]. The 3D network of CFC not only prevent the agglomeration of WS_2 nanoplates but also enhances the ion transport efficiency due to low charge transfer resistance (R_{ct}) of 0.1 Ω. The as fabricated WS_2/CFC electrode exhibited a high specific capacitance of 399 F g^{-1} at 1 A g^{-1} current density with cyclic retention of 99% over charge-discharge 500 cycles, which is higher than compared with bare WS_2. In addition, developing such composite of WS_2 with the carbon fibre helps for fabricating wearable SCs which are in demand for wearable electronics. Yang *et al.* fabricated WS_2@CNT hybrid film electrode by incorporating conducting CNTs into WS_2. The WS_2@CNT hybrid film with a unique skeleton structure showed a maximum specific area capacitance of 752.53 mF cm^{-2} at a scan rate 20 mV s^{-1} with very good cyclic stability by only loss of 1.28% capacitance after 10,000 cycles. In addition, a quasi-solid-state flexible SC made by WS_2@CNT hybrid film exhibited excellent bendability under bending to 135 10, 000 times with the loss of 23.12% at scan rate of 100 mV s^{-1} [53]. Tu *et al.* have been synthesized WS_2/RGO hybrid material by using a simple molten salt process, which showed a high specific capacitance of 2508.07 F g^{-1} at 1 mV s^{-1} scan rate with excellent capacitance retention of 98.6% over 5000 cycles, due to synergic effect of highly conducting RGO and large charge-accumulating sites of WS_2 networks. Likewise, Xu *et al.* demonstrated 3D composite of WS_2 nanoflakes and quantum dots on N and S co-doped reduced graphene oxide (WS_2/N,S-rGO) crumpled nanosheets through a rapid solution combustion synthesis of the precursor and subsequent gas-solid phase sulfurization process, which presented a significant specific capacitance of 1562.5 F g^{-1} at 1 A g^{-1} current density, and a rate capability of 780 F g^{-1} at 40 A g^{-1} (**Figure 8(a-c)**) [78]. The high specific capacitance of WS_2/N,S-rGO hybrids is because of synergistic effect between WS_2 and N,S-rGO, where N,S-rGO provides larger contact surface area, excellent charge transport, and shorter ion diffusion path. Hierarchical $MoSe_2$/C hybrid was successfully fabricated by facile one-step hydrothermal strategy, which composed of few-layered $MoSe_2$ nanosheets and amorphous carbon obtained from the decomposition of the triethylene glycol. As fabricated hierarchical $MoSe_2$/C electrode exhibited high specific capacitance of 878.6 F g^{-1} in comparison with the bare $MoSe_2$ at current density of 1 A g^{-1} and maintained 98% of initial capacitance over 2000 cycles without obvious decrease. The superior electrochemical performances of $MoSe_2$/C hybrid can be ascribed to hierarchical structure of $MoSe_2$ and conducting nature of carbon, which help for providing large surface area for electrochemical reactions and enhancing charge carriers transfer at the electrolyte/electrode interface [79]. Liu *et al.* fabricated VACNTF@$MoSe_2$/NF composite electrode through a combined chemical vapor deposition method and solvothermal methods by growing $MoSe_2$ nanoflakes on the vertically aligned carbon nanotube array film (VACNTF) with binder-free nickel foam as current collector [80]. The as fabricated VACNTF@$MoSe_2$/NF composite electrode exhibited high specific capacitance of 435 F g^{-1} at a current density of 1 A g^{-1} with outstanding cycling stability

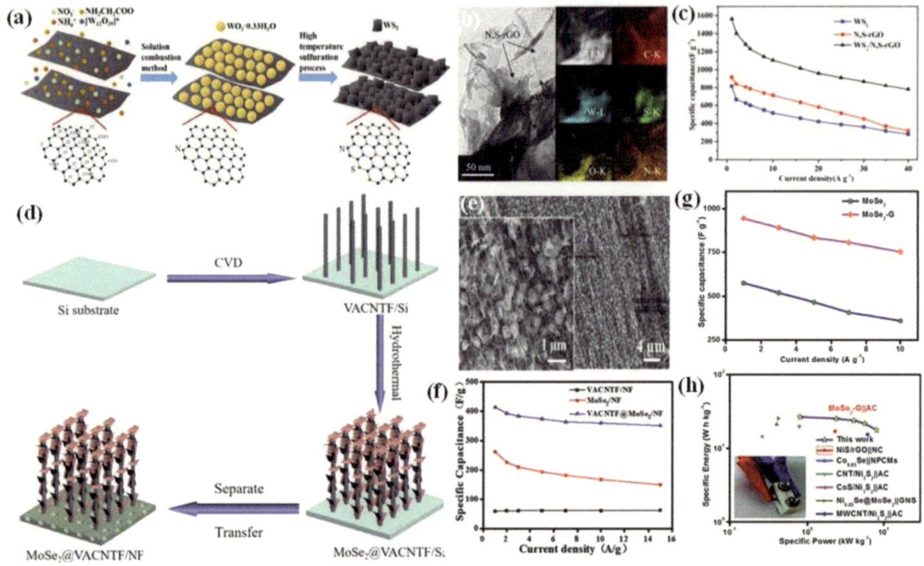

Figure 8.
(a) Schematic illustration of synthetic processes of WS_2/N,S-rGO hybrid, (b) HRTEM, STEM and EDS elemental mapping images of WS_2/N,S-rGO hybrid, and (c) The specific capacitances of the WS_2, N,S-rGO and WS_2/N,S-rGO hybrid at different current densities [78]. (d) Schematic illustration of the synthesis process of the VACNTF@$MoSe_2$/NF composite electrode, (e) SEM images of the VACNTF@$MoSe_2$ composites (inset: high magnification), and (f) The specific capacitance comparison of the $MoSe_2$/NF, VACNTF/NF and VACNTF@$MoSe_2$/NF electrodes at various current densities [80]. (g) Specific capacitance of the $MoSe_2$ NS and $MoSe_2$/G nanohybrid based electrodes as a function of current density, and (h) Ragone plot for the $MoSe_2$G||AC ASC device (inset: photograph of ASC device) [81].

of 92% after 5000 cycles (**Figure 8(d-f)**). In addition, the VACNTF@$MoSe_2$/NF composite based ASC displays a high energy density with 22 W h kg^{-1} for a power density of 330 W kg^{-1}. Kirubasankar *et al.* $MoSe_2$/graphene nanohybrid based electrode prepared by a simple and facile sonochemical route, which showed higher specific capacitance (945 F g^{-1}) as compared to $MoSe_2$ nanosheets (576 F g^{-1}) at 1 A g^{-1} current density. Further, as fabricated ASC device based on $MoSe_2$/graphene nanohybrid retains 88% of its capacitance over 3000 cycles and delivers an energy density of 26.6 W h kg^{-1} at a power density of 0.8 kW kg^{-1} (**Figure 8(g, h)**) [81]. The high specific capacitance with better rate capability is due to the effective penetration and migration of electrolyte, reduction of the contact resistance and shortness of the diffusion path of ions between the electrode-electrolyte interface, which enhances the redox kinetics and provide maximum utilization of the electroactive area, so providing a high structural stability during charge-discharge processes. Similarly, Huang *et al.* demonstrated $MoSe_2$/graphene on flexible Ni electrode, which could deliver a specific capacitance of 1422 F $g^{-1}$and fully retention of initial capacitance over 1500 cycles [82]. Wei *et al.* first time fabricated free-standing SC anode based on 3D $MoSe_2$ nanoflowers ($MoSe_2$ NFs) and hierarchically porous anisotropic carbonized delignified wood (CDW), which exhibited ultrahigh capacitance of 1043 mF cm^{-2} at a current density of 1 mA cm^{-2} and excellent cycling stability less than 5% capacitance loss over 5000 cycles. The ASC device was made by integration of 3D $MoSe_2$ NFs@CDW anode and a common MnO_2-based cathode, which exhibited a high capacitance of 415 mF cm^{-2} at a current density of 2.5 mA cm^{-2} with high energy density of 147 mW h cm^{-2} at power density of 2 mW cm^{-2}. These results confirm that 3D $MoSe_2$ NFs@CDW based anode can be used as a potential anode for the development of high-performance SCs [83].

7. Conclusion

In the past few decades, SCs have been extensively studied as energy storage devices and more focusing area in the multidisciplinary science over the world. The selection of high performance SC electrode materials based on high specific capacitance, low internal resistance and good stability. In this article, we have reviewed the carbon-based composite materials (*i.e.*, metal oxide, metal hydroxide, TMDs composited with carbon materials) as promising SC electrode materials due to the synergic effect of the composite materials such as high surface area, interconnected porous structure, high electrical conductivity, excellent wettability towards the electrolyte, and presence of electrochemically active surface functionalities of the carbon supports which improves the EDL capacitance while metal oxide or metal hydroxide or TMDs enhances electrochemical performance through pseudocapacitive/faradaic charge-storage process. The carbon-based composite materials demonstrated herein usually possesses high specific capacity, impressive energy density and maintain long term stability with better mechanical flexibility. We also observe the microstructural changes in the carbon-based composite materials would be more favorable for fabrication of high performance supercapacitor. We also explained how the composite materials overcome the traditional obstacles while formulating the standard electrode designs as compare to individual components.

Acknowledgements

This work was supported by the Ministry of Science and Technology (MOST) of Taiwan, under grant numbers 107-2113-M-845-001-MY3.

Author details

Prasanta Kumar Sahoo[1], Chi-Ang Tseng[2], Yi-June Huang[3] and Chuan-Pei Lee[4]*

1 Department of Mechanical Engineering, Siksha 'O' Anusandhan, Deemed to be University, Bhubaneswar, Odisha, India

2 Department of Chemistry, National Taiwan University, Taipei, Taiwan

3 Graduate Institute of Nanomedicine and Medical Engineering, College of Biomedical Engineering, Taipei Medical University, Taipei, Taiwan

4 Department of Applied Physics and Chemistry, University of Taipei, Taiwan

*Address all correspondence to: CPLee@utaipei.edu.tw

References

[1] Yu Z., Duong B., Abbitt D., Thomas J., Highly ordered MnO_2 nanopillars for enhanced supercapacitor performance. Adv Mater. 2013; 25; 3302-3306, DOI: 10.1002/adma.201300572.

[2] Etacheri Vinodkumar, Marom Rotem, Elazari Ran, Salitra Gregory, Aurbach Doron, Challenges in the development of advanced Li-ion batteries: a review. Energy & Environmental Science. 2011; 4; 3243, DOI: 10.1039/c1ee01598b.

[3] Miller John R., Simon Patrice, Electrochemical Capacitors for Energy Management. Science. 2008; 321; 651-652, DOI: DOI: 10.1126/science.1158736.

[4] Naoi Katsuhiko, Nao Wako, Aoyagi Shintaro, Miyamoto Jun-ichi, Kamino Takeo, New Generation "Nanohybrid Supercapacitor". Accounts of Chemical Research. 2013; 46; 1075-1083, DOI: https://doi.org/10.1021/ar200308h.

[5] Choudhary N., Li C., Moore J., Nagaiah N., Zhai L., Jung Y., Thomas J., Asymmetric Supercapacitor Electrodes and Devices. Adv Mater. 2017; 29, DOI: 10.1002/adma.201605336.

[6] Meng C., Gall O. Z., Irazoqui P. P., A flexible super-capacitive solid-state power supply for miniature implantable medical devices. Biomed Microdevices. 2013; 15; 973-983, DOI: 10.1007/s10544-013-9789-1.

[7] Zhang Li Li, Zhou Rui, Zhao X. S., Graphene-based materials as supercapacitor electrodes. Journal of Materials Chemistry. 2010; 20; 5983, DOI: 10.1039/c000417k.

[8] Ke Qingqing, Wang John, Graphene-based materials for supercapacitor electrodes – A review. Journal of Materiomics. 2016; 2; 37-54, DOI: 10.1016/j.jmat.2016.01.001.

[9] Wang G., Zhang L., Zhang J., A review of electrode materials for electrochemical supercapacitors. Chem Soc Rev. 2012; 41; 797-828, DOI: 10.1039/c1cs15060j.

[10] Zou Yong, Han Bu-Xing, High-Surface-Area Activated Carbon from Chinese Coal. Energy & Fuels. 2001; 15; 1383-1386, DOI: https://doi.org/10.1021/ef0002851.

[11] Zhu Y., Murali S., Cai W., Li X., Suk J. W., Potts J. R., Ruoff R. S., Graphene and graphene oxide: synthesis, properties, and applications. Adv Mater. 2010; 22; 3906-3924, DOI: 10.1002/adma.201001068.

[12] Portet C., Yushin G., Gogotsi Y., Electrochemical performance of carbon onions, nanodiamonds, carbon black and multiwalled nanotubes in electrical double layer capacitors. Carbon. 2007; 45; 2511-2518, DOI: 10.1016/j.carbon.2007.08.024.

[13] Lota Grzegorz, Fic Krzysztof, Frackowiak Elzbieta, Carbon nanotubes and their composites in electrochemical applications. Energy & Environmental Science. 2011; 4; 1592, DOI: 10.1039/c0ee00470g.

[14] Geim A. K., Novoselov K. S., The rise of graphene. Nature Materials. 2007; 6; 183-191, DOI: https://doi.org/10.1038/nmat1849.

[15] Simon PATRICE, Gogotsi Yury, Materials for electrochemical capacitors. Nature Materials. 2008; 7; 845-854, DOI: https://doi.org/10.1038/nmat2297.

[16] Simon P., Gogotsi Y., Capacitive Energy Storage in Nanostructured Carbon-Electrolyte Systems. Accounts of Chemical Research. 2013 46; 1094-1103, DOI: https://doi.org/10.1021/ar200306b.

[17] Trasatti Sergio, Buzzanca Giovanni, Ruthenium dioxide: A new interesting electrode material. Solid state structure and electrochemical behaviour. Journal of Electroanalytical Chemistry and Interfacial Electrochemistry. 1971 29; A1-A5, DOI: https://doi.org/10.1016/S0022-0728(71)80111-0.

[18] Lee Hee Y., Goodenough J. B., Supercapacitor Behavior with KCl Electrolyte. Journal of Solid State Chemistry. 1999; 144; 220-223, DOI: https://doi.org/10.1006/jssc.1998.8128.

[19] Wu Nae-Lih, Nanocrystalline oxide supercapacitors. Materials Chemistry and Physics. 2002 75; 6-11, DOI: https://doi.org/10.1016/S0254-0584(02)00022-6.

[20] Reddy A. Leela Mohana, Ramaprabhu S., Nanocrystalline Metal Oxides Dispersed Multiwalled Carbon Nanotubes as Supercapacitor Electrodes. The Journal of Physical Chemistry C. 2007; 111; 7727-7734, DOI: https://doi.org/10.1021/jp069006m.

[21] Ho M. Y., Khiew P. S., Isa D., Tan T. K., Chiu W. S., Chia C. H., A Review of Metal Oxide Composite Electrode Materials for Electrochemical Capacitors. Nano. 2014; 09; 1430002, DOI: 10.1142/s1793292014300023.

[22] Chen Yan-Li, Hu Zhong-Ai, Chang Yan-Qin, Wang Huan-Wen, Zhang Zi-Yu, Yang Yu-Ying, Wu Hong-Ying, Zinc Oxide/Reduced Graphene Oxide Composites and Electrochemical Capacitance Enhanced by Homogeneous Incorporation of Reduced Graphene Oxide Sheets in Zinc Oxide Matrix. The Journal of Physical Chemistry C. 2011; 115; 2563-2571, DOI: 10.1021/jp109597n.

[23] Rolison D. R., Long J. W., Lytle J. C., Fischer A. E., Rhodes C. P., McEvoy T. M., Bourg M. E., Lubers A. M., Multifunctional 3D nanoarchitectures for energy storage and conversion. Chem Soc Rev. 2009; 38; 226-252, DOI: 10.1039/b801151f.

[24] Wu Z., Li L., Yan J. M., Zhang X. B., Materials Design and System Construction for Conventional and New-Concept Supercapacitors. Adv Sci (Weinh). 2017; 4; 1600382, DOI: 10.1002/advs.201600382.

[25] Fischer Anne E., Pettigrew Katherine A., Rolison Debra R., Stroud Rhonda M., Long Jeffrey W., Incorporation of Homogeneous, Nanoscale MnO_2 within Ultraporous Carbon Structures via Self-Limiting Electroless Deposition: Implications for Electrochemical Capacitors. Nano Letters. 2007; 7; 281-286, DOI: https://doi.org/10.1021/nl062263i.

[26] Hwang Jee Y., El-Kady Maher F., Wang Yue, Wang Lisa, Shao Yuanlong, Marsh Kristofer, Ko Jang M., Kaner Richard B., Direct preparation and processing of graphene/RuO_2 nanocomposite electrodes for high-performance capacitive energy storage. Nano Energy. 2015; 18; 57-70, DOI: 10.1016/j.nanoen.2015.09.009.

[27] Suktha Phansiri, Phattharasupakun Nutthaphon, Sawangphruk Montree, Transparent supercapacitors of 2 nm ruthenium oxide nanoparticles decorated on a 3D nitrogen-doped graphene aerogel. Sustainable Energy & Fuels. 2018; 2; 1799-1805, DOI: 10.1039/c8se00177d.

[28] Zhang K., Han X., Hu Z., Zhang X., Tao Z., Chen J., Nanostructured Mn-based oxides for electrochemical energy storage and conversion. Chem Soc Rev. 2015; 44; 699-728, DOI: 10.1039/c4cs00218k.

[29] Wei W., Cui X., Chen W., Ivey D. G., Manganese oxide-based materials as electrochemical supercapacitor electrodes. Chem Soc Rev. 2011; 40; 1697-1721, DOI: 10.1039/c0cs00127a.

[30] Gao Peng-Cheng, Lu An-Hui, Li Wen-Cui, Dual functions of activated carbon in a positive electrode for MnO_2-based hybrid supercapacitor. Journal of Power Sources. 2011; 196; 4095-4101, DOI: 10.1016/j.jpowsour.2010.12.056.

[31] Huang Huajie, Zhang Wenyao, Fu Yongsheng, Wang Xin, Controlled growth of nanostructured MnO_2 on carbon nanotubes for high-performance electrochemical capacitors. Electrochimica Acta. 2015; 152; 480-488, DOI: 10.1016/j.electacta.2014.11.162.

[32] Yan Jun, Fan Zhuangjun, Wei Tong, Qian Weizhong, Zhang Milin, Wei Fei, Fast and reversible surface redox reaction of graphene–MnO_2 composites as supercapacitor electrodes. Carbon. 2010; 48; 3825-3833, DOI: 10.1016/j.carbon.2010.06.047.

[33] Zhang Zheye, Xiao Fei, Wang Shuai, Hierarchically structured MnO_2/graphene/carbon fiber and porous graphene hydrogel wrapped copper wire for fiber-based flexible all-solid-state asymmetric supercapacitors. Journal of Materials Chemistry A. 2015; 3; 11215-11223, DOI: 10.1039/c5ta02331a.

[34] Fang Cuiqin, Zhang Dong, A large areal capacitance structural supercapacitor with a 3D rGO@MnO_2 foam electrode and polyacrylic acid–Portland cement–KOH electrolyte. Journal of Materials Chemistry A. 2020; 8; 12586-12593, DOI: 10.1039/d0ta03109g.

[35] He Guangyu, Li Jianghua, Chen Haiqun, Shi Jian, Sun Xiaoqiang, Chen Sheng, Wang Xin, Hydrothermal preparation of Co_3O_4@graphene nanocomposite for supercapacitor with enhanced capacitive performance. Materials Letters. 2012; 82; 61-63, DOI: 10.1016/j.matlet.2012.05.048.

[36] Xia Xin-hui, Tu Jiang-ping, Zhang Yong-qi, Mai Yong-jin, Wang Xiu-li, Gu Chang-dong, Zhao Xin-bing, Freestanding Co_3O_4 nanowire array for high performance supercapacitors. RSC Advances. 2012; 2; 1835, DOI: 10.1039/c1ra00771h.

[37] Iqbal Muhammad Zahir, Haider Syed Shabhi, Zakar Sana, Alzaid Meshal, Afzal Amir Muhammad, Aftab Sikandar, Cobalt-oxide/carbon composites for asymmetric solid-state supercapacitors. Materials Research Bulletin. 2020; 131; 110974, DOI: 10.1016/j.materresbull.2020.110974.

[38] Sun Juan, Man Ping, Zhang Qichong, He Bing, Zhou Zhenyu, Li Chaowei, Wang Xiaona, Guo Jiabin, Zhao Jingxin, Xie Liyan, Li Qingwen, Sun Jing, Hong Guo, Yao Yagang, Hierarchically-structured Co_3O_4 nanowire arrays grown on carbon nanotube fibers as novel cathodes for high-performance wearable fiber-shaped asymmetric supercapacitors. Applied Surface Science. 2018; 447; 795-801, DOI: 10.1016/j.apsusc.2018.03.244.

[39] Tan Hui Yun, Yu Bao Zhi, Cao Lin Li, Cheng Tao, Zheng Xin Liang, Li Xing Hua, Li Wei Long, Ren Zhao Yu, Layer-dependent growth of two-dimensional Co_3O_4 nanostructure arrays on graphene for high performance supercapacitors. Journal of Alloys and Compounds. 2017; 696; 1180-1188, DOI: 10.1016/j.jallcom.2016.12.050.

[40] Tseng C. A., Sahoo P. K., Lee C. P., Lin Y. T., Xu J. H., Chen Y. T., Synthesis of CoO-Decorated Graphene Hollow Nanoballs for High-Performance Flexible Supercapacitors. ACS Appl Mater Interfaces. 2020; 12; 40426-40432, DOI: 10.1021/acsami.0c12898.

[41] Zomorodian Esfahani Majid, Aghaei Alireza, Khosravi Mohsen, Bagheri Narjes, Khakpour Zahra, Javaheri Masoumeh, Pore structure improvement of carbon aerogel and investigation of the supercapacitive behavior of a Co_3O_4 nanoball/carbon

aerogel composite. New Journal of Chemistry. 2017; 41; 11731-11741, DOI: 10.1039/c7nj02897k.

[42] Zhu Yun Guang, Wang Ye, Shi Yumeng, Huang Zhi Xiang, Fu Lin, Yang Hui Ying, Phase Transformation Induced Capacitance Activation for 3D Graphene-CoO Nanorod Pseudocapacitor. Advanced Energy Materials. 2014; 4; 1301788, DOI: 10.1002/aenm.201301788.

[43] Fan Shi, Lu Li, Xiu-li Wang, Chang-dong Gu, Jiang-ping Tu, Metal oxide/hydroxide based materials for supercapacitors. RSC Adv. 2014;4; 41910-41921, DOI: 10.1039/C4RA06136E.

[44] Yufei Zhang, Laiquan Li, Haiquan Su, Wei Huang, Xiaochen Dong, Binary metal oxide: advanced energy storage materials in supercapacitors. J. Mater. Chem. A 2015; 3; 43-59, DOI: 10.1039/C4TA04996A.

[45] Niraj Kumar, P. K. Sahoo, H. S. Panda, Tuning the electro-chemical properties by selectively substituting transition metals on carbon in Ni/Co oxide–carbon composite electrodes for supercapacitor devices. NewJ.Chem. 2017; 41; 3562-3573, DOI: 10.1039/C6NJ04123J.

[46] Ganesh Kumar Veerasubramani, Karthikeyan Krishnamoorthy, Sang JaeKim, Improved electrochemical performances of binder-free $CoMoO_4$ nanoplate arrays@Ni foam electrode using redox additive electrolyte, J.PowerSources 2016; 306; 378-386, DOI: 10.1016/j.jpowsour.2015.12.034.

[47] Ganesh Kumar Veerasubramani, M. S. P. Sudhakaran, Nagamalleswara Rao Alluri, Karthikeyan Krishnamoorthy, Young Sun Mok, Sang Jae Kim, Effective use of an idle carbon-deposited catalyst for energy storage applications, J. Mater. Chem. A 2016; 4 ;12571-12582, DOI: 10.1039/C6TA05082D.

[48] Ankur Soam, Rahul Kumar, Prasanta Kumar Sahoo, C. Mahender, Balwant Kumar, Nitin Arya, Mamraj Singh, Smrutiranjan Parida, Rajiv O Dusane, Synthesis of Nickel Ferrite Nanoparticles Supported on Graphene Nanosheets as Composite Electrodes for High Performance Supercapacitor, ChemistrySelect 2019; 4; 9952-9958, DOI: 10.1002/slct.201901117.

[49] You Zhou, Zhengyong Huang, Huijun Liao, Jian Li, Hanxiang Wang, Yu Wang, 3D porous graphene/$NiCo_2O_4$ hybrid film as an advanced electrode for supercapacitors, Appl. Surf. Sci. 2020;534;147598, DOI: 10.1016/j.apsusc.2020.147598.

[50] Shi Fan, Li Lu, Wang Xiu-li, Gu Chang-dong, Tu Jiang-ping, Metal oxide/hydroxide-based materials for supercapacitors. RSC Adv. 2014; 4; 41910-41921, DOI: 10.1039/c4ra06136e.

[51] Yusin S. I., Bannov A. G., Synthesis of composite electrodes for supercapacitors based on carbon materials and the metal oxide/metal hydroxide system. Protection of Metals and Physical Chemistry of Surfaces. 2017; 53; 475-482, DOI: 10.1134/S2070205117030261.

[52] Patil U., Lee S. C., Kulkarni S., Sohn J. S., Nam M. S., Han S., Jun S. C., Nanostructured pseudocapacitive materials decorated 3D graphene foam electrodes for next generation supercapacitors. Nanoscale. 2015; 7; 6999-7021, DOI: 10.1039/c5nr01135c.

[53] Yang X., Li J., Hou C., Zhang Q., Li Y., Wang H., Skeleton-Structure WS_2@CNT Thin-Film Hybrid Electrodes for High-Performance Quasi-Solid-State Flexible Supercapacitors. Front Chem. 2020; 8; 442, DOI: 10.3389/fchem.2020.00442.

[54] Tang Zhe, Tang Chun-hua, Gong Hao, A High Energy Density Asymmetric Supercapacitor from

Nano-architectured $Ni(OH)_2$/Carbon Nanotube Electrodes. Advanced Functional Materials. 2012; 22; 1272-1278, DOI: 10.1002/adfm.201102796.

[55] Ma Yufang, Chen Wanjun, Zhang Peng, Teng Feng, Zhou Jinyuan, Pan Xiaojun, Xie Erqing, $Ni(OH)_2$ nanosheets grown on a 3D graphene framework as an excellent cathode for flexible supercapacitors. RSC Adv. 2014; 4; 47609-47614, DOI: 10.1039/c4ra08786k.

[56] Jagadale Ajay D., Guan Guoqing, Du Xiao, Hao Xiaogang, Li Xiumin, Abudula Abuliti, Cobalt hydroxide [$Co(OH)_2$] loaded carbon fiber flexible electrode for high performance supercapacitor. RSC Advances. 2015; 5; 56942-56948, DOI: 10.1039/c5ra11366k.

[57] Chen Chao, Cho Misuk, Lee Youngkwan, Electrochemical preparation and energy storage properties of nanoporous $Co(OH)_2$ via pulse current deposition. Journal of Materials Science. 2015; 50; 6491-6497, DOI: 10.1007/s10853-015-9207-6.

[58] Cheng Q., Tang J., Shinya N., Qin L. C., $Co(OH)_2$ nanosheet-decorated graphene-CNT composite for supercapacitors of high energy density. Sci Technol Adv Mater. 2014; 15; 014206, DOI: 10.1088/1468-6996/15/1/014206.

[59] Zhang Jing, Wang Xiuchun, Ma Jie, Liu Shuo, Yi Xibin, Preparation of cobalt hydroxide nanosheets on carbon nanotubes/carbon paper conductive substrate for supercapacitor application. Electrochimica Acta. 2013; 104 110-116, DOI: 10.1016/j.electacta.2013.04.052.

[60] Chen Li-Feng, Yu Zi-You, Wang Jia-Jun, Li Qun-Xiang, Tan Zi-Qi, Zhu Yan-Wu, Yu Shu-Hong, Metal-like fluorine-doped β-FeOOH nanorods grown on carbon cloth for scalable high-performance supercapacitors. Nano Energy. 2015; 11; 119-128, DOI: 10.1016/j.nanoen.2014.10.005.

[61] Owusu K. A., Qu L., Li J., Wang Z., Zhao K., Yang C., Hercule K. M., Lin C., Shi C., Wei Q., Zhou L., Mai L., Low-crystalline iron oxide hydroxide nanoparticle anode for high-performance supercapacitors. Nat Commun. 2017; 8; 14264, DOI: 10.1038/ncomms14264.

[62] Shen Baoshou, Guo Ruisheng, Lang Junwei, Liu Li, Liu Lingyang, Yan Xingbin, A high-temperature flexible supercapacitor based on pseudocapacitive behavior of FeOOH in an ionic liquid electrolyte. Journal of Materials Chemistry A. 2016; 4; 8316-8327, DOI: 10.1039/c6ta01734g.

[63] Sun C., Pan W., Zheng D., Zheng Y., Zhu J., Liu C., Low-Crystalline FeOOH Nanoflower Assembled Mesoporous Film Anchored on MWCNTs for High-Performance Supercapacitor Electrodes. ACS Omega. 2020; 5; 4532-4541, DOI: 10.1021/acsomega.9b03869.

[64] Liu Jiaqi, Zheng Mingbo, Shi Xiaoqin, Zeng Haibo, Xia Hui, Amorphous FeOOH Quantum Dots Assembled Mesoporous Film Anchored on Graphene Nanosheets with Superior Electrochemical Performance for Supercapacitors. Advanced Functional Materials. 2016; 26; 919-930, DOI: 10.1002/adfm.201504019.

[65] Wei Y., Ding R., Zhang C., Lv B., Wang Y., Chen C., Wang X., Xu J., Yang Y., Li Y., Facile synthesis of self-assembled ultrathin alpha-FeOOH nanorod/graphene oxide composites for supercapacitors. J Colloid Interface Sci. 2017; 504; 593-602, DOI: 10.1016/j.jcis.2017.05.112.

[66] Wang Fuxin, Zeng Yinxiang, Zheng Dezhou, Li Cheng, Liu Peng, Lu Xihong, Tong Yexiang, Three-dimensional iron oxyhydroxide/reduced graphene oxide composites as advanced electrode for

electrochemical energy storage. Carbon. 2016; 103; 56-62, DOI: 10.1016/j.carbon.2016.02.088.

[67] Li Xiao, Zhu Hongwei, Two-dimensional MoS_2: Properties, preparation, and applications. Journal of Materiomics. 2015; 1; 33-44, DOI: 10.1016/j.jmat.2015.03.003.

[68] Li Ming-Yang, Chen Chang-Hsiao, Shi Yumeng, Li Lain-Jong, Heterostructures based on two-dimensional layered materials and their potential applications. Materials Today. 2016; 19; 322-335, DOI: 10.1016/j.mattod.2015.11.003.

[69] Ge Yu, Jalili Rouhollah, Wang Caiyun, Zheng Tian, Chao Yunfeng, Wallace Gordon G., A robust free-standing MoS_2/poly(3,4-ethylenedio xythiophene):poly(styrenesulfonate) film for supercapacitor applications. Electrochimica Acta. 2017; 235; 348-355, DOI: 10.1016/j.electacta.2017.03.069.

[70] Soon Jia Mei, Loh Kian Ping, Electrochemical Double-Layer Capacitance of MoS_2 Nanowall Films. Electrochemical and Solid State Letters. 2007; 10; A250-A254, DOI: 10.1149/1.2778851兴.

[71] Huang Ke-Jing, Wang Lan, Zhang Ji-Zong, Wang Ling-Ling, Mo Yan-Ping, One-step preparation of layered molybdenum disulfide/multi-walled carbon nanotube composites for enhanced performance supercapacitor. Energy. 2014; 67; 234-240, DOI: 10.1016/j.energy.2013.12.051.

[72] Ali Gomaa A. M., Thalji Mohammad R., Soh Wee Chen, Algarni H., Chong Kwok Feng, One-step electrochemical synthesis of MoS_2/graphene composite for supercapacitor application. Journal of Solid State Electrochemistry. 2020; 24; 25-34, DOI: 10.1007/s10008-019-04449-5.

[73] Ali B. A., Metwalli O. I., Khalil A. S. G., Allam N. K., Unveiling the Effect of the Structure of Carbon Material on the Charge Storage Mechanism in MoS_2-Based Supercapacitors. ACS Omega. 2018; 3; 16301-16308, DOI: 10.1021/acsomega.8b02261.

[74] Sun G., Zhang X., Lin R., Yang J., Zhang H., Chen P., Hybrid fibers made of molybdenum disulfide, reduced graphene oxide, and multi-walled carbon nanotubes for solid-state, flexible, asymmetric supercapacitors. Angew Chem Int Ed Engl. 2015; 54; 4651-4656, DOI: 10.1002/anie.201411533.

[75] Zhang Y., He T., Liu G., Zu L., Yang J., One-pot mass preparation of MoS_2/C aerogels for high-performance supercapacitors and lithium-ion batteries. Nanoscale. 2017; 9; 10059-10066, DOI: 10.1039/c7nr03187d.

[76] Sun Tianhua, Li Zhangpeng, Liu Xiaohong, Ma Limin, Wang Jinqing, Yang Shengrong, Facile construction of 3D graphene/MoS_2 composites as advanced electrode materials for supercapacitors. Journal of Power Sources. 2016; 331; 180-188, DOI: 10.1016/j.jpowsour.2016.09.036.

[77] Shang Xiao, Chi Jing-Qi, Lu Shan-Shan, Gou Jian-Xia, Dong Bin, Li Xiao, Liu Yan-Ru, Yan Kai-Li, Chai Yong-Ming, Liu Chen-Guang, Carbon fiber cloth supported interwoven WS_2 nanosplates with highly enhanced performances for supercapacitors. Applied Surface Science. 2017; 392; 708-714, DOI: 10.1016/j.apsusc.2016.09.058.

[78] Xu Yunpeng, Wang Lizhong, Xu Qiang, Liu Liyun, Fang Xiaochen, Shi Ce, Ye Bin, Chen Lingyun, Peng Wenyi, Liu Zongjian, Chen Weifan, 3D hybrids based on WS_2/N, S co-doped reduced graphene oxide: Facile fabrication and superior performance in supercapacitors. Applied Surface Science. 2019; 480; 1126-1135, DOI: 10.1016/j.apsusc.2019.02.217.

[79] Ma Lin, Xu Limei, Zhou Xiaoping, Xu Xuyao, Zhang Lingling, Synthesis of a hierarchical $MoSe_2$/C hybrid with enhanced electrochemical performance for supercapacitors. RSC Advances. 2016; 6; 91621-91628, DOI: 10.1039/c6ra16157j.

[80] Liu Y., Li W., Chang X., Chen H., Zheng X., Bai J., Ren Z., $MoSe_2$ nanoflakes-decorated vertically aligned carbon nanotube film on nickel foam as a binder-free supercapacitor electrode with high rate capability. J Colloid Interface Sci. 2020; 562; 483-492, DOI: 10.1016/j.jcis.2019.11.089.

[81] Kirubasankar Balakrishnan, Vijayan Shobana, Angaiah Subramania, Sonochemical synthesis of a 2D–2D $MoSe_2$/graphene nanohybrid electrode material for asymmetric supercapacitors. Sustainable Energy & Fuels. 2019; 3; 467-477, DOI: 10.1039/c8se00446c.

[82] Huang Ke-Jing, Zhang Ji-Zong, Cai Jia-Lin, Preparation of porous layered molybdenum selenide-graphene composites on Ni foam for high-performance supercapacitor and electrochemical sensing. Electrochimica Acta. 2015; 180; 770-777, DOI: 10.1016/j.electacta.2015.09.016.

[83] Wei S., Wan C., Jiao Y., Li X., Li J., Wu Y., 3D nanoflower-like $MoSe_2$ encapsulated with hierarchically anisotropic carbon architecture: a new and free-standing anode with ultra-high areal capacitance for asymmetric supercapacitors. Chem Commun (Camb). 2020; 56; 340-343, DOI: 10.1039/c9cc07362k.

Chapter 7

3D Ionic Networked Hydrophilic-Hydrophobic Nano Channeled Triboelectric Nanogenerators

Ravi Kumar Cheedarala

Abstract

The power demand is increasing day by day owing to the diminishing of fossil fuel reserves on the globe. To overcome the future energy crises, there is a strong need to fulfill the energy loophole by novel technologies such as triboelectric nanogenerators to harvest miniature resources from renewable natural resources. Here, I discussed the synthesis and fabrication of novel triboelectric nanogenerators (TENGs) using highly reproducible power generators as electropositive surfaces from the monomers of naphthalene tetracarboxylic dianhydride, benzdiene diamine, and sulfonated polyimide (Bno-Spi), and modified nonwoven carbon fibers (Wcf) and polytetrafluoroethylene (PTFE) and polyvinylidene difluoride (PVDF) as electronegative TENG electrodes, respectively. Here, novel double characteristic hydrophilic and hydrophobic nano-channels concerned with Bno-Spi films were proposed through contact electrification process through ion and electron transfer by an electron-donor-acceptor complex mechanism. The proposed Bno-Spi-TENG system High triboelectric open circuit voltage 75 V (Voc) and short circuit current 1 μA (Jsc) have been achieved from Bno-Spi-TENGs, in particular, and for SO_3H.Bno-Spi-TENG at 6 Hz. Besides that, we used improved knitted woven carbon fiber composite (wcf-COOH), as one of the TENGs to generate a greater open-circuit voltage (*Voc*), and short circuit current (*Isc*). Also, I aimed the contact and separation mode TENG which is using spring structure through oxidation of Wcf into Wcf-COOH followed by coupling of aniline through and one-step oxidative polymerization to get woven carbon fiber-polyaniline emraldine salt (Wcf-Pani.Es). The Wcf-PANI.Es composite film (thickness ~ 100 nm) shows the surface resistivity of 0.324 Ω m, and functions as a rubbing surface to produce charges through harvesting of energy using vertical contact-separation mode TENG. The vibrant exchanges of novel Wcf-Pani.Es, and PVDF membrane produced higher *Voc* of 95 V, and *Isc* of 180 μA, correspondingly. In specific, Wcf-Pani.Es -TENG is shown an enhancement of 498% of *Voc* concerning Wcf-COOH-TENG due to the availability of the Pani.Es layer. The novel Bno-Spi-TENGs and Wcf-Pani.Es are the potential candidates for fulfilling the need for improved energy harvesting devices as an alternate substantial choice for contact-separation mode TENGs.

Keywords: triboelectric nanogenerator, contact-electrification, Bno-Spi, Wcf-TENG, ionic electrets

1. Introduction

The urbanization process is quickening in the recent past due to speedy economic development and population growth in megalopolises considerably leads to the upsurge of resource demand, especially energy. Energy demand is gradually rising due to global warming and ecological degradation to overcome Technologists/academic scientists who are considering potential substitute energies that can desperately exchange the traditional sources for example solar, wind, and tidal energies, etc. [1–4]. Numerous kinds of energy harvesting methods were developed for accumulating energies which are thermoelectric, electromagnetic devices piezoelectric, photoelectric, electrostatic, and triboelectric devices [5, 6]. On the other hand, more robust, eco-friendly, economically viable, miniature and easy to handle, and highly reproducible energy harvesting systems with advanced technologies are desirable to satisfy the tight requirement of manufacturing demand. The energy produced from mechanical devices is a common renewable energy resource that is achieved using various modes such as humanoid motion, including the body's pulsation, and rotation, etc. [7, 8]. Freshly, triboelectric nanogenerators (TEGs) have received worldwide attention for the collecting of feasible green energy from ambient resources. Classical TEGs were designed and established based on a combination of synthetic polymers for contact and separation electrification, and electrostatic induction for generating mechanical energy [9, 10]. The appropriate choice of triboelectric paired polymers and their coherent design can upsurge the rate of energy collection and conversion efficiency [11, 12]. The oppositely charged material surfaces when they contact each other at regular intervals, consequently, the ions or electrons should be motivated to flow over the external load and, create a continuous voltage, and currents, respectively. Whiteside et al. and others have studied the ion-transfer technique by integrating ionic functional groups on the solid surface such as polystyrene (PS), glass, and silicone to produce ionic electrets on the active surface [13–19]. Microspheres with internal cross-linking's that contain mobile ions and counter ions which transfer some of them in the air through another material upon contact [20].

2. Different working mechanisms of triboelectric nanogenerators (TENGs)

Generally, TENGs produced energy by contact electrification through the coupling effect of two oppositely charged materials. While contact electrification, the dissimilar materials becomes charged after contacting each other and generate opposites triboelectric charge from the surface of two dissimilar materials with different electron affinities. When the external mechanical motion is driven, the materials were separated resulted in the potential difference between the two opposite electrodes on the backside of the resource materials. To continue the electrostatic equilibrium, the free electrons from the electrodes were flown in the external circuit to balance the induced potential difference, consequently, the mechanical energy converted into electrical energy.

Depending on the different construction strategies of potential electrodes which show four different modes of TENGs have been constructed, as explained as follows.

2.1 Vertical contact-separation style

The mechanistic approach of vertical contact-separation mode is described typically by an example. The simplest construction of TENG includes two metal electrodes, and dielectric surface, in which two Aluminum layers work as a top electrode and bottom electrode attached to a dielectric film, respectively [21, 22]. When the mechanical system is vertically functions, the top electrode and dielectric film will be contacted each other, and therefore, the dielectric layer and electrode become positively charged, and negatively charged, respectively, owing to the triboelectrification. Once they were separated by a small detachment, the potential difference among the two electrodes could be convinced, which drive electrons to flow from the posterior-electrode to the top-electrode, subsequent in a pulse current with an external circuit coupled. If the two electrodes are brought into exchange again, the electrons will be flown back, and the current will be upturned (**Figure 1a**).

2.2 Lateral-sliding style

The basic construction of TENG in lateral sliding mode is the same as vertical contact-separation mode, but the difference is the top electrode will be moved over the bottom surface electrode as shown in **Figure 1b**. Next, the top electrode, and dielectric surface are fully overlap, and intimate interaction with each other, prominent to the oppositely charged surfaces. With the top electrode descending outward, the contact surface-area is gradually decreased until the wide-ranging departure of two surfaces. The departed surface will create a potential difference across the two TENG electrodes that generate a current movement from the upper electrode to the lower electrode. When the electrode moves backward, then there is reversed current flow to equilibrium the potential variance [23, 24]. The four fundamental methods of triboelectric nanogenerators: (a) vertical contact departure style, (b) in-plane contact-descending style, (c) single-electrode style, and (d) free standing triboelectric layer style.

2.3 Single-electrode style

The single-electrode mode TENG shows a bottom electrode is connected with the ground and the top surface is connected to the upper surface to get charged

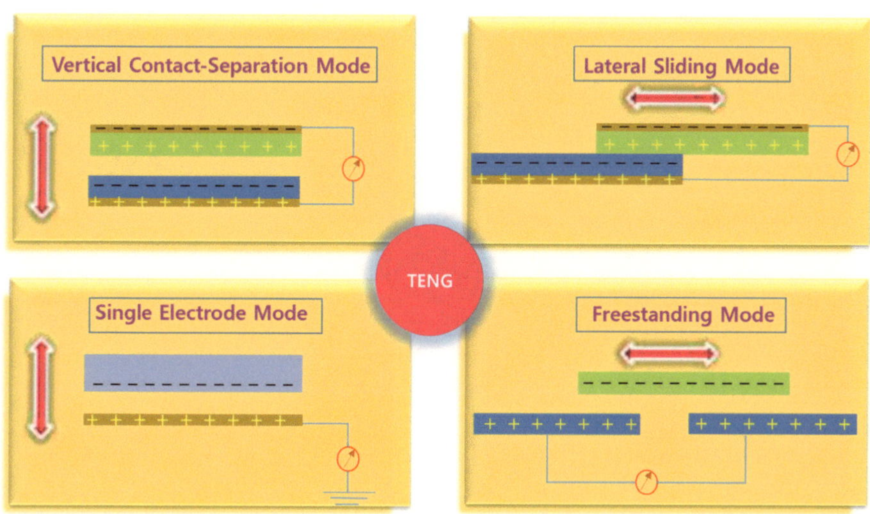

Figure 1.
The four basic triboelectric nanogenerator styles: (a) vertical contact- departure style, (b) in-plane contact-descending style, (c) single electrode style, and (d) free-standing triboelectric-layer style.

then the triboelectric effect is triggered as shown in **Figure 1c**. While approaching and leaving the top surface, the generated electric field is distributed through charged surfaces when they change. Then, the change in potential difference occurs between the bottom electrode and the ground. Subsequently, electrons can exchange between them to maintain the potential change [25, 26].

2.4 Freestanding triboelectric-layer style

Figure 1d shows the moving electrode surface which is a dielectric layer, and the two electrodes were positioned in the similar horizontal direction. The distance between the two symmetric electrodes is lesser than the length of the dielectric layer. The state of the dielectric layer and electrode are the same as in the lateral-sliding mode. Once the movement starts, simultaneously, the dielectric layer and bottom electrodes are charged oppositely as mentioned earlier. During movement the dielectric layer is sliding forward and backward, the potential difference is triggered between the two electrodes owing to the change of the affected area, and drives the electron exchanges between them [27, 28].

3. Experimental and methods of fabrication

3.1 Construction of the contact and separation mode Bno-Spi (or) Wcf-Pani.ES TENG devices

The construction and the functioning principle of the contact and separation Bno-Spi (or) Wcf-Pani.ES TENGs were discussed [20–21]. A methodical understanding of Bno-Spi (or) Wcf-Pani.ES TENGs are designated in diverse studies. Here, the building of the typical TENG models are depicted in **Figure 2**. First, the SO_3H.Bno-Spi-TENG was developed by attributing the SO_3H.Bno-Spi membrane with the sizes of 2 cm x 2 cm = 4 cm^2 on an Aluminum (Al) conductor. Next, the SO_3H.Bno-Spi-Al conductor was glued to soft sponge to reduce the reflecting strength during experiment. First, a load cell was linked to the top of the Al

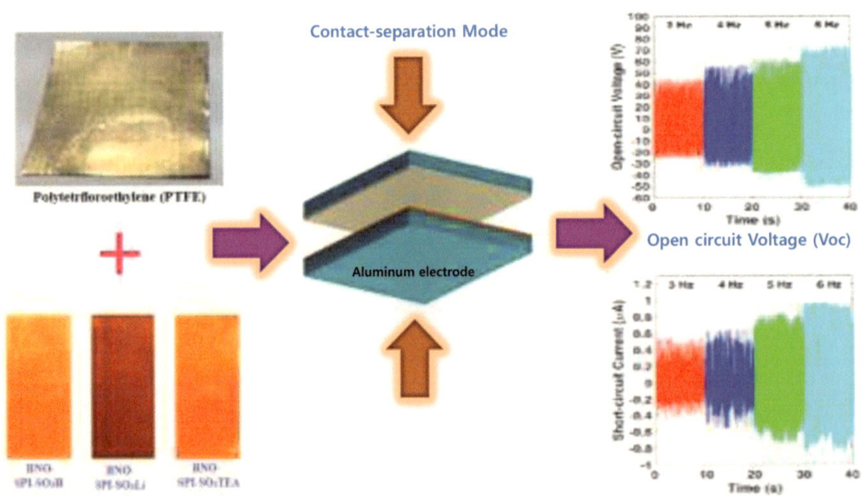

Figure 2.
Schematic illustration of a dual demonstrative Bno-Spi-TENGs aimed at robust contact electrification through vertical contact and separation style.

conductor. Secondly, the Al conductor was placed on the PTFE (or) PVDF film with similar dimensions along with soft sponge and a linear oscillator connected to a DC motor with an eccentric arrangement steadily fluctuated on a linear slider. The extreme swinging amplitude is 4 cm. The higher portion of the Bno-Spi (or) wCF-PANI.ES films (load cell, and Al) was then postponed by a cantilever style shaft of light that is connected to the linear slider. The cautious setting of the complete arrangement lead to in a slender contact between the upper and lower films while the slider oscillation is consistent. The similar protocol was followed for the all designated TENGs such as SO_3Li. Bno-Spi-Al, SO_3H.TEA.Bno-Spi-Al, and wCF-PANI.ES-Al TENGs.

4. Results and discussion

4.1 Dual demonstrative Bno-Spi-TENGs for strong contact-electrification using a vertical contact-separation approach through ions and electrons charge transfer

In this study, for the first time, we motivated to use, a Bno-Spi-TENG as a real ion, and electron-transfer route with a counter electronegative Polytetrafluoroethylene (PTFE) surface for the contact-separation electrification process [29, 30]. The anticipated novel Bno-Spi-TENG shown superior characteristics which have a special π-π stacked layer-on-layer oligomer morphology with an alternate hydrophobic and hydrophilic network with representative regular nano-channels that are comprising with -SO_3H or SO_3Li ionic electrets for active ions transfer, and inter-connected merged aromatic sextets with imides bridges for electrons transfer, respectively. The robust coordination can empower the Bno-Spi-TENG to endure the time-honored electrostatic potential on the contact surface which displays an inequality between the number of protons (cations), and electron on the targeted surface. Moreover, Bno-Spi film displays an ions hopping mechanism at hydrophilic -SO_3H or SO_3Li centers through ion charge electrets, and at the same time, the hydrophobic π-π stacking network can prompt the triboelectric open-circuit voltage Voc, and short circuit currents Jsc, individually. The induced charges on the Bno-Spi surface are comparative to its surface area and are close to the theoretical limit levied by the dielectric breakdown by air [30]. However, a noteworthy claim was shown to enhance the triboelectric polarity by fluctuating their surface morphologies, chemical construction, and interpenetration of ionic groups within the polymer network. The projected novel polymeric Bno-Spi-TENGs might show robust chemical steadiness, stretchable modulus, and strength to improve the triboelectric current [31]. The electric out-puts through altered frequencies of contact-separation manner have shown the increased Voc and Jsc of 75 V, and 1 μA at 6 Hz (**Figure 2**).

4.2 Mechanism of the dual demonstrative Bno-Spi-TENGs for strong contact-electrification through hydrophilic and hydrophobic nano-channels

In this study, for the first time, we motivated to use, a Bno-Spi-TENG is an effective ion and electron-transfer root with a counter electronegative PTFE film for the contact-separation electrification process [29, 30]. The anticipated novel Bno-Spi-TENG shown superior characteristics which have a superior π-π stacked layer-on-layer oligomeric morphology with an alternate hydrophobic and hydrophilic network with representative regular nano-channels that are comprising with -SO_3H, or SO_3Li ionic electrets for active ions transfer, and inter-connected merged aromatic sextets with imides bridges for electrons transfer, respectively. The vigorous arrangement could allow the Bno-Spi-TENG to bear the enduring electrostatic potential on the contact surface which shows an imbalance between the numbers of electrons. For the first time, we have examined the mechanism of contact electrification procedure in two methods among the Bno-Spi films (i.e. SO_3H.Bno-Spi, SO_3Li.Bno-Spi, and SO_3H.TEA.Bno-Spi) as a positive layer, and PTFE as a negative TENG layer. At this point, the projected Bno-Spi-TENGs have been fabricated with interchanged hydrophilic, and hydrophobic nano-channels for the generation of high-throughput Voc, and Isc [32].

The mechanistic approach of sulfonic acid (SO_3H) group was attached to the backbone of Bno, during the triboelectric process, when they interact with an adjacent fluorocarbon ($-CF_2$) of PTFE polymer chain has comprised the splitting of the $-SO_3H$, or $-SO_3Li$ into positive H^+ protons or Li^+ ions, and negative SO_3^- ions. Consequently, the H^+ protons or Li^+ ions attract momentarily on the C–F to form a temporary chemical bond by the transition state of [C^+---F----- H^+ or Li^+---SO_3^-----C] to transfer the charges through an ionic mechanism between two surfaces. In certain, the charge-transfer application was approved in three steps (**Figure 3**) [33–36]. In the Step 1, the Bno-Spi, and PTFE surfaces have generated initial

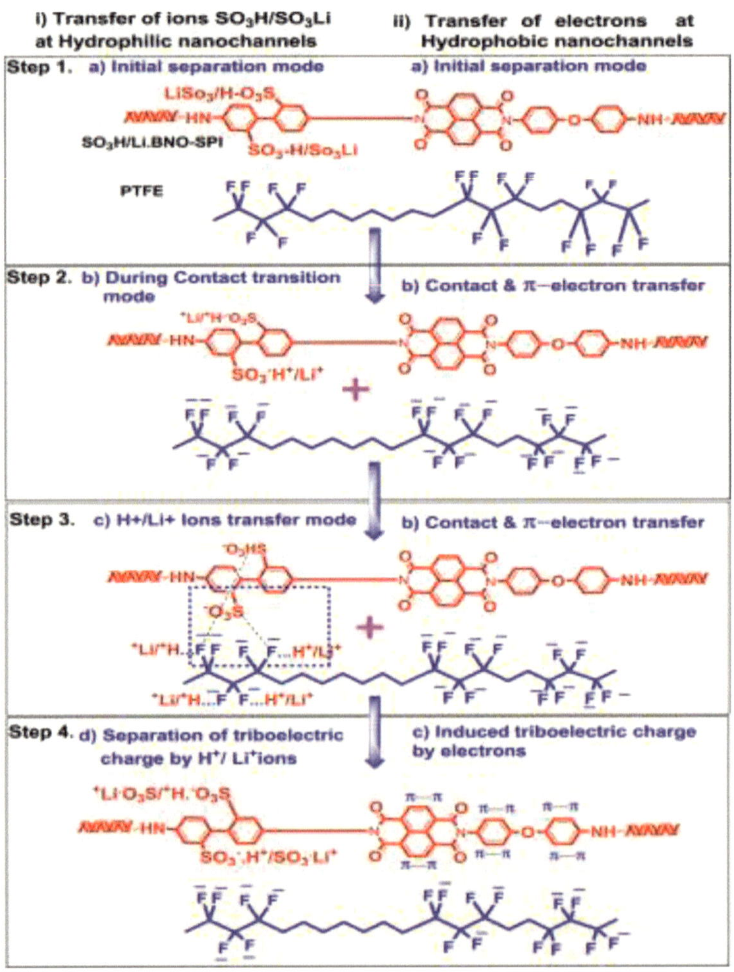

Figure 3.
The mechanistic approach of Bno-Spi-TENGs through chemical reaction pathways for the ions and electrons transfer at the hydrophilic and hydrophobic nano-channels, respectively, by contact-separation mode TENG.

charges on their surfaces where the ions from -SO_3H or -SO_3Li of the Bno-Spi, and CF_2 groups on PTFE. The Bno-Spi can produce the temporary charge-transfers through ion transfer mechanism at hydrophilic sites through nano-channels, and electron transfer at the hydrophobic nano-channels. In Step-2, when the PTFE was brought into contact, the H^+ protons or Li^+ ions were at hydrophilic nano-channels, and π-electrons at hydrophobic nano-channels of SO_3H.Bno-Spi- or SO_3Li.Bno-Spi was induced by electrostatic field effect. Thus, the projected Bno-Spi-TENG was produced electric charges through ions, and electrons from both surfaces. In Step-3, during transporting of H^+ protons or Li^+ ions from the Bno-Spi surfaces into PTFE wherein transition state, and forms a momentary ion bridge. In the four, while detaching of electrodes, the generated electric TENG charges were excited through π-π bonds in aromatic ring systems through hydrophobic nano-channels were moved into Bno-Spi-TENGs, and the net negative charges are remain the same on PTFE surface. This TENG process is continued during the contact and separation process [20].

The SO_3H.Bno-Spi-TENG, SO_3Li.Bno-Spi-TENG, and SO_3H.TEA-Bno-Spi-TENGs have shown the *Voc* and *Jsc* of 75 V, and 1 μA, 43 V, and 0.6 μA, and 9 V, and 0.13 μA at applied frequency of 6 Hz, correspondingly. The *Voc*, and *Jsc* of SO_3H.Bno-Spi-TENGs have shown upto 733%, and 669% concerning SO_3H.TEA-Bno-Spi-TENGs since the movement of H^+ ions remains very high on the device surface. Therefore, the maximum instantaneous power of SO_3H.Bno-Spi-TENGs, SO_3Li.Bno-Spi-TENGs, and SO_3H.TEA-Bno-Spi-TENGs were reached to 71.4 μW, 18.07 μW, and 10.89 μW at 20 MΩ conforming to the power density of 17.85 $\mu W/cm^2$ (0.1785 W/m^2), 4.515 $\mu W/cm^2$ (0.0045 W/m^2), and 2.72 $\mu W/cm^2$ (0.0272 W/m^2), correspondingly. The numerical characterization of the output performance has presented from SO_3H.Bno-Spi-TENG is 8 folds higher than SO_3H.TEA-Bno-Spi-TENG, and 1.8 folds higher than that of SO_3Li.Bno-Spi-TENG since the ion sizes were enlarged from $H^+ > Li^+ > SO_3H$.TEA. It was strongly recommended that the competence of the SO_3H.Bno-Spi-TENGs is significantly larger over the corresponding SO_3H.TEA-Bno-Spi-TENGs and SO_3Li.Bno-Spi-TENGs (**Figure 4**) [36–39].

4.3 Electric impulse spring-assisted contact separation mode TENG

In this study, we established a self-effacing and movable self-powered contact-separation approach that includes coil-aided Wcf-Pani.Es-TENG such as positive interaction superficial surface, and PVDF membrane as a negative triboelectric electrode. The established Wcf-Pani.Es-TENG presented special appearances such as inner π-π stacking's network, and amidic connections together with quaternary anilinium ions that are linking between each monomer of aniline blocks. The width, and resistivity of the Wcf-Pani.Es deposition are 0.65 μm, and 0.324 Ω which are determined by four-point probe method [21]. Owing to this morphology, the Wcf-Pani.Es is showed a huge superficial zone which is increasing the output presentation of the TENG. The established innovative Wcf-Pani.Es-TENG is revealed a short circuit current (*Isc*) of ~180 μA, and the open-circuit voltage (*Voc*) of 95 V (**Figure 5**) [40–45].

4.4 Ions transfer mechanism of contact-separation of Wcf-Pani.Es TENG

Figure 6 showed the mechanistic approach between Pani.Es and PVDF where the electric charges were reorganized when the electrification happens. During the triboelectrification process, the formation of H^+ protons from Pani.Es, and adjacent F- ions from PVDF interact to induce opposite charges. Successively, when they are in full contact mode, the H^+ ions are attracted temporarily on the C-F to form a transition bond of PVDF---F-----H^+ ---Pani.Es to transmit the charges through an

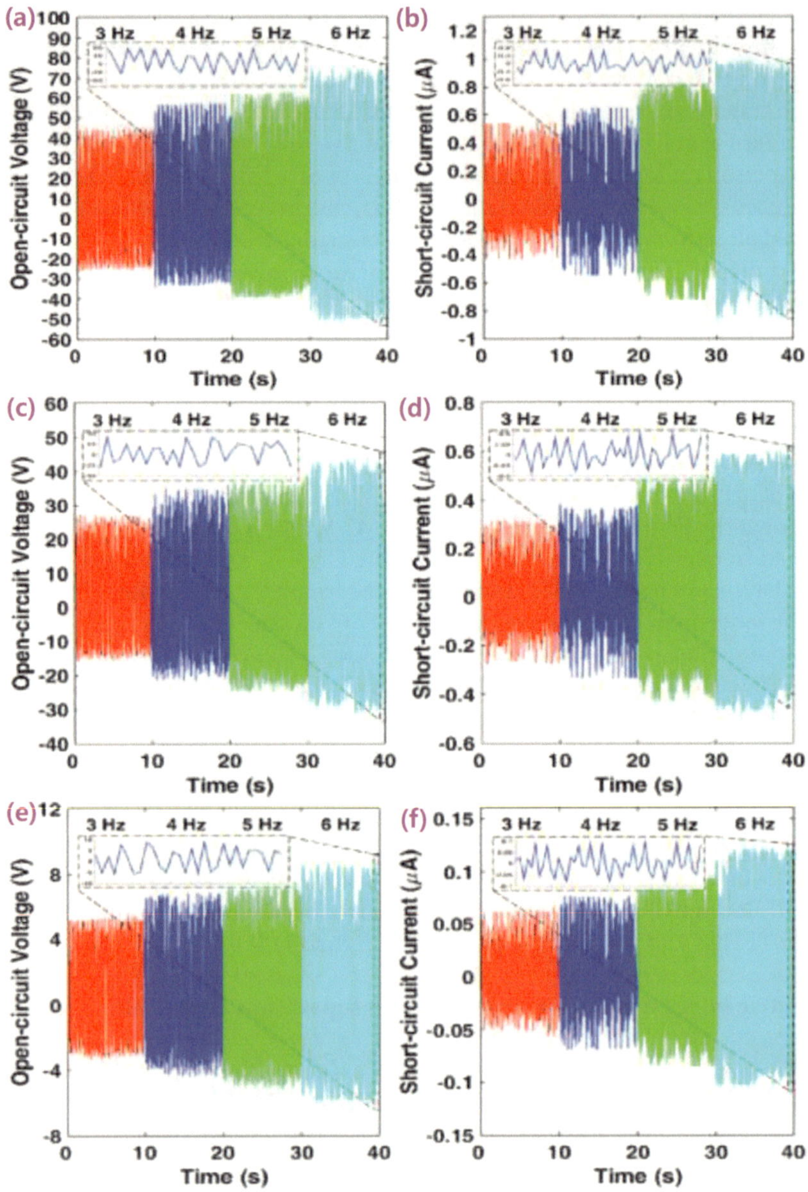

Figure 4.
(a), (c), and (e) open circuit voltages Voc and (b), (d), and (f) short-circuit currents Jsc of Bno-Spi-TENG, SO_3H.Bno-Spi-TENG, SO_3Li.Bno-Spi-TENG, and SO_3H.TEA.Bno-Spi-TENG in contradiction of PTFE film at 3 Hz, 4 Hz, 5 Hz, and 6 Hz, respectively. Inset: An enlarged view of the signals when the Bno-Spi-TENGs were interacts with PTFE surface.

ionic passage or temporary chelation between the two films [46]. The charge-transfer mechanism is carried out in four steps. Step 1, it represents the Wcf-Pani.Es, and PVDF membrane are display an early charges on their surfaces through NH^+ and F^- positioned on the Wcf-Pani.Es and PVDF, respectively. Step 2, when the PVDF membrane was carried into interaction, the H^+ protons of Wcf-Pani.Es are induced by the electrostatic field effect. Therefore, the electric charges by H^+ protons and F^- ions can generate in both films. Step 3, shows the transferring of H^+ protons is occur from the Wcf-Pani.Es surface into PVDF during the transition state, and form a temporary ion bridge between them. Step 4, during the separation process, the two oppositely charged surfaces induced a potential variance, and to minimize

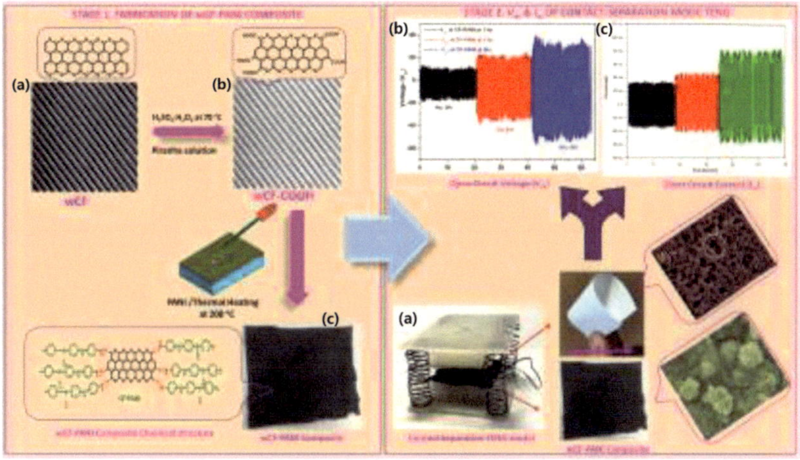

Figure 5.
Schematic illustration of electric impulse coil-aided contact separation style TENG. Stage 1. Chemical alteration of Wcf-Pani.Es film (a) woven carbon fiber mat (Wcf); b) chemically oxidized woven carbon fiber mat (Wcf-COOH); c) construction of Wcf-Pani.Es composite through electrostatic connections with aniline monomer, and in-situ oxidative polymerization using $(NH_4)_2.S_2O_8$. Stage 2. a) the actual archetypal of coil-aided TENG, (inset nanoporous PVDF membrane (upper) and variable Wcf-Pani.Es nano-pillared composite (lower) and their inset SEM pictures. b) Voc, and c) Isc.

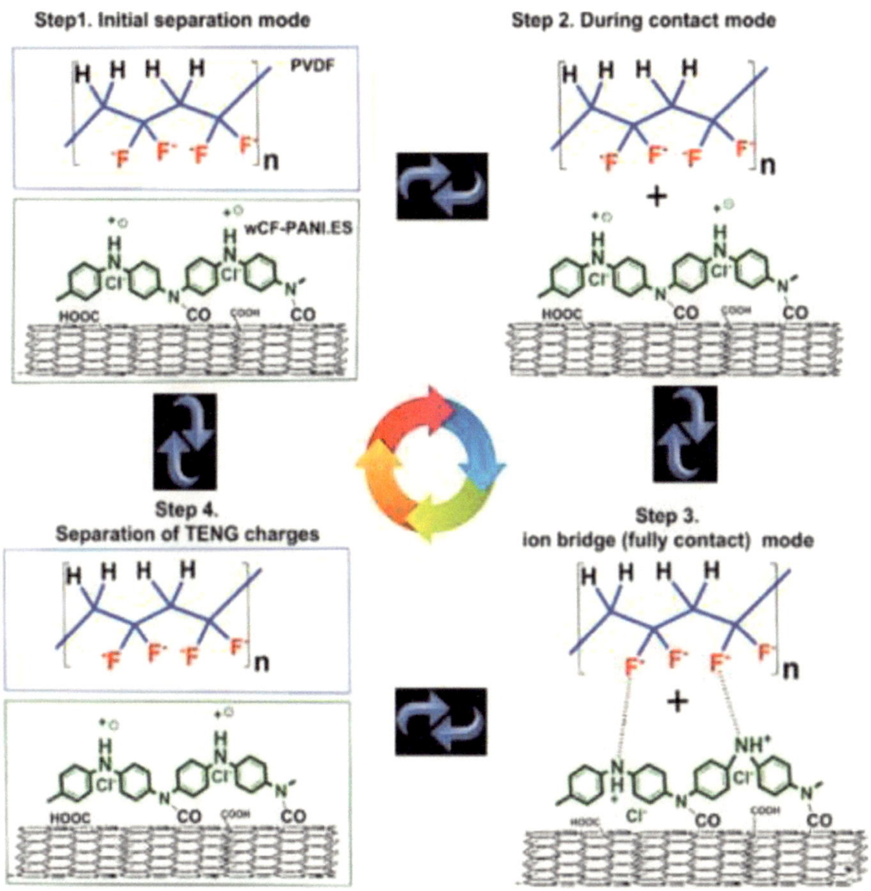

Figure 6.
The mechanism pathway of ions that are prompted between negative PVDF, and wCF-PANI.ES surfaces when they contact separated each other.

these differences, the flow of electrons emerged between two electrodes. For the validation of the automatic investigations, we achieved a measureable analysis of the out-put presentation of Wcf-Pani.Es [47–49].

4.5 Demonstration of output presentation of Wcf, Wcf-COOH, and Wcf-Pani.Es-TENGs

Initially, we inspected the performance of the Wcf, Wcf-TENG, and Wcf-COOH-TENG in contradiction of PVDF membrane through the contact-separation style technique at numerous applied regularities of 1 Hz, 3 Hz, and 5 Hz, correspondingly, is depicted in **Figure 7**. **Figure 7a** and **b** showed the *Voc,* and *Isc* of Wcf-TENG were at −2.5 V to 2.7 V, and 170 nA to −171 nA, −2.3 V to +3.1 V, and 225 nA to −221 nA, and 2.4 V to −3.7 V, and 326 nA to −328 nA at 1 Hz, 3 Hz, and 5 Hz, correspondingly, upon regular contact and separation of electrodes. Subsequently, the examination remained discovered the *Voc,* and *Isc* of Wcf-COOH-TENG have shown 3.7 V to −4.1 V, and 0.2 μA to −0.6 μA, 4.6 V to −6.4 V and 0.5 μA to −1.3 μA, and

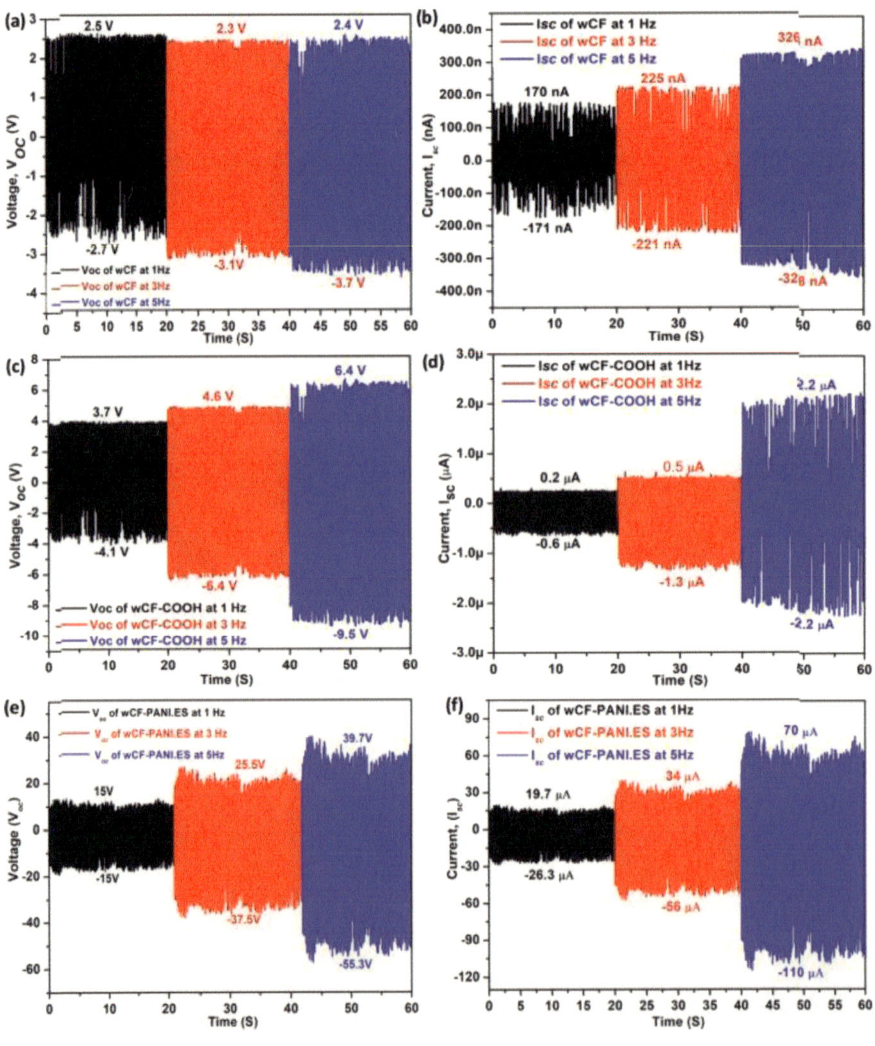

Figure 7.
The Voc, and Isc in different input circumstances in open circuit arrangements. The Voc, and Isc of (a, b) Wcf-TENG, (c, d) Wcf-COOH-TENG, and (e, f) Wcf-Pani.Es-TENG.

6.4 V to −9.5 V and 2.2 μA to −2.2 μA at 1 Hz, 3 Hz, and 5 Hz, correspondingly, as depicted in **Figure 7c** and **d**. The V*oc,* and I*sc* of Wcf-Pani.Es-TENG showed of 15 V to −15 V, and 19.7 μA to −26.3 μA, 25.5 V to −37.5 V, and 34 μA to −56 μA, and 39.7 V to −55.3 V, and 70 μA to −119 μA at 1 Hz, 3 Hz and 5 Hz, correspondingly (**Figure 7e** and **f**) [21, 50].

5. Conclusions

To develop novel technologies for the harvesting of energy, TENG is an alternative mode of technology by collecting trillions of electrons combining. These electrons are collectively obtained from various smart materials that contain high characteristic features such as flexibility, thinness and durability, long self-life, high power density, and reproducibility to harvest clean energy. Besides, TENGs can be used to transform physical characteristics such as pressure contact mode, sliding mode, and single electrode mode of features for the accumulation of energies at sub-molecular levels. Based on our novel technologies, the self-powered energy systems have given the higher out-put performance of voltage and currents. The proposed Bno-Spi-TENG and Wcf-Pani.Es-TENG are highly durable and can be used with a lower speed of contact separation modes to generate the desired amount of voltage and I*sc*. The systematic method created on the ionic electrets mechanism on the superficial electric potential of the polymeric surface has been maintained the generation of Voc and Isc from our developed novel TENGs. The established Bno-Spi-TENGs have been transported together ions, and electrons through ion, and electron transfer device when they communicated each other through PTFE. The H^+ protons or Li^+ ions attract temporarily on the C–F of PTFE surface to form a transition state of [C^+----F---- H^+ or Li^{+-}---SO_3----C] bond to transfer the charges through ionic mechanism between two active TENG films. In particular, the rate of transmission of H^+ protons, and Li^+ ions from the SO_3H.Bno-Spi-TENG, and SO_3Li. Bno-Spi-TENG surfaces have achieved huge voltage, and currents owing to the presence of the SO_3H and SO_3Li ionic clusters involved to hydrophilic nanochannels, and effective electron transfer arisen at the hydrophobic nanochannels. Also, Wcf-Pani.Es-TENG showed extraordinarily vigorous, and reliable energy gathering presentations owing to mechanically strong material assets of woven carbon fibers chemically changed by the Pani nano-flakes using a simple chemical process, this Wcf-Pani.Es-TENG has shown a great potential for self-powered TENGs even under numerous strict surroundings, and in distinct medical applications without harmful effects.

Acknowledgements

There is no funding support for this book chapter.

Conflict of interest

The authors declare no conflict of interest.

Author details

Ravi Kumar Cheedarala
School of Mechanical Engineering, Ulsan National Institute of Science and Technology (UNIST), Ulsan, South Korea

*Address all correspondence to: rkchidrala@gmail.com

References

[1] Wiles JA, Grzybowski BA, Winkleman A, Whitesides GM. Analytical Chemistry. **2003;75**:4859

[2] Diaz AF, Guay J. IBM Journal of Research and Development. **1993;37**:249

[3] Horn RG, Smith DT, Grabbe A. Nature. **1993;366**:442

[4] Horn RG, Smith DT. Science. **1992;256**:362

[5] Fan FR, Tian ZQ, Wang ZL. Nano Energy. **2012;1**:328

[6] Fan FR, Lin L, Zhu G, Wu W, Zhang R, Wang ZL. Nano Letters. **2012;12**:3109

[7] Zhu G, Pan C, Guo W, Chen CY, Zhou Y, Yu R, et al. Nano Letters. **2012;12**:4960

[8] Wang S, Lin L, Wang ZL. Nano Letters. **2012;12**:6339

[9] Zhu G, Lin ZH, Jing Q, Bai P, Pan C, Yang Y, et al. Nano Letters. **2013;13**:847

[10] Lowell J, Rose-Innes AC. Advances in Physics. **1980;29**:9947

[11] Zhao D, Duan LT, Xue MQ, Ni W, Cao TB. Angewandte Chemie, International Edition. **2009;48**:6699

[12] Bandodkar AJ, Jeerapan I, Wang J. ACS Sensors. **2016;1**:464

[13] Soh S, Kwok SW, Liu H, Whitesides GM. Journal of the American Chemical Society. **2012;134**:2015

[14] Diaz AF. The Journal of Adhesion. **1998;67**:11

[15] Diaz AF, Alexander DF. Langmuir. **1993;9**:1009

[16] Liu C, Bard AJ. Nature Materials. **2008**;7:505

[17] McCarty LS, Winkleman A, Whitesides GM. Journal of the American Chemical Society. **2007;129**:4075

[18] McCarty LS, Whitesides GM. Angewandte Chemie (International Ed. in English). **2008;47**:2188

[19] Soh S, Liu H, Cademartiri R, Yoon HJ, Whitesides GM. Journal of the American Chemical Society. **2014;136**:13348

[20] Cheedarala RK, Duy LC. K. K. Ahn[a]. Nano Energy. **2018;44**:430

[21] Cheedarala RK, Parvez AN, Ahn KK. Nano Energy. **2018;53**:362

[22] B. J. Akle, M. D. Bennett, D. J. Leo, *Sen. Actuator, A* **2006**, *126*, 173.

[23] V. Panwar K. Cha, J. Park, S. Park, *Sen. Actuators, B,* **2012**, *161, 460.*

[24] Rajagopalan M, Oh IK. ACS Nano. **2011**;*5*:2248

[25] Jeon JH, Cheedarala RK, Kee CD, Oh IK. Advanced Functional Materials. **2013;23**:6007

[26] Cheedarala RK, Jeon JH, Kee CD, Oh IK. Advanced Functional Materials. **2014**;**25**:6005

[27] J. W. Lee, Y. T. Yoo, *Sen. Actuator, B* **2009**, *137*, 539.

[28] J. W. Lee, S. M. Hong, J. Kim, C. M. Koo, *Sen. Actuator, B* **2012**, *162*, 369.

[29] Jo C, Pugal D, Oh IK, Kim KJ, Asaka K. Progress in Polymer Science. **2013**;*38*:1037

[30] Imaizumi S, Kokubo H, Watanabe M. Macromolecules. **2013**;*45*:401

[31] Park S, An J, Suk JW, Ruoff RS. Small. **2010**;*6*:210

[32] Lee JW, Kim JH, Goo NS, Lee JY, Yoo YT. Journal of Bionic Engineering. **2010**;*7*:19

[33] C. A. Dai, C. J. Chang, A. C. Kao, W. B. Tsai, W. S. Chen, W. M. Liu, W. P. Shih, C. C. Ma, *Sensors and actuators, A* **2009**, *155*, 152.

[34] Rajagopalan M, Jeon JH, Oh IK. Sensors and Actuators B: Chemical. **2011**;*151*:198

[35] Jeon JH, Kang SP, Lee S, Oh IK. *Sensors and actuators*. B. **2009**;*143*:357

[36] Mirfakhrai T, Madden JDW, Baughman RH. Materials Today. **2007**;*10*:30

[37] Lee JW, Yu S, Hong SM, Koo CM. Journal of Materials Chemistry C. **2012**;*1*:3784

[38] Imaizumi S, Kokubo H, Watanabe M. Macromolecules. **2013**;*45*:401

[39] K. Cui, X. Feng, Y. Huang, Q. Zhao, Z. Huang, W. Zhang, *Proc. SPIE*, DOI: 10.1117/2.1200710.0901.

[40] Li J, Ma W, Song L, Niu Z, Cai L, Zeng Q, et al. Nano Letters. **2011**;*11*:4636

[41] Yoonessi M, Shi Y, Scheiman DA, Lebron-Colon M, Tigelaar DM, Weiss RA, et al. ACS Nano. **2012**;*6*:7644

[42] Pyshkina OA, Panova TV, Boeva ZA, Lezov AA, Polushina GE, Lezov AV, et al. Nanotechnologies in Russia. **2012**;*7*:629

[43] Cui K, Zheng Y, Liang J, Wang D. Nano Research. **2018;11**:1873

[44] Liu J, Rinzler AG, Dai H, Hafner JH, Bradley RK, Boul PJ, et al. Fuller. Pipes Sci. **1998;280**:1253

[45] Lu J, Kim SG, Lee S, Oh IK. Advanced Functional Materials. **2008**;*18*:1290

[46] Cheedarala RK, Song JI. RSC Advances. **2019;9**:31735

[47] S A Nahian, *R. K. Cheedarala,* A. N. Parvez, K. K. Ahn, Nano Energy, **2017**, 38, 447.

[48] Cheedarala RK, Kim GH, Cho S, Lee JH, Kim J, Song HK, et al. **Journal of Materials Chemistry**. **2011;21**:843

[49] *R. K. Cheedarala,* J. I. Song, Int. J. of Smart and Nano Materials, **2020**, 11, 38.

[50] Cheedarala RK, Song JI, Journal of Mechanics Engineering and Automation, 2019, 9, 225-229

Section 3

Nanomaterials for Biosensors

Chapter 8

Novel Two-Dimensional Siloxene Material for Electrochemical Energy Storage and Sensor Applications

Rajendran Ramachandran, Zong-Xiang Xu and Fei Wang

Abstract

After discovering graphene, the two-dimensional materials have gained considerable interest in the electrochemical applications, especially in energy conversion, storage, and bio-sensors. Siloxene, a novel two-dimensional low-buckled structure of Si networks with unique properties, has received the researcher's attention for a wide range of applications. Though the electronic and optical properties of siloxene have been explored in detail previously, there is a lack of electrochemistry studies of siloxene as the result of material degradation, and the investigation is still open-ended to enhance the electrochemical application. Recently, siloxene has been used for supercapacitor, lithium-ion batteries, and dopamine bio-marker detections. This chapter highlights the recent development of siloxene synthesis and its electrochemical properties in energy and sensor applications. The plannar Si structure with Si_6 rings interconnected with different oxygen, hydroxyl functional groups, and large interlayer spacing of siloxene sheets can promote the active sites for enhanced electrochemical performance. This chapter provides the current state-of-the-art in the field and a perspective for future development in the electrochemistry field of siloxene.

Keywords: siloxene, electrochemistry, functional groups, active sites

1. Introduction

After discovering graphene in 2004, the two-dimensional (2D) materials have drawn significant attention in broad applications due to their unique physico-chemical properties. The 2D materials such as transition metal dichalcogenides (TMDs), phosphorene and bismuthine, etc., which consists of a one-atom-thick monolayer network can exhibit different chemical and physical properties, including the electrical and thermal conductivity, magnetic, photonic and catalytic properties when compared to their bulk counterparts [1]. Over the past few years, the enormous 2D family materials like graphene [2, 3], molybdenum disulphide (MoS_2) [4, 5], tungsten disulphide (WS_2) [6, 7], graphitic carbon nitride (g-C_3N_4) [8] and recently MXene [9, 10] have been investigated for various applications in electronic, energy, catalysis and electrochemical applications. However, the electrochemistry investigation of those materials is yet to be explored in detail.

IntechOpen

The limitation in the bandgap of these materials has hindered their performance in practical applications. Therefore, exploring a new novel 2D material is highly recommended, especially for the future electrochemical energy conversion, storage, and biosensors applications. Recently, silicon (Si) based one-atom-thick layered material named siloxene has been investigated for electrochemical energy and sensing applications, including supercapacitors, batteries, and dopamine sensors [1, 11–13].

Siloxene is a direct bandgap material that was discovered by Wohler in 1863. It can be obtained through the deintercalation of calcium and exfoliation from the Zintl phase of calcium silicide ($CaSi_2$) powder [14–16]. Different from the graphene planner structure, siloxene possesses a low-buckled structure due to its double band role. As a result of the surface-terminated functional groups with Si chain and the mixed sp^2 and sp^3 hybridization, siloxene can provide several advantages in the electrochemical energy and sensor applications [1, 11, 12, 17].

1.1 Synthesis of siloxene and its structural types

Siloxene is prepared by deintercalation of Ca^{2+} from $CaSi_2$ under concentrated hydrochloric acid. Briefly, the required amount of $CaSi_2$ powder and HCl acid stirred in the ice-cold condition under the inert gas atmosphere for 2-4 days (**Figure 1**). During this reaction, the deintercalation of Ca layers and functionalization of Si

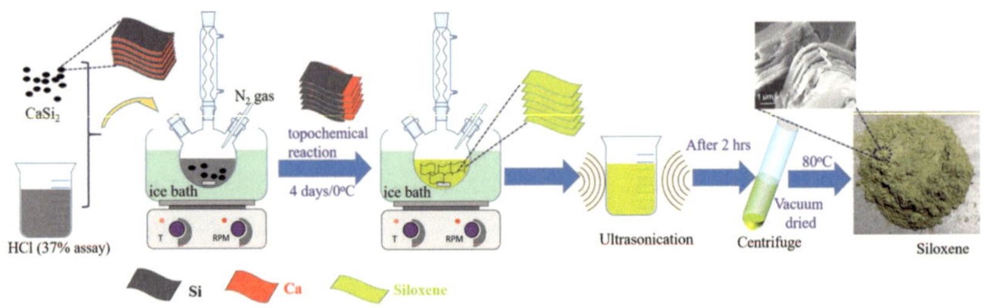

Figure 1.
Siloxene synthesis process (reproduced from [11] with permission from Elsevier).

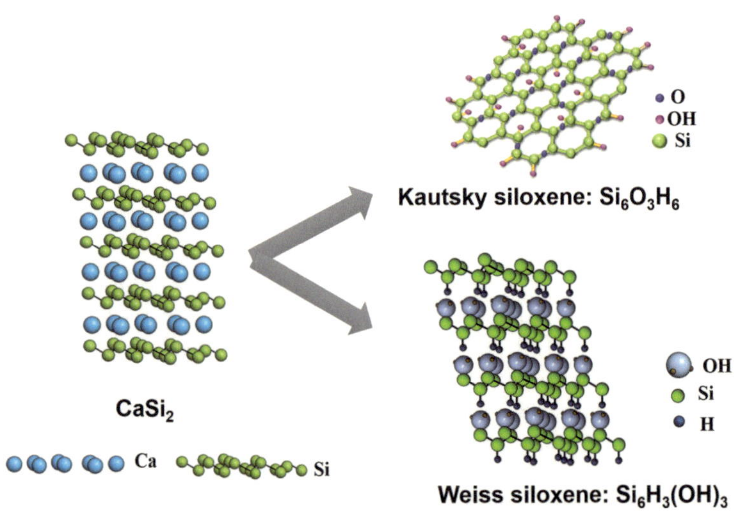

Figure 2.
Different types of Siloxene structure [19].

sheets can be occurred simultaneously and formed the siloxene structure. The following equation can describe the common formation mechanism of siloxene from $CaSi_2$ [11].

$$3CaSi_2 + 6HCl + 3H_2O \rightarrow Si_6O_3H_6 + 3CaCl_2 + 3H_2\uparrow \quad (1)$$

In general, the siloxene stoichiometric ratio of Si:H:O is 2:2:1. Based on the exfoliation and deintercalation conditions such as reaction time, the concentration of the acidic medium, and temperature, siloxene can be classified into two major types. (1) Weiss, and (2) Kautsky type siloxene structures [18]. In Weiss type siloxene ($Si_6(OH)_3H_3$), the six-membered Si_6 rings connected with alternative Si-H and Si-OH bonds, whereas Kautsky type siloxene ($Si_6O_3H_6$), the Si_6 rings connected by Si-O-Si bridge (**Figure 2**). It is noteworthy that the crystalline silicon (common impurity) in $CaSi_2$ may affect the siloxene structure formation [19], which deviates from the structures mentioned above.

2. Electrochemical application of siloxene

Due to the unique 2D structure and the abundant functional groups of siloxene, it can be applied in various applications such as optoelectronics, catalysis, water splitting, etc. Theoretical investigations of the siloxene have shown the high possibilities in different electrochemical applications [20]. However, because of limited knowledge of siloxene's electrochemistry, only a few works have been reported on the electrochemical application of siloxene so far. The siloxene has been mainly employed in supercapacitors and batteries as an electrode material and detection of biomarkers in electrochemical biosensors.

2.1 Supercapacitors

2.1.1 Siloxene based supercapacitors

Though siloxene was discovered in 1863, it has recently received considerable attention in the electrochemical energy storage application. The researchers have been focused on siloxene based electrode materials for energy storage and conversion application. Due to the increases in energy consumption and the non-renewable sources decreasing gradually, the development of high-efficiency energy storage devices is highly demanded. Electrochemical or supercapacitors are the perfect choice for high-performance devices as the results of its high-power density and long cyclic lifetime [21]. Compared with the commercial activated carbon-based supercapacitors, the integration of Si-based materials with the current microelectronic technology can lead to higher performance in energy storage devices because of its high theoretical capacity (3579 mA hg^{-1}). However, Si-based materials such as silicon carbide (SiC), Si nanowire, porous silicon have been employed as electrode materials in supercapacitor application, the functionalization of the one-atom-thick Si layers with interconnected Si_6 rings can accommodate the better performance in supercapacitors [1].

Krishnamoorthy et al. have reported the siloxene based symmetric supercapacitor (SSC) application in 2018 [1]. The Kautsky-type of siloxene structure prepared by deintercalation of calcium from $CaSi_2$ and confirmed its Si-O-Si bridges Si_6 rings interconnection by Fourier transform infrared spectroscopy.

The capacitance behavior of the siloxene has been studied in tetraethylammonium tetrafluoroborate ($TEABF_4$) electrolyte under optimal conditions. Interestingly, the operating potential window (OPW) of the siloxene-SSC device was determined from 0 to 3.0 V. This result confirms the excellent electrochemical stability of the SSC device even at a higher voltage window. The fabricated SSC device showed unique capacitance behavior with an energy density of 5.08 W h kg^{-1} (areal energy density of 9.82 mJ cm^{-2}) and about 98% of capacitance retention even after 10 k cycles (**Figure 3**). The ion diffusion and the electron transfer rate were significantly enhanced by the conductive hexagonal Si frameworks in the siloxene during the electrochemical redox reactions. Also, the high surface area and the larger interlayer spacing between the siloxene sheets were enabled fast ion transport and improved the electrochemical performance of the SSC device.

It is well known that the reduced graphene oxide (rGO) can increase the electroactive sites for the electrochemical reactions than bare graphene oxide (GO) because of its higher electronic conductivity [22]. The electrical conductivity of siloxene sheets may decrease when a higher amount of the oxygen functional groups is attached on its edge/basal surface; thus, the reduction of oxygen functional groups in siloxene enhances the active sites for electrochemical redox reactions due to its better conductivity. In this scenario, Parthiban et al. have investigated the removal of oxygen functional groups in pristine siloxene (p-siloxene) at high temperatures and obtained reduced siloxene sheets (denoted as HT-siloxene). Calcinating siloxene sheets removed the functional groups at edge/basal planes of siloxene at 900°C, which led to the formation of reduced siloxene sheets [23]. Interestingly, the calcination process has decomposed the oxygen functional groups at edge/basal planes of siloxene and preserved the Si_6 rings' connection with oxygen atom without affecting the 2D layer structure. The obtained HT-siloxene possessed a higher electrical conductivity than p-siloxene resulting in improved

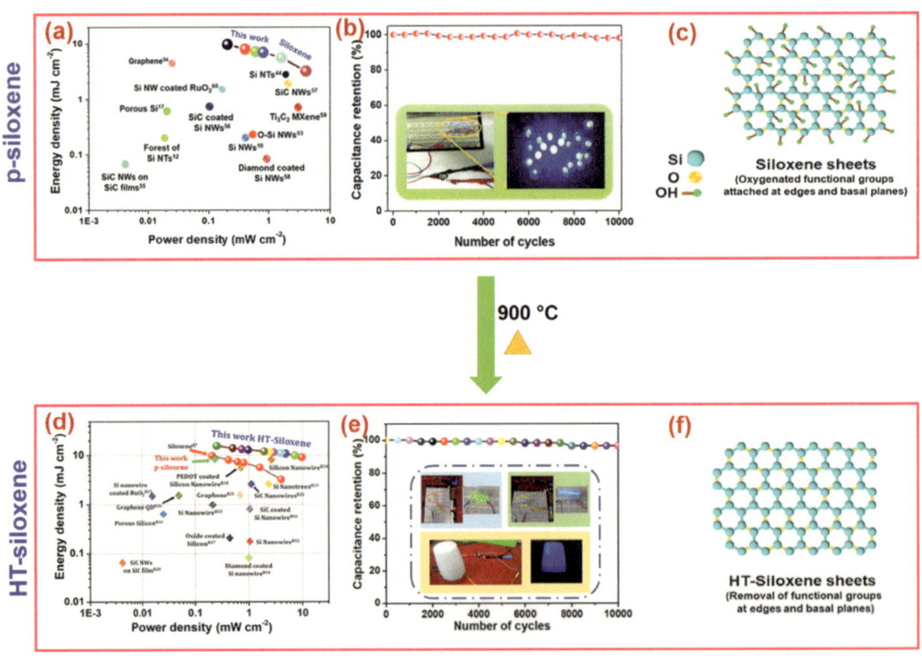

Figure 3.
(a, b) Ragone plot and cyclic stability of p-siloxene SSC device; (c) structure of p-siloxene [1]; (d, e) Ragone plot and cyclic stability of HT- siloxene SSC device; (f) structure of HT-siloxene (reproduced from [23] with permission from ACS).

electrochemical performance. The specific capacitance of the HT-siloxene increased almost 1.71 times higher than that of p-siloxene. The maximum energy density of the HT-siloxene SSC device has been achieved by about 6.64 Wh kg^{-1}, higher than p-siloxene (3.89 Wh kg^{-1}) due to its lower equivalent series resistance and better electrical conductivity. The complete removal of the oxygen functional groups in p-siloxene enhanced the energy density of SSC. It also increased the cyclic stability of the SSC (96.3% after 10000 cycles), as shown in **Figure 3**.

Another fascinating strategy has been developed by Kim and co-workers recently that dry reforming methane (DRM) recycled siloxene/Ni foam catalyst towards supercapacitor applications. The siloxene coated Ni foam was initially utilized for DRM reactions for producing H_2 and CO gas by CO_2 reduction. After the DRM reaction, the siloxene/Ni foam catalyst has employed as electrode material in SSC [24]. The regeneration of carbon during the DRM reactions deposited on the siloxene/Ni foam catalyst and could improve the electrochemical performance. Compared to the p-siloxene and HT-siloxene, the carbon-coated siloxene/Ni foam exhibited superior performance in the supercapacitor. A maximum energy density of 30.81 Wh kg^{-1} was achieved for carbon/siloxene/Ni foam-based SSC, indicates the remarkable performance enhancement. Thus, utilizing spent siloxene catalysts to supercapacitor can be an effective approach for waste-to-energy applications. Besides, the direct use of the siloxene in a supercapacitor, siloxene was also confirmed as a flexible template for fabricating silicon oxy-carbide (SiOC). Carbothermal conversion of siloxene to SiOC has been proposed by Pazhamali and co-works [12]. Mixing siloxene and sodium alginate at 900°C led to the formation of SiOC. Since the SiC-based electrodes can intensify the cycling stability and areal capacitance in supercapacitors, the SiOC electrodes were expected to improve the stability of the SSC device than siloxene based SSC. The SiOC based SSC device delivered an excellent electrochemical performance with an energy density of 20.89 Wh kg^{-1}, which is higher than that of p-siloxene. However, the cyclic stability of SiOC supercapacitor decreased to 92.8% after 5000 cycles. As pointed out in the previous paragraph, the removal of oxygen functional groups can improve the SSC performance; the complete reduction of oxygen in SiOC may help to facilitate the fast ion transport and wettability of the electrode during the long cyclic time.

2.1.2 Siloxene composite supercapacitor

Making composite electrodes is an efficient approach to increase the supercapacitor's electrochemical performance due to its synergistic behavior [25]. The specific capacitance of siloxene is restricted because of its aggregation effect; consequently, the layers agglomerations generate poor utilization of the pores and the lower specific surface area. Thus, introducing a spacer material such as metal oxides or carbon between the siloxene sheets can enhance the accessible sites for the electrochemical reactions. Meng and co-works have reported the construction of a three-dimensional (3D) architecture of siloxene-reduced graphene oxide hydrogel (SGH) through a simple hydrothermal method (**Figure 4**) [26].

The hybrid structure of SGH has increased the specific surface area and facilitated the electrolyte ions transportation, resulting in improved capacitive performance. As compared to bare siloxene electrode specific capacitance (23 F g^{-1}), the SGH with 1:3 ratio composite electrode exhibited a maximum specific capacitance of 520 F g^{-1} at a current density of 1 A g^{-1}. However, the EDLC of the graphene in SGH has contributed significantly to the capacitive enhancement of siloxene-graphene composite. Though graphene could facilitate the capacitance performance, the surface oxygen-functional groups of siloxene provided pseudocapacitance and improved the wettability of the electrode, results in an excellent rate capability and outstanding cyclic stability.

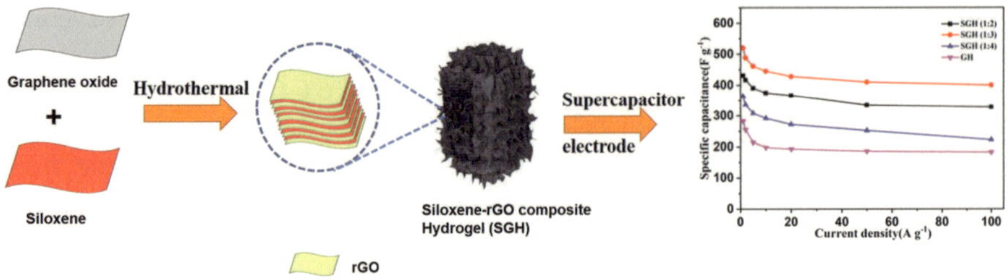

Figure 4.
Synthesis of siloxene-reduced graphene oxide hydrogel and its specific capacitance plot [26].

2.2 Siloxene application in batteries

Like supercapacitors, rechargeable batteries (e.g., lithium-ion batteries, sodium-ion batteries, lead-acid, etc.) are primary power sources for large-scale portable and wearable electronic devices. They have received significant consideration due to their high energy density and long cyclic stability [27, 28]. However, the current battery technologies cannot meet the advanced application requirement as the result of confined energy storage capacity. Thus, the development of commercial electrodes in the existing technologies is highly needed.

The theoretical capacity of silicon (Si) is 4200 mA hg^{-1} [29], which is higher than the capacity of graphite (372 mA hg^{-1}), has considered being an active anode material for the future lithium-ion batteries (LIBs). However, the severe capacity degradation and the high-volume change during the lithiation-delithiation process may lead to lower Coulombic efficiency. Making 2D Si nanosheets with oxygen functional groups provides a high specific surface area, resulting in fast lithium storage and preventing volume changes. As mentioned in the previous section, the siloxene oxidation level can be controlled by the various synthesis conditions such as temperature, oxidants, concentrations, etc. The oxidation level may influence the lithiation-delithiation process. Xu and co-workers have demonstrated the siloxene preparation with different oxidation levels in the various oxidants and the temperature [30]. Three types of siloxene oxidation level have been achieved by altering the oxidants and temperature: (i) $CuCl_2$ aqueous solution used to prepare fully oxidized siloxene nanosheet (FO-SNS) at room temperature; (ii) partially oxidized siloxene nanosheet (PO-SNS) made in $SnCl_2$ ethanol solution at 60°C and (iii) hardly oxidized siloxene nanosheet (HO-SNS) synthesized in a LiCl-KCl molten salt at 400°C (**Figure 5(i)**). The FO-SNS, PO-SNS, and HO-SNS electrodes delivered the lithiation capacity of 298, 1218, and 1450 mA hg^{-1}. Besides, the HO-SNS presented a higher Coulombic efficiency of 66%, which is higher than FO-SNS (24%) and PO-SNS (56%). The improved performance of HO-SNS has associated with the presence of a higher atomic percentage (64%) of bulk Si (Si0) and the lower percentage (7%) of SiO_2 (Si^{4+}) in HO-SNS, which were estimated from the XPS analysis (**Figure 5(ii)**). Besides, the hierarchical nanostructure of HO-SNS could buffer the volume expansion and contribute to the good rate performance. Fu and co-workers have remarked that bare siloxene is an unsuitable anode material for LIBs due to its inadequate electrochemical capacity, resulting in the lower Coulombic efficiency. However, Si- derivatives such as silicon suboxides (SiO_x), carbon-coated SiO_2, etc., from siloxene can meet higher capacity requirements with satisfactory Coulombic efficiency. Fu et al., have demonstrated the carbon-coated 2D SiO_x nanocomposites (nano-Si/α-SiO_2) from siloxene to moderate the volume expansion during the electrochemical lithiation-delithiation process [31]. The carbon-coated nano-Si/α-SiO_2 anode materials showed the limited volume change, fast electrons

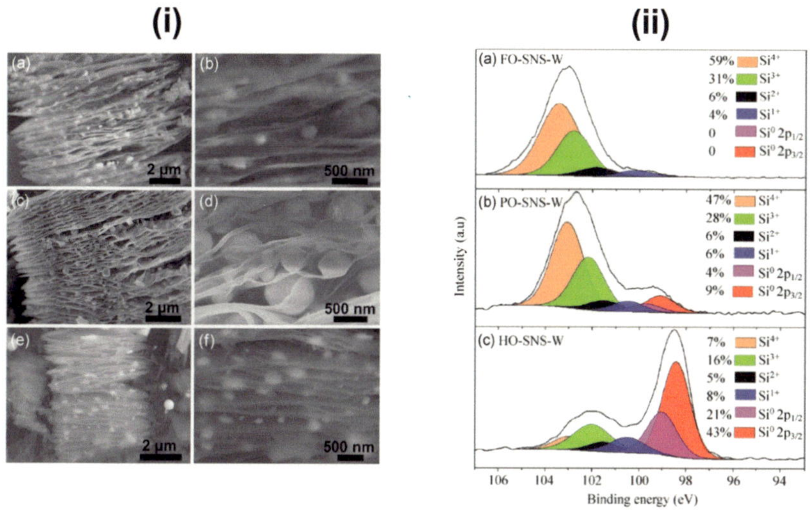

Figure 5.
(i) SEM images of (a, b) FO-SNS, (c, d) PO-SNS, (e, f) HO-SNS; (ii) XPS spectrum of siloxene samples at different oxidation (reproduced from [30] with permission from Springer).

transport, and more significant Li-ion kinetics, resulting in high initial Coulombic efficiency (72.5%) with a capacity of 946 mA hg^{-1}. On the other hand, the value of x in SiO_x can influence Li storage's electrochemical performance. Thus, controlling the oxidation level of the SiO_x is a crucial process to achieve higher capacity than bare Si structures. Many previous studies showed that SiO_x with x = 1.0 presented the specific capacity value higher than 1000 mA hg^{-1} [32]. However, unsatisfactory cyclic life has limited its practical usage. Thus, turning the oxygen content in SiO_x is a proper way to improve the electrochemical performance in LIBs. After investigating carbon-coated nano-Si/α-SiO_2, Fu and co-workers have prepared siloxene with different levels of oxidation in SiOx and used as anode material for LIB. They controlled the SiOx oxidation level in the siloxene via stepwise oxidizing of the siloxene precursor at various times [32]. SiO_x with four different oxidation levels, such as $SiO_{1.01}$, $SiO_{1.25}$, $SiO_{1.47}$, and $SiO_{1.78}$ has been tailored through siloxene oxidation and investigated their Li-storage capacity. The sample $SiO_{1.47}$ exhibited optimal electrochemical behavior due to the synergistic effect of electrical conductivity and Li-ion diffusivity. The higher oxygen level in SiO_x caused a larger polarization effect, resulting in the poor Coulombic efficiency and smaller reversible capacity.

Similar to the graphene-siloxene composite electrode in supercapacitors, the incorporation of siloxene sheets between the graphene layers enhances the specific surface area, facilitating the fast Li-storage. In the siloxene-graphene (SiG) composite, the siloxene sheets have provided higher Li-storage, and the encapsulated graphene sheets prevented the volume expansion during lithium insertion-extraction process. SiG anode material exhibited the initial cycle charge and discharge capacities of 3016 mA hg^{-1} and 3880 mA hg^{-1} with a capacity decay of 78%, which were higher than the bare siloxene and graphene electrodes. The synergistic effect of graphene and siloxene and the excellent electrical conductivity of graphene in the composite contributed to the higher electrochemical performance for LIBs [29].

2.3 Siloxene based electrochemical sensor

The 2D siloxene sheets not only possessed the excellent electrochemical characteristics towards electrochemical energy application. Besides, due to the large surface area and the unique 2D structure of siloxene, the heterogeneous electron transfer (HET) is high, which beneficial for selective electrochemical bio-marker detections. We have recently demonstrated the siloxene-based novel electrochemical dopamine sensor and obtained remarkable achievements in dopamine detection by the siloxene modified sensor [11]. Dopamine (DA) is an important neurotransmitter that plays a crucial role in the central nervous system and cardiovascular systems. A variety of materials have been employed for electrochemical DA detection in the past decades. However, the high selectivity of DA is limited to the existing materials. As a result of high HET rates, large surface area, and improved mass transportation, siloxene possessed high selectivity for DA detection (**Figure 6**). Siloxene modified glassy carbon electrode showed a well-defined redox peak in the cyclic voltammetry technique towards DA detection. Excellent linearity has been achieved for the siloxene electrode in the presence of a different concentration of DA, and the modified electrode exhibited a detection limit of 0.327 μM. Besides, the proposed sensor revealed a wide linear range from 10 to 1100 μM (**Figure 6(b)**).

The DA detection performance by the 2D siloxene sheets is remarkably higher than that of other reported 2D graphene and g-C_3N_4 modified electrodes. Siloxene sheets owned a higher response for the detection limit and showed high selectivity for DA detection. The stronger π-π interaction between the siloxene planar structure and the dopamine phenyl structure enables faster electron transportation during the DA oxidation process, making the high selectivity characteristic of the siloxene modified electrode. On the other hand, the π-π interaction of the siloxene structure with other biomolecules such as ascorbic acid, uric acid, etc., is weak, resulting in

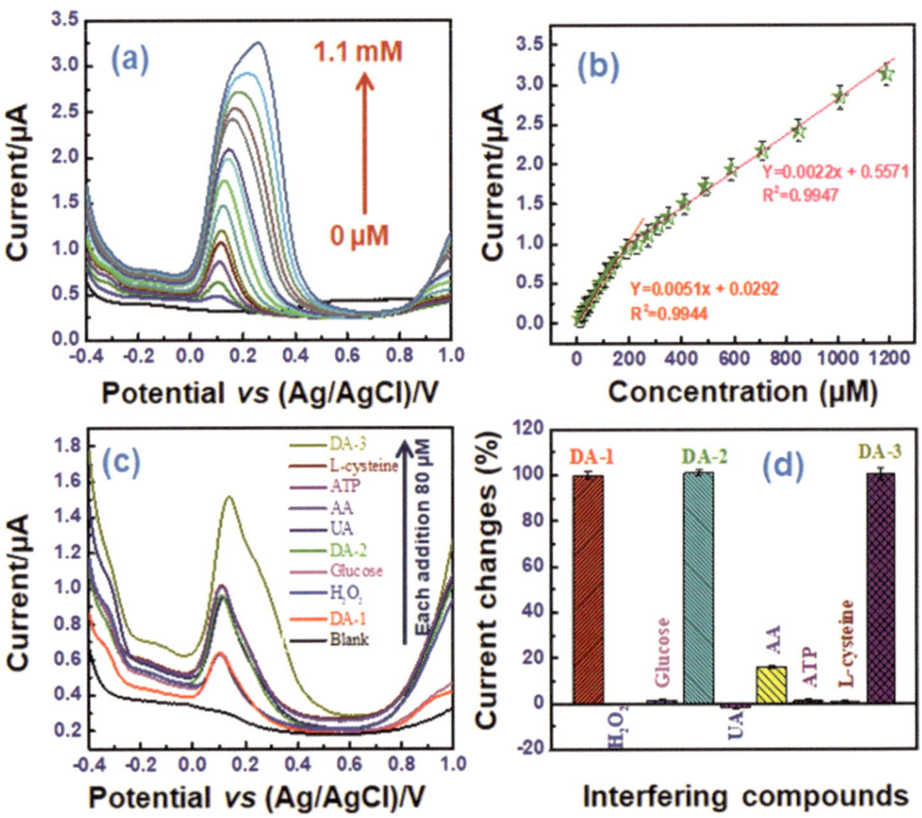

Figure 6.
(a-d) Electrochemical differential pulsed voltammetry response and linear range of siloxene modified electrode for DA detections (reproduced from [11] with permission from Elsevier).

the inactive oxidation. However, the thickness of the siloxene sheets can affect the electron conduction during the electrochemical reactions similar to graphene [33]. Reducing the size and the layer thickness of siloxene could tremendously enhance its performance for DA detection.

3. Summary and future research direction

In conclusion, this chapter deals with the comprehensive review of the raising star 2D siloxene based electrochemical energy and sensor applications. The siloxene synthesis process and the siloxene structure affecting parameters have been reviewed in detail. The functional groups in siloxene and the oxidation level can be varied at different synthesis times and the annealing temperature. Compared to pristine siloxene, high temperature treated siloxene possessed an excellent performance in the electrochemical supercapacitors because of its reduced functional groups. Besides, the siloxene and its composite have been used as anode materials for LIBs and showed a significant capacity and Coulombic efficiency. Li-storage has influenced by the oxidation level in siloxene due to the presence of different atomic percentages of Si functional groups. However, both supercapacitors and LIBs applications, siloxene derivatives such as SiOx, SiOC showed improved performance as the results of its better electrical conductivity and Li-ion diffusivity compared to the bare siloxene. The reported siloxene works have focused on the performance of siloxene in supercapacitors and LIBs. But many works failed to investigate the insight of the electrochemistry of siloxene and its derivatives for better energy density, capacity, and cyclic stability. Thus, the research direction should be focused more on the study of electrochemistry of siloxene. On the other hand, the 2D siloxene sheets proved as a novel electrochemical sensor for highly selective dopamine detection. Moreover, the size and thickness of the layer can influence the HET rate, specific surface area, and active sites for DA detection, which need to be optimized in the near future.

Acknowledgements

This work was supported in part by the National Natural Science Foundation of China (Project No. 51950410598), in part by Shenzhen Science and Technology Innovation Committee (Projects No. JCYJ20170412154426330), and in part by Guangdong Natural Science Funds (Project No.: 2016A030306042 and 2018A050506001). Also, this work was supported by the Major Program of Guangdong Basic and Applied Research (No. 2019B030302009).

Author details

Rajendran Ramachandran[1,2,3], Zong-Xiang Xu[3*] and Fei Wang[2,4*]

1 SUSTech Academy for Advanced Interdisciplinary Studies, Southern University of Science and Technology, Shenzhen, China

2 School of Microelectronics, Southern University of Science and Technology, Shenzhen, China

3 Department of Chemistry, Southern University of Science and Technology, Shenzhen, China

4 Engineering Research Center of Integrated Circuits for Next-Generation Communications, Ministry of Education, Shenzhen, China

*Address all correspondence to: xu.zu@sustech.edu.cn and wangf@sustech.edu.cn

References

[1] Krishnamoorthy K, Pazhamalai P, Kim S-J. Two-dimensional siloxene nanosheets: novel high-performance supercapacitor electrode materials, Energy Environ. Sci., 2018:11:1595-1602. DOI: https://doi.org/10.1039/C8EE00160J

[2] Ramachandran R, Felix S, Joshi GM, Raghupathy BPC, Jeong SK, Grace AN. Synthesis of graphene platelets by chemical and electrochemical route, Mater. Res. Bull., 2013:48:3834-3842. DOI: http://dx.doi.org/10.1016/j.materresbull.2013.05.085

[3] Ramachandran R, Saranya M, Kollu P, Raghupathy BPC, Jeong SK, Grace AN. Solvothermal synthesis of Zinc sulfide decorated Graphene (ZnS/G) nanocomposites for novel Supercapacitor electrodes, Electrochim. Acta 2015:178:647-657. DOI: http://dx.doi.org/10.1016/j.electacta.2015.08.010

[4] Kumuthini R, Ramachandran R, Therese HA, Wang F. Electrochemical properties of electrospun MoS_2@C nanofiber as electrode material for high-performance supercapacitor application, J. Alloys Compd, 2017:705:624-630. DOI: http://dx.doi.org/10.1016/j.jallcom.2017.02.163

[5] Pazhamalai P, Krishnamoorthy K, Manoharan S, Kim S.-J. High energy symmetric supercapacitor based on mechanically delaminated few-layered MoS_2 sheets in organic electrolyte, J. Alloys Compd, 2019:771:803-809. DOI: https://doi.org/10.1016/j.jallcom.2018.08.203

[6] Kumar KS, Choudhary N, Pandey D, Ding Y, Hurtado L, Chung h-S, Jung Y, Thomas J. Investigating 2D WS_2 supercapacitor electrode performance by Kelvin probe force microscopy, J. Mater. Chem. A 2020:8:12699-12704. DOI: https://doi.org/10.1039/D0TA03383A

[7] Liu S, Zeng Y, Zhang M, Xie S, Tong Y, Cheng F, Lu X. Binder-free WS_2 nanosheets with enhanced crystallinity as a stable negative electrode for flexible asymmetric supercapacitors, J. Mater. Chem. A. 2017:5:21460-21466. DOI: https://doi.org/10.1039/C7TA07009H

[8] Sakthivel T, Ramachandran R, Kirubakaran K. Photocatalytic properties of copper-two dimensional graphitic carbon nitride hybrid film synthesized by pyrolysis method, J. Environ. Chem. Engg, 2018:6: 2636-2642. DOI: https://doi.org/10.1016/j.jece.2018.04.009.

[9] Ramachandran R, Hu Q, Rajavel K, Zhu P, Zhao C, Wang F, Xu Z-X. Non-peripheral octamethyl-substituted copper (II) phthalocyanine nanorods with MXene sheets: An excellent electrode material for symmetric supercapacitor with enhanced electrochemical performance, J. Power Sources, 2020:471:228472. DOI: https://doi.org/10.1016/j.jpowsour.2020.228472

[10] Ramachandran R, Rajavel K, Xuan W, Lin D, Wang F. Influence of $Ti_3C_2T_x$ (MXene) intercalation pseudocapacitance on electrochemical performance of Co-MOF binder-free electrode, Ceram. Int., 2018:44:14425-14431. DOI: https://doi.org/10.1016/j.ceramint.2018.05.055

[11] Ramachandran R, Leng X, Zhao C, Xu Z-X, Wang F. 2D siloxene sheets: A novel electrochemical sensor for selective dopamine detection, Appl. Mater. Today 2020:18:100477. DOI: https://doi.org/10.1016/j.apmt.2019.100477

[12] Pazhamalai P, Krishnamoorthy K, Sahoo S, Mariappan VM, Kim S-J.

Carbothermal conversion of siloxene sheets into silicon-oxy-carbide lamellae for high-performance supercapacitors, Chem. Engg. J., 2020:387:123886. DOI: https://doi.org/10.1016/j.cej.2019.123886

[13] Loaiza LC, Monconduit L, Seznec V. Siloxene: A potential layered silicon intercalation anode for Na, Li and K ion batteries, J. Power Source, 2019:417:99-107. DOI: https://doi.org/10.1016/j.jpowsour.2019.02.030

[14] Nakano H, Ishii M, Nakamura H. Preparation and structure of novel siloxene nanosheets, Chem. Commun., 2005:2945-2947. DOI: https://doi.org/10.1039/B500758E

[15] Li S, Wang H, Li D, Zhang X, Wang Y, Xie J, Wang J, Tian Y, Ni W, Xie Y. Siloxene nanosheets: a metal-free semiconductor for water splitting, J. Mater. Chem. A 2016:4:15841-15844. DOI: 10.1039/c6ta07545b

[16] Kong X, Liu Q, Zhang C, Peng Z, Chen Q. Elemental two-dimensional nanosheets beyond graphene, Chem. Soc. Rev., 2017:46:2127-2157. DOI: 10.1039/c6cs00937a

[17] Zhang W, Sun L, Vianney JM, Nsanzimana, Wang X. Lithiation/Delithiation Synthesis of Few Layer Silicene Nanosheets for Rechargeable Li–O2 Batteries, Adv. Mater., 2018:30:1705523. DOI: 10.1002/adma.201705523

[18] Yamanaka S, Matsu-ura H, Ishikawa M. New deintercalation reaction of calcium from calcium disilicide synthesis of layered polysilane, Mater. Res. Bull., 1996:31:307-316. DOI: https://doi.org/10.1016/0025-5408(95)00195-6

[19] Dahn JR, Way BM, Fuller E. Structure of siloxene and layered polysilane (Si_6H_6), Phys. Rev. B 1993:48(24):17872-17877. DOI: http://dx.doi.org/10.1103/PhysRevB.48.17872

[20] Rosli NF, Rohaizad N, Sturala J, Fisher AC, Webster RD, Pumera M. Siloxene, Germanane, and Methylgermanane: Functionalized 2D Materials of Group 14 for Electrochemical Applications, Adv. Funct. Mater. 2020:30:1910186. DOI: https://doi.org/10.1002/adfm.201910186.

[21] Ramachandran R, Lan Y, Xu Z-X, Wang F. Construction of NiCo-Layered Double Hydroxide Microspheres from Ni-MOFs for High-Performance Asymmetric Supercapacitors, ACS Appl. Energy Mater. 2020:3:6633-6643. DOI: https://dx.doi.org/10.1021/acsaem.0c00790

[22] Ramachandran R, Saranya M, Velmurugan V, Raghupathy BPC, Jeong SK, Grace AN. Effect of reducing agent on graphene synthesis and its influence on charge storage towards supercapacitor applications, Appl. Energy 2015:153:22-31. DOI: http://dx.doi.org/10.1016/j.apenergy.2015.02.091

[23] Pazhamalai P, Krishnamoorthy, Sahoo S, Mariappan VK, Kim S-J. Understanding the Thermal Treatment Effect of Two-Dimensional Siloxene Sheets and the Origin of Superior Electrochemical Energy Storage Performances, ACS Appl. Mater. Interfaces 2019:11:624-633. DOI: 10.1021/acsami.8b15323

[24] Krishnamoorthy K, Sudhakaran MSP, Pazhamalai P, Mariappan VK, Mok YS, Kim S-J. A highly efficient 2D siloxene coated Ni foam catalyst for methane dry reforming and an effective approach to recycle the spent catalyst for energy storage applications, J. Mater. Chem. A 2019:7:18950-18958. DOI: https://doi.org/10.1039/C9TA03584B

[25] Ramachandran R, Saranya M, Grace AN, Wang F. MnS nanocomposites based on doped

graphene: simple synthesis by a wet chemical route and improved electrochemical properties as an electrode material for supercapacitors, RSC Adv. 2017:7:2249-2257. DOI: 10.1039/c6ra25457h

[26] Meng Q, Du C, Xu Z, Nie J, Hong M, Zhang X, Chen J. Siloxene-reduced graphene oxide composite hydrogel for supercapacitors, Chem. Engg. J. 2020:393:124684. DOI: https://doi.org/10.1016/j.cej.2020.124684

[27] An Y, Tian Y, Wei C, Jiang H, Xi B, Xiaong S, Feng J, Qian Y. Scalable and Physical Synthesis of 2D Silicon from Bulk Layered Alloy for Lithium-Ion Batteries and Lithium Metal Batteries, ACS Nano. 2019:13:13690-13701. DOI: 10.1021/acsnano.9b06653

[28] Zhang X, Li L, Fan E, Xue Q, Bian Y, Wu F, Chen R. Toward sustainable and systematic recycling of spent rechargeable batteries, Chem. Soc. Rev. 2018:47:7239-7302. DOI: https://doi.org/10.1039/C8CS00297E

[29] Kumar KT, Reddy MJK, Sundari GS, Raghu S, Kalaivani RA, Ryu SH, Shanmugharaj AM. Synthesis of graphene-siloxene nanosheet based layered composite materials by tuning its interface chemistry: An efficient anode with overwhelming electrochemical performances for lithium-ion batteries, J. Power Sources 2020:450:227618. DOI: https://doi.org/10.1016/j.jpowsour.2019.227618

[30] Xu K, Ben L, Li H, Huang X. Silicon-based nanosheets synthesized by a topochemicalreaction for use as anodes for lithium ion batteries, Nano Res. 2015:8:2654-2662. DOI: 10.1007/s12274-015-0772-4

[31] Fu R, Zhang K, Zaccaria RP, Huang H, Xia Y, Liu Z. Two-dimensional silicon suboxides nanostructures with Si nanodomains confined in amorphous SiO_2 derived from siloxene as high performance anode for Li-ion batteries, Nano Energy 2017:39:546-553. DOI: http://dx.doi.org/10.1016/j.nanoen.2017.07.040

[32] Fu R, Li Y, Wu Y, Shen C, Fan C. Controlling siloxene oxidization to tailor SiO_x anodes for high performance lithium ion batteries, J. Power Sources 2019:432:65-72. DOI: https://doi.org/10.1016/j.jpowsour.2019.05.071

[33] Hu Y, Li X, Geng D, Cai M, Li R, Sun X. Influence of paper thickness on the electrochemical performances of graphene papers as an anode for lithium ion batteries, Electrochim. Acta 2013:91:227-233. DOI: https://doi.org/10.1016/j.electacta.2012.12.106

Chapter 9

The Novel Nanomaterials Based Biosensors and Their Applications

Kübra Gençdağ Şensoy and Mihrican Muti

Abstract

Since the development of the first biosensor reported, biosensor has received considerable attention due to its high selectivity and sensitivity. Biosensors are highly pursued in order to meet the growing demands and challenges in a large number of analytic applications such as medical diagnosis, food safety control, environmental monitoring, or even military defense. Due to the unique physical, chemical, mechanical and electrical properties, nanomaterials have been widely investigated for their ability and used to fabricate sensors. High surface to volume ratio, good stability, excellent electrocatalytic properties of the nanomaterials plays an important role in the sensitive and selective detection of biomolecules. The synthesis of new nanomaterials with different properties is increasingly common in order to improve these counted properties of nanomaterials. This chapter gives an overview of the importance of the development of novel nanomaterials based biosensors technologies. The use of different funtionalized carbon nanomaterilas, metal oxide nanoparticles, metal nanoparticles, polymeric nanoparticles, quantum dots, graphene sheets and other novel nanomaterials in biosensor technology, and their innovations and advantages are discussed.

Keywords: novel nanomaterials, biosensor, biorecognition, nanosensors

1. Introduction

A biosensor device is defined as a biological or bio-inspired receptor unit with unique specificities for analytes. These analytes are generally of biological origin. One of the challenges in biosensor development is that efficient signal capture can be achieved with biological recognition. Novel nanomaterials represent a rapidly developing field in bioanalysis applications. The sensitivity and performance of biosensors can be improved by using nanomaterials. Typical schemeatic presentation of a biosensor is illuatrated in **Figure 1**.

With the development of nanotechnology, many new nanomaterials such as gold nanostructure, magnetic nanoparticles, nanozymes, and carbon-based nanomaterials have been synthesized [1]. Nanomaterials have been widely applied in the areas of invivo imaging [2], cancer treatment [3], drug delivery [4], catalysis [5], bacteriostasis [6], and so on. Due to the outstanding physical and chemical properties of nanomaterials, nanomaterial-based biosensors have been developed [7].

In this chapter, synthesis, properties and possible applications of these materials in biosensors were examined. The high sensitivity and selectivity of nanomaterial-based biosensors have led to major advances in the development of new methodologies for early detection. Due to its submicron dimensions, it allows simple and fast

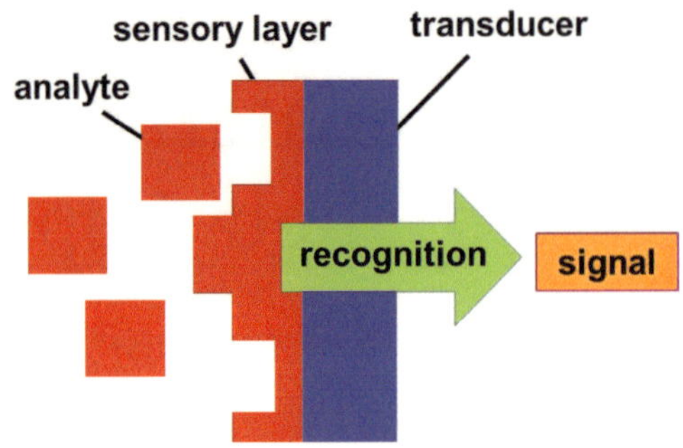

Figure 1.
Schematic presentation of a biosensor.

analysis in vivo. Their reactivity, toughness and other properties are also dependent on their unique shape, size and structure. In addition, the application of nanomaterials to biosensors provides different detection limits depending on the samples to be analyzed and facilitates the adjustment of the sensitivity level according to the needs.

2. Experimental

2.1 Synthesize of the novel nanomaterials

2.1.1 Graphdiyne

The synthesis of the GDY was reported in the literature as follows:

Graphdiyne (GDY) was prepared on the copper surface by a cross-linking reaction using hexaetynylbenzene (HEB) as a monomer. Firstly, hexakis [(trimethylsilyl) ethynyl] benzene (HEB-TMS) was prepared using Negishi cross-linking reaction. Then HEB monomer was obtained by the addition of tetrabutylammonium fluoride into tetrahydrofuran solution of HEB-TMS with stirring at 0°C for 10 minutes. Finally, GDY in the presence of pyridine was successfully grown on the surface of copper foils by a cross-coupling reaction of the HEB monomer for 72 hours at 60°C under a nitrogen atmosphere. After the reaction was completed, GDY grown on copper foils was removed by ultrasonic treatment and concentrated by rotary evaporator, and then washed with heated acetone and N,N-dimethylformamide. The GDY powder was refluxed with dilute hydrochloric acid and sodium hydroxide, respectively at 80°C for 3 hours. It was then washed repeatedly and centrifuged. Finally, black GDY powder was obtained by centrifugation and drying vacuum [8].

2.1.2 Gold nanostructures

According to the modified El-Sayed method the synthesis of the gold nanorods (GNRs) was performed as follows:

Ttwo solutions were prepared as seed solution and growth solution. For the seed solution, ice-cold sodium borohydride ($NaBH_4$) (0.3 mL, 0.01 M) was added to the solution mixture containing hydrogen tetrachloroaurate ($HAuCl_4$) (0.5 mM) and cetyltrimethylammonium bromide (CTAB) (0.2 M) in a volume ratio of 1:1, and the entire reaction mixture was incubated at room temperature for 3 hours. For the growth solution, a 200 mL solution containing $HAuCl_4$ (0.5 mM) and CTAB (0.1 M) was made and 6 mL of silver nitrate ($AgNO_3$) (4 mM) was added to it. Following this, 0.5 M sulfuric acid (H_2SO_4) (1 mL) and 0.0788 M ascorbic acid (1.4 mL) were added and mixed gently. In the last step, the seed solution (0.24 mL) was added to the growth solution mixture and left at room temperature for a period of 12 hours. The brownish colored solution was centrifuged at 9000 rpm (2 times) for 30 minutes to remove unbound CTAB and stored at room temperature (28°C) [9].

2.1.3 Inorganic nanomaterials

The synthesis of Zirconium Phosphate Nanoparticle (ZrP-NP) is described in this section as one of the inorganic nanomaterials.

Zirconium Phosphate (ZrP), one of the inorganic nanomaterials, has been synthesized by hydrothermal process. First, 1.6 g of $ZrOCl_2.8H_2O$ was added to 30 mL of DI water and stirred continuously. Then 15 M H_3PO_4 (10 mL) was added to this prepared solution and stirred continuously for 30 minutes. The solution was transferred to a hydrothermal autoclave (50 mL) and heated in an air oven at 200°C for 24 hours. The products obtained were collected by centrifugation and washed several times with ethanol and deionized water. In the last step, the purified ZrP powder was dried in an air oven at 50°C [10].

2.1.4 Nanozymes

The synthesis of the core-shell Au@Co-Fe hybrid nanoparticles is described as peroxidase mimetic nanozyme.

In the synthesis of the core-shell Au@Co-Fe hybrid nanoparticles as the peroxidase mimetic nanozyme, gold nanoparticles (AuNPs) with the average diameter of 22 nm were synthesized by citrate reduction of $HAuCl_4$. Briefly, 1.5 mL of 1% (w/v) sodium citrate solution was added to 21 mL of 0.8 mM $HAuCl_4.3H_2O$ solution at boiling point while the solution was stirred vigorously. After hanging its color from pale yellow to deep red, the mixture was stirred for 15 min and let to cool to room temperature and, then stored at 4°C until use. In the second step, 1 μL tween 20 was added to 1.5 mL of the synthesized AuNPs. Then, 100 μL of $FeSO_4$ 0.18 M and 180 μL of $CoCl_2$ 0.1 M were added to the mixture, and incubated at the room temperature for 24 h. After that, the mixture was centrifuged and washed with deionized water [11].

2.1.5 Hybrid nanocomposites

One of the hybrid nanocomposites is reduced graphene oxide-magnetite nanoparticle (RGO-Fe_3O_4 NP) and its synthesis is described below according to the literature [12].

Reduce graphene oxide magnetite nanoparticle (RGO-Fe_3O_4 NP) hybrid was synthesized by alkaline reduction. For this purpose, the powder was redispersed in the 0.5 mg mL^{-1} graphene oxide (GO) suspension. Citric and ascorbic acids were added and the mixture was stirred at 55°C (12 hours). 1 M NaOH was added and the mixture It was stirred again at 95°C (6 hours). After centrifugation at 10000 rpm (RCF = 1118 x g), the solid is filtered, washed, and dried in vacuum during 24 hours [12].

2.1.6 DNA nanomaterials

Y-DNA was prepared by mixing equimolar amounts of three single stranded DNA (ssDNA), two long and one short. The two long sequences have regions that hybridize to the shorter one. One of the fields is not completely linked to the corresponding fragment. Thus, the target miRNA became able to replace this fragment and remove the Y-DNA nanostructure. ssDNAs were dissolved in hybridization buffer at 10 μM final concentration per sequence and annealed to form the desired Y-shaped DNA: annealed at 95°C for 2 minutes, cooled to 65°C and incubated for 5 minutes, followed by 2 minutes while its temperature dropped to 60°C and cooled to 20°C at a rate of 1° per minute. The final products were stored at 4°C. Double stranded substrates were formed by mixing in the hybridization buffer. The mixture was heated to 95°C for 5 minutes and slowly cooled to 4°C, then allowed to stand at room temperature for 20 minutes to form a specific double stranded substrate [13].

2.1.7 DNAzyme

DNA phosphorylation was made by incubating 200 pmol of FS1 with 20 units of T4 polynucleotide kinase (PNK) at 37°C for 30 min in a 100 μL reaction mixture containing 50 mM Tris–HCl (pH 7.6 at 25°C), 10 mM $MgCl_2$, 5 mM 1,4-Dithiothreitol (DTT), 0.1 mM spermidine and 1 mM adenosine 5′-triphosphate (ATP). The reaction was stopped by heating the mixture at 90° C for 5 minutes. RFT1 (100 μM) and 2 μL RFS1 (100 μM) were then added to the solution, and the mixture was heated to 90°C for 40 seconds and cooled to room temperature for 10 minutes. In the last step, 10 units of T4 DNA ligase were added for DNA ligation at 25°C for 2 hours. The ligation mix contains 10 mM $MgCl_2$, (150 μL) 40 mM Tris–HCl (pH 7.6 at 25°C), 10 mM DTT and 0.5 mM ATP. The products were concentrated by standard ethanol precipitation and further purified by polyacrylamide gel electrophoresis [14].

2.1.8 Carbon Nanodots

The syntheses of carbon nanodots (CDs) will describe according to the literature [15].

CDs were synthesized hydrothermally with citric acid and ethylenediamine (EDA). Initially citric acid (3.0 g) and ethylenediamine (1875 μL) were dissolved in 30 mL of distilled water. The solution was then transferred to a 500 mL round bottom flask and heated at 150°C for 5 hours. The product was dialyzed against ddH_2O to obtain CDs. CDs powder was obtained by evaporating, redispersed in deionized water, and stored at 4°C for later use [15].

2.1.9 Carbon black nanomaterials

Carbon black (CB) is produced by the reaction of a hydrocarbon fuel such as gas or oil with a limited supply of combustion air at temperatures of 1320 to 1540°C. The hydrocarbons which were degraded from polyethylene (PE) or high density polyethylene (HDPE) at the pyrolysis step were injected into decomposing chamber. They were introduced to pass through dc-plasma jet, and were decomposed into the carbon particles. The carbon particles were cooled down in the stream of nitrogen and they were deposited on the surface of outer graphite chamber after decomposition by the plasma jet. As-synthesized carbon black samples were characterized by the analytical instrument without further purification in the case of carbon black synthesis. Two major processes are the oil furnace process and the thermal process. The oil furnace process accounts for about 90 percent of production, and the thermal, about 10 percent. Two other processes are, the lamp for production of lamp black and the cracking of acetylene to produce acetylene black. However, these are small-volume specialty black operations that constitute less than 1 percent of total production in this country [16].

2.1.10 Nanodiamonds

For the nanodiamond synthesis the graphitic C_3N_4 (g-C_3N_4) used for the starting material which prepared by a benzenethermal reaction between $C_3N_3Cl_3$ and $NaNH_2$ at 220°C for 12 hours. For the synthesis of the C_3N_4, 1.10 g (6.0 mmol) $C_3N_3Cl_3$(1,3,5-trichlorotriazine) and 0.70 g (18.0 mmol) $NaNH_2$ (sodium amide) powders were put into a 50 mL teflon-lined autoclave, which was then filled with benzene up to 90% of the total volume. The autoclave was sealed and maintained at 180–220°C for 8–12 h, then allowed to cool to room temperature naturally. The mixed product was washed three times with distilled water, acetone and again distilled water to remove NaCl impurities, some organic-like impurities. The g-C_3N_4 obtained in such a way is a light yellowish brown powder of amorphous-like, poorly crystalline particles [8]. The resulting yellow powders was dried in vacuum at 50°C for several hours. The sample was compressed to a desired pressure at room temperature, heated to 800–2000°C for 5–30 min, and then quenched and decompressed to ambient condition [17].

2.1.11 Magnetic nanoparticles

Magnetic nanoparticles (MNP) were prepared by chemical co-precipitation and then processed under hydrothermal conditions. Briefly, iron (II) chloride and iron (III) chloride (1:2) were chemically precipitated at room temperature (25°C) by adding 30% ammonium hydroxide at pH=10.0–10.4. The precipitates were heated at 80°C for 35 minutes with continuous stirring and washed in deionized water and ethanol [18].

3. Result and discussion

3.1 Graphdiyne

Graphdiyne (GDY) is a new two-dimensional all-carbon allotrope composed of benzene rings and alkyne unites.

The carbon based nanomaterials are usually used to build electrochemical biosensors because of their physical and chemical properties. According to conventional carbon nanomaterials, GDY possesses richer carbon chemical bonds, which are of great importance for their practical applications. More importantly, GDY has a typical 2D structure similar to graphene, but also has the properties of three-dimensional materials such as a hard carbon network and uniformly distributed pores that can greatly increase active bonding areas [19, 20].

Figure 2 illustrates surface characterization of GDY [21].

As can be seen from this figure it is clear that GDY has a porous structure which is very important in sensor design to the effective diffusion of the analyte to the sensor surface.

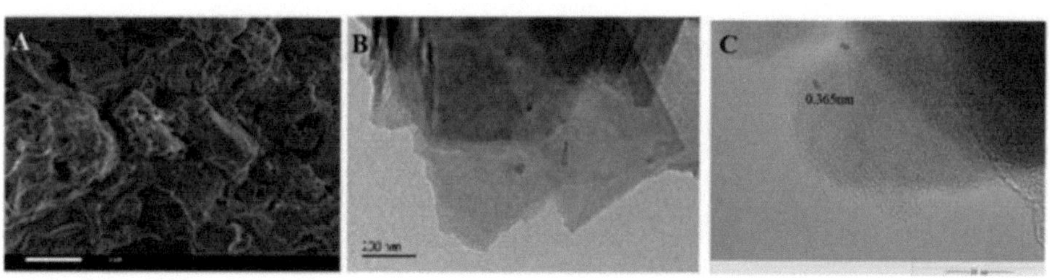

Figure 2.
A) SEM, B) TEM and C) HRTEM of GDY.

There are studies in which GDY has been used in the preparation of electrochemical enzyme biosensors [21], for microRNA testing [22] and in the determination of bacterium [23]. GDY was investigated as matrix for tyrosinase (a model enzyme) immobilization to create a mediator-free GDY based biosensor for rapid detection of bisphenol A (BPA). In this study between different carbon nanomaterial based biosensors including carbon nanotube and graphene was compared and it was reported, GDY-based tyrosinase biosensor performed better analytical for BPA detection than CNTs and graphene-based biosensors [21]. A new photoactive material has been synthesized that integrates the properties of MoS_2 and GDY to implement ultra-sensitive detection of microRNA [22]. Controllable synthesis of two-dimensional graphite nanosheet (GDY NS) is of great importance for the clinical diagnosis and treatment of tuberculosis [23].

It is though that as a new promising 2D all-carbon nanomaterial after graphene, graphdiyne with intriguing properties would inevitably attract the general interest of scientists.

3.2 Gold nanostructures

Metal nanoparticles (NPs) such as gold and silver NPs have gained immense recognition in nanosensing and diagnostic applications [24, 25]. Therefore, ease of synthesis, versatile surface functionalization and long term stability of gold nanomaterials increases their potential as efficient detection probes [26].

Gold nanostars modified with biotin were used for streptavidin determination [27]. Sensing applications using other shapes of gold nanomaterials include the use of gold nanowires and nanocubes for detection of bacteria in human kidney infection and catechol, respectively [28, 29].

Gold nanorods have also employed as a SERS substrate where in they have achieved highly sensitive and selective detection of DNA [30].

It has been reported that the nanosensor based on gold nanorodes is highly reproducible and has excellent selectivity. It was also reported the nanosensing platform is reliable, facile, cost-effective and less labor intensive. The nanomaterial with aspect ratio tunable property can be possibly used for several biomedical applications.

Figure 3 illusrates TEM and SEM images of some kind of gold nanosructures [27–30].

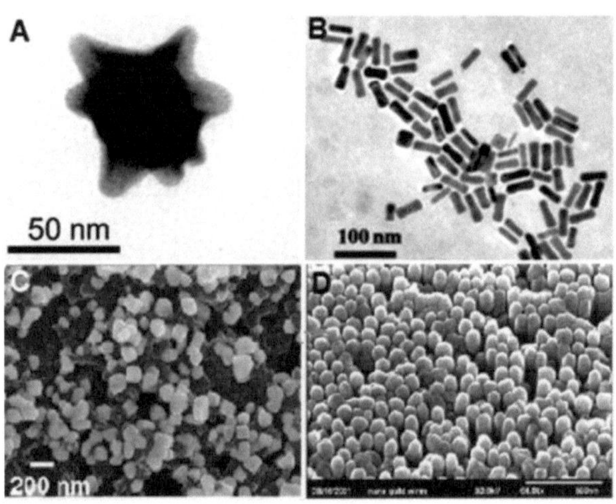

Figure 3.
TEM (A,B) and SEM (C,D) images of gold nanostar (A) [27], gold nanorods (B) [28], gold nanoparticle (C) [29] and gold nanowire (D) [30].

3.3 Inorganic novel nanomaterials

Recently, inorganic nanostructured materials have gained widespread attention as potential electrode materials of electrochemical sensors with excellent structural adjustability and other properties [31, 32].

In the past few years, binary metal oxides (denoted BMOs) are considered as one of the state-of-the-art electrocatalyst materials for various electrochemical applications [33, 34]. Among the different categories of BMOs, transition-metal phosphates/phosphides (denoted TMPs) have attracted increasing attention as a promising electrocatalyst [35–37]. Ultrathin cobalt phosphate-based modified electrode was used for the non enzymatic electrochemical determination of glucose [38]. α-zirconium phosphate (α-ZrP) based electrocatalysts have been recognized as crucial for numerous electrochemical applications [39]. The sensitive electrochemical sensing probe using the ZrP nanoplates was successfully applied for Furazolidone detection [10].

Figure 4 illustrates surface characterization of ZrP [10].

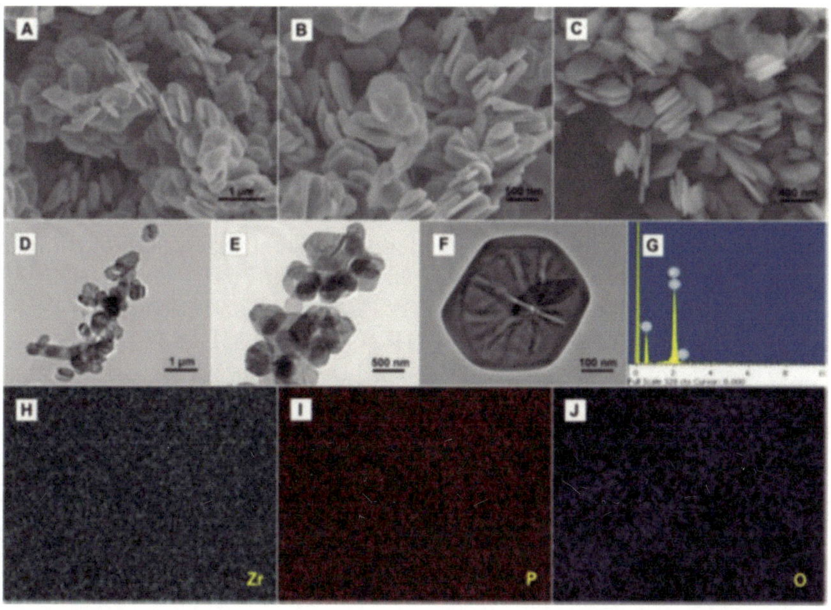

Figure 4.
(A–C) FEG-SEM image, (D–F) TEM images, (G) EDX spectrum, and (H–J) elemental mapping of ZrP.

3.4 Nanozymes

In the last decade, artificial nanomaterials, which exhibit properties similar to enzymes, have been shown as highly stable and low-cost alternatives to enzymes in electrochemical biosensing.

Nanozymes, combining the advantages of chemical catalysts and enzymes [40, 41], outperform natural enzymes because they are usually synthetized using low-cost, simple, and mass-production methods and offer high operational stability and self-life, robust catalytic performance [42–45]. Moreover, the smooth surface modification of nanomaterials provides more room for modifications than the natural enzymes. In addition, their inherent nanomaterial properties impart them both tunable and tenable catalytic activity [46, 47].

Figure 5 illustrates the schematic presentation of the enzyme-based and nanozyme-based immunoassay.

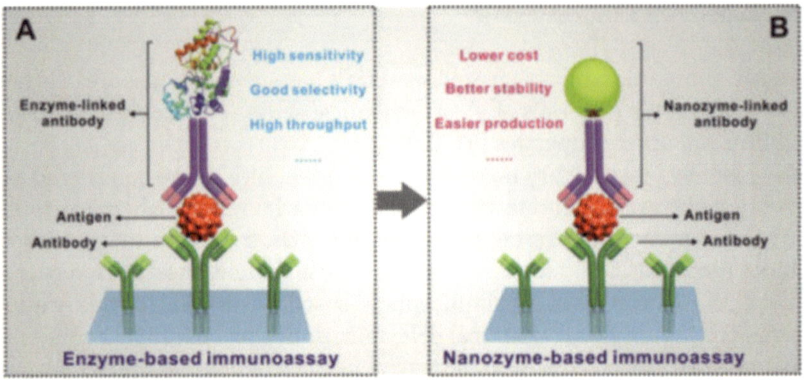

Figure 5.
Comparison of (A) natural enzyme-based immunoassays and (B) nanozyme-based immunoassays [46].

The lack of selectivity of nanozymes is compensated for by using specific bioreceptors. However, it is important to be aware of the current lack of bio-ligands for emerging analytes and that their use compromises both stability and the low cost of nanozymes [48].

Affinity ligand-based electrochemical biosensors using nanozymes have been successfully developed and exhibit some excellent merits such as higher selectivity and sensitivity, lower cost, shorter detection time, and better signal readout [49].

Nanozymes, being a special type of nanomaterial, can be exploited in electrochemical affinity biosensing as electrode modifiers, nanocarriers, and/or catalytic labels. These multi functional nanozymes, which include PtNPs/CoTPP/rGO [49], Pd/APTES-MCeO_2-GS [50], rGO-NR-Au@Pt [51], Mn_3O_4 and Pd@Pt nanoflowers [52], Fe_3O_4/PDDA/Au@Pt [53], MWCNTs/ GQDs [54, 55], and FeS_2-AuNPs [56], have been decorated with detector antibody (Ab_2) [49, 50], detector antibody (Ab_2) + HRP [51, 54, 55], AuNPs + Ab_2 [56], detector aptamer (Apt_2) + HRP [47], or (Apt_2) + HRP + G-quadruplex/hemin DNAzyme [46]. It is important to note that these nanozymes are often dressed with the natural enzyme to further enhance the sensitivity [51, 54, 55].

The combination of nanozyme-based electrochemical affinity biosensors with personalized equipment such as smartphones and/or portable low-cost devices will also be exciting to move forward in point-of-care testing. This nanozymes development to achieve catalytic activity and efficiency comparable or even better than natural enzymes will bring a revolution to conventional electrochemical biosensing and more practical applications in other expectation fields.

3.5 Hybrid nanocomposites

Hybrid sensing materials, which are organized by interaction of organic molecules onto inorganic supports, have been developed as a novel and hopeful class of hybrid sensing probes. Magnetic silica hybrid rather than other hybrid materials such as polymer, titania, and selfassembled monolayers [57–60] provides low toxicity, simple separation via external magnetic field, stability, biocompatibility and thermally stable advantages [61–63].

Biosensors prepared using hybrid materials were used to detect biological materials by thermal, electrical or optical signals. Examples of various applications of biosensors can be mentioned as environmental monitoring [64], forensic science [65–67], water characteristic testing [68], defense and the military [69], biomedicine, food industry and medical diagnosis [70].

Magnetic silica hybrids were reported as fluorescent, colorimetric, electrochemical and Surface-enhanced raman spectroscopy (SERS) sensing probes [71]. Inorganic mesoporous material is one of the best materials as molecular catalysts due to its thermal stability, easy production and modification. It used in the fields of biomedicine, electronics, and physicochemistry. Silica-coated Fe_3O_4 nanoparticle (Fe_3O_4@SiO_2 NPs), have good excellent conductivity, electrochemical transducers, biocompatibility, catalytic activity, separation ability and low toxicity properties to produce "electronic wires" to increase the electron transfer between redox centers and electrode surfaces in proteins [72].

Figure 6 illustrates the surface characterization of Fe_3O_4@ SiO_2 nanoparticles performed by transmission electron microscopy (TEM) [57].

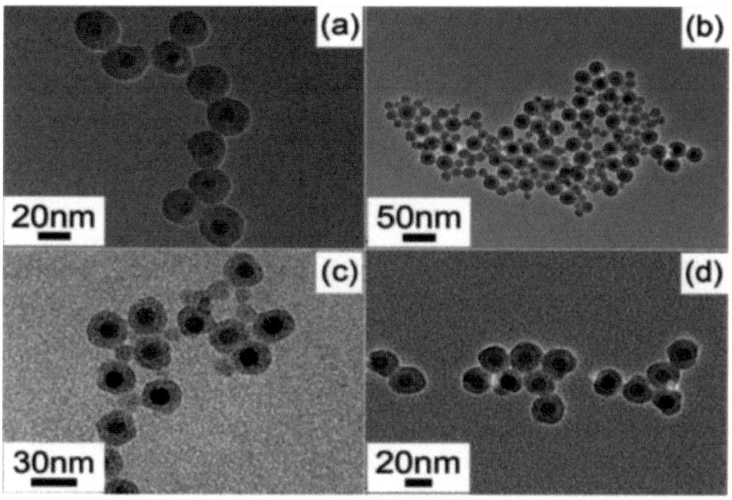

Figure 6.
TEM images of Fe_3O_4@ SiO_2 NPs with Fe_3O_4 sizes of (a) 8.8 nm and (b–d) 12.2 nm.

3.6 DNA nanomaterials

DNA nanomaterials have been widely used in bioassays due to their promising properties for sensitive and specific detection of biomolecules. The electrochemical biosensor has received greater attention in clinical diagnosis due to its high sensitivity, easy controllability and low cost [72]. For this reason, the biomolecular recognition and signal amplification based on electrochemical platform to achieve miRNAs detection still need to be considered.

In recent years, legion nucleic acid nanostructures have been applied to biological detection, including DNA tetrahedron, DNA gels, DNA dendrimers, and so on [73–75]. Y-shaped DNA (Y-DNA), as a constant nanostructure with high selectivity, provides an effective method for completely measuring target molecules [76]. Y-DNA consists of three oligonucleotides that are partially hybridized to each other. Some older biosensors used this feature to perform DNA detections where one DNA stand is fixed to the surface, another DNA stand and target DNA are added to form a specific structure [77].

Numbers of signal amplification strategies have been developed, including hybridization chain reaction (HCR), strand displacement amplification (SDA), catalytic hairpin assembly (CHA) and rolling circle amplification (RCA) [78–81]. The HCR consists of a trigger sequence and two partially complementary hairpin probes. Once triggered, the two hairpin probes can autonomously hybridize continuously [82].

Compared to HCR, this reaction consists of more complex components, including a trigger sequence, two double stranded substrates with bridging loops in the middle, and two helper sequences [83]. Thus, non-linear HCR can achieve higher rates of amplification and molecular weights [84].

To join non-linear HCR and Y-DNA nanostructures, the Y-DNA's terminals were designed as triggers that could initiate the amplification reaction. As a result, the new biosensing method can provide high-precision and selective detection of biological molecules. An unlabeled DNA nanostructured electrochemical biosensor was designed to detect miRNA-25, which is reported to be a potential molecular biomarker for non-small cell lung cancer and heart failure [85, 86].

Expanding the application of DNA nanomaterials to bioassays in the future may enable early and effective detection of various diseases.

3.7 DNAzyme

DNAzymes are single-stranded (ss) DNA sequences are able to catalyze a number of reactions, including cleavage of the phosphodiester backbone at a ribonucleotide or deoxyribonucleotide site [87]. It has been shown that metal ions play an important role in the catalytic process and are essential for the catalytic activity of most known DNAzymes [88].

The ability to select a DNAzyme with metal ion specific activity without previous chemical knowledge of the DNAzyme structure, and then to subsequently modify DNAzyme binding arms and other insignificant nucleotides with minimal to no effect on sensitivity and selectivity has made DNAzymes ideal metal-selective components for new metal ion sensing technologies. RNA-cleaving DNAzyme is a very useful biomaterial for the determination of metal ions, but some parts of DNAzymes can be cleaved by several metal ions, which makes different concentrations of metal ions difficult to distinguish [89].

In the last two decades, the rapid development of nanomaterials and biomaterials [90] offers more opportunities to improve electrochemical sensor performance. For the determination of Cu (II) and Hg (II), many highly sensitive sensors are manufactured using small molecules, peptides, proteins and antibodies at low cost.

The ligand sites of proteases composed of nitrogen, oxygen or sulfur can combine with heavy metal ions to form a stable complex [91]. Cu(II) is a small ion that has to be chelated first and then bind to the antibody recognition [92]. Both antibody and enzyme work best under physiological conditions that limit application in real environment. DNA is not only the genetic material of most living organisms, but also an excellent biological functional material [93].

Metal ions can be specifically bound with a single-stranded DNA to form a stable metal-mediated DNA, and this mechanism is applied to detect metal ions [94, 95]. Therefore, numerous studies have focused on the newly discovered biosensor using different DNA-based aptamers functionalized with nanomaterials to increase sensitivity. DNAzymes that break down RNA as DNA-based catalysts are obtained through in vitro selection, which turned out to be a very useful platform for the identification of metal ions. After binding with heavy metal ions, many biochemical and biophysical studies have been conducted on DNAzymes due to their high metal ion selectivity and high catalytic efficiency [96]. Therefore, DNAzymes have been applied in various biosensors (colorimetric, electrochemical and fluorescent) that realize the detection of various metal ions such as Mg(II) [97], Ag(I) [98], Pb(II) [99], Zn(II) [100], Hg(II) [101], UO_2(II) [102].

The field of DNAzyme-based metal ion sensing is continuing to develop for future cellular and portable detection technologies.

3.8 Carbon nanodots

Carbon dots (CDs) are nanomaterials less than 10 nm in size and became the new potential material for the electrode modifier [103]. Formerly, CDs have been applied in electrochemical sensing platforms, mainly focusing on their electrocatalytic properties toward analytes of interest [104, 105] rather than electrode modifiers. Thus, the studies on carbon dots owing a noticeable potential to be used as electrode modifiers in electrochemical techniques to increase the sensitivity of the electrochemical sensor has been exploited.

Recently, a new member of CDs, have gained attention because of their water solubility, fine properties, high luminescence, low cytotoxicity and good conductivity [106]. Depending on the precursors employed in their synthesis, CNDs are surrounded by different functional groups including, among others, hydroxyl, amide groups and carboxyl which facilitate the immobilization of biomolecules. Hence, due to their ability to be modified with a wide variety of biomolecules, and in conjunction with the excellent properties mentioned above, CNDs have been employed in many biological applications such as solar cell development and photocatalysis [107, 108]. Concerning the employment of CNDs for electrochemical biosensors, it should be highlighted that despite the previously mentioned advantages, very few attempts to incorporate CNDs into electrodes are reported. Reporting the application of CNDs in electrochemical sensors are focused on the electrocatalytic properties of this nanomaterial toward oxygen reduction [109], biomedical application [110], exploited for glucose biosensing [111] and DNA sensing [112].

Transmission electron microscopy (TEM) of carbon nanodots in different scale from 20 nm to 2 nm are illustrate in **Figure 7** [110].

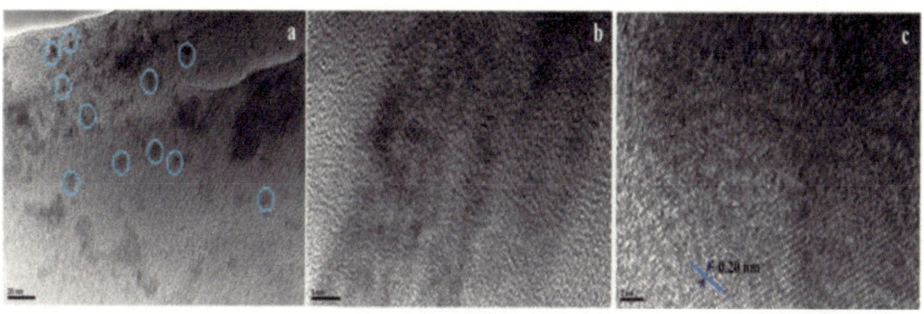

Figure 7.
High resolution transmission electron microscopic images of fish scale derived carbon nanodots (a-c).

3.9 Carbon black nanomaterials

Since the discovery of carbon nanotubes, carbon-based nanomaterials being researched in various disciplines including electrochemistry. An old and cost-effective material recently called carbon black (CB) reinvented. CB has good electrical conductivity, dispersible in solvents, possibility of easy functionalization and has a large number of defect areas and fast electron transfer kinetics [113–116].

Previously, CB's main application in the electrochemical field was based on the design of sensors for analyte detection in fuel cell and gas phases for lithium and sodium batteries [117, 118]. However, until 2009, only a few CB-based electrochemical sensors were reported for analyte detection.

Among nanomaterials, CB demonstrated high potential in customizing all from the oldest carbon paste to glassy carbon and printed electrodes thanks to their fascinating electrochemical properties combined with cost effectiveness.

One of the main properties of CB is its ability to produce easily stable dispersions in a variety of solvents such as ethanol, acetonitrile, a mixture of dimethylformamide water [119], chitosan [120], or di hexadecylphosphate water solution [121], usually at a concentration of 1 mg/mL.

CB is widely used in the design of biosensors with a variety of biological recognition elements including enzymes, DNA and antibodies. The main potential of the enzyme combination with CB is based on the outstanding advantages this nanomaterial has in enhancing the biosensor sensitivity. CB can increase both conductivity and enzyme loading areas, thus causing increased signals and hence higher sensitivity. Some examples have shown that CB is a compatible substrate for the immobilization of enzymes in the design of amperometric biosensors [122].

Immunosensors have attracted great attention for specific, sensitive, cost-effective and in-field analysis. Examples of CB-based immunosensors in unlabeled configuration have been reported in the literature [123, 124].

Alongside traditional bioreceptors such as enzymes, antibodies, and nucleic acids, CB also demonstrated the ability to improve their analytical performance by combining with alternative biological recognition elements or molecularly imprinted polymers [125].

Besides the biosensor application, CB was used in sensor design for both single analyte detection and multiple analysis, showing increased sensitivity thanks to its high conductivity, number of defective areas and surface area. Nowadays, most CB-based detection systems are mainly sensors, but in recent years there has been a sharp increase in publications in the development of enzymatic, immuno, and DNA biosensors [126].

CB is a new generation material due to its environmentally friendly properties in terms of costs and environmental impact.

The morphological properties of the synthesized carbon black by using commercial and waste polystyrene (PS) and high density polyethylene in different pyrolysis conditions were illustrated in **Figure 8** [16].

Figure 8.
FE-SEM images of the carbon black obtained from: A commercial PS pyrolyzed at 500°C; b commercial PS pyrolyzed at 900°C; c waste PS pyrolyzed at 500°C; d waste PS pyrolyzed at 900°C; e high density polyethylene (HDPE) pyrolyzed at 500°C; f HDPE pyrolyzed at 900°C.

3.10 Nanodiamonds

Nanodiamonds (ND), a new member of the carbon nanoparticle class, has recently received much attention in drug delivery, bio-imaging, and biosensor applications due to its physical and chemical properties [127].

Nanodiamond (ND) is of great interest in various fields of material science due to its various functional groups. An electrochemical biosensor containing copper, nano-diamond (ND) and carbon nanotube (CNT) was built to detect the amino acids of Parkia Seeds (PS). Electrochemical reaction of PS was carried out with composite electrodes prepared using nanodiamond [128].

The AFM and SEM characterization of nanocrystalline diamond (NCD) and boron doped nanocrystalline diamond (BDND) were illustrated in **Figure 9** respectively [129, 130].

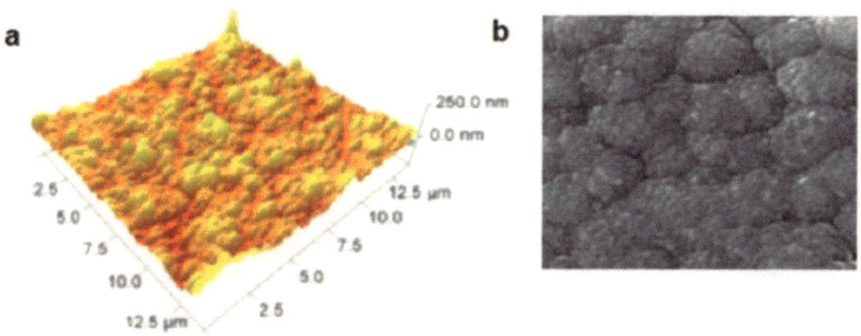

Figure 9.
(a) AFM topographic images of NCD films and (b) SEM image of BDND film grown on a Si substrate.

3.11 Magnetic nanoparticles

Nanomaterials provide high surface areas and a biocompatible environment for enzyme loading. In the last decade, research of magnetic particles has resulted in their use in a large number of nano-sensing devices, providing ease of separation in solution.

Various iron magnetic nanoparticles (MNPs) have proven to be an excellent nanomaterial for electrochemical biosensing applications due to their electroconductivity, biocompatibility and ease of synthesis properties. They make important contributions to the development of electrochemical nanobiosensors. Functionalized magnetic nanoparticles can be directed by the external magnetic field to site-specific drug delivery targets. Iron and iron oxide nanoparticles have been studied as signal amplification elements in biosensing [131]. Among these materials, magnetite (Fe_3O_4), a Fe^{2+} and Fe^{3+} complex oxide, is one of the most studied super paramagnetic nanoparticles. It has unique mesoscopic mechanical and physical properties and has many potential applications in various fields such as cell separation [132] and microwave absorption [133]. Fe_3O_4 nanoparticles have been widely used for in vivo examination [134]. The direct binding of cholesterol oxidase to Fe_3O_4 magnetic nanoparticles was investigated and the kinetic behavior, stability and activity of bound cholesterol were investigated [135]. Due to its easy preparation process, low toxicity, strong superparamagnetism and good biocompatibility, Fe_3O_4 has recently been used in biosensors for glucose, ethanol and acetaminophen. Prepared biosensors showed fast response and high sensitivity with a wide linear range [136, 137]. Fe_3O_4 - Au nanoparticles, have been successfully used for the first time in the dual-mode detection of carcinoembryonics antigens (CEA) and have correctly confirmed the presence of antigens [138].

Figure 10 illustrates TEM images of Fe_3O_4, Au and Fe_3O_4-Au nanoparticles [138]. **Table 1** illustrates the studies based novel nanomaterials.

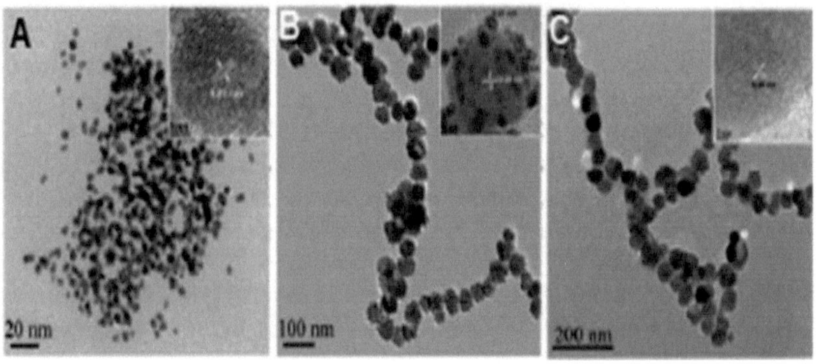

Figure 10.
TEM images of (A) Fe_3O_4, (B) Au and (C) Fe_3O_4–Au nanoparticles; the corresponding HRTEM images are inserted.

Nanomaterial	Analyzed	Detection limit	Linear range	Method	Ref.
Graphdiyne	Bisphenol A	1,0 × 10^{-7}- 3,5 × 10^{-6} mol/L	24 nmol/L	CV[1]	[21]
Hybrid Nanocomposite	Metronidazole	0,001–2444 μM	0,8 nM	CV and EIS[2]	[34]
Inorganic nanomaterial	Furazolidone (FZD)	0,009–339 μM	1,2 nM	CV, EIS and Amperometry	[10]
Noble metal nanoparticles	Alpha fetoprotein (AFP	0,1 pg./mL to 50 ng/mL	0,033 pg./mL	CV	[50]
Bimetallic Pt-Au/multi-walled carbon nanotubes	Organophosphorous pesticides	50 to 500 nmol/L	29,7 nmol/L	CV, Amperometric i-t curve and EIS	[64]
Quantum dots	Dopamine	0,375–450 μM	100 nM	Electro-chemiluminescence	[66]
DNAzyme-functionalized single-walled carbon nanotubes	Cu(II) and Hg(II)	Cu(II) 0,01–10,000 nM Hg(II) 5–10,000 nM	Cu(II) 6,7 pM Hg(II) 3,43 nM	EIS	[89]
DNAzyme Functionalized Single-Walled Carbon Nanotube	Silver Ion	10 pM to 10^6 pM	5 pM	UV–Vis Spectrometry	[98]
Carbon nanodots	Gene mutation	0,001–20 μM.	0,16 nM	CV and DPV[3]	[112]
Carbon-coated nickel magnetic nanoparticles	Acetaminophen	2,0 × 10^{-6} to 2,3 × 10^{-4} mol/L.	6,0 × 10^{-7} mol/L	DPV	[137]
Carbon black	Bisphenol A	0,03 μ M	0,1–0,9 μM 1–50 μM	SVW[4]	[139]

Nanomaterial	Analyzed	Detection limit	Linear range	Method	Ref.
3D DNA nanonet structure	MicroRNA	36,083 fM	10 fM-1 nM	CV, DPV and EIS	[140]
Carbon nanodot	17ß-Estradiol	$0,5 \times 10^{-12}$ M	$1,0 \times 10^{-7}$ - $1,0 \times 10^{-12}$ M	CV and EIS	[141]
Carbon black	Photosynthetic herbicide	0,1–5 mu M	1 nM	Amperometric measurement	[142]
Metal-polymer hybrid nanomaterial	Human papillomavirus	1–100 pg. mu/L	2,74 pg. mu/L	CV and EIS	[143]
Nanozymes (magnetic metal organik framework)	Hydrogen peroxide (H_2O_2)	5 mu M-120 mM	0,9 mu M	CV, EIS and Amperometry	[144]
Gold nanorod	Aflatoxin	0,25–10 ng/mL	0,11 ng/mL	SPR[5]	[145]
Nanodiamond	Urea	0,1–0,9 mg/mL	0,005 mg/mL	Direct current voltage	[146]
Carbon dots, chitosan, gold nanoparticles	Patulin	1×10^{-12} - 1×10^{-9} mol/L	$7,57 \times 10^{-13}$ mol/L	CV and DPV	[147]

1: Cyclic voltammetry, 2: Electrochemical impedance spectrometry, 3: Differential pulse voltammetry 4: Square wave voltammetry, 5: Surface plasmon resonance.

Table 1.
Biosensor applications based navel nanomaterials.

4. Conclusion

Nanomaterials offer significant advantages, especially in sensor technology, due to their large surface area. When biocompatible nanomaterials are used as biorecognition layers, it enables the design of highly sensitive biosensors. Many nanomaterials, which are widely used today, are now being replaced by novel nanomaterials due to their physical stability, easy synthesis, easy fabrication, and cheapness.

Nanomaterials became important components in bioanalytical devices since they clearly increase the performances in the sense of detection limits and sensitivity down to single molecules detection.

Over content of this chapter aims to evaluate developments in the fields of new nanomaterial-based biosensors. Their production and potential applications for the direct and reliable detection of bioanalytes are described. In addition, research interests for the production of nanomaterial-based biosensors were encouraged with examples.

Author details

Kübra Gençdağ Şensoy[1] and Mihrican Muti[2*]

1 Department of Food Processing, Köşk Vocational High School, Aydın Adnan Menderes University, Aydin, Turkey

2 Department of Chemistry, Faculty of Arts and Sciences, Aydın Adnan Menderes University, Aydin, Turkey

*Address all correspondence to: mihricanmuti@adu.edu.tr

References

[1] Wu, L., Xiong, E. H., Zhang, X., Zhang, X. H., & Chen, J. H. (2014). Nanomaterials as signal amplification elements in DNA-based electrochemical sensing. Nano Today, 9(2), 197-211.

[2] Michalet, X., Pinaud, F. F., Bentolila, L. A., Tsay, J. M., Doose, S., Li, J. J., Weiss, S. (2005). Quantum dots for live cells, in vivo imaging, and diagnostics. Science, 307(5709), 538-544.

[3] .Cheng, L., Wang, X. W., Gong, F., Liu, T., & Liu, Z. (2020). 2D nanomaterials for cancer theranostic applications. Advanced Materials, 32(13), 1902333.

[4] Tsou, Y. H., Zhang, X. Q., Zhu, H., Syed, S., & Xu, X. Y. (2018). Drug delivery to the brain across the blood–brain barrier using nanomaterials. Small, 14(25).

[5] Zhu, W., Chen, Z., Pan, Y., Dai, R. Y., Wu, Y., Zhuang, Z. B., Li, Y. D. (2019). Functionalization of hollow nanomaterials for catalytic applications: Nanoreactor construction. Advanced Materials, 31(38).

[6] Xin, Q., Shah, H., Nawaz, A., Xie, W. J., Akram, M. Z., Batool, A., Gong, J. R. (2019). Antibacterial carbon-based nanomaterials. Advanced Materials, 31(45), 1804838.

[7] Chen, Y., Zhou, S. W., Li, L. L., & Zhu, J. J. (2017). Nanomaterialsbased sensitive electrochemiluminescence biosensing. Nano Today, 12, 98-115.

[8] Arduini, F., Cinti, S., Mazzaracchio, V., Scognamiglio, V., Amine, A., Moscone, D. (2020). Carbon black as an outstanding and affordable nanomaterial for electrochemical (bio)sensor design. Biosensors and Bioelectronics, 156, 112033.

[9] Nikoobakht, B., El-Sayed, M. A. (2003). Preparation and growth mechanism of gold nanorods (NRs) using seed-mediated growth method. Chem. Mater., 15, 1957-1962.

[10] Kokulnathan, T., Wang, T. J., Kumar, E. A., Suvina, V., Balakrishna, R. G. (2020). Development of an Electrochemical Platform Based on Nanoplate like Zirconium Phosphate for the Detection of Furazolidone. ACS Appl. Nano Mater., 3, 4522-4529.

[11] Mirhosseini, M., Shekari-Far, A., Hakimian, F., Haghiralsadat, B. F., Fatemi, S. K., Dashtestani, F. (2020) Core-shell Au@Co-Fe hybrid nanoparticles as peroxidase mimetic nanozyme for antibacterial application, Process Biochemistry 95, 131-138.

[12] Fernandes, P. M. V., Campiña, J. M., Silva, A. F. (2020). A layered nanocomposite of laccase, chitosan, and Fe3O4 nanoparticles-reduced graphene oxide for the nanomolar electrochemical detection of bisphenol A. Microchimica Acta, 187, 262.

[13] Zhoua, L., et al. (2019). A label-free electrochemical biosensor for microRNAs detection based on DNA nanomaterial by coupling with Y-shaped DNA structure and non-linear hybridization chain reaction. Biosensors and Bioelectronics, 126, 657-663.

[14] Liu, M., Zhang, Q., Brennan, J. D., Li, Y. (2018). Graphene-DNAzyme-based fluorescent biosensor for *Escherichia coli* Detection. MRS Communications, 8, 687-694.

[15] Zhu, S., Meng, Q., Wang, L., Zhang, J., Song, Y., Jin, H., Zhang, K., Sun, H., Wang, H., Yang, B. (2013). Highly photoluminescent carbon dots for multicolor patterning, sensors, and bioimaging. Angew. Chem. Int. Ed. Engl., 52, 3953-3957.

[16] Guo, X. F., Kim, G. J. (2009). Synthesis of Ultrafine Carbon Black

by Pyrolysis of Polymers Using a Direct Current Thermal Plasma Process. Plasma Chemistry and Plasma Processing, 30, 75-90.

[17] Fang, L., Ohfuji, H., Irifune, T. (2013). A Novel Technique for the Synthesis of Nanodiamond Powder, Journal of Nanomaterials, 2013, 201845,

[18] Cornell, R. M., Schwertmann, U. (2006). The iron oxides structures, properties, reactions, occurrences and uses. 2nd, Completely Revised and Extended Edition, p 139-146.

[19] Li, Y., Guo, C., Li, J., Liao, W., Li, Z., Zhang, J., et al. (2017). Pyrolysis-induced synthesis of iron and nitrogen-containing carbon nanolayers modified graphdiyne nanostructure as a promising core-shell electrocatalyst for oxygen reduction reaction. Carbon, 119, 201-210.

[20] Wang, K., Wang, N., He, J., Yang, Z., Shen, X., Huang, C. (2017). Preparation of 3D architecture graphdiyne nanosheets for high-performance sodium-ion batteries and capacitors. ACS Appl. Mater, Interfaces 9 (46), 40604-40613.

[21] Wu, L., Gao, J., Lu, X., et al. (2020). Graphdiyne: A new promising member of 2D all-carbon nanomaterial as robust electrochemical enzyme biosensor platform, Carbon, 156, 568-575.

[22] Li, X., Li, Y., Zhang, J., et al. (2019). Molybdenum disulfide/graphdiyne-based photoactive material derived photoelectrochemical strategy for highly sensitive MicroRNA assay. Sensors and Actuators B-Chemical, 297, 126808.

[23] Chang, F., Huang, L., Guo, C., et al. (2019). Graphdiyne-Based One-Step DNA Fluorescent Sensing Platform for the Detection of *Mycobacterium tuberculosis* and Its Drug-Resistant Genes. ACS Applied Materials & Interfaces, 11(39), 35622-35629.

[24] Sau, T.K., Rogach, A.L., J€ackel, F., Klar, T.A., Feldmann, J. (2010). Properties and applications of colloidal nonspherical noble metal nanoparticles. Adv. Mater, 22(16), 1805-1825.

[25] Doria, G., Conde, J., Veigas, B., Giestas, L., Almeida, C., Assunç~ao, M., Rosa, J., Baptista, P. V. (2012). Noble metal nanoparticles for biosensing applications. Sensors, 12(2),1657-1687.

[26] Emrani, A.S., Danesh, N.M., Lavaee, P., Ramezani, M., Abnous, K., Taghdisi, S.M. (2016). Colorimetric and fluorescence quenching aptasensors for detection of streptomycinin blood serum and milk based on double-stranded DNA and gold nanoparticles. Food Chem., 190, 115-121.

[27] Dondapati, S.K., Sau, T.K., Hrelescu, C., Klar, T.A., Stefani, F.D., Feldmann, J. (2010) Label free biosensing based on single gold nanostars as plasmonic transducers. ACS Nano, 4 (11), 6318-6322.

[28] Basu, M., Seggerson, S., Henshaw, J., Jiang, J., del A. Cordona, R., Lefave, C., Boyle, P.J., Miller, A., Pugia, M., Basu, S. (2004). Nano-biosensor development for bacterial detection during human kidney infection: use of glycoconjugate-specific antibody bound gold nano wire arrays (GNWA). Glycoconj. J. 21(8-9), 487-496.

[29] Karim, M.N., Lee, J.E., Lee, H.J. (2014). Amperometric detection of catechol using tyrosinase modified electrodes enhanced by the layer-by-layer assembly of gold nanocubes and polyelectrolytes. Biosens. Bioelectron, 61, 147-151.

[30] Parab, H.J., Jung, C., Lee, J.-H., Park, H.G. (2010). A gold nanorod-based optical DNA biosensor for the diagnosis of pathogens, Biosens. Bioelectron. 26 (2), 667-673.

[31] Kokulnathan, T., Chen, S. M. (2019). Rational Design for the Synthesis of

Europium Vanadate-Encapsulated Graphene Oxide Nanocomposite: An Excellent and Efficient Catalyst for the Electrochemical Detection of Clioquinol. ACS Sustainable Chem. Eng., 7, 4136-4146.

[32] Li, X., Zhu, J., Wei, B. (2016). Hybrid nanostructures of metal/ two dimensional nanomaterials for plasmon-enhanced applications. Chem. Soc. Rev., 45, 3145-318.

[33] He, P., Huang, Q., Huang, B., Chen, T. (2017). Controllable synthesis of Ni–Co–Mn multi-component metal oxides with various morphologies for high-performance flexible supercapacitors. RSC Adv., 7, 24353-24358.

[34] Kokulnathan, T.; Chen, S. M. Praseodymium Vanadate Decorated Sulfur-doped Carbon Nitride Hybrid Nanocomposite: The Role of Synergistic Electrocatalyst for the Detection of Metronidazole. ACS Appl. Mater. Interfaces 2019, 11, 7893-7905.

[35] Lin, R.; Ding, Y. A review on the synthesis and applications of mesostructured transition metal phosphates. Materials 2013, 6, 217-243.

[36] Liu, J., Meyns, M., Zhang, T., Arbiol, J., Cabot, A., Shavel, A. (2018). Triphenyl phosphite as the phosphorus source for the scalable and cost-effective production of transition metal phosphides. Chem. Mater., 30, 1799-1807.

[37] Zhang, Y., Xiao, J., Lv, Q., Wang, S. (2018). Self-supported transition metal phosphide based electrodes as high-efficient water splitting cathodes. Front. Chem. Sci. Eng., 12, 494-508.

[38] Tomanin, P. P., Cherepanov, P. V., Besford, Q. A., Christofferson, A. J., Amodio, A., McConville, C. F., Yarovsky, I., Caruso, F., Cavalieri, F. (2018). Cobalt Phosphate Nanostructures for Non- Enzymatic Glucose Sensing at Physiological pH. ACS Appl. Mater. Interfaces, 10, 42786-42795.

[39] Pessoa, C. A., Gushikem, Y., Kubota, L. T., Gorton, L. (1997). Preliminary electrochemical study of phenothiazines and phenoxazines immobilized on zirconium phosphate. J. Electroanal. Chem., 431, 23-27.

[40] Singh, S. (2019). Nanomaterials exhibiting enzyme-like properties (nanozymes): current advances and future perspectives. Front Chem., 7, 46.

[41] Jiang, B., Fang, L., Wu, K., Yan, X., Fan, K. (2020). Ferritins as natural and artificial nanozymes for theranostics. Theranostics, 10, 687-706.

[42] Niu, X., Cheng, N., Ruan, X., Du, D., Lin, Y. (2020). Review—Nanozyme-based immunosensors and immunoassays: recent developments and future trends. J Electrochem Soc., 167, 037508.

[43] Liu, J. (2019). Special topic: nanozyme-based analysis and testing. J Anal Test, 3, 189-190.

[44] Liu, B., Liu, J. (2017). Surface modification of nanozymes. Nano Res., 10, 1125-1148.

[45] Chatterjee, B., Das, S. J., Anand, A., Sharma, T. K. (2020) Nanozymes and aptamer-based biosensing. Mater Sci Technol., 3, 127-135.

[46] Sun, D., Line, X., Lu, J., Wei, P., Luo, Z., Lu, X., et al. (2019). DNA nanotetrahedron-assisted electrochemical aptasensor for cardiac troponin I detection based on the co-catalysis of hybrid nanozyme, natural enzyme and artificial DNAzyme. Biosens Bioelectron, 142, 111578.

[47] Bazin, I., Tria, S. A., Hayat, A., Marty, J. L. (2017). New biorecognition molecules in biosensors for the

detection of toxins. Biosens Bioelectron, 87, 285-298.

[48] Zhang, X., Wu, D., Zhou, X., Yu, Y., Liu, J., Hu, N., et al. (2019). Recent progress on the construction of nanozymes-based biosensors and their applications to food safety assay. TrAC Trends Anal Chem., 121, 115668.

[49] Shu, J., Qiu, Z., Wei, Q., Zhuang, J., Tang, D. (2015). Cobalt-porphyrinplatinum- functionalized reduced graphene oxide hybrid nanostructures: a novel peroxidase mimetic system for improved electrochemical immunoassay. Sci Rep., 5, 15113.

[50] Wei, Y., Li, Y., Li, N., Zhang, Y., Yan, T., Man, H., Wei, Q. (2016). Sandwich-type electrochemical immunosensor for the detection of AFP based on Pd octahedral and APTES-M-CeO2-GS as signal labels. Biosens Bioelectron, 79, 482-487.

[51] Zhu, F., Zhao, G., Dou, W. (2018). Electrochemical sandwich immunoassay for *Escherichia coli* O157:H7 based on the use of magnetic nanoparticles and graphene functionalized with electrocatalytically active Au@Ptcore/shell nanoparticles. Microchim Acta, 185, 455.

[52] Golchin, J., Golchin, K., Alidadian, N., Ghaderi, S., Eslamkhah, S., Eslamkhah, M., Akbarzadeh, A. (2017). Nanozyme applications in biology and medicine: an overview. Artif Cells Nanomed Biotechnol., 45, 1-8.

[53] Serafín, V., Valverd,e A., Garranzo-Asensio, M., Barderas, R., Campuzano, S., Yáñez-Sedeño, P., Pingarrón, J. M. (2019). Simultaneous amperometric immunosensing of the metastasis related biomarkers IL-13Rα2 and CDH-17 by using grafted screen-printed electrodes and a composite prepared from quantum dots and carbon nanotubes for signal amplification. Microchim Acta, 186, 411.

[54] Serafín, V., Valverde, A., Martínez-García, G., Martínez-Periñán, E., Comba, F., Garranzo-Asensio, M., Barderas, R., Yáñez-Sedeño, P., Campuzano, S., Pingarrón, J. M. (2019). Graphene quantum dots functionalized multi-walled carbon nanotubes as nanocarriers in electrochemical immunosensing. Determination of IL-13 receptor α2 in colorectal cells and tumor tissues with different metastatic potential. Sensors Actuators B Chem., 284, 711-722.

[55] Zhang, L., Xie, X., Yuan, Y., Chai, Y., Yuan, R. (2019). FeS_2-AuNPs nanocomposite as mimicking enzyme for constructing signal-off sandwich-type electrochemical immunosensor based on electroactive nickel hexacyanoferrate as matrix. Electroanalysis, 31, 1019-1025.

[56] Sun, D., Line, X., Lu, J., Wei, P., Luo, Z., Lu, X., et al. (2019). DNA nanotetrahedron-assisted electrochemical aptasensor for cardiac troponin I detection based on the co-catalysis of hybrid nanozyme, natural enzyme and artificial DNAzyme. Biosens Bioelectron, 142, 111578.

[57] Ding, H., Zhang, Y., Wang, S., Xu, J., Xu, S., Li, G. (2012). Fe_3O_4@ SiO_2 core/shell nanoparticles: the silica coating regulations with a single core for different core sizes and shell thicknesses. Chem mater., 24: 4572-4580.

[58] Cheng, G., Zhang, J. L., Liu, Y. L., Sun, D. H., Ni, J. Z. (2011). Synthesis of novel Fe_3O_4@ SiO_2@ CeO_2 microspheres with mesoporous shell for phosphopeptide capturing and labeling. Chem commun., 47, 5732-5734.

[59] Shao, M., Ning, F., Zhao, J., Wei, M., Evans, D. G, Duan, X. (2012). Preparation of Fe_3O_4@ SiO_2@ layered double hydroxide core–shell microspheres for magnetic separation of proteins. J am Chem Soc., 134, 1071-1077.

[60] Wang, L., Cole, M., Li, J., Zheng, Y., Chen, Y. P., Miller, K. P., Decho, A. W., Benicewicz, B. C. (2015). Polymer grafted recyclable magnetic nanoparticles, Polym Chem., 6, 248-255.

[61] Peng, X., Wang, Y., Tang, X., Liu, W. (2011). Functionalized magnetic core–shell Fe_3O_4@ SiO_2 nanoparticles as selectivity enhanced chemosensor for Hg (II). Dyes pigments, 91, 26-32.

[62] Wang, Y., Peng, X., Sh,i J., Tang, X., Jiang, J., Liu, W. (2012). Highly selective fluorescent chemosensor for Zn2+ derived from inorganic organic hybrid magnetic core/shell Fe_3O_4@ SiO_2 nanoparticles. Nanoscale Res. Lett., 7, 86.

[63] Zhao, Y., Li, J., Zhao, L., Zhang, S., Huang, Y., W,u X., Wang, X. (2014). Synthesis of amidoxime-functionalized Fe_3O_4@ SiO_2 core–shell magnetic microspheres for highly efficient sorption of U (VI). Chem Eng., 235, 275-283.

[64] Miao, S. S., Wu, M. S., Ma, L. Y., He, H. J., Yang, H. (2016). Electrochemiluminescence biosensor for determination of organophosphorous pesticides based on bimetallic Pt- Au/ multi-walled carbon nanotubes modified electrode. Talanta, 158, 142-151.

[65] Gao, W., Wang, C., Muzyka, Kitte, S. A., Li, J., Zhang, W., Xu, G. (2017). Artemisinin-luminol chemiluminescence for forensic bloodstain detection using a smart phone as a detector. Anal Chem., 89, 6160-6165.

[66] Stewart, A. J., Hendry, J., Dennany, L. (2015). Whole blood electrochemiluminescent detection of dopamine. Anal Chem, 87, 11847-11853.

[67] Takahashi, F., Nitta, S., Shimizu, R., Jin, J. (2018). Electrochemiluminescence and voltammetry of tris (2, 2-bipyridine) ruthenium (II) with amphetaminetype stimulants as coreactants: An application to the discrimination of methamphetamine. Forensic Toxicol., 36, 185-191.

[68] Zhang, J. J., Kang, T.F., Hao, Y. C., Lu, L. P., Cheng, S.Y. (2015). Electrochemiluminescent immunosensor based on CdS quantum dots for ultrasensitive detection of microcystin- LR, Sens. Actuators B Chem., 214, 117-123.

[69] Li, G., Yu, X., Liu, D., Liu, X., Li, F., Cui, H. (2015). Label-free electrochemiluminescence aptasensor for 2, 4, 6-trinitrotoluene based on bilayer structure of luminescence functionalized graphene hybrids. Anal Chem., 87, 10976-10981.

[70] Kitte, S. A., Gao, W., Zholudov, Y. T., Qi, L., Nsabimana, A., Liu, Z., Xu, G. (2017). Stainless steel electrode for sensitive luminol electrochemiluminescent detection of H_2O_2, glucose, and glucose oxidase activity. Anal Chem., 89, 9864-9869.

[71] Bagheri, E., Ansari, L., Abnous, K., Taghdisi, S. M., Naserifar, M., Ramezani, M., Alibolandi, M. (2020). Silica -magnetic inorganic hybrid nanomaterials as versatile sensing platform. Nanomed. J., 7(3), 183-193.

[72] Xu, T., Song, Y., Gao, W., Wu, T., Xu, L.P., Zhang, X., Wang, S. (2018). Superwettable Electrochemical Biosensor toward Detection of Cancer Biomarkers. ACS Sens., 3, 72-78.

[73] Brown 3rd, C. W., Buckhout-White, S., Díaz, S. A., Melinger, J. S., Ancona, M. G., Goldman, E. R., Medintz, I. L. (2017). Evaluating Dye-Labeled DNA Dendrimers for Potential Applications in Molecular Biosensing. ACS Sens., 2, 401-410.

[74] Zhou, L., Morel, M., Rudiuk, S., Baigl, D. (2017). Intramolecularly Protein-Crosslinked DNA Gels: new biohybrid nanomaterials with controllable size and catalytic activity. Small, 13.

[75] Zhou, X., Zhao, M., Duan, X., Guo, B., Cheng, W., Ding, S., Ju, H. (2017). Collapse of DNA tetrahedron nanostructure for "Off-On" fluorescence detection of DNA methyltransferase activity. ACS Appl. Mater. Interfaces, 9, 40087-40093.

[76] Li, Y., Chang, Y., Yuan, R., Chai, Y. (2018). Highly Efficient Target Recycling-Based Netlike Y-DNA for Regulation of Electrocatalysis toward Methylene Blue for Sensitive DNA Detection. ACS Appl. Mater. Interfaces, 10, 25213-25218.

[77] Wang, K., Lei, Y., Zhong, G.X., Zheng, Y.J., Sun, Z.L., Peng, H.P., Chen, W., Liu, A.L., Chen, Y.Z., Lin, X.H. (2015). Dual-probe electrochemical DNA biosensor based on the "Y" junction structure and restriction endonuclease assisted cyclic enzymatic amplification for detection of double-strand DNA of PML/RARα related fusion gene. Biosens. Bioelectron. 71, 463-469.

[78] Peng, X., Zhu, J., Wen, W., Bao, T., Zhang, X., He, H., Wang, S. (2018). Integrated amplified aptasensor with in-situ precise preparation of copper nanoclusters for ultrasensitive electrochemical detection of microRNA 21. Biosens. Bioelectron. 118, 174-180.

[79] Miao, P., Jiang, Y., Zhang, T., Huang, Y., Tang, Y. (2018). Electrochemical sensing of attomolar miRNA combining cascade strand displacement polymerization and reductant-mediated amplification. Chem. Commun., 54, 7366-7369.

[80] Liu, C., Chen, C., Li, S., Dong, H., Dai, W., Xu, T., Liu, Y., Yang, F., Zhang, X. (2018). Target-Triggered Catalytic Hairpin Assembly-Induced Core-Satellite Nanostructures for High-Sensitive "Off-to-On" SERS Detection of Intracellular MicroRNA. Anal. Chem., 90, 10591-10599.

[81] Peng, X., Liang, W.B., Wen, Z.B., Xiong, C.Y., Zheng, Y.N., Chai, Y.Q., Yuan, R. (2018). Ultrasensitive Fluorescent Assay Based on a Rolling-Circle-Amplification-Assisted Multisite-Strand-Displacement-Reaction Signal-Amplification Strategy. Anal. Chem., 90, 7474-7479.

[82] Bi, S., Yue, S., Zhang, S. (2017). Hybridization chain reaction: a versatile molecular tool for biosensing, bioimaging, and biomedicine. Chem. Soc. Rev., 46, 4281-4298.

[83] Chang, C .C., Chen, C. Y., Chuang, T. L., Wu, T.H., Wei, S. C., Liao, H., Lin, C. W. (2016). Aptamer-based colorimetric detection of proteins using a branched DNA cascade amplification strategy and unmodified gold nanoparticles. Biosens. Bioelectron. 78, 200-205.

[84] Ding, X., Cheng, W., Li, Y., Wu, J., Li, X., Cheng, Q., Ding, S. (2017). An enzyme-free surface plasmon resonance biosensing strategy for detection of DNA and small molecule based on nonlinear hybridization chain reaction. Biosens. Bioelectron. 87, 345-351.

[85] Xu, F.X., Su, Y.L., Zhang, H., Kong, J.Y., Yu, H., Qian, B.Y. (2014). Prognostic implications for high expression of MiR-25 in lung adenocarcinomas of female non-smokers. Asian Pac. J. Cancer Prev., 15, 1197-1203.

[86] Wahlquist, C., Jeong, D., Rojas-Muñoz, A., Kho, C., Lee, A., Mitsuyama, S., van Mil, A., Park, W.J., Sluijter, J.P., Doevendans, P.A., Hajjar, R.J., Mercola, M. (2014). Inhibition of miR-25 improves cardiac contractility in the failing heart. Nature, 508, 531-535.

[87] Breaker, R. R., Joyce, G. F. (1994). A DNA enzyme that cleaves RNA. Chem. Biol., 1, 223-229.

[88] Ihms, H. E., Lu, Y. (2012). In vitro selection of metal ion-selective DNAzymes. In Ribozymes: Methods and Protocols. Edited by Hartig JS. Totowa, NJ: Humana Press, 297-316.

[89] Wang, H., Liu, Y., Wang, J., Xiong, B., Hou, X. (2020). Electrochemical impedance biosensor array based on DNAzyme-functionalized single-walled carbon nanotubes using Gaussian process regression for Cu(II) and Hg(II) determination. Microchimica Acta, 187, 207.

[90] Oh, W. K., Kwon, O. S., Jang, J. (2013). Conducting polymer nanomaterials for biomedical applications: cellular interfacing and biosensing. Polym Rev., 53(3), 407-442.

[91] Syshchyk, O., Skryshevsky, V. A., Soldatkin, O. O., Soldatkin, A. P. (2015). Enzyme biosensor systems based on porous silicon photoluminescence for detection of glucose, urea and heavy metals. Biosens Bioelectron, 66, 89-94.

[92] Ouyang, H., Shu, Q., Wang, W., Wang, Z., Yang, S., Wang, L., Fu, Z. (2016). An ultra-facile and label-free immunoassay strategy for detection of copper (II) utilizing chemiluminescence self-enhancement of Cu (II)-ethylenediaminetetraacetate chelate. Biosens Bioelectron, 85, 157-163.

[93] Seeman, N. C., Sleiman, H. F. (2017). DNA nanotechnology. Nat Rev Mater, 3(1), 1-23.

[94] Wang, H., Liu, Y., Liu, G. (2018). Reusable resistive aptasensor for Pb(II) based on the Pb(II)-induced despiralization of a dna duplex and formation of a G-quadruplex. Microchim Acta, 185(2), 142.

[95] Liu, L., Wu, H. C. (2016). Dna-based nanopore sensing. Angew Chem Int Ed., 55(49), 15216-15222.

[96] Liang, G., Man, Y., Li, A., Jin, X., Liu, X., Pan, L. (2017). Dnazyme-based biosensor for detection of lead ion: a review. MicrochemJ, 131, 145-153.

[97] Liu, S., Cheng, C., Gong, H., Wang, L. (2015). Programmable Mg^{2+} dependent DNAzyme switch by the catalytic hairpin DNA assembly for dual-signal amplification toward homogeneous analysis of protein and DNA. Chem Commun., 51(34), 7364-7367.

[98] Wang, H., Liu, Y., Liu, G. (2018). Label-free biosensor using a silver specific RNA-cleaving DNAzyme functionalized single-walled carbon nanotube for silver ion determination. Nanomaterials, 8, 258.

[99] Wang, H., Yin, Y., Gang, L. (2019). Single-gap microelectrode functionalized with single-walled carbon nanotubes and Pbzyme for the determination of Pb^{2+}. Electroanal., 31(6), 1174-1181.

[100] Shen, W., Li, Y., Qi, T., Wang, S., Sun, J., Deng, H., Tang, S. (2018). Fluorometric determination of zinc (II) by using DNAzymemodified magnetic microbeads. Microchim Acta, 185(10), 447.

[101] Wang, H., Liu, Y., Liu, G. (2018). Electrochemical biosensor using DNA embedded phosphorothioate modified RNA for mercury ion determination. ACS Sens, 3(3), 624-631.

[102] Gupta, V. K., Singh, A. K., Kumawat, L. K., Mergu, N. (2016). An easily accessible switch-on optical chemosensor for the detection of noxious metal ions Ni (II), Zn (II), Fe (III) and UO_2 (II). Sensor Actuat BChem., 222, 468-482.

[103] Wang, X., Feng, Y., Dong, P., Huang, J. (2019). A Mini Review on Carbon Quantum Dots: Preparation, Properties, and Electrocatalytic Application. Front. Chem., 7, 671.

[104] Peng, Z., Han, X., Li, S., Al-Youbic, A., O Bashammakh, A. S., SEl-Shahawi M. Leblan, R. M. (2017). Carbon dots: Biomacromolecule interaction, bioimaging and nanomedicine. Coord. Chem. Rev., 343, 256-277.

[105] Wang, T., Wang, A., Wang, R., Liu, Z., Sun, Y., Shan, G., Chen, Y., Liu,

Y. (2019). Carbon dots with molecular fluorescence and their application as a "turn-o" fluorescent probe for ferricyanide detection. Sci. Rep., 9, 10723.

[106] Baker, S. N., Baker, G. A. (2010). Luminescent Carbon Nanodots: Emergent Nanolights. Angew. Chem. Int. Ed., 49, 6726-6744.

[107] Margraf, J. T., Lodermeyer, F., Strauss, V., Haines, P., Walter, J., Peukert, W., Costa, R. D., Clark, T., Guldi, D. M. (2016). Using Carbon Nanodots as Inexpensive and Environmentally Friendly Sensitizers in Mesoscopic Solar Cells. Nanoscale Horiz., 1, 220-226.

[108] Hutton, G. A. M., Martindale, B. C. M., Reisner, E. (2017). Carbon Dots as Photosensitisers for Solar-Driven Catalysis. Chem. Soc. Rev., 46, 6111-6123.

[109] Martinez-Perinan, E., Bravo, I., Rowley-Neale, S. J., Lorenzo, E., Banks, C. E. (2018). Carbon Nanodots as Electrocatalysts Towards the Oxygen Reduction Reaction. Electroanalysis, 30, 436-444.

[110] Athinarayanan, J., Periasamy, V. S., Alshatwi, A. A. (2020) Simultaneous fabrication of carbon nanodots and hydroxyapatite nanoparticles from fish scale for biomedical applications, Materials Science and Engineering C 117 111313.

[111] Li, H., Chen, L., Wu, H., He, H., Jin, Y. (2014). Ionic Liquid-Functionalized Fluorescent Carbon Nanodots and their Applications in Electrocatalysis, Biosensing, and Cell Imaging. Langmuir, 30, 15016-15021.

[112] Garcia-Mendiola, T., Bravo, I., Maria Lopez-Moreno, J., Pariente, F., Wannemacher, R., Weber, K., Popp, J., Lorenzo, E. (2018). Carbon Nanodots Based Biosensors for Gene Mutation Detection. Sens. Actuators B Chem., 256, 226-233.

[113] Arduini, F., Zanardi, C., Cinti, S., Terzi, F., Moscone, D., Palleschi, G., Seeber, R. (2015). Effective electrochemical sensor based on screen-printed electrodes modified with a carbon black-Au nanoparticles composite. Sens. Actuat. B, 212, 536-543.

[114] Vicentini, F.C., Ravanini, A.E., Figueiredo-Filho, L.C., Iniesta, J., Banks, C.E., Fatibello- Filho, O. (2015). Imparting improvements in electrochemical sensors: evaluation of different carbon blacks that give rise to significant improvement in the performance of electroanalytical sensing platforms. Electrochim. Acta, 157, 125-133.

[115] Cinti, S., Arduini, F., Carbone, M., Sansone, L., Cacciotti, I., Moscone, D., Palleschi, G. (2015). Screen-Printed Electrodes Modified with Carbon Nanomaterials: A Comparison among Carbon Black, Carbon Nanotubes and Graphene. Electroanal, 27, 2230-2238.

[116] Mazzaracchio, V., Tomei, M.R., Cacciotti, I., Chiodoni, A., Novara, C., Castellino, M.,Scordo, G., Amine, A., Moscone, D., Arduini, F. (2019). Inside the different types of carbon black as nanomodifiers for screen-printed electrodes. Electrochim. Acta, 317, 673-683.

[117] Alcantara, R., Jimenez-Mateos, L. M., Lavela, P., Tirado, J. L. (2001). Carbon black: a promising electrode material for sodium-ion batteries. Electrochem. Comm., 3, 639-642.

[118] Drofenik, J., Gaberscek, M., Dominko, R., Poulsen, F. W., Mogensen, M. B., Pejovnik, S., Jamnik, J. (2003). Cellulose as a binding material in graphitic anodes for Li ion batteries: A performance and degradation study. Electrochim. Acta, 48, 883-889.

[119] Mazzaracchio, V., Tomei, M. R., Cacciotti, I., Chiodoni, A., Novara, C., Castellino, M., Scordo, G., Amine, A., Moscone, D., Arduini, F. (2019). Inside

the different types of carbon black as nanomodifiers for screen-printed electrodes. Electrochim. Acta, 317, 673-683.

[120] Talarico, D., Arduini, F., Amine, A., Cacciotti, I., Moscone, D., Palleschi, G. (2016). Screen-printed electrode modified with carbon black and chitosan: a novel platform for acetylcholinesterase biosensor development. Anal. Bioanal. Chem., 408, 7299-7309.

[121] Silva, T. A., Fatibello-Filho, O. (2017). Square-wave adsorptive anodic stripping voltammetric determination of ramipril using an electrochemical sensor based on nanostructured carbon black. Anal. Methods, 9, 4680-4687.

[122] Arduini, F., Cinti, S., Mazzaracchio, V., Scognamiglio, V., Amine, A., Moscone, D. (2020). Carbon black as an outstanding and affordable nanomaterial for electrochemical (bio)sensor design. Biosensors and Bioelectronics, 156, 112033.

[123] Aydın, E. B., Aydın, M., Sezgintürk, M. K. (2018). Electrochemical immunosensor based on chitosan/conductive carbon black composite modified disposable ITO electrode: An analytical platform for p53 detection. Biosensors and Bioelectronics,121, 80.

[124] Aydın, M,, Aydın, E. B., Sezgintürk, M. K. (2018). A highly selective electrochemical immunosensor based on conductive carbon black and star PGMA polymer composite material for IL-8 biomarker detection in human serum and saliva, Biosensors and Bioelectronics, 117, 720.

[125] Messaoud, N.B., Lahcen, A.A., Dridi, C., Amine, A. (2018). Ultrasound assisted magnetic imprinted polymer combined sensor based on carbon black and gold nanoparticles for selective and sensitive electrochemical detection of Bisphenol A. Sensors and Actuators B Chemical 276, 304.

[126] Arduini, F., Cinti, s., Mazzaracchio, V., Scognamiglio, V., Amine, A., Moscone, D. (2020). Carbon black as an outstanding and affordable nanomaterial for electrochemical (bio)sensor design. Biosensors and Bioelectronics, 156, 112033. Doi: 10.1016/j.bios.2020.112033

[127] Zhang, X. Q., Lam, R., Xu, X., Chow, E. K., Kim, H. J., Ho, D. (2011). Multimodal Nanodiamond Drug Delivery Carriers for Selective Targeting, Imaging, and Enhanced Chemotherapeutic Efficacy. Advanced Materials, 23, 4770-4775.

[128] Babadia, F. E., Hosseinia, S., Shavandib, A., Moghaddasc, H., Shotipruka, A., Kheawhoma, S. (2019). Electrochemical investigation of amino acids Parkia seeds using the composite electrode based on copper/carbon nanotube/nanodiamond. Journal of Environmental Chemical Engineering, 7, 102979.

[129] Azevedo, A. F., Matsushima, J. T., Vicentin, F. C., Baldan, M R., Ferreira, N. G. (2009). Surface characterization of NCD films as a function of sp2/sp3 carbon and oxygen content, Applied Surface Science 255, 6565-6570.

[130] Azevedo, A. F., Baldan, M. R., and Ferreira, N. G. (2012). Nanodiamond Films for Applications in Electrochemical Systems, International Journal of Electrochemistry, 508453, Doi: 10.1155/2012/508453

[131] Hasanzadeh, M., Shadjou, N., And de la Guardia, M. (2015). Iron and iron-oxide magnetic nanoparticles as signal-amplification elements in electrochemical biosensing. Trends Anal Chem., 72, 1-9.

[132] Sieben, S., Bergemann, C., Lubbe, A., Brockmann, B., Rescheleit, D. (2001). Comparison of different particles and methods for magnetic isolation of circulating tumor cells. J Magn Mater., 225, 175.

[133] Pinho, M. S., Gregori, M. L., Nunes, R. C. R, Soares, B. G. (2001). Aging effect on the reflectivity measurements of polychloroprene matrices containing carbon black and carbonyl-iron powder. Polym Degrad Stab., 73,1-5.

[134] Abdelhamid, H. N. (2019). Nanoparticle based surface assisted laser desorption ionisation mass spectrometry: a review. Microchim Acta, 186, 682.

[135] Kouassi, G. K., Irudayaraj, J., McCarty, G. (2005). Examination of cholesterol oxidase attachment to magnetic nanoparticles. J Nanobiotechnol, 3, 1.

[136] Lu, B. W., Chen, W. C. (2006). A disposable glucose biosensor based on drop- coating of screen-printed carbon electrodes with magnetic nanoparticles. J Magn Magn Mater, 304, 400-402.

[137] Wang, S. F., Xie, F., Hu, R. F. (2007). Carbon-coated nickel magnetic nanoparticles modified electrodes as a sensor for determinationof acetaminophen. Sens Act B, 123, 495-500.

[138] Lou, L., Yu, K., Zhang, Z., Huang, R., Zhu, J., Wang, Y., Zhu, Z. (2012). Dual-mode protein detection based on Fe_3O_4– Au hybrid nanoparticles. Nano Res., 5, 272-282.

[139] Dhouha, J., Eleonora, M., Danila, M., et al. (2020). Highly sensitive paper-based electrochemical sensor for reagent free detection of bisphenol A. Talanta, 216, 120924.

[140] Wenqing, Z., Huan, X., Xianxian, Z. (2020). 3D DNA nanonet structure coupled with target -catalyzed hairpin assembly for dual -signal synergistically amplified electrochemical sensing of circulating microRNA, Analytica Chimica Acta, 1122, 39-47.

[141] Mat Zaid, M. H., Abdullah, J., Rozi, N., Rozlan, A. A. M., Hanifah, S. A. (2020). A Sensitive Impedimetric Aptasensor Based on Carbon Nanodots Modified Electrode for Detection of 17ß-Estradiol. Nanomaterials , 10, 1346.

[142] Attaallah, R., Antonacci, A., Mazzaracchio, V., et al. (2020). Carbon black nanoparticles to sense algae oxygen evolution for herbicides detection: Atrazine as a case study. Biosensors & Bioelectronics, 159, 112203.

[143] Avelino, K. Y. P. S., Oliveira, L. S., Lucena-Silva, N., et al. (2020). Metal-polymer hybrid nanomaterial for impedimetric detection of human papillomavirus in cervical specimens. Journal of Pharmaceutical and Biomedical Analysis, 185, 113249.

[144] Lu, J., Hu, Y., Wang, P., et al. (2020). Electrochemical biosensor based on gold nanoflowers-encapsulated magnetic metal-organic framework nanozymes for drug evaluation with in-situ monitoring of H_2O_2 released from H_9C_2 cardiac cells. Sensors and Actuators B-Chemical, 311, 127909.

[145] Fang, B., Xu, S., Huang, Y., et al. (2020). Gold nanorods etching-based plasmonic immunoassay for qualitative and quantitative detection of aflatoxin M1 in milk. Food Chemistry, 329, 127160.

[146] Kumar, V., Kaur, I., Arora, S., et al. (2020). Graphene nanoplatelet/ graphitized nanodiamond-based nanocomposite for mediator-free electrochemical sensing of urea. Food Chemistry, 303, 125375.

[147] Guo, W., Pi, F., Zhang, H., Sun, J., Zhang, Y., Sun, X. (2017). A Novel Molecularly Imprinted Electrochemical Sensor Modified with Carbon Dots, Chitosan, Gold Nanoparticles for the Determination of Patulin. Biosens. Bioelectron, 98, 299-304.

Chapter 10

Perspectives of Nano-Materials and Nanobiosensors in Food Safety and Agriculture

Sivaji Mathivanan

Abstract

Nanobiosensor is one type of biosensor made up with usage of nanomaterials i.e., nanoparticles and nanostructures. Because of the nanomaterials' unique properties such as good conductivity, and physicochemical, electrochemical, optical, magnetic and mechanical properties, Nanobiosensors are highly reliable and more sensitive in biosensing approaches over conventional sensors which is having various limitation in detection. Quantum dots, nanotubes, nanowires, magnetic and other nanoparticles enhance sensitivity and lower limit of detection by amplifying signals and providing novel signal transduction mechanisms enable detection of a very low level of food contaminants, pesticides, foodborne pathogens, toxins and plant metabolites. Nanobiosensors are having a lot of scope in sustainable agriculture because of its detecting ability i.e., sensing changes occurred in molecular level. So it can be utilized to find out the variations or modification of plant metabolities, volatiles, gas exchange, hormonal and ion concentration etc. which are the indicators of various harsh environmental stresses (abiotic), biotic and physiological stress. Identification of the stress in the starting stage itself will help us to avoid intensive plant damage and prevent yield losses created by the stress. Nanosensors can be used in smart farming, in which all the environmental factors related to plant growth like temperature, water, pH, humidity, nutritional factor etc. are measured and precaution taken to control the factors which reduce the crop production with the help of IOT platform, thereby enhance the productivity. In this review, discussed about nanobiosensors for detection of food contaminants and various application and its potential in agriculture.

Keywords: biosensor, nano material, nano-biosensor, food contaminants, agriculture and smart farming

1. Introduction

Agriculture and food industry are a main source of income and employment for major section of population. Agriculture sector plays a strategic role in the self-sustaining economic development by providing basic ingredients to mankind and raw material for industrialisation. Global estimates indicate, the people engaged in agriculture are about 2.5 billion [1]. Agriculture is much diversified field, but continuing with technological growth at brisk pace. Many advanced technologies are introduced in agriculture to increase the yield by reducing the direct and indirect factors which affect the crop yield. Major yield reduction factors are insect, pathogens and weeds which can be controlled by human beings through application of insecticide, fungicide and herbicide respectively. In order to control these biotic factors and serious intention to increase yield heavy dosage of chemical pesticides and fertilizers are applied to the growing crops. As reciprocation of this residual pesticides and chemicals are contaminating soil food and water. When the contaminated food got consumed, it produce so many serious ill effects to the consumer [2]. Other than chemicals many bacterial pathogens and the toxin produced by the microorganism also major food contaminants and creating more health complications. One of the fine solution for avoiding contaminated food related health issues is effective detection of the food contaminants whether chemical or biological before consumption will ensure the food safety. So many conventional and advanced methods such as culture plate technique, chromatography, spectroscopy, immunology and molecular biology technique etc. are available for detection of biological and chemical contaminants in food sample, but these all are either time consuming, or more expensive and low sensitivity [3]. Agriculture is the primery food source for both human and livestock. So many biotic and abiotic factors are challenging the agriculture production and productivity. To maintain the food security for the fast growing population need to increase the food production and productivity with enormous level [4, 5]. There is need of some advance technology to increase the production in to maximum with higher quality assurance, risk identification, diagnosis and prevention to achieve goal of regional and global food security. Thus to improve consumer livelihood and optimal utilization of resources, rapid, real-time, portable, and cost effective technologies are desired in agriculture and food industry [6–8]. Recently many technologies are developed and revolutionized the agriculture sector, among that, the most promising one is the nanotechnology [9]. Eventhough its practical application is negligible at present moment, it has a lot of scope in near future to improve agricultural practices over conventional farming at various stages from crop production to post harvest, there by flourishing the agriculture sector by enhancing food production and crop productivity. Normally crop productivity or yield enhancement is possible in two ways 1. By reducing the yield loss caused by various factors at different crop stages such as insect and diseases (Biotic stress), various adverse environmental factors (Abiotic stress) i.e. water stress, high temperature stress, salt stress, cold stress, harmful radiation and nutritional deficiency during the crop production stage and avoiding losses after harvesting of farm product (Post harvest stages) 2. Enhancing the yield by adopting highly improved advance crop production techniques there by reducing the cost of inputs and increasing yield with high cost benefit ratio. Nanotechnology can be applied in both of the strategies to improve the production and productivity in agriculture and food sector [10]. Nanobiosenors i.e. biosensors with nanomaterials, is one of the major application of nanotechnology, are synthesized with the help of various departments like, bioelectronics, material science, miniaturization techniques, electrode design, fabrication technology, nanolithography and microfluidics [11–13]. Biosensor is a self contained integrated tool for sensing and characterization of biological materials. Improvement in basic characteristics of biosensors will lead to widespread application in major challenging areas in food and agriculture [14–16]. In this review, role and applications of nanobiosensors in agriculture and food industry at present are explored and also discussed the potential of nanobiosensor, possible application with brisk development and benefits in future.

2. Biosensor

A **sensor** is a device, module, machine, or subsystem whose purpose is to detect events or changes in its environment like temperature, humidity, water flow, intensity of light etc. and send the information to other electronic circuit or electrical instrument that can be measured and/or analyzed. Biosensor is a one type of sensing technology consisting of biological component, such as a cell, enzyme or antibody, connected to a tiny transducer, a device changing one form of signal in to another form so that it can be easily measured by other system. The biosensors enable to sense changes that happen in the cells in cells and molecules even in very low concentration of the tested material. When the substance binds with the biological component, the transducer produces a signal proportional to the quantity of the substance [17–20]. For example if there is a more number of bacteria in a particular food, the biosensor will produce a strong signal indicating that the food is unsafe to eat. With this technology, mass amounts of food can be readily checked for their safety of consumption [21–23].

The biological element of a biosensor contains a biosensitive layer, which can either contain bioreceptors or be made of bioreceptors covalently attached to the transducer. The different types of biosensors are classified based on the bioreceptor and transducer present in the biosensor [24–27]. Based upon biorecepting molecules majorly it is divided in to five categories.

1. Protein based mainly Antibody/antigen based,

2. Enzymes based,

3. DNA based.

4. Based on cellular interactions either whole cells or cell organelles,

5. Employing biomimetic materials (e.g., synthetic bioreceptors).

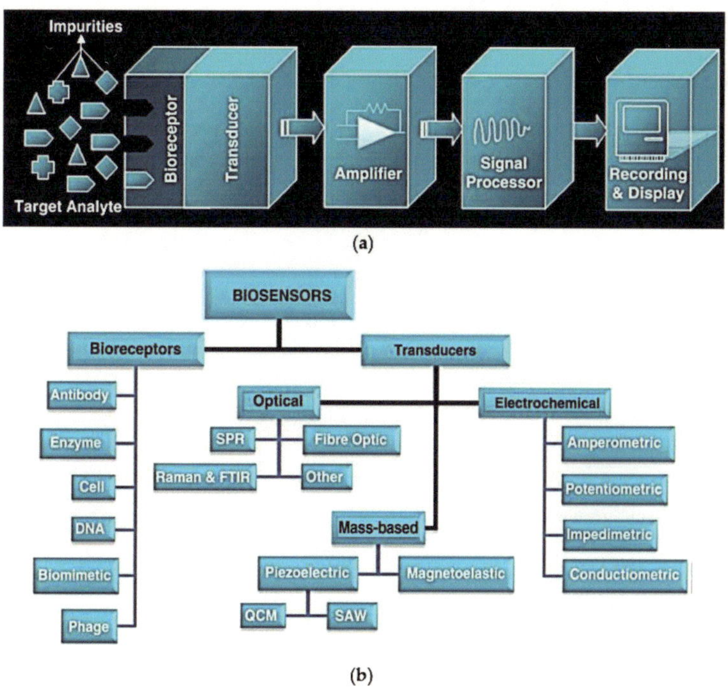

Figure 1.
Various components of a typical Biosensor (a) and its classification (b) [25].

By using of the transducer, mainly it is divided in to three types, those are, 1. Electrochemical, 2. Mass based and 3. Optical Biosensor.

In these major types is divided in to many subtypes based upon the mechanism of signal detection (**Figure 1**).

Biosensors especially nanobiosensors can overcome all the disadvantages of conventional detection methods by offering a rapid, non-destructive and affordable methods for quality control [28, 29].

3. Nanobiosensor

Biosensors are synthesized using nanomaterial is called nanobiosensor. This type of biosensor can able to detect the changes happen in the atomic level with more accuracy. In normal biosensor s receptor and transducer materials are made up of normal micro and macro sized material, but in nanobiosensor either receptor or transducer or both are made up of nano material i.e. at least any one dimension is less than 100 nm [30, 31]. Nanomaterials are very small size, so it is having unique physical, mechanical, optical, electrical and magnetic characters when compared to the conventional material. This is the added advantage of the nanobiosensor and reason for all the superior sensing qualities. Researchers have used various nanomaterials and nanocomposites to enhance the sensitivity, shelf life and get the precision in the biosensing results [32, 33]. Mostly four types of nanomaterials i.e., carbon based, metal based, polymer based and nanocomposites are used for various field application [34]. Perhaps among four, dendrimer is not used that much as frequently as other nanomaterial in the bio sensing field. Carbon nanotubes (CNTs), quantum dots (QDs), gold (Au), silica, silver (Ag), graphene and other nanocomposites are synthesized in such a way having large surface area to volume ratio to improve electrochemical parameters (**Figure 2**). The molecular binding is a subject of the biological surface science, which is strongly related to the research on modification of nanostructures properties by controlling their structure and surface

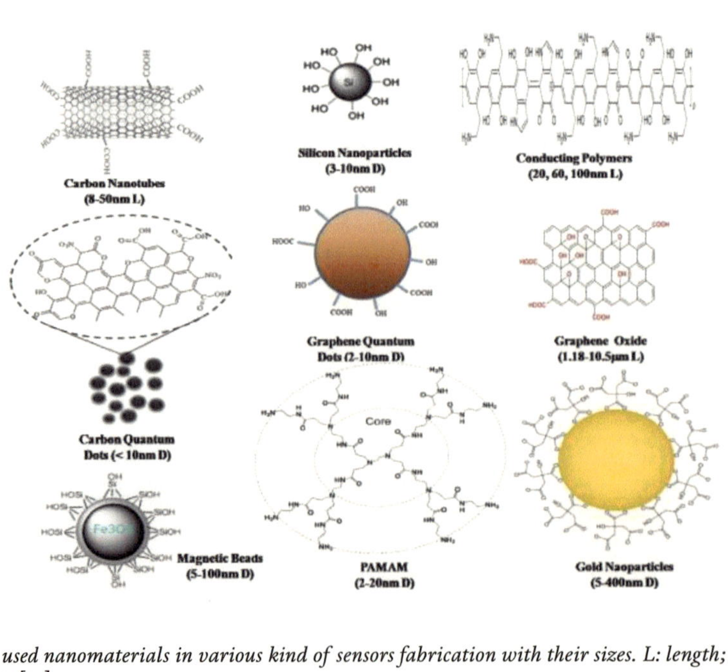

Figure 2.
Commonly used nanomaterials in various kind of sensors fabrication with their sizes. L: length; D: Diameter [35].

at a nanoscale level [36, 37]. Recent technological advancement in nanotechnology enable to synthesize nanomaterials with extraordinary optical and electronic properties for electronics and sensing applications [38]. The efficiency of biosensor can be improved by increase the detection limit and the overall performance through using nanomaterials. Nanomaterials provide friendly platform for the assembly of bio-recognition element, the high surface area, high electronic conductivity that increase the limit of detection [39].

4. Nanobiosensor application in food

Harmful microorganisms such as pathogenic bacteria, viruses, or parasites and its toxin and excessive use of agrochemicals (pesticides, herbicides and food preservative) are the common causes of food contamination. The primary contaminants leading to foodborne illness are pathogenic microorganisms including: *Bacillus cereus, Clostridium botulinum, Escherichia coli (E. coli), Listeria monocytogenes, Salmonella* and *Staphylococus aureus* [40]. Currently, there are limited methods for field detection of toxins and foodborne pathogens, making early identification of a possible contamination is much difficult because of the low efficient conventional methods. The biosensors integration with various nanostructures like thin films, nanorods, nanoparticles and nanofibers, in the analysis methods for detection of food contaminants has improved the detection sensitivity and increased portability. In food microbiology, nanosensors or nanobiosensors are used for the detection of pathogens in processing plants or in food material, quantification of available food constituents, alerting consumers and distributors on the safety status of food [26, 41].

4.1 Nanobiosensors for food pathogen detection

The conventional method for food pathogen detection is colony counting (CFU) on an agar plate which takes 2–3 days for initial results, and up to 1 week for confirming pathogen specificity [42]. These conventional method is not suitable for highly perishable food products. Polymerase chain reaction (PCR) and enzyme-linked immunosorbent assay-based (ELISA) can be used as alternative to traditional CFU methods [43]. But these methods are labour-intensive and costly. So using nanobiosensor, that can be adapted on portable platforms to enable rapid testing of wide range of pathogens with potential for on-site analysis [44]. A dimethylsiloxane microfluidic immunosensor integrated with specific antibody immobilized on an alumina nanoporous membrane was developed for rapid detection of foodborne pathogens *Escherichia coli O157:H7* and *Staphylococcus aureus* with electrochemical impedance spectrum [45]. Due to good electrical conductivity and ample functional groups present on surface area carbon nanotubes have been used to develop biosensors for detection of foodborne pathogenic bacteria (*Staphylococcus aureus*) in fresh meat [46]. Miranda et al. developed a hybrid colorimetric enzymatic nanocomposite biosensor for the detection of *E. coli* in aqueous solutions based on enzyme amplification. The efficiency of the method was demonstrated in both solution and test strip format [47]. β-galactosidase an anionic enzyme is electrostatically attached to the cationic Gold nanoparticles (AuNPs) featuring quaternary amine head groups by this way it inhibit the activity of enzyme. When AuNPs binds with bacteria, the attached β-galactosidase is released restoring its activity and this binding activity and colour formation because of the enzymatic reaction was measured by colorimetric means. Using this method, bacteria can be detected at the concentrations of 1×10^2 bacteria/mL in solution [48]. In an effort to ensure food safety,

Escherichia coli (E. coli) O157:H7 has been detected using synthesized attractive 3D architecture of silver (Ag) nanoparticles as nanoflowers as the interface material for electrochemical biosensing [49]. Bovine serum albumin (BSA) has been used as stabilizing agent for proper conjugation of Ag nanoflowers. *Salmonella* spp. a food pathogenic organism responsible for causing salmonellosis diseases, [50] millions of people are naffected by this disease annually in all over the World [50–52]. Magnetic nanoparticles and TiO2 nanocrystals are used to detect the *Salmonella* in milk [53]. In this method the pathogen is first captured by antibody-immobilized magnetic NPs. Then antibody-conjugated TiO2 binds the MNP–*Salmonella* complexes and this was monitored by absorbance measurement. This method sensitive enough to detect of 100 CFU/mL for Salmonella in milk sample [53].

4.2 Detection of mycotoxins

Mycotoxin is the toxic metabolites secreted by fungus during its growth and development. Some of the major example for mycotoxin are ochratoxin synthesized by penicillium, aflatoxin secreted by *Aspergillus.* Many agricultural and horticultural food crops are easily affected by different group of pathogenic fungi so invariably food material produced from the affected plants contaminated with mycotoxins. The importance of the mycotoxin is, these create severe health complication even at in small concentrations [54, 55]. Many type of nanomaterial are used to synthesize the nanobiosensor, among all carbon based nanomaterials graphene and its derivatives have great oppurtunity to detect the mycotoxins from various food sample. A nanocomposite of graphene oxide and gold nanocomposites (GO/AuNCs) has high sensitivity detection of aflatoxin B1 (AFB1) in peanut samples [56, 57]. The improvement in sensitivity of the biosensor is due to better quenching ability of nanocomposite. Moreover low detection limit with wide linear range leads to better reliability and wider applicability of such biosensors. Gold nanoparticle (AuNP) based aptasensor is used to detect the aflatoxin B1 with a detection limit of 7 nM [58, 59].

4.3 Detection of pesticides

After green revolution the application and usage of chemical pesticide shoots very high in order to get more yield. In this farmers might have got success in the sense of productivity but at the same time contamination of food material with chemical pesticide is also unavoidable, this issue creates insecurity in the food safety. Organophosphorus (OP) and carbamates (C) are the pesticides used mostly representing ~40% of the world pesticide market [60, 61]. Acetylcholinesterase (AChE) is one of the important enzyme in our body, this catalyses the hydrolysis of neurotransmitter acetylcholine. Primarily the pesticide molecule inhibit this enzyme activity, hence pesticide toxin presence in food ultimately affect the human body dangerously [62, 63]. In order to avoid this the food material should be analysed and detected the amount of pesticide residues exist before its consumption. Commercially many techniques such as chromatographic techniques (GC and HPLC) and coupled chromatographic-spectrometric procedures such as GC–MS and HPLC-MS are available, perhaps these methods are more costlier and not able to do the real time analysis [64]. Mostly pesticide detection is done by measuring AChE activity before and after exposure to the pesticides through colorimetric Ellman assay [65]. AuNPs (3 nm) nanoparticles based biosensors are used to detect the many pesticide molecule like paraoxon, dimethoate, carbaryl, chlorpyrifos, carbofuran, etc. at a concentration of 24 µg/mL [66–69]. The Lum-AgNPs were used in conjunction with a H_2O_2 based CL detection to generate a CL "fingerprinting" related to each specific pesticide. A highly reproducible and stable biosensor based on chitosan-TiO_2 graphene nanocomposite has been developed recently for detection of organophosphate pesticides in cabbage. The porosity of the nanocomposite provides additional stability to the biosensor via efficient enzyme immobilization [70].

4.4 Detection of metal contaminants

Heavy metal ions like mercury, lead, cadmium, arsenic etc., are present in the environment. When food crops are cultivated in such environment, heavy metal residues are accumulated in food. Consumption of this heavy metal contaminated food will create various disorders and health issues such as neurological, reproductive, cardiovascular problems [71]. AuNP-based sensors working based upon colorimetric detection was used to detect the metal ions in water with the detection limits of 30 ppb for Pb^{2+} and 89 ppb for Al^{3+} [72]. AuNPs based sensors also used to detect the other heavy metals like Hg^{2+} [73–77], Cu^{2+} and Ag^{+} [78], Mn^{2+} [79], Cd^{2+} [80, 81], Fe^{3+}, Pb^{2+}, Al^{3+}, Cu^{2+}, and Cr^{3+} [77].

4.5 Nanobiosensors for intelligent food packaging or smart packaging

Normally packaging of food helps to maintain the nutrient content and increasing the shelf life. But smart packaging is have an added option, it indicate the temporal and spatial changes occur in the food constituents that contains with in it. Intelligent tags and stickers are combined with nanobiosensing material inside the packaging and nanobiosensor connected with consumers through electronic devises will indicate the real time sensing data about the food material present whether normal or deteriorated from time to time [82–84]. These work by sensing through a nanobiosensor integrated with polymer film or polymer matrix and radio frequency identification (RFID) components are used by intelligent tags to sensing [85, 86]. Biosensors for food packaging can function in particular physico-chemical conditions in the packaged microenvironment. Zeolite-molybdate tablets are prepared by placing Ammonium molybdate in to the zeolite nanopores used to detect and measure ethylene in avocados packages. In ten day old package, the zeolite-molybdate tablet changes its colur from yellow to blue because of the redction of Mo(VI) to Mo(V) [87–89].

4.6 Nanobiosensors in E-nose technology

Most common factors for food rotting and developing of foul odor in food is pathogenic bacteria. Above certain level of odors can be sensed by the human nose, but sometimes it may not be useful to prevent of food poisoning. Therefore, rapid assessment of odor at earlier stage should be most useful, in this regard nanobiosensors can be used for the detection of these odors with high sensitivity. Nanoparticles help in better absorption of gas on sensor surface due to more surface area than macroscopic particles [90, 91].

Electronic nose (E-nose) is used to identify different types volatile organic compounds present in food to ensure good quality, uniformity and consistency of raw material during mixing, cooking and of final product during packaging and storage processes [92]. Gas sensors composed of nanoparticles e.g. ZnO nanowires are used to detect the gas. More amount of ethylene gas fruits and vegetables deteriorates its quality, Tungsten oxide–tin oxide nanocomposites have been employed for ethylene sensing [93]. SWCNT field-effect transistor functionalized with human olfactory receptor 2AG1 protein has been employed for sensing fruit odorant amyl butyrate in apricot [94, 95]. Olfactory receptors-functionalized carbon nanotubes-based transistor has been documented for the selective detection of hexanal as olfactory indicator of spoiled milk and oxidized food [96].

5. Application of nanobiosensor in agriculture

Nanotechnology is basically dealing with smallest particles which plays important role in fixing problems exist in agriculture that cannot be solved through existing approaches. The development in synthesis of new nanomaterial and nano-devices depict novel applications in agriculture. One of the applications of nanotechnology is formation of superior biosensors that leads to development of miniature structures known as nanobiosensors, that are greater efficient and well organized when compared to traditional biosensors. Nanobiosensors can be used effectively in agriculture for sensing soil pH, moisture, wide variety of pathogens, plant hormones, plant metabolites, pesticide, herbicide, fertilizers, and metal ions. Appropriate and controlled use of nanobiosensor can support sustainable agriculture for improving crop productivity. It can also help in controlled use of agricultural inputs there by control pollution and lowering cost of cultivation [97].

5.1 Nanobiosensor in crop stress management

Crop growth and production undergo for various stress such as biotic, abiotic and nutritional stress. The plant hormones also called as phytohormones play an important role in control and regulation of physiological processes of development and much importantly involved in the stress response and regulation in plants. Abiotic and biotic stress are inducing various unusual chemical metabolites and different plant hormones in the plants, in order to make necessary arrangements to face the adverse condition in the surrounding environment. For addressing problems related to imbalance of phytohormones and related consequences, nanobiosensors have played a pivotal role in term of detection. So that further recovery action of the plant can be taken very quickly, it prevent the considerable amount of plant damage and yield reduction. Nanobiosensors have most significantly contributed to achieve the ever existing goal of precision in agriculture. For abscisic acid detection, a label-free electrochemical impedance immunosensor has been developed using an anti-ABA antibody that is adsorbed on to a porous nanogold film [98]. With the same biosensing technique different matrices have been tried to obtain desired output. The results indicated that the abscisic acid in plants can be detected successfully using an antobody based nanobiosensor. Gibberellic acid (GA) is detected in the hybrid rice grain samples by electrochemical impedance spectroscopy fabricated with successful grown porous nanogold film and consequently modification of the glassy carbon electrode [99]. Simple amperometric biosensor developed by graphite coated with polypyrrole (PPy) for the determination of salicylic acid in samples of plasma and milk [100]. A real time highly selective nanobiosensor developed for determination of cytokinins and auxion concentrations in tomato xylem sap exudates [101, 102]. It is recommended that precautions are better than cure or remedies in case of plant stresses otherwise at the calamity stage it will cause huge crop loss and poor quality of the produce. Presently, it has been observed that the operational stability of the biosensor is limiting its technological implications in agricultural sector. But this early detection of the alarming stress conditions (Biotic and abiotic stress) of plants will be possible in future with complete growth of nanobiosensing technology. In this direction, some nanobiosensors have been developed to detect indicators or signalling compounds of stress conditions of plants. Recently, to analyse the jasmonate signalling in plants, a fluorescence based biosensor has been developed that exclusively provides data about hormone distribution in conditions of plant abiotic and biotic stresses [103]. Water is important not only for photosynthesis, but also for flow of nutrients and many microelements necessary for healthy crop. An optical fluorescence biosensor for plant water stress detection has been reported [104]. Deficiency of macro and micro nutrients will affect the plant growth, physical strength, grain formation and yield. Detection of the deficient nutrient by nanosensor will help to improve the plant growth and yield; also prevent excess unwanted nutrient application. Enzyme based phosphate biosensors with fluorescent nanoparticles using fluorescence transduction mechanism for specific and rapid detection of phosphate and other nutrients such as nitrogen, calcium and zinc [105, 106].

5.2 Nanobiosensors in smart farming

Nanosensors are very minute and it can sense the soil condition, irrigation requirement, pH of the water and soil, nutrient requirement, disease and pest incidence, soil temperature like many important paramaters by scattering all over the field. Based upon the parameter recorded by the nanosensor, need based action will increase the crop yields and reduces the unwanted manpower resourses like fertilizer, pesticide etc. This nanosensor concept fitting well with the precision farming or smart farming goal. Nanosensors, made up of non-biological materials, such as carbon nanotubes, have ability to sense and signal, acting as wireless nanoantennas, because of their small dimensions, can collect information from numerous different points [107]. External devices can then integrate the data to automatically generate incredibly detailed report and respond to potentially devastating changes in their environment. For instance, connected nanosensors for monitoring soil or plant conditions can alert automatically according to conditions detected by sensors and therefore influence more efficient usage of the fertilizers, herbicide, pesticide, insecticide, etc. Nanobiosensors are now developed with all integrated devices such as power source, sensing unit, detector and display unit in a single chip for detecting the plant stress indicators [108, 109]. For that various indicative signals such as increase in sucrose content [110], change in concentration of nutrients [111, 112] and hormones [103] etc. can be used to further transform these in to visual indication through biosensing technology of processing signals [111, 113].

Nanoscale devices are envisioned that would have the capability to detect and treat diseases, nutrient deficiencies or any other maladies in crops long before symptoms were visually exhibited. This is the future of agriculture, an army of nano-sensors will be scattered like dust across the farms and fields, working like the eyes, ears and noses of the farming world. These tiny wireless sensors are capable to communicate the information they sense. These are programmed and designed to respond various parameters like variation in temperature, humidity and nutrients. The distributed intelligence of smart particles can be networked to respond immediately to any change in environment, hence giving an alert in advance to devise ways and means to deal with environmental variations. By smart dust and gas sensors it is possible to evaluate the amount of pollutants in the environment. The most efficacious approach in this sense is real time detection of parameters by the use of autonomous sensors connected to global positioning system (GPS) [114–117].

5.3 Nanobiosensors to maintain seed purity

Seed production is very difficult process particularly in wind pollinated crops since pollen can fly for long distance. Humidity, wind speed, temperature are some of the factor affect the pollen plight Very effective method to ensure the genetic purity is the detecting pollen load that cause contamination. Bionanosensors can be used to identify the specific contaminating pollen and thus reduces contamination. Novel genes are being incorporated into/seeds and sold in the market. Tracking of sold seeds could be done with the help of nanobarcodes [118] that are encodable, machine - readable, durable and sub-micron sized taggants [119, 120].

5.4 Nanobiosensor for disease detection

Frequently recurring diseases are considered as one of the major factors limiting the crop productivity. Early prediction of the occurrence is the only prevention to eradicate diseases at the root. Such devices can diagnose plant health issues before these actually get visible to the farmer. Antibody conjugated nanoparticles are used to detect *Xanthomonas axonopodis* that causes bacterial spot disease [121]. Optical immunosensors based on Gold nanoparticle and antibody conjugated Fluorescent silica nanoparticles (FSNPs) are being used to detect the karnal bunt disease in wheat and bacterial spot diseases (*Xanthomonas axonopodis* pv. *vesicatoria*) in *Solanaceae* plant respectively [121, 122]. Due to the unique optical properties of Quantum dots (QD) [123] they are used for detection of witches' broom disease of lime (WBDL) caused by *Candidatus* Phytoplasma aurantifolia (*Ca.* P. aurantifolia) using fluorescence resonance energy transfer (FRET) mechanism [124].

Many novel sensor fabricated with nanomaterials have been explored in order to obtain high sensitivity and low limits of detection [125–129]. Methyl salicylate is a volatile compound synthesized more by plant during infection stage, so detection of methyl salicylate or other volatile organic compounds specific for particular diseases will be more helpful to identify the diseases before forming symptoms and to take proper control measures in initial stage itself. Gold nanoparticle and semi-conductive metal oxide nanoparticles based amperometric biosensors are used to detect diseases causing different types of bacteria, viruses and fungi [130–134].

5.5 Assessment of harvest index by nanobiosensor

As harvesting of proper mature fruits and vegetables ensure good quality and consumer acceptability, sensing maturity of agricultural produce is very important for good post-harvest quality and enhanced shelf life. The overmature fruits/ vegetables will over ripe and have to be discarded, while immature fruits/vegetables will lead to inferior eating quality for consumers. Thus only proper mature fruits/ vegetables should be harvested at proper time. Physio chemical properties of horticultural crops like fruits and vegetables are linked with their maturity. Changes of these characters are linked to the maturity and real time measurement of these characters will guide the harvesting of crops at proper mature stage only. Nanobiosensors are used for measurement of intrinsic quality attributes of horticultural crops such as ascorbic acid [135], total phenolic compounds [136] and L-arginine [137]. As the crop attains maturity, it is harvested and stored for further processing.

5.6 Biosensors for heavy metal deposits in soil and water

Plants need many elements in the trace level for their healthy growth and metabolism. Soil, water and air are the major source of these elements and plants obtained nutrients from these sources. Accumulation of these heavy metals and elements more than the optimum level cause serious problem to the plants there by to the human up on the food consumption. Heavy metals such as nickel, mercury [138], arsenic [139], lead [140], chromium [141], cadmium and copper [142] are commonly found in contaminated soils [143]. The presence of heavy metals increases oxidative stress in plants, which further triggers synthesis of pigments like chlorophyll and βcarotene [141, 144].

In a research study for simultaneous detection of mercury (Hg^{2+}) and silver (Ag^{+}) ions in drinking water, serum and cell lysate, tungsten disulfide-nanosheets (WS2) based biosensing platform has been developed. The study has implications in environmental monitoring and diagnosis [145]. Heavy metals such as Pb, Ni, Cd, Zn, Co, and Al are detected effectively by cantilever nanobiosensors with phosphatase alkaline in water [146].

5.7 Nanosensors in storage

Inexpensive sensors, cloud computing and intelligent software, hold the potential to transform the agri-food sector. Internet of Things (IoT), is an emerging field in which many more instruments are interconnected to the user agricultural field and internet will increase the precision of the agriculture there by maintain the sustainability and cost effectiveness of agricultural production. The joint application of nanotechnology in IoT creates a new things, namely the Internet of Nano Things (IoNT). Nanosensors, because of their small dimensions, can collect information from numerous different points [107]. External devices can then integrate the data to automatically generate incredibly detailed report and respond to potentially devastating changes in their environment. For instance, connected nanosensors for monitoring soil or plant conditions can alert automatically according to conditions detected by sensors and therefore influence more efficient usage of the fertilizers, herbicide, pesticide, insecticide, etc. Involving nanosensors in the design of smart or intelligent packaging, enable the transfer of information regarding product conditions during distribution or storage. The response generated due to changes related to internal or external environmental factor, are recorded through specific sensors [147, 148] and data are stored in the database. So at any time, from any where those data and parameters related to soil and plant health can be accessed via Internet. Rapid response and detections of unusual parameters' values, are enabled to increase the food quality and safety, that directly influence to human health.

6. Conclusion

Latest improvements in nanotechnology and information and communications technology (ICT) exhibit its great potential towards the agri-food sector. Increased fertilizer efficiency, enhancing the plant nutrient absorbtion, Improved quality of the soil, stimulation of plant growth, the use of precise farming, enhancing the food safety, food processing and pakage, distribution and storage, are some of the benefits of nano-based technology in agricultural and food industry. Efficiency and accuracy of the biosensor on detection of agricultural and food safety parameters will be enhanced by the integration of the nanomaterials. The application of biosensor and its efficiency can also be improved further in future by developing of novel nanomaterials that will boost the agriculture and food sector. Therefore, the full potential of nanotechnology in the agri-food sector is yet to be realized. In the coming decades the convergence among nanotechnology, agriculture engineering, and plant science will lead the path towards food security, sustainability, precision agriculture, automation, robotic farming and cost effective technologies. This innovation in technology has important consequences in agriculture.

Author details

Sivaji Mathivanan
Agricultural College and Research Institute, Tamil Nadu Agricultural University, Vazhavachanur, Tiruvannamalai, India

*Address all correspondence to: sivajibt@gmail.com

References

[1] Food and Agriculture FAO statistical yearbook 2013. Food & agricultural organization, 2013.

[2] Chauhan, N., Narang, J., Jain, U. Amperometric acetylcholinesterase biosensor for pesticides monitoring utilising iron oxide nanoparticles and poly (indole-5-carboxylic acid). Journal of Experimental Nanoscience. 2016; 11 111-122.

[3] Fang, Y., Ramasamy, R.P. Current and prospective methods for plant disease detection. Biosensors, 2015; 5 537-561.

[4] Gruere, G. Narrod, C., Abbott, L. Agriculture, Food, and Water Nanotechnologies for the Poor: Opportunities and Constraints, IFPRI Policy Brief, Vol. 19, 2011.

[5] Joseph, T., Morrison, M. Nanotechnology in agriculture and food, Nanoforum Report, 2006; 2 2-3.

[6] Mustafa, F., Andreescu, S. Chemical and biological sensors for food-quality monitoring and smart packaging, Foods, 2018; 7(10) 168.

[7] Gallocchio, F., Belluco, S., Ricci, A. Nanotechnology and food: brief overview of the current scenario, Procedia Food Sci., 2015; 5 85-88.

[8] Farahi, R.H., Passian, A., Tetard, L., Thundat, T. Critical issues in sensor science to aid food and water safety. ACS Nano 2012; 6 4548-4556.

[9] Sekhon, B.S. Nanotechnology in agri-food production: an overview, Nanotechnol Sci Appl. 2014; 7 31-53.

[10] Dasgupta, N., Ranjan, S., Mundekkad, D., Ramalingam, C., Shanker, R., Kumar, A. Nanotechnology in agro-food: from field to plate. Food Res Int. 2015; 69 381-400.

[11] Bhatia, S. Nanoparticles types, classification, characterization, fabrication methods and drug delivery applications. Natural polymer drug delivery systems Springer. 2016; 33-93.

[12] Ettenauer, J., Zuser, K., Kellner, K., Posnicek, T., Brandl, M. Development of an automated biosensor for rapid detection and quantification of E. coli in water. Procedia Engineering. 2015; 120 376-379.

[13] Sangadevan, S., Periasamy, M. Recent trends in Biosensors and their Application. Revised Advance Material Science. 2014; 62-69

[14] Reddy, M.N.K., Ratna, N.P. Nanobiosensors. Advance in Electronic and Electric Engineering. 2013; 3(3) 321-326.

[15] Cui, D., Guo, Q., Zhang X. Recent Advances in Nanotechnology Applied to Biosensors. Sensors. 2015; 15 30736-30758.

[16] Weiss, P.S. Nanoscience and nanotechnology: present and future. ACS Nano 2010; 4 1771-1772.

[17] You, C., Bhagawati, M., Brecht, A., Piehler, J. "Affinity Capturing for Targeting Proteins into Micro and nanostructures," Analytical and Bioanalytical Chemistry. 2009; 360 (6-7) 1563-1570.

[18] Velasco, M.N. "Optical Biosensors for Probing at the Cellular Level: A Review of Recent Progress and Future Prospects," Seminars in Cell & Developmental Biology. 2009; 20 (1) 27-33.

[19] Fan, X., White, I.M., Shopova, S.I., Zhu, H., Suter, J.D., Sun, Y. "Sensitive Optical Biosensors for Unlabeled Targets: A Review," Analytica Chimica Acta. 2008; 620 (1-2) 8-26.

[20] Khanna, V.K. "New-Generation Nano-Engineered Biosensors, Enabling Nanotechnologies and Nanomaterials," Sensor Review, Vol. 28, No. 1, 2008, pp. 39-45.

[21] A. Shana and K. R. Rogers, "Biosensors," Measurement Science and Technology, 1994; 5 461-472.

[22] Sharma, H., Agarwal, M., Goswami, M., Sharma, A., Roy, S.K. Biosensors: tool for food borne pathogen detection. Vet World. 2013; 6: 968-973.

[23] Bhalla, N., Jolly, P., Formisano, N., Estrela, P. Introductrion to bisensors. Essay Biochem. 2016; 60 1-8.

[24] Su, L., Jia, W., Hou, C., Lei, Y. Microbial biosensors: A review. Biosens Bioelectron. 2011; 26 1788-1799.

[25] Velusamy, V., Arshak, K., Korostynska, O., Oliwa, K., Adley, C. An overview of foodborne pathogen detection: In the perspective of biosensors. Biotechnol Adv. 2010; 28 232-254.

[26] Meshram, B.D., Agrawal, A.K., Adil, S., Ranvir, S., Sande, K.K. Biosensor and its application in food and dairy ındustry: A review. Int J Curr Microbiol App Sci. 2018; 7: 3305-3324.

[27] Singh, R., Mukherjee, M.D., Sumana, G., Gupta, R.K., Sood, S. Biosensors for pathogen detection: A smart approach toward clinical diagnosis. Sens Actuators B. 2014; 197 385-404.

[28] Ozoglu, O., Unal, M.A., Altuntaş, E.G. Biyosensörler: gıda ve sağlık alanında laktat biyosensorleri. Turk J Life Sci, 2017; 2 180-193.

[29] Neethirajan, S., Ragavan, V., Weng, X., Chand, R. Biosensors for sustainable food engineering: challenges and perspectives. Biosensors. 2018; 8 23

[30] Doria, G., Conde, J., Veigas, B., Giestas, L., Almeida, C., Assuncao, M., Rosa, J., Baptista, P.V. Noble Metal nanoparticles for Biosensing Applications. Sensors. 2012; 12 1657-1687.

[31] SRackauskas, S., Barbero, N., Barolo, C., Viscardi, G. ZnO Nanowire Application in Chemoresistive Sensing: A Review. Nanomaterials (Basel). 2017; 7(11) 381.

[32] Anam, M., Yori, O., Romana, S., Muhammad, A., Waheed, S., Sadia, Z. Nanosensors for diagnosis with optical, electric and mechanical transducers RSC Adv. 2019; 9 6793.

[33] Maduraiveeran, G., Jin, W. Nanomaterials based electrochemical sensor and biosensor platforms for *environmental* applications. Trends Environ. Anal. Chem. 2017; 13 10.

[34] Khan, I., Saeed, K., Khan, I. Nanoparticles: Properties, applications and toxicities Arabian Journal of Chemistry. 2019; 12 908-931.

[35] Kumar, H., Kuca, K., Bhatia, S. K., Saini, K., Kaushal, A. , Verma, R., Bhalla, T.C. Kumar, D. Applications of Nanotechnology in Sensor-Based Detection of Foodborne Pathogens. Sensors. 2020; *20*(7), 1966.

[36] Khanna, V.K. in Nanosensors, Physical Chemical and Biological (CRC Press, India) 2011; 1st ed., Chap. 2.

[37] Lim, T.C., Ramakrishna, S. A Conceptual Review of Nanosensors. Z. Naturforsch., 2006; 61, 402.

[38] Shipway, A.N., Katz, E., Willner, I. Nanoparticle arrays on surfaces for electronic, optical, and sensor applications. Chem. Phys. Chem. 2000; 1 18-52.

[39] Saini, R.K., Bagri, L.P., Bajpai, A.K. New Pesticides and Soil Sensors, ed. A.

Grumezescu Elsevier, Amsterdam. 2017; Chap. 14.

[40] Inbaraj, B.S., Chen, B. Nanomaterial-based sensors for detection of foodborne bacterial pathogens and toxins as well as pork adulteration in meat products. J. Food Drug Anal. 2015; 24 15-28.

[41] Thakur, M.S, Ragavan, K.V. Biosensors in food processing. J Food Sci Technol. 2013; 50: 625-641.

[42] Poltronieri, P., Mezzolla, V., Primiceri, E., Maruccio, G. Biosensors for the Detection of Food Pathogens. Foods. 2014; *3* 511-526.

[43] Hossain, S.Z., Ozimok, C., Sicard, C., Aguirre, S.D., Ali, M.M., Li, Y., Brennan, J.D. Multiplexed paper test strip for quantitative bacterial detection. Anal. Bioanal. Chem. 2012; *403* 1567-1576.

[44] Tan, L., Schirmer, K. Cell culture-based biosensing techniques for detecting toxicity in water. Current Opinion in Biotechnology. 2017; 45, 59-68.

[45] Abdalhai, M.H., Maximiano Fernandes, A., Bashari, M., Ji, J., He, Q., Sun, X. Rapid and sensitive detection of foodborne pathogenic bacteria (staphylococcus aureus) using an electrochemical DNA genomic biosensor and its application in fresh beef. Journal of Agricultural and Food Chemistry, 2014; 62 12659-12667.

[46] Huang, H., Liu, M., Wang, X., Zhang, W., Yang, D.-P., Cui, L. Label-free 3D Ag nanoflower-based electrochemical immunosensor for the detection of *Escherichia coli* O157: H7 pathogens. Nanoscale Research Letters, 2016; 11, 507.

[47] Miranda, O.R., Li, X., Garcia-Gonzalez, L., Zhu, Z.J., Yan, B., Bunz, U.H., Rotello, V.M. Colorimetric Bacteria Sensing Using a Supramolecular Enzyme-Nanoparticle Biosensor. J. Am. Chem. Soc. 2011; *133* 9650-9653.

[48] Jodi Woan-Fei, L., Ab Mutalib, N.S., Chan, K.G., Lee, L.H. Rapid Methods for the Detection of Foodborne Bacterial Pathogens: Principles, Applications, Advantages and Limitations. *Front. Microbiol*. 2015; *5*, doi:10.3389/fmicb.2014.00770.

[49] Amiri, M., Bezaatpour, A., Jafari, H., Boukherroub, R., Szunerits, S. Electrochemical Methodologies for the Detection of Pathogens. ACS Sensors, American Chemical Society. 2018; 3 (6) 1069-1086.

[50] Stephen Inbaraj, B., Chen, B.H. Nanomaterial-based sensors for detection of foodborne bacterial pathogens and toxins as well as pork adulteration in meat products. J. Food Drug Anal. 2016; 24(1) 15-28.

[51] Hendriksen, R.S., Vieira, A.R., Karlsmose, S., Lo Fo Wong, D.M., Jensen, A.B., Wegener, H.C., Aarestrup, F.M. Global monitoring of Salmonella serovar distribution from the World Health Organization Global Foodborne Infections Network Country Data Bank: Results of quality assured laboratories from 2001 to 2007. *Foodborne Pathog. Dis.* 2011; *8* 887-900.

[52] Amagliani, G., Brandi, G., Schiavano, G.F. Incidence and role of Salmonella in seafood safety. Food Res. Int. 2012; *45* 780-788.

[53] Joo, J., Yim, C., Kwon, D., Lee, J., Shin, H.H., Cha, H.J., Jeon, S. A facile and sensitive detection of pathogenic bacteria using magnetic nanoparticles and optical nanocrystal probes. Analyst. 2012; *137* 3609-3612.

[54] Cheli, F., Pinotti, L., Campagnoli, A., Fusi, E., Rebucci, R., Baldi, A. Mycotoxin Analysis, Mycotoxin-Producing Fungi Assays and Mycotoxin

Toxicity Bioassays in Food Mycotoxin Monitoring and Surveillance. Ital. J. Food Sci. 2008; *20* 447-462.

[55] Hussein, H.S., Brasel, J.M. Toxicity, metabolism, and impact of mycotoxins on humans and animals. Toxicology. 2001; *167*, 101-134.

[56] Fernandez-Cruz, M.L., Mansilla, M.L., Tadeo, J.L. Mycotoxins in fruits and their processed products: Analysis, occurrence and health implications. J. Adv. Res. 2010; *1* 113-122.

[57] Li, Z., Xue, N., Ma, H., Cheng, Z., Miao, X. An ultrasensitive and switch-on platform for aflatoxin B 1 detection in peanut based on the fluorescence quenching of graphene oxide-gold nanocomposites. Talanta, 2018; 181, 346-351.

[58] Hosseini, M., Khabbaz, H., Dadmehr, M., Ganjali, M.R., Mohamadnejad, J. Aptamer-Based Colorimetric and Chemiluminescence Detection of Aflatoxin B1 in Foods Samples. Acta Chim. Slov. 2015; *62* 721-728.

[59] Luan, Y., Chen, J., Xie, G., Li, C., Ping, H., Ma, Z., Lu, A. Visual and microplate detection of aflatoxin B2 based on NaCl-induced aggregation of aptamer-modified gold nanoparticles. Microchim. Acta. 2015; *182* 995-1001.

[60] Singh, B.K., Walker, A. Microbial degradation of organophosphorus compounds. FEMS Microbiol. Rev. 2006; *30* 428-471.

[61] Kumar, S.V., Fareedullah, M., Sudhakar, Y., Venkateswarlu, B., Kumar, E.A. Current review on organophosphorus poisoning. Arch. Appl. Sci. Res. 2010; *2* 199-215.

[62] Andreescu, S., Marty, J.L. Twenty years research in cholinesterase biosensors: From basic research to practical applications. Biomol. Eng. 2006; *23* 1-15.

[63] Pope, C.N. Organophosphorus pesticides: Do they all have the same mechanism of toxicity? J. Toxicol. Environ. Health B Crit. Rev. 1999; *2* 161-181.

[64] Valdes, M., Valdes Gonzalez, A., García Calzón, J., Diaz-García, M. Analytical nanotechnology for food analysis. Microchim. Acta. 2009; *166* 1-19.

[65] Perez-Lopez, B., Merkoci, A. Nanomaterials based biosensors for food analysis applications. Trends Food Sci. Technol. 2011; *22* 625-639.

[66] Ellman, G.L., Courtney, K.D., Andres, V., Featherstone, R.M. A new and rapid colorimetric determination of acetylcholinesterase activity. Biochem. Pharmacol. 1961; *7* 88-95.

[67] Hossain, S.M.Z., Luckham, R.E., Smith, A.M., Lebert, J.M., Davies, L.M., Pelton, R.H., Filipe, C.D., Brennan, J.D. Development of a Bioactive Paper Sensor for Detection of Neurotoxins Using Piezoelectric Inkjet Printing of Sol-Gel-Derived Bioinks. Anal. Chem. 2009; *81*, 5474-5483.

[68] Hossain, S.M.Z. Luckham, R.E., McFadden, M.J., Brennan, J.D. Reagentless Bidirectional Lateral Flow Bioactive Paper Sensors for Detection of Pesticides in Beverage and Food Samples. Anal. Chem. 2009; *81* 9055-9064.

[69] Luckham, R.E., Brennan, J.D. Bioactive paper dipstick sensors for acetylcholinesterase inhibitors based on sol-gel/enzyme/gold nanoparticle composites. Analyst 2010; *135* 2028-2035.

[70] Cui, H.F., Wu, W.W., Li, M.M., Song, X., Lv, Y., Zhang, T.T. A highly stable acetylcholinesterase biosensor

based on chitosan-TiO2-graphene nanocomposites for detection of organophosphate pesticides. Biosensors and Bioelectronics. 2018; 99, 223-229.

[71] Kim, H.N., Ren, W.X., Kim, J.S., Yoon, J. Fluorescent and colorimetric sensors for detection of lead, cadmium, and mercury ions. Chem. Soc. Rev. 2012; *41* 3210-3244.

[72] Zhao, C., Zhong, G., Kim, D.E., Liu, J., Liu, X. A portable lab-on-a-chip system for gold-nanoparticle-based colorimetric detection of metal ions in water. *Biomicrofluidics.* 2014; *8* 052107.

[73] Du, J., Wang, Z., Fan, J., Peng, X. Gold nanoparticle-based colorimetric detection of mercury ion via coordination chemistry. Sens. Actuat. B Chem. 2015; *212* 481-486.

[74] Chansuvarn, W., Tuntulani, T., Imyim, A. Colorimetric detection of mercury(II) based on gold nanoparticles, fluorescent gold nanoclusters and other gold-based nanomaterials. TrAC Trends Anal. Chem. 2015; *65* 83-96.

[75] Zhou, Y.; Dong, H.; Liu, L.; Li, M.; Xiao, K.; Xu, M. Selective and sensitive colorimetric sensor of mercury (II) based on gold nanoparticles and 4-mercaptophenylboronic acid. Sens. Actuat. B Chem. 2014; *196*; 106-111.

[76] Chen, G.H. Chen, W.Y. Yen, Y.C., Wang, C.W., Chang, H.T., Chen, C.F., Detection of Mercury (II) Ions Using Colorimetric Gold Nanoparticles on Paper-Based Analytical Devices. Anal. Chem. 2014; *86*, 6843-6849.

[77] Sener, G., Uzun, L., Denizli, A. Colorimetric Sensor Array Based on Gold Nanoparticles and Amino Acids for Identification of Toxic Metal Ions in Water. ACS Appl. Mater. Interfaces. 2014; *6* 18395-18400.

[78] Alizadeh, A. Khodaei, M.M. Hamidi, Z. Shamsuddin, M.B. Naked-eye colorimetric detection of Cu2+ and Ag+ ions based on close-packed aggregation of pyridines-functionalized gold nanoparticles. Sens. Actuat. B Chem. 2014; *190* 782-791.

[79] Annadhasan, M., Muthukumarasamyvel, T., Sankar Babu, V.R., Rajendiran, N. Green Synthesized Silver and Gold Nanoparticles for Colorimetric Detection of Hg2+, Pb2+, and Mn2+ in Aqueous Medium. ACS Sustain. Chem. Eng. 2014; *2* 887-896.

[80] Sung, Y.M. Wu, S.P. Colorimetric detection of Cd(II) ions based on di-(1H-pyrrol-2-yl) methanethione functionalized gold nanoparticles. Sens. Actuat. B Chem. 2014; *201* 86-91.

[81] Liana, D.D., Raguse, B., Wieczorek, L., Baxter, G.R., Chuah, K., Gooding, J.J., Chow, E. Sintered gold nanoparticles as an electrode material for paper-based electrochemical sensors. RSC Adv. 2013; *3*, 8683-8691.

[82] Kuswandi, et al., A novel on-package sticker sensor based on methyl red for real-time monitoring of broiler chicken cut freshness, Packag. Technol. Sci., 2014; 27(1), 69-81.

[83] Kuswandi, B., Nurfawaidi. On-package dual sensors label based on pH indicators for real-time monitoring of beef freshness, Food Control, 2017; 82 91-100.

[84] Schaude, C. et al., Developing a sensor layer for the optical detection of amines during food spoilage, Talanta, 2017; 170 481-487.

[85] El-Nour, K. et al., Gold nanoparticles as a direct and rapid sensor for sensitive analytical detection of biogenic amines, Nanoscale Res. Lett., 2017; 12(1), 231.

[86] Putri, V. et al., Application Nano Zeolite-Molybdate for Avocado Ripeness Indicator, in IOP Conference Series:

Earth and Environmental Science, IOP Publishing, 2019.

[87] Lang, C., Hubert, T. A colour ripeness indicator for apples, Food Bioprocess Technol., 2012; 5(8) 3244-3249.

[88] Mills, A. et al., An O-2 smart plastic film for packaging, Analyst, 2012; 137(1) 106-112.

[89] de Chiara, M.L.V. et al., Photocatalytic degradation of ethylene on mesoporous TiO2/SiO2 nanocomposites: Effects on the ripening of mature green tomatoes, Biosyst. Eng., 2015; 132 61-70.

[90] Ranjan, S., Dasgupta, N., Chakraborty, A.R, Samuel, S.M., Ramalingam, C., Shanker, R., Kumar, A. Nanoscience and nanotechnologies in food industries: opportunities and research trends. J Nanopart Res. 2014; 16 2464.

[91] Dasgupta, N., Ranjan, S., Mundekkad, D., Ramalingam, C., Shanker, R., Kumar, A. Nanotechnology in agro-food: from field to plate. Food Res Int. 2015; 69 381-400.

[92] Wilson, A.D., Baietto, M. Applications and advances in electronic-nose technologies. Sensors. 2009; 9 5099-5148.

[93] Pimtong-Ngam, Y., Jiemsirilers, S., Supothina, S. Preparation of tungsten oxide–tin oxide nanocomposites and their ethylene sensing characteristics. Sensor Actuators A-Phys. 2007; 139:7-11

[94] Kim, T.H., Lee, S.H., Lee, J., Song, H.S., Oh, E,H., Park, H.T., Hong, S. Single-carbon-atomic-resolution detection of odorant molecules using a human olfactory receptor-based bioelectronic nose. Adv Mater.2009; 21 91-94.

[95] Jin, H.J., Lee, S.H., Kim, T.H., Park, J., Song, H.S., Park, T.H., Hong, S. Nanovesicle-based bioelectronic nose platform mimicking human olfactory signal transduction. Biosens Bioelectron. 2012; 35:335-341.

[96] Park, J., Lim, J.H., Jin, H.J., Namgung, S., Lee, S.H., Park, T.H., Hong, S. A bioelectronic sensor based on canine olfactory nanovesicle carbon nanotube hybrid structures for the fast assessment of food quality. Analyst. 2012; 137 3249-3254.

[97] Ghaffar, N., Farrukh, M.A., Naz, S. Applications of Nanobiosensors in Agriculture. In: Javad S. (eds) Nanoagronomy. Springer, Cham. 2020; 10 179-196.

[98] Li, Y.-W., Xia, K., Wang, R.-Z., Jiang, J.-H., & Xiao, L.-T. (2008). An impedance immunosensor for the detection of the phytohormone abscisic acid. Analytical and Bioanalytical Chemistry, 391, 2869-2874

[99] Li, J., Xiao, L.T., Zeng, G.M., Huang, G.H., Shen, G.L., and Yu, R.Q. Immuno-sensor for rapid detection of gibberellin acid in the rice grain. Journal of Agricultural and Food Chemistry, 2005; 53, 1348-1353.

[100] Zavar, M., Heydari, S., & Rounaghi, G. Electrochemical determination of salicylic acid at a new biosensor based on polypyrrole-banana tissue composite. Arabian Journal for Science and Engineering. 2013; 38 29-36.

[101] Tian, F., Greplova, M., Frebort, I., Dale, N., & Napier, R. A highly selective biosensor with nanomolar sensitivity based on cytokinin dehydrogenase. PLoS One, 2014; 9 1-7.

[102] Wend, S., Bosco, C.D., Kampf, M.M., Ren, F., Palme, K., Weber, W. A quantitative ratiometric sensor for time-resolved analysis of auxin dynamics. Scientific Reports, 2013; 3, 1-7.

[103] Larrieu, A., Champion, A., Legrand, J., Lavenus, J., Mast, D., Brunoud, G. A fluorescent hormone biosensor reveals the dynamics of jasmonate signalling in plants. Nature Communications. 2015; 6, 1-8.

[104] Chong, J.P.C., Liew, O.W., Li, B.Q., Asundi, A.K. Optical fluorescence biosensor for plant water stress detection. Saratov fall meeting 2006: Optical technologies in biophysics and medicine: Vol. International Society for Optics and Photonics. 2007; 65350-65358.

[105] Kahveci, Z., Martinez-Tome, M.J., Mallavia, R., Mateo, C.R. Fluorescent biosensor for phosphate determination based on immobilized polyfluorene-liposomal nanoparticles coupled with Alkaline Phosphatase. ACS Applied Materials & Interfaces. 2016; 9(1), 136-144

[106] Barney, B.M., Eberhart, L.J., Ohlert, J.M., Knutson, C.M., Plunkett, M.H. Gene deletions resulting in increased nitrogen release by azotobacter vinelandii: Application of a novel nitrogen biosensor. Applied and Environmental Microbiology. 2015; 81, 4316-4328

[107] Garcia-Martinez, J. The Internet of Things Goes Nano, http://www.scientificamerican.com/article/the-internet-of-things-goes-nano/.2016.

[108] Kuswandi, B., Wicaksono, Y., Jayus, A., Abdullah, A., Heng, L.Y., Ahmad, M. Smart packaging: sensors for monitoring of food quality and safety. Sens. Instrum. Food Qual. Saf. 2011; 5 (3-4), 137-146.

[109] Sassolas, A., Prieto-Simon, B., Marty, J.L. Biosensors for pesticide detection: New trends. American Journal of Analytical Chemistry, 2012; 3 210-232.

[110] Soldatkin, O.O., Peshkova, V.M., Saiapina, O.Y., Kucherenko, I.S., Dudchenko, O.Y., Melnyk, V.G., Development of conductometric biosensor array for simultaneous determination of maltose, lactose, sucrose and glucose. Talanta, 2013; 115 200-207.

[111] Chiou, T.J., Lin, S.I. Signaling network in sensing phosphate availability in plants. Annual Review of Plant Biology. 62; 185-206.

[112] Raul, R.G., Irineo, T.P., Gerardo, G.G.R., Miguel, C.M.L. Biosensors used for quantification of nitrates in plants. Journal of Sensors. 2016; 1-12.

[113] Liew, O.W., Chong, P.C.J., Li, B., Asundi, A.K. Signature Optical Cues: Emerging Technologies for Monitoring Plant Health. Sensors. 2008; 8 3205-3239;

[114] Gruere G.P. Implications of nanotechnology growth in food and agriculture in OECD countries. Food Policy. 2012; 37 191-198.

[115] Hajirostamlo, B., Mirsaeedghazi, N., Arefnia, M., Shariati, M.A., Fard E.A. The role of research and development in agriculture and its dependent concepts in agriculture [Short Review]. Asian J. Appl. Sci. Eng. 2015; 4 1010-1016.

[116] Mukhopadhyay, S.S. Nanotechnology in agriculture: prospects and constraints. Nanotechnol. Sci. Appl. 2014; 7 63-71.

[117] Prasad, R., Kumar, V., Prasad, K.S. Nanotechnology in sustainable agriculture: present concerns and future aspects. Afr. J. Biotechnol. 2014; 13 705-713.

[118] Nicewarner-Pena, S.R., Carado, A.J., Shale, K.E., Keating, C.D. Barcoded Metal Nanowires: Optical Reflectivity and Patterned Fluorescence. The Journal of Physical Chemistry B. 2003; *107* (30) 7360-7367.

[119] Li, Y., Cu, H.Y.T., Luo, D. Multiplexed detection of pathogen DNA with DNA-based fluorescence nanobarcodes. Nat Biotechnol. 2005; 23 885-889.

[120] Lin, C., Jungmann, R., Leifer, A.M., Li, C., Levner, D., Church, G.M., Shih, W.M., Yin, P. Submicrometre geometrically encoded fluorescent barcodes self-assembled from DNA. Nat Chem. 2012; 4 832-839.

[121] Yao, K.S., Li, S.J., Tzeng, K.C., Cheng, T.C., Chang, C.Y., Chiu, C.Y., Liao, C.Y., Hsu, J.J., Lin, Z.P. Fluorescence silica nanoprobe as a biomarker for rapid detection of plant pathogens. Adv. Mater. Res. 2009; *79* 513-516.

[122] Singh, S., Singh, M., Agrawal, V.V., Kumar, A. An attempt to develop surface plasmon resonance based immunosensor for Karnal bunt (*Tilletia indica*) diagnosis based on the experience of nano-gold based lateral flow immuno-dipstick test. Thin Solid Films. 2010; *519* 1156-1159.

[123] Frasco, M.F., Chaniotakis, N. Semiconductor quantum dots in chemical sensors and biosensors. Sensors. 2009; *9* 7266-7286.

[124] Algar, W.R., Krull, U.J. Quantum dots as donors in fluorescence resonance energy transfer for the bioanalysis of nucleic acids, proteins, and other biological molecules. Anal. Bioanal. Chem. 2008; *391* 1609-1618.

[125] Rad, F., Mohsenifar, A., Tabatabaei, M., Safarnejad, M.R., Shahryari, F., Safarpour, H., Foroutan, A., Mardi, M., Davoudi, D., Fotokian, M. Detection of Candidatus Phytoplasma aurantifolia with a quantum dots fret-based biosensor. J. Plant Pathol. 2012; *94* 525-534.

[126] Safarpour, H., Safarnejad, M.R., Tabatabaei, M., Mohsenifar, A., Rad, F., Basirat, M., Shahryari, F., Hasanzadeh, F. Development of a quantum dots FRET-based biosensor for efficient detection of Polymyxa betae. Can. J. Plant Pathol. 2012; *34* 507-515.

[127] Kuila, T., Bose, S., Khanra, P., Mishra, A.K., Kim, N.H., Lee, J.H. Recent advances in graphene-based biosensors. Biosens. Bioelectron. 2011; *26* 4637-4648.

[128] Perez-Lopez, B., Merkoçi, A. Nanoparticles for the development of improved (bio) sensing systems. Anal. Bioanal. Chem. 2011; *399* 1577-1590.

[129] Shiddiky, M.J., Torriero, A.A. Application of ionic liquids in electrochemical sensing systems. Biosens. Bioelectron. 2011; *26* 1775-1787.

[130] Cao, X., Ye, Y., Liu, S. Gold nanoparticle-based signal amplification for biosensing. Anal. Biochem. 2011; *417* 1-16.

[131] Mandler, D., Kraus-Ophir, S., Self-assembled monolayers (SAMs) for electrochemical sensing. J. Solid State Electrochem. 2011; *15* 1535-1558.

[132] Umasankar, Y., Ramasamy, R.P. Highly sensitive electrochemical detection of methyl salicylate using electroactive gold nanoparticles. Analyst. 2013; *138* 6623-6631.

[133] Boonham, N., Glover, R., Tomlinson, J., Mumford, R. Exploiting generic platform technologies for the detection and identification of plant pathogens. In *Sustainable Disease Management in a European Context*; Springer: Berlin, Germany, 2008; 355-363.

[134] Chartuprayoon, N., Rheem, Y., Chen, W., Myung, N. Detection of plant pathogen using LPNE grown single conducting polymer Nanoribbon. In *Meeting Abstracts*; The Electrochemical Society: Pennington, NJ, USA, 2010.

[135] Vermeir, S., Nicolai, B.M., Verboven, P., Van Gerwen, P., Baeten, B., Hoflack, L. Microplate differential calorimetric biosensor for ascorbic acid analysis in food and pharmaceuticals. Analytical Chemistry. 2007; 79, 6119-6127.

[136] Lupetti, K.O., Vieira, I.C., Fatibello-Filho, O. Jack fruit-capric acid biosensor for total phenols determination in wastewaters. Analytical Letters 2004; 37 1833-1846.

[137] Verma, N., Singh, A.K., Kaur, P. Biosensor based on ion selective electrode for detection of L-arginine in fruit juices. Journal of Analytical Chemistry. 2015; 70 1111-1115.

[138] Bontidean, I., Mortari, A., Leth, S., Brown, N.L., Karlson, U., Larsen, M.M. Biosensors for detection of mercury in contaminated soils. Environmental Pollution. 2004; 131 255-262.

[139] Srivastava, S., Ali, M.A., Umrao, S., Parashar, U.K., Srivastava, A., Sumana, G. Graphene oxide-based biosensor for food toxin detection. Applied Biochemistry and Biotechnology. 2014; 174 960-970.

[140] Davis, H.T., Aelion, C.M., Liu, J., Burch, J.B., Cai, B., Lawson, A.B. Potential sources and racial disparities in the residential distribution of soil arsenic and lead among pregnant women. The Science of the Total Environment. 2016; 551-552,

[141] Oliveira, H. Chromium as an environmental pollutant: Insights on induced plant toxicity. Journal of Botany. 2012; 1-8

[142] Wyszkowska, J., Borowik, A., Kucharski, M., Kucharski, J. Effect of cadmium, copper and zinc on plants, soil microorganisms and soil enzymes. Journal of Elementology. 2013; 18 769-796.

[143] Bhalerao, S.A., Sharma, A.S., Poojari, A.C. Toxicity of nickel in plants. International Journal of Pure and Applied Bioscience. 2015; 3 345-355.

[144] Tekaya, N., Saiapina, O., Ouada, H.B., Lagarde, F., Namour, P., Ouada, H.B. Bi-Enzymatic conductometric biosensor for detection of heavy metal ions and pesticides in water samples based on enzymatic inhibition in arthrospira platensis. Journal of Environmental Protection. 2014; 5 1-13.

[145] Zuo, X., Zhang, H., Zhu, Q., Wang, W., Feng, J., Chen, X. A dual-color fluorescent biosensing platform based on WS2 nanosheet for detection of Hg2+ and Ag+. Biosensors and Bioelectronics. 2016; 85 464-470.

[146] Rigo, A.A., Cezaro, A.M.D., Muenchen, D.K., Martinazzo, J., Manzoli, A., Steffens, J., Steffens, C. Heavy metals detection in river water with cantilever nanobiosensor. Journal of Environmental Science and Health. 2020; 36 (4) 1429-1437.

[147] Chen, H., Yada, R. Nanotechnologies in agriculture: new tools for sustainable development. Trends Food Sci Technol. 2011; 22 585-594.

[148] Berekaa, M. Nanotechnology in Food Industry; Advances in Food processing, Packaging and Food Safety, Int.J. Curr. Microbiol. App. Sci. 2015; 4(5) 345-357.

Chapter 11

Ti_3C_2 MXene-Based Nanobiosensors for Detection of Cancer Biomarkers

Lenka Lorencova, Kishor Kumar Sadasivuni, Peter Kasak and Jan Tkac

Abstract

This chapter provides information about basic properties of MXenes (2D nanomaterials) that are attractive for a design of various types of nanobiosensors. The second part of the chapter discusses MXene synthesis and various protocols for modification of MXene making it a suitable matrix for immobilization of bioreceptors such as antibodies, DNA aptamers or DNA molecules. The final part of the chapter summarizes examples of MXene-based nanobiosensors developed using optical, electrochemical and nanomechanical transducing schemes. Operational characteristics of such devices such as sensitivity, limit of detection, assay time, assay reproducibility and potential for multiplexing are provided. In particular MXene-based nanobiosensors for detection of a number of cancer biomarkers are shown here.

Keywords: MXene, nanomaterials, biosensors, cancer, biomarkers

1. Introduction

1.1 MXenes: their precursors, characterization, unique properties and applications

Nanomaterials of the 2D kind are in the research spotlight due to their superior properties like ultrathin structure and intriguing physico-chemical properties [1–3]. Graphene has made researchers believing in extracting single layer transition metal dichalcogenides, which in turn has led to extensive research dedicated towards 2D nanomaterials [4, 5]. Since their inception, 2D nanomaterials have been characterized to have exceptional electronic, mechanical, and optical properties. These outstanding characteristics have driven research to use them in almost all fields of materials science and nanotechnology [6–8]. Rather recently in 2011 and 2012, Gogotsi, Barsoum, and colleagues have successfully prepared a new kind of 2D nanomaterial - MXenes, composed of a large group of transition metal carbides and carbonitrides [9–13]. These 2D nanomaterials are found to possess many striking properties and boost attraction in applications such as energy storage [14–16], electromagnetic shielding [17, 18], water treatment [19, 20], disease treatment [21] and (bio)sensing [22, 23], MXenes are made up of atomic layers of different materials like transition metal carbides, nitrides, or carbonitrides. All MAX phases consist of two-dimensional slabs of close-packed alternating layers of M and A, where M is a transition metal, A is an A-group element and X is C and/or N [23].

The selective chemical etching of “A” in “MAX” phases have led to successful synthesis of MXenes. MAX phases are found to have elusive properties like stiff elasticity, good thermal and electrical conductivity, as well as relatively low thermal expansion coefficients and resistance towards chemical attack. There is a general formula for MXene synthesis where, “MAX” phases have a formula of $M_{n+1}AX_n$, with “M” meaning early d-transition metal, “A” representing the main group sp-element, and “X” indicates C and/or N [24]. Hence, with this analogy, more than 70 different kinds of MXenes with different M and X are theoretically possible to synthesize. Out of these theoretical MXene types, 20 different combinations of MXenes have been synthesized successfully [25]. MXenes can conduct heat and electricity like metals and are strong and brittle like ceramics with high surface area. Exfoliated MXene exhibits higher pseudocapacitance than most capacitive materials [26]. Additionally, the MXene-have properties like a clay. Furthermore, Ca^{2+}, Mg^{2+} and Al^{3+} ions (intercalated polyvalent cations) have all shown a huge storage power capacity [27–29]. It needs to be stressed out that energy storage capacity, high conductivity, photochemical properties, modulated surface chemistry and tunable composition make MXene and their derivatives very perspective to (bio)sensing applications.

2D MXenes are candidates for energy storage [30] (Li-ion batteries, supercapacitors) and electromagnetic interference shielding applications [31–35] and in the form of composites become ever more useful for sensing as *e.g.* gas sensing devices [36, 37], pressure sensor [38, 39] and sensors for various analytes [40–43]. Number of other biomedical applications (such as biosensor, biological imaging, photothermal therapy, drug delivery, theranostic nanoplatforms and antibacterial agents) have become a challenge for MXenes [44]. The antibacterial properties making them potentially appealing for nanomedicine were proved for ($Ti_3C_2T_x$) MXene quantum dots [45], MXene-hybridized silane film [46], Cu_2O/MXene [47] and MXene-gold nanoclusters [48] *etc*. The multifunctional MXenes have attracted attention in biosensing [49, 50] with the aim at the ultrasensitive determination of cancer diseases related biomarkers. Examples include biosensors based on Ti_3C_2 MXenes-Au NPs hybrids, delaminated $Ti_3C_2T_x$ MXene@AuNPs, nanohybrid of $Ti_3C_2T_x$ MXene and phosphomolybdic acid (PMo_{12}) embedded with polypyrrole, MXene-TiO_2/$BiVO_4$ hybrid and AuNPs/Ti_3C_2 MXene three-dimensional nanocomposite for detection of carcinoembryonic antigen [51], prostate specific antigen [52], osteopontin [53], CD44 [54] and microRNA-155 [55], respectively.

2. MXene synthesis

Generally, top-down selective etching process is used for the synthesis of MXenes [56]. Strong etching solutions containing a fluoride ion (F^-) such as hydrofluoric acid (HF), ammonium difluoride (NH_4HF_2), and a mixture of hydrochloric acid (HCl) and lithium fluoride (LiF) are used for production of MXene in such processes [57]. Since typically, the etching process results in replacement of the M-A bond by M-O, M-OH, M-H, and M-F bonds on the surface of MXenes, the structure of MXenes can be expressed as $M_{n+1}X_nT_x$ or $M_{n+1}X_n$ (M and X are in same form as the MAX phase and T is =O, -OH, -H, or -F) [58, 59].

A single and/or few layers of MXene can be synthetized by exfoliation or delamination of a multilayer structure of a MAX phase. The composition and electrochemical properties of MXene strongly depend on the conditions used during etching procedure [60]. As an example, application of LiF/HCl as etchant led to production of MXene with interlayers intercalated with Li^+ ions. Exfoliation can be done by a simple shaking or by sonication and prolonged sonication time results in production of MXene with small size of nanosheets and high density of defects [61]. An alternative to use of highly corrosive and harmful HF is to employ small organic molecules or ions such as urea [62], dimethyl sulphoxide (DMSO) [12] (only for $Ti_3C_2T_x$ MXene) or isopropylamine as etchants [63]. MAX phase containing Si can be also exfoliated using tetrabutylammonium hydroxide (TBAOH) and tetramethylammonium hydroxide (TMAOH) [64].

3. MXene characterization

Since introduction of nanolayered and machinable MXenes in 2011 by Gogotsi and co-workers through wet-etching process with HF to obtain multilayered flakes of $Ti_3C_2T_x$ [13], few improvements in MXene synthesis and MXene-nanocomposite preparation resulted in various elemental composition and surface functionality [65]. In last few years the single layers of MXene were isolated adding salts or organic solvents (NH_4HF_2, tetrabuthylammonium hydroxide, isopropylamine) during synthesis process and resulted in delaminated MXene layers. The significant breakthrough for MXene synthesis named as "clay method" in 2014 was based on *in situ* formation of HF (LiF/HCl). The lattice *c* parameter increased to a value of $\approx$40 Å by applying LiF-HCl as an etchant to produce $Ti_3C_2T_x$ instead of HF etchant with a lattice *c* parameter of 20 Å [60]. The battery of techniques were employed to observe variations in the composition of $Ti_3C_2T_x$ MXene produced either by HF or LiF-HCl method including nuclear magnetic resonance (^1H, ^{13}C and ^{19}F NMR), scanning electron microscopy (SEM), X-ray diffraction method (XRD), energy-dispersive X-ray spectroscopy (EDS) techniques [59]. The most suitable combination presented utilization of LiF/HCl as an etchant with minimally intensive layer delamination "MILD" method instead of sonication to produce huge MXene flakes with minimum of defects [66]. $Ti_3C_2T_x$ MXene has become an attractive subject of interest due its high capacitance ($\sim$1500 F cm^{-3}) in supercapacitors and an excellent high metallic conductivity ($\sim$15,000 S cm^{-1}). On the other hand there is still demand to improve stability of MXene flakes with a poor resistance in aerated aqueous suspensions resulting in oxidized form with loss of its activity for potential applications [67]. The optimization of etching process is cardinal to access single- to few-layer Ti_3C_2 MXene flakes. SEM technique providing information about flake size and distribution revealed formation of aggregates on the surface varying in size i.e. having few μm in size or with size larger than 10 μm in a lateral dimension. It was found out by atomic force microscopy (AFM), that thickness of single MXene monolayer was (1.1 $\pm$ 0.1) nm for $Ti_3C_2T_x$ [68]. Platinum nanoparticles with average diameter of 3 nm were homogeneously distributed on the MXene sheets surface, that was found out by transmission electron microscopy (TEM) [69]. MXene and oxidized MXene were analyzed and differentiated by applying Raman spectroscopy method providing more detailed information about the characteristic vibrational bands and the dependence thickness of Ti_3C_2Tx layers on Raman signal enhancement [68–71].

The electrochemical behavior employing methods like cyclic voltammetry (CV), chronoamperometry (CA), differential pulse voltammetry (DPV) and electrochemical impedance spectroscopy (EIS) revealed significant findings related to the electrochemical activity of MXene. The electrochemical investigation of $Ti_3C_2T_x$ MXene to detect significant analytes (O_2, H_2O_2 and NADH) was performed by applying cyclic voltammetry and chronoamperometry techniques, whereas $Ti_3C_2T_X$ demonstrated electrocatalytic activity towards H_2O_2 reduction with LOD at nanomolar level [68]. Unfortunately, formation of TiO_2 layer or domains with subsequent TiO_2 dissolution caused by F^- ions was observed during oxidation process at anodic potential window in a plain phosphate buffer electrolyte pH 7.0 leading to the decrease in electrochemical activity of $Ti_3C_2T_x$ MXene.

The improvement of stability and redox behavior was achieved by further modification of MXene with nanoparticles of platinum ($Ti_3C_2T_x$/Pt) [69, 72]. The electrocatalytically active sensor based on $Ti_3C_2T_x$/Pt nanocomposite successfully determined H_2O_2 by CA, and moreover small organic molecules (acetaminophen, dopamine, ascorbic acid, uric acid) were selectively determined by DPV [72].

In addition electrochemical study confirmed significant differences in a negative charge density on the MXene surface as well electrocatalytic activity depending on the etchant (HF, LiF/HCl) used during MXene synthesis with preference towards utilization of LiF/HCl [60].

Aryldiazonium salts were utilized in modification of $Ti_3C_2T_x$ MXene either spontaneously by free electrons or electrochemically. Electrochemical modification of $Ti_3C_2T_x$ MXene by aryldiazonium-based grafting with derivatives bearing a SB- or CB- betaine pendant moiety was performed by cyclic voltammetry in a potential window from 0 V to −1 V with a sweep rate of 0.25 V s^{-1} and 48 cycles. The electrochemical grafting resulted in denser CB or SB layer on MXene interface, lower interfacial resistance and an electrochemically active surface area for SB layer in comparison to CB layer [73].

In the following years the exponential increase in the number of affinity-based MXene biosensors can be expected, though it is necessary to develop advanced strategies for modification of MXene interfaces with an effort to eliminate non-specific binding of proteins, bring in anti-fouling behavior and immobilize target biomolecules. Electrochemical methods can be employed as a useful tools for interfacial patterning, characterization of MXene-based biosensors and furthermore ultrasensitive detection of cancer related biomarkers [23].

4. MXene functionalization

4.1 Covalent modification of Ti_3C_2 MXenes with biomolecules

Functionalization and various methods for synthesis of MXenes can result in production of the nanomaterial with a diverse range of properties. This is why, it is very important to describe synthesis of MXenes in full details. Another point to focus on is to properly describe delamination conditions since the flake size and density of defects governs MXene's surface properties and stability. It is important to know the molecular structure of MXenes in order to decide the best application of such nanomaterial for catalysis, (bio)sensing or for chemical adsorption of various compounds.

Due to presence of -OH groups on surface, functionalization of MXene employing silylation reagents was developed by a simple reaction with triethoxysilane derivatives [74–76]. Such modification led to production of nanosheets of Ti_3C_2-MXene uniformly patterned by aminosilane moieties allowing NHS/EDC-based amine coupling for covalent immobilization of bioreceptors such as anti- carcinoembryonic antigen (CEA) antibodies [77].

Another viable surface modification of MXenes can be done by applying zwitterions. It was observed that spontaneous grafting of sulfobetaine (SB) and carboxybetaine (CB) derivatives onto $Ti_3C_2T_x$ MXene is feasible [73]. The approach is similar to spontaneous grafting of diazonium salt modified zwitterions to gold nanoshell modified particles by consuming surface plasmons (free electron cloud) present within Au nanoshells [78]. Even though spontaneous grafting of diazonium

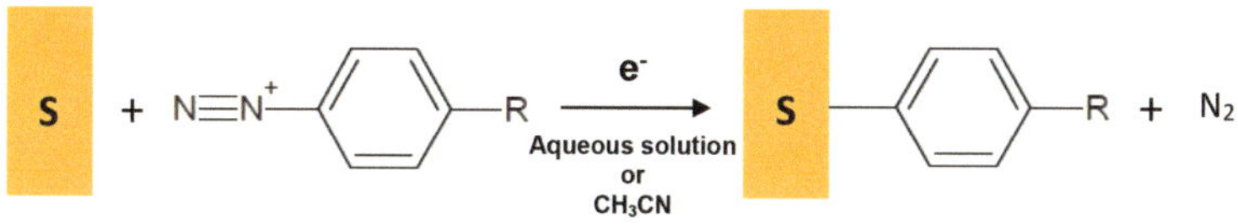

Figure 1.
Electrochemically triggered grafting of diazonium salt-containing compounds to conductive surfaces. Electrochemical reduction of diazonium salt-containing compounds is feasible via freely available clouds of electrons (plasmons) present in metallic nanoparticles, but also in MXene.

salt modified zwitterions to MXene was feasible, electrochemically triggered grafting of diazonium salts bearing zwitterionic pendants was more effective (**Figure 1**) [73]. Electrochemical characterization tools confirmed a much quicker spontaneous SB grafting compared to spontaneous CB grafting. Zwitterionic modification is considered as a benchmark to design antibiofouling interfaces with such modification offering to reduce dramatically non-specific protein binding compared to an unmodified MXene interface [73]. It is worth mentioning that grafting of a mixed layer composed of CB and SB can be applied to tune density of carboxylic groups and by amine coupling chemistry it is possible to finely tune density of immobilized bioreceptors for effective and efficient recognition of an analyte *via* affinity interactions [79]. Diazonium salts can be utilized in order to achieve stable modification of all surfaces (radical reaction providing most often disordered oligomers ("multilayers")) [80].

Diazonium salts can be easily synthetized from aromatic amines that are commercially available. Modification can be performed by applying different grafting methods like electrochemistry, spontaneous reduction, by reducing surfaces and reagents, photochemistry etc.

Besides application of APTES there are other strategies for modification of MXene such as self-initiated photo-grafting and photopolymerization not requiring an anchor layer, self-assembled monolayer (SAM) and initiator, applying a nature polymer, soy phospholipid (SP) improving permeability, stable cycling, and retention and PEGylation of MXene improving the water dispersibility of MXene by electrostatic adsorption [81].

Recently, a novel MXene modification approach was developed by substitution or elimination reactions in molten inorganic salts. Such modification allowed to synthetize MXenes containing = O, -NH, =S, -Cl, -Se, -Br, and -Te surface terminations [82].

4.2 Preparation of hybrid nanoparticles based on MXene

The hydrothermal method run in a Teflon-lined stainless steel autoclave (150°C, 5 h; aqueous solution of vitamin C and Fe^{3+} salt) allowed preparation of composite of MXene with small magnetic Fe_3O_4 nanoparticles with an average size of ~4.9 nm (TiO_2/$Ti_3C_2T_x$/Fe_3O_4). These hybrid magnetic nanoparticles show a great promise for selective enrichment of various biomolecules/antigens based on affinity interactions [83].

Other promising nanocomposite option is represented by MXene sheets combined with metallic NPs [84–87], which can be further effectively modified by crosslinkers due to their high affinity towards MXene or by other biomolecules for final detection of target molecules/biomarkers. MXene/metallic nanoparticles (NPs) based nanocomposites can be prepared by spontaneous reduction of salts of precious metals or by applying an external reducing agent such as $NaBH_4$. A simple spontaneous reduction of metallic salts to form Ag, Au, and Pd nanoparticles onto the $Ti_3C_2T_x$ MXene sheets was applied for formation of particles exhibiting surface-enhanced Raman spectroscopy (SERS) phenomenon [85]. Moreover, an AuNP/MXene composite boosts sensitivity of detection of oncomarker such as microRNA [88]. Similarly, the composite consisting of $Ti_3C_2T_x$ MXene and PtNPs was prepared by means of *in-situ* reduction of Pt precursor (spontaneously or by external reducing agents) on MXene surface. Composite was used for electrochemical catalysis [69] and sensing of important small bioactive compounds [72]. The negatively charged acetylcholinesterase (AChE) was electrostatically deposited on the hybrid nanocomposite of MXene/AgNPs/chitosan from a mixture of the enzyme and chitosan onto MXene/AuNPs for detection of organophosphate pesticide [86].

Graphite oxide as another 2D material was used to form composite together with MXene and such a composite led to a stable and efficient electrochemical detection of H_2O_2 and maintained hemoglobin biological activity even after ink jet printing applied for a sensor-based application [89].

4.3 Electrostatic and other interactions

MXene surface can be patterned *via* electrostatic interactions between MXene and chitosan making a nanocomposite from a negatively charged $Ti_3C_2T_x$ MXene and positively charged biopolymer. Chitosan due to beneficial properties i.e. biocompatibility, nontoxicity and film-forming ability was successfully applied in numerous studies for preparation of MXene/chitosan bionanocomposites or hybrid MXene/chitosan-based nanoparticles. Such a bionanocomposite was used for attachment of an enzyme sarcosine oxidase for detection of sarcosine as a potential prostate cancer biomarker. The biosensor could detect the analyte from LOD of 18 nM up to 7.8 μM. The device responded to the analyte in an extremely short time of 2 s and the analyte was detected in a complex sample with recovery index of 102.6% (**Figure 2**) [90]. Glutaraldehyde was not needed for immobilization of the enzyme [90]. Such approach with glutaraldehyde was also used in other studies [91, 92]. In addition Nafion was proved to be an effective "adhesive" to deposit MXene onto the surface, e.g. for a final electrostatic immobilization of glucose oxidase (GOx) [84], to deposit MXene with adsorbed hemoglobin [93, 94], MXene/$Mn_3(PO_4)_2$ hybrid particles [95] or MXene/TiO_2 mixed with hemoglobin on GCE [96].

DNA aptamer activated through EDC/NHS chemistry was covalently immobilized onto MXene electrostatically modified with polyethyleneimine (PEI) [97].

Zheng *et al.* [98] described *in situ* adsorption of DNA on MXene surface through aromatic hydrophobic bases and in further step modified Ti_3C_2/DNA interface was patterned by PdNPs and PtNPs deposited using $NaBH_4$ as a reducing agent to obtain Ti_3C_2/DNA/Pd/Pt nanocomposite.

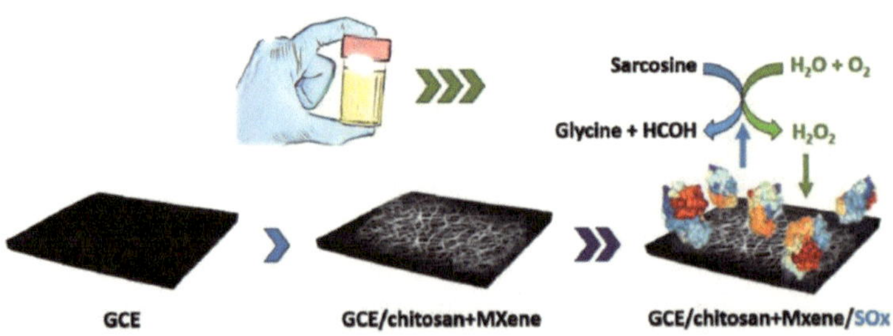

Figure 2.
A graphical presentation of a glassy carbon electrode (GCE) modified using a MXene/chitosan nanocomposite as a support for sarcosine oxidase (SOx) immobilization and indirect sarcosine detection in urine, based on hydrogen peroxide electrochemical reduction. SOx structure is adapted from the protein data Bank (code 1EL5). Figure taken from Ref. [90].

Any conductive interface can be patterned by MXene by a simple casting of a MXene dispersion on untreated electrodes with formation of MXene layer after drying [73]. Alternatively, the electrodes can be pretreated in order to make them more adhesive for formation of MXene layer. To make surface of screen-printed electrodes (SPEs) hydrophilic for subsequent deposition of MXene, SPEs were electrochemically activated in 0.1 M NaOH by CV in a potential range from −0.6 V to 1.3 V [50]. SPEs patterned with delaminated MXene suspension as signal enhancer were applied for quantifying acetaminophen (ACOP) and isoniazid (INZ) in blood serum samples [99]. The presence of abundant highly active surface sites due to functional groups (=O, -F and -OH) offers additional opportunity for MXene to interact with various positively charged functional groups of molecules.

Besides electrostatic modification of MXene by a modifier applied as glue for subsequent attachment of bioreceptors, electrostatic interactions could be applied also to modify MXene by redox molecules. Methylene blue as a redox probe due to its positive charge can be electrostatically deposited on MXene layer with a final immobilization of the enzyme urease on the surface using glutaraldehyde [50]. Moreover electrostatic interaction was utilized for deposition of positively (CTA^+) charged cetyltrimethylammonium chloride (CTAC) on the negatively (OH^-) charged Nb_2C nanosheets resulting in CTAC-anchored Nb_2C nanosheets and subsequently *in situ* formation of mesoporous silica layer (a pore size of 2.9 nm) by co-deposition of CTAB and tetraethyl orthosilicate (TEOS) in the next step [100].

Rich surface chemistry of MXenes can be also applied for interaction with a number of molecules. High applicability of exfoliated MXene (e-MXene) has been investigated as a matrix due to its high laser energy absorption, electrical conductivity and photothermal conversion for laser desorption/ionization time-of-flight mass spectrometry (LDI-MS) analysis of various analytes (saccharides - glucose, sorbitol, sucrose, and mannitol, amino acids - Arg, Phe, His, and Pro, peptide - leu-enkephalin and antibiotics - sulfamerazine and norfloxacin, benzylpyridinium salt (BP), environmental pollutants). Before LDI-MS measurement 1 μL of each small molecule solution was spotted on a target plate, mixed with 1 μL of e-MXene suspension and dried under ambient condition. The e-MXenes exhibiting a high resolution and salt-tolerance demonstrated a strong potential for the development of an efficient analytical platform based on LDI-MS analysis [101]. In addition, Ti_3C_2 MXene assisted LDI-LIFT-TOF/TOF was utilized for differentiation and relative quantitative analysis of three types of glycan isomers resulting in higher sensitivity, better homogeneity and stable relative peak intensity for glycan analysis. Moreover nine disaccharides, two trisaccharides, three heptasaccharides and ten natural product extractions were resolved by applying MXene with LDI-LIFT-MS/MS. The enhanced sensitivity and background-free nature of the fragment profile obtained by LDI-LIFT-TOF/TOF opens up a new realm for nanomaterial assisted glycan structural analysis and/or enrichment either through MXenes themselves or in combination with other functionalized magnetic nanoparticles [102].

> *Glycans are biomolecules, both simple and complex carbohydrates playing important roles in molecular recognition, protein conformation, cell proliferation and differentiation. The analysis of glycans and their structure has gained considerable attention because of their close relationship with disease occurrence and progression. Major types of glycans include N-linked glycans attached to the nitrogen atom in the asparagine side chain within a consensus amino acid sequence Asn-X-Ser/Thr (X should not be proline), and O-linked glycans attached to the oxygen atom of several amino acid residues including serine and threonine. Other types of glycans include glycosaminoglycans usually found attached to the proteins (proteoglycans) and also lipid chains as in glycolipids. MXenes can play an important role in the hydrogen-bonding interactions with glycans and in metal ions (Na^+/K^+) enrichment and transfer leading to improvement of the ionization efficiency of glycans.*

5. Advanced 2D MXenes-based nanobiosensors as ultrasensitive detection tools

The link between progressive detection and daily/routine tests is fostered by (bio)sensing platforms employing nanomaterials/nanostructures with outstanding electronic, electrocatalytic, magnetic, mechanical, and optical properties. Novel multifunctional nanometer-sized structures combine advantageous large surface-to-volume ratio, controlled morphology and structure that would allow immobilizing bioreceptors with preserved biocompatibility, biostability and biodistribution [103]. Compared to other 2D materials (graphene, graphitic carbon nitride, MoS_2), MXenes nanomaterials carry a unique combination of excellent electrical conductivity, complete metal atomic layers, ease of functionalization, high stability, hydrophilicity, large surface area, ultrathin 2D sheet-like morphology, excellent mechanical properties and good bio-compatibility [49, 77].

Bioreceptor's intrinsic characteristics including its affinity towards the analyte, structural stability during biosensor's operation and a methodology deployed for bioreceptor immobilization onto the transducing surface can significantly affect sensitivity, selectivity and robustness (reproducibility, stability *etc.*) of a biosensor. The biorecognition element is usually grafted onto a surface, *i.e.*, in the close vicinity of the transducer. Additional specifications, which need to be optimized for advanced biosensing performance, are the accessibility of the analyte to the biorecognition site of the bioreceptor, the distance between the bioreceptor and the transducer (surface) and bioreceptor's interfacial density.

Both enhanced biocompatibility and increase of the transducing surface area of the (bio)sensors related to enhanced catalytic activity drive a design of 2D MXene nanomaterial-based biosensors utilizing aptamers, antibodies, enzymes and protein molecules [23, 60, 68]. Ultrathin 2D sheet-like morphology with potential for high density incorporation of a number of functional groups as well as excellent ion intercalation behavior also show up as promising features for (bio)sensing applications [104]. On the other hand the implementation of MXenes as next-generation detection devices will require a substantial improvement of the stability of MXenes towards oxidation.

"Detect-to-protect" biosensors are compact analytical devices converting the biochemical reaction into an analytical and measureable signal. Due to their high specificity which is directly dependent on the receptor used (biomolecules or synthetic compounds), their sensitivity, compact size and simple operation, biosensors are the tool of choice for detection of chemical and biological components. Principally, biosensors are formed by two components, a biorecognition part consisting of a biological or synthetic receptor (enzymes, antibodies, nucleic acids, organelles, plant and animal tissue, whole organism, or organs) that utilizes a specific biochemical or chemical reaction mechanism with an analyte and a transducer where the interaction between a bioreceptor and an analyte is transformed into a measurable signal. There are two major obstacles in biosensor development; incorporation/immobilization of (bio)receptors in suitable matrix and monitoring/quantifying the interactions between the analytes and these receptors [105].

In order to allow for a rapid screening of analytes/antigens from human samples a real-time analysis is the preferred approach. The corresponding biosensor should be cheap, small, portable and user-friendly.

The key part of a biosensor is the transducer, which screens a physical change accompanying the bioaffinity reaction (*amperometric biosensors, calorimetric biosensors, optical biosensors, piezo-electric biosensors, potentiometric biosensors*).

A typical biosensor consists of:
1. *a bioreceptor that specifically bind to the analyte;*
2. *an interface architecture where a specific biological event takes place and gives rise to a signal screened by such an interface;*
3. *a transducer element converting a biorecognition event into a measurable signal;*
4. *a computer software able to further process and store measured signal;*
5. *an interface to the human operator.*

The morphology of ultrathin 2D Ti_3C_2 MXene single or few layered nanosheets with high density of functional groups offers improved biomolecule loading and rapid access to the analyte. The covalent immobilization of biorecognition elements (DNAs, enzymes, proteins, *etc.*) leads not only to improved uniformity and accessibility of immobilized bioreceptors, but also to higher density of bound bioreceptors, all resulting in an enhanced biosensor performance.

Jastrzębska *et al.* observed that 2D Ti_3C_2 MXene superficially oxidized into titanium (III) oxide i.e., Ti_2O_3 by sonication of MXene flakes followed by a mild thermal oxidation in water at 60°C for 24 h resulted in "fine-tuning" of the toxicity of the flakes to cancerous cell lines. The authors found out, that thermally oxidized samples showed the highest cytotoxic effect, moreover they were selectively toxic towards all cancerous cell lines with increasing concentration of nanomaterial up to 375 mg L^{-1} [106].

5.1 State-of-the-art approaches of MXenes-based nanobiosensors for cancer biomarkers detection

Cancer is one of the deadliest diseases worldwide, and acquiring cancer-specific data by quantitative analysis of cancer-associated biomarkers is crucial to monitor cancer progression and for the early treatment [107]. As reported by the World Health Organization, the year of 2030 should be marked by approximately 12 million cancer related deaths, making cancer a major public health problem and one of the most prominent death-causing factors worldwide. The number of new cases of cancer (cancer incidence) is presently around 439 *per* 100,000 *per* capita *per* year [108]. Early-stage diagnostics of various types of cancer diseases is important since it offers opportunities to extend life expectation of patients. Tumor markers exist in tumor cells themselves or are secreted by tumor cells. In either case the presence of these tumor markers above a set threshold may suggest the existence and/or growth of a tumor. The phrase "tumor marker" is often transposed for the term "biomarker" [109] and *vice versa*. Biomarkers can be applied as an early diagnostic tool, to monitor disease progression, as a prognostic tool and as means for prediction and monitoring of clinical response to an intervention.

According to the National Institute of Health, a biological marker (biomarker): is a characteristic that is objectively measured and evaluated as an indicator of normal biological processes, pathogenic processes, or pharmacologic responses to a therapeutic intervention.

A tumor/cancer marker is a substance produced by a tumor or by the host in response to a cancer cell that can be objectively measured and evaluated as an indicator of cancerous processes within the body. The term tumor marker was firstly coined in 1847 and presently there are more than 100 known different tumor markers [110]. Biomarkers have a great potential for screening and diagnostics because they are present in blood and provide information about the health condition [111]. In healthy individuals, the tumor marker concentration is comparatively low level or even absent, while increased values can reveal development and/or progression of a disease [112]. Serum biomarkers providing key information about the disease are important for management of cancer patients since blood aspiration is only a moderately invasive procedure. There is clear need for early-stage cancer diagnostics, efficient treatment and posttreatment monitoring to avoid progress of the disease into advanced stages. Therefore there is an enormous demand for efficient less-invasive investigation *i.e.* analysis of cancer biomarkers in plasma/ serum samples at low limit of detection [113].

Limit of Detection (LOD): the level of analyte that leads to a sensor signal which is statistically significantly different from the background signal obtained in the absence of the analyte. A frequently used definition of LOD is a concentration that gives a signal greater than three times the standard deviation of a blank sample consisting entirely of a matrix (S/N) = 3.

The unique physico-chemical properties of MXenes make them a significant tool that can be employed in the cancer therapy (photothermal therapy, photodynamic therapy, radiation therapy, chemotherapy), cancer imaging (CT/MRI/PA imaging) as well as cancer theranostic applications [21].

5.2 MXene-based electrochemical nanobiosensors

Electrochemical biosensors are prospective tool of choice for an early-stage diagnostics of cancer diseases [114]. Electrochemical methods such as CV, CA, DPV, EIS, square wave voltammetry (SWV) provide a number of advantages. They are reliable, easy-to-use, affordable and highly sensitive and reliable [107, 115, 116]. Lab-on-chip biosensors are compact and portable miniaturized devices that can be employed in cancer biomarkers research leading to potential clinical applications. Biosensors employing surface nanoarchitectures with this type of detection offer attractive features including straightforward miniaturization, excellent LODs, robustness, small analyte volumes and the ability to be applied in turbid biofluids with optically absorbing and fluorescing compounds.

Single/few-layered MXene (Ti_3C_2) nanosheets were functionalized with (3-aminopropyl)triethoxysilane (APTES) to enable covalent attachment of bio-receptor onto *f*-Ti_3C_2-MXene for electrochemical detection of carcinoembryonic antigen (CEA) as a widely used tumor marker [77]. The ultrathin 2D nanosheets of single/multilayer MXene Ti_3C_2 with high density of functional groups brought in improved antibodies anchoring and faster access to analyte. The label-free aminosilane and bio-functionalized *f*-Ti_3C_2-MXene-based biosensor (BSA/anti-CEA/f-Ti_3C_2-MXene/GC) demonstrated LOD of 0.000018 ng mL^{-1} with sensitivity of 37.9 μA ng^{-1} mL cm^{-2} *per* decade (a linear detection range of 0.0001–2000 ng mL^{-1}) for CEA determination using hexaammineruthenium ($[Ru(NH_3)_6]^{3+}$) as a preferable redox probe and CV as a detection technique [77].

Carcinoembryonic antigen (CEA, molecular mass of 180–200 kDa) is a highly glycosylated cell surface protein consisting of approx. 60% carbohydrates, which attains elevated levels in a number malignancies, such as colorectal, breast and ovarian, gastric, liver and pancreatic cancer. Serum CEA is used in clinical research to identify early stages of disease, monitor tumor recurrence and metastatic disease. The normal range of serum CEA in healthy adults of non-smokers is below 2.5 ng mL^{-1} and in the serum of smokers below 5.0 ng mL^{-1}, but increases rapidly when normal cells become cancerous.

Due to their excellent electrical conductivity and large specific surface area with a large number of potential attachment binding sites, 2D MXenes are also applied as a conductive support for immobilization of aptamer probes. Wang *et al.* modified electrode surface with MXene for development of a MUC1 biosensor [117]. The ferrocene-labeled complementary DNA was bound onto MXene nanosheets to design a detection probe for electrochemical signal amplification. GCE was modified by electrodeposited AuNPs with MUC1 aptamer attached to the modified electrode *via* Au-S bonds. The modified electrode was blocked using bovine serum albumin (BSA) in order to resist non-specific interactions. Next, a detection probe was attached to the modified electrode *via* hybridization between complementary DNA and a MUC1 aptamer. Upon interaction of MUC1 with such an electrode, the detection probe was detached from the working electrode resulting in a decrease of an electrochemical signal (a signal-off response). This competitive aptasensor detected MUC1 with LOD of 0.33 pM with a linear range up to 10 mM. The relative standard deviation (RSD) of the peak current difference response was 1.43%, indicating that the aptasensor had good reproducibility. The peak current difference response of the aptasensor did not change much in ten days, indicating its acceptable stability [117].

Currently, there are more than 20 known types of mucins. They are encoded by MUC genes and represent high molecular weight glycoproteins expressed on epithelial cells. Aberrantly glycosylated mucins are expressed in cancer cells and serve as oncogenic molecules.

MicroRNAs (miRNAs) overexpression is a biomarker for a number of diseases including cardiovascular disorders, cancer, rheumatic diseases, diabetes, neurological disorders, liver diseases, kidney diseases, and immune dysfunction. The microRNAs (miRNAs) are biomolecules composed of 18–24 nucleotides and they play a key role in biological processes such as cell proliferation, apoptosis and tumorigenesis. Abnormal expression has been monitored in breast cancer as well as in other cancer types with observed blood stability. The miRNA-182 demonstrates tissue specificity and sequential expression in the different stages during lung cancer development or evolution. The miRNA-155 is overexpressed in human breast cancers.

The label-free strategy for the ultrasensitive detection of miRNA-182 was based on glassy carbon electrode (GCE) modified step-by-step by van der Waals forces and electrostatic interactions with MoS_2/Ti_3C_2, AuNPs, ssRNA [118]. BSA was used to block unbound gold particles surface and avoid nonspecific adsorption. The biosensor was able to determine miRNA-182 with LOD of 0.43 fM (a linear range of 1 fM - 0.1 nM) by DPV method [118]. The recovery was 105%, 95.3% and 93.0% for the concentration of 10^{-10} M, 10^{-12} M and 10^{-14} M of the analyte respectively, manifesting its effective detection of miRNA-182 in real sample [118].

Duan with co-workers [119] developed an impedimetric aptasensing strategy based on a novel zero dimensional (0D)/2D nanohybrid of $Ti_3C_2T_x$ nanosheets decorated with FePc QDs (denoted as $Ti_3C_2T_x$@using iron phthalocyanine quantum dots (FePcQDs)) for miRNA-155 detection. The miRNA-155 was established by applying impedimetric aptasensor with LOD of 4.3 aM (S/N = 3, a linear concentration range from 0.01 fM to 10 pM). The observed relative standard deviation (RSD) of the five aptasensors for detection of miRNA-155 was as low as 2.98%, demonstrating good reproducibility of the proposed aptasensor. Moreover, the signal remained 104% of the original signal after 15 days of storage, revealing a satisfactory stability of the present aptasensor [119].

Multiple (miRNA-21 and miRNA-141) and rapid (80 min) analysis of onco microRNAs in total plasma was carried out with combination of AuNPs (5 nm) decorated MXene as an electrode interface and a duplex-specific nuclease (DSN) as an amplification system applied onto home-made screen-printed gold electrode

(SPGE) [88]. As the initial step functionalization of two magnetic particles (MPs) with two different single-stranded DNAs (ssDNAs) was performed through labeling with methylene blue (MB) and ferrocene (Fc) that were partially complementary to the target miRNA. After the invasion of targets and amplification cycle, the released uncleaved DNA sequences harboring redox labels were hybridized with the electrochemical sensor platforms for subsequent measurements. To enhance the electrochemical signal, the SPGE was modified with the synthesized MXene-$Ti_3C_2T_x$ and patterned with AuNPs and further loaded with abundant ssDNAs (base) to provide a significantly higher electrochemical signal compared to the AuNP/Au electrodes (almost 4 orders of magnitude increase). The LODs of the biosensor exhibiting multiplex ability, antifouling activity and single mutation recognition for microRNA-21 and microRNA-141 detection reaching low LOD levels down to 204 aM and 138 aM (a wide linear range up to 50 nM), respectively. The synergic effect of combining MXene based electrochemical amplification and DSN target recycling, resulted in a short assay time of 80 min, a good assay reproducibility (RSD ≈ 4.7%) and stability of 95.2% and 97.1% of its initial signal values assigned to MB and Fc, respectively, after 4 weeks of storage [88].

Xu *et al.* [120] treated Ti_3C_2 MXene with NaOH and hydrogen peroxide in a Teflon lined stainless-steel autoclave by a simultaneous oxidation and alkalization resulting in the synthesized 3D sodium titanate nanoribbons (M-NTO) in order to overcome restacking of MXenes flakes. Such a composite offered fast electron transfer ability, high specific surface area and excellent biocompatibility by connection of 3D M-NTO with conductive poly(3,4-ethylenedioxythiophene) (PEDOT). AuNPs were electrodeposited in the next step on the surface of M-NTO-PEDOT for immobilization of antibodies against prostate specific antigen (PSA) for PSA detection. Assay reproducibility was high with RSD of 1.89% with satisfactory biosensor stability (84.2% of its original response after 2 weeks storage at 4°C). The label-free immunosensor could detect PSA with LOD of 0.03 pg. L^{-1} (S/N = 3) by DPV [120].

The prostate-specific antigen (PSA, 28.4 kDa) belongs to the tissue kallikrein-related family of peptidases and is also known as g-seminoprotein, kallikrein-3 or KLK3. PSA presenting a single-chain glycoprotein containing approximately 8% (by mass) of N-glycan with a single glycosylation site is produced by vesicles in prostate epithelial cells. Prostate cancer (PCa, adenocarcinoma or glandular cancer of the prostate gland) is the 2nd most abundant cancer type in men worldwide, with an estimated 1.1 million cases diagnosed in 2012 alone. The PSA level in health body is lower than 4 ng mL^{-1}.

In addition, PSA was sensitively detected with capacitance-based enzyme immunosensor [121] based on enzymatic biocatalytic precipitation of precipitate on interdigitated micro-comb electrode (IDE). AuNPs heavily functionalized with HRP and detection antibodies (HRP-Au-Ab_2) were utilized as the signal generating probe. Firstly, MXene dispersion in 1.0 wt % Nafion ethanol solution was dropped onto IDE to modify it. Next anti-PSA capture antibodies (Ab_1) were physically adsorbed onto the nanosheets. Subsequently PSA, HRP-Au-Ab_2 conjugates, H_2O_2 and HRP-tyramine conjugates were incubated step-by-step with the immunosensor at room temperature. The target PSA was determined with LOD of 0.031 ng mL^{-1} in a linear range up to 50 ng mL^{-1} with RSD of 10.7%, indicating good reproducibility [121].

Liu *et al.* [122] designed a "signal-on" photoelectrochemical (PEC) biosensor employing a $Ti_3C_2/BiVO_4$ Schottky junction for a signal generation for ultrasensitive detection of vascular endothelial growth $factor_{165}$ ($VEGF_{165}$) with LOD of 3.3 fM (a linear range of 10 fM - 100 nM). First, *in situ* synthesized

$Ti_3C_2/BiVO_4$ nanocomposite covered the surface of the electrode to produce an initial photocurrent signal. The T7 Exonuclease (T7 Exo)-assisted dual signal amplification strategy was applied to achieve improved sensitivity of the PEC sensor. With the target $VEGF_{165}$, the hairpin DNA (HP2), containing the aptamer of $VEGF_{165}$ can be specifically identified and opened to specifically recognize the exposed toehold of S1 on the magnetic bead, releasing the output DNA S2, S3, and S4. Further, T7 Exo was used to digest the recessed 5′ termini of double-stranded DNA (dsDNA). $VEGF_{165}$-HP2 complex was released for the next cycle, which can be converted to multiple output DNAs. Next, the output DNA hybridized with hairpin DNA (HP1) on the electrode to form a double-stranded structure, which provided a wonderful platform for the intercalation of methylene blue. Methylene blue effectively increased light absorption and promoted the electron transfer along the dsDNA, resulting in an enhanced PEC signal. The inter-assay and intra-assay RSD values were calculated to be 2.42% and 2.26%, respectively, illustrating the outstanding reproducibility of the biosensor [122].

The vascular endothelial growth factor (VEGF) is a biomarker with a molecular mass of 18–27 kDa and can be related to various cancer types for example brain, lung, gastrointestinal, hepatobiliary, renal, breast, ovarian. Normal level of VEGF in serum is ~ 220 pg mL^{-1}.

An impedimetric aptasensor based on the nanostructured multicomponent hybrid of $Ti_3C_2T_x$ nanosheets and phosphomolybdic acid (PMo_{12}) nanoparticles integrated by embedding within the polypyrrole (PPy) matrix ($PPy@Ti_3C_2T_x/PMo_{12}$) was utilized for detection of osteopontin (OPN) [53]. The $PPy@Ti_3C_2T_x/PMo_{12}$-based aptasensor estimated OPN with LOD of 0.98 fg mL^{-1} in a linear range of 0.05–10,000 pg. mL^{-1}. The biosensor exhibited low RSD of the assays of around 1.7% and during the biosensor offered also good operational stability [53].

Osteopontin (OPN, 41–75 kDa) known as a phosphoprotein regulates tumor metastasis and leads to cancer progression (breast, colon, liver, lung, ovarian, prostate). OPN plays an important role in tumor invasion, growth, angiogenesis, and metastasis by upregulating several signaling pathways. Normal level in serum is 16 ng mL^{-1}.

5.3 MXene-based optical nanobiosensors

Surface plasmon resonance (SPR) is a principal technique for *in situ* bioaffinity assays of various target (bio)molecules without a need for fluorescent or enzymatic labeling. SPR (bio)sensors could be developed with improved operational parameters by applying nanomaterials [123]. SPR detection platform offers beneficial advantages for the biosensing including label-free and real-time detection, high sensitivity and selectivity, ease of miniaturization and rapid detection making the technique well suited for bioassays.

Surface plasmon resonance (SPR) is optical sensing technology which can be used for health monitoring, early disease diagnosis, and environment safety. It has become a valuable tool for biological, chemical, and biomedical applications. It has been widely used in various biochemical and biosensing applications, particularly for enzyme detection, drug diagnostic, dsDNA hybridization, and applied as an immune sensor. SPR sensors are refractive index based sensors that can be experimentally implemented for real-time biosensing without the labeling of the analytes or bioreceptors.

It is an established high sensitivity platform for measuring minute concentrations of analyte and kinetics of biomolecular interactions. SPR generates an evanescent wave at the interface of two materials, when properly polarized incident light excites charge density oscillation (also called surface plasmons, SPs) supported by thin metal film deposited on the prism. However, SPR condition is established only after proper coupling of p-polarized incident wave with surface plasmon wave (SPW), when the frequency of evanescent wave matches the natural frequency of the SPW. SPR in reflection mode measures the resonance angle at a dip in reflectivity and a complete energy transfer from evanescent wave to SPW is achieved. Resonance angle is very sensitive to alteration of sensing medium refractive index (RI), i.e. on adsorption of analytes, which changes SPR condition. The sensitivity of the sensor is directly related to the resonance angle shift, which is sensitive to the modification of the RI of the sensing medium. In conventional SPR, a thin film of noble metal is used for SPs generation as well as adsorption of biomolecules or other analytes. Gold is a preferred interface, as it is non-oxidizing, corrosion-free, with substantial chemical stability, and shows stable adsorption of analytes with high sensitivity. However it shows a broad resonance curve causing reduction in detection accuracy. Silver (Ag) on the other hand shows higher accuracy through sharper reflectance curve but a poor chemical stability, as it is assumed to oxidize quickly on direct exposure to the atmosphere. Ag can be utilized efficiently in the SPR sensor if its oxidation can be avoided by using some protective layer over it. Conventional SPR sensor utilizing prism, metal layer and sensing medium offers smaller sensitivity.

Due to its absorption, a few-layer $Ti_3C_2T_x$ MXene can contribute to the improved sensitivity of SPR biosensors. Enhanced sensitivity by 16.8%, 28.4%, 46.3% and 33.6% was achieved for the proposed SPR biosensors based on Au with 4 layers, Ag with 7 layers, Al with 12 layers and Cu with 9 layers of $Ti_3C_2T_x$, respectively [124].

The platform based on prism/gold layer/MXene/WS_2/black phosphorus using monolayer of each nanomaterial was proved as a novel SPR sensing material with enhanced sensitivity of 15.6% compared to a bare metal film [125]. MXene-based composite, g-C_3N_4/MXene-AgNPs, including g-C_3N_4 as a photocatalyst, MXene as a co-catalyst and AgNPs as an electron mediator offered enhanced photocatalytic activity. The increased optical absorption and reduced band-gap energy due to the SPR effect of AgNPs deposited on such nanocomposite modified interface was observed [126].

Wu with co-workers [49] took advantage of hydrophilic and biocompatible Ti_3C_2 surface as a platform for making a nanohybrid consisting of multi-walled carbon nanotubes (MWCNTs)-polydopamine (PDA)-Ag nanoparticles (AgNPs) as a signal probe to develop SPR biosensor, that is easy to prepare, convenient to operate, and provides high sensitivity and selectivity. In order to obtain good orientation and immobilization of monoclonal anti-CEA antibody (Ab_1), synthesized Ti_3C_2/AuNPs composite was firstly decorated with staphylococcal protein A (SPA) to which Ab_1 was captured by affinity interaction through its Fc region. Polyclonal anti-CEA antibodies (Ab_2) were conjugated with a nanohybrid through Schiff-base reaction between amino residues and quinone groups of PDA. By introducing a MWPAg-Ab_2 conjugate to form a sandwich format, LOD of 0.07 fM was achieved for CEA detection (a dynamic range of 2×10^{-16} - 2×10^{-8} M). However, there are some limitations of such biosensing platform including time-consuming fabrication of the interfacial layer *prior* to analysis, but the biosensor exhibited good assay reproducibility with RSD below 5%. The stability of the fabricated sensing platform was also investigated by measuring the SPR responses to 10^{-12} M CEA concentration over the period of 7 days, during which the sensing platform was stored at 4°C. The developed biosensor lost 13% of its initial activity after storing for 7 days [49].

Wu *et al.* [127] utilized amino-functionalized N-Ti_3C_2-MXene-hollow gold nanoparticles (HGNPs)-staphylococcal protein A (SPA) complexes as a signal enhancer for CEA detection with LOD of 0.15 fM (a linear range of 0.001–1000 pM) at SPR (**Figure 3**). The SPR biosensor was stable (80% of the initial response

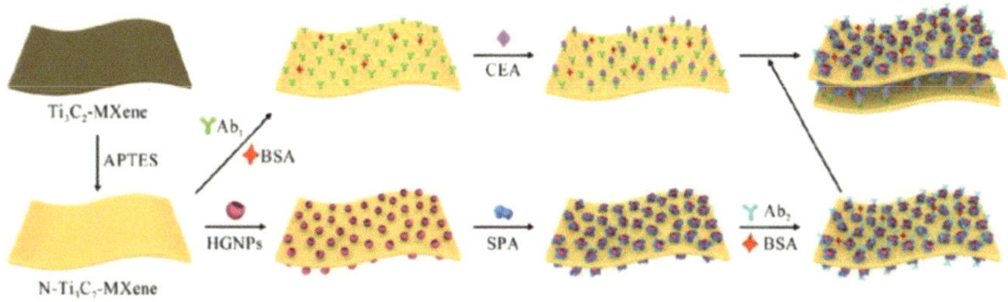

Figure 3.
Schematic of detection procedure of the prepared SPR biosensor. Reprinted with permission from ref. [127]. Copyright ACS, 2020.

after storage for 28 days), reproducible (average assay RSD less than 5%) and offering operational stability (84% of its initial activity after five reuse cycles) [127].

Among various investigation methods for detection of cancer biomarkers, fluorescence analysis methods, especially fluorescent nanoprobes based on "turn on" mechanism, are regarded as sensitive and reliable analytical tools for cancer diagnosis. The nanoprobes can be ideally stabilized in both extracellular and intracellular microenvironment and respond to multi-biomarkers with different spatial distributions to achieve multilayer information of diverse biomarkers range from cell membrane to the cytoplasm at a cellular level [128]. Wang with colleagues [128] investigated fluorescence quenching capacity of Ti_3C_2 MXenes for biosensing of dual biomarkers in single (MCF-7) living cells. A chimeric DNA-functionalized Ti_3C_2 probe was employed for real-time and multilayer simultaneous fluorescent imaging of plasma membrane glycoprotein MUC1 and cytoplasmic microRNA-21 at nM concentration *in vitro (***Figure 4***)*. Ti_3C_2 MXene was decorated with polyacrylic acid to achieve high stabilization and dispersion of MXene with delivering functional groups were employed for covalent linkage of the bioreceptor (a dual signal-tagged chimeric DNA probe (dcDNA)) [128].

Guo *et al.* [129] fabricated Ti_3C_2 QDs (~4.2 nm in diameter) by a hydrothermal treatment with beneficial and excellent salt tolerance, anti-photobleaching and dispersion stability in aqueous solution. Ti_3C_2 QDs were applied as the fluorescent markers for fluorescent signal readout without and for sensitive fluorimetric analysis of alkaline phosphatase (ALP) activity with LOD of 0.02 U L^{-1}. Moreover, an accurate analysis of ALP by applying Ti_3C_2 QDs-based strategy for assays of AFP in the lysates of embryonic stem cells was also achieved by such a biosensor device [129].

Alkaline phosphatase (ALP), as an essential enzyme in phosphate metabolism, responsible for catalysis of the dephosphorylation of a variety of substrates. The abnormal level of serum ALP, as a crucial biomarker for clinical diagnostics, is closely related to various diseases, such as diabetes, hepatitis and prostatic cancer. In addition a high ALP activity is the traditional biomarker of pluripotent embryonic stem cells.

A strand displacement dual amplification (SDDA) strategy was developed by Chen *et al.* [130] for simultaneous detection of multiple miRNAs analytes in a cell lysate using a unique single strand-double strand-single strand DNA (sdsDNA) probe, which was generated by the target recognition probe hybridizing with the site region probe. The fluorescence resonance energy transfer (FRET) assay was used for highly photostable, specific and sensitive detection of miRNAs with LOD

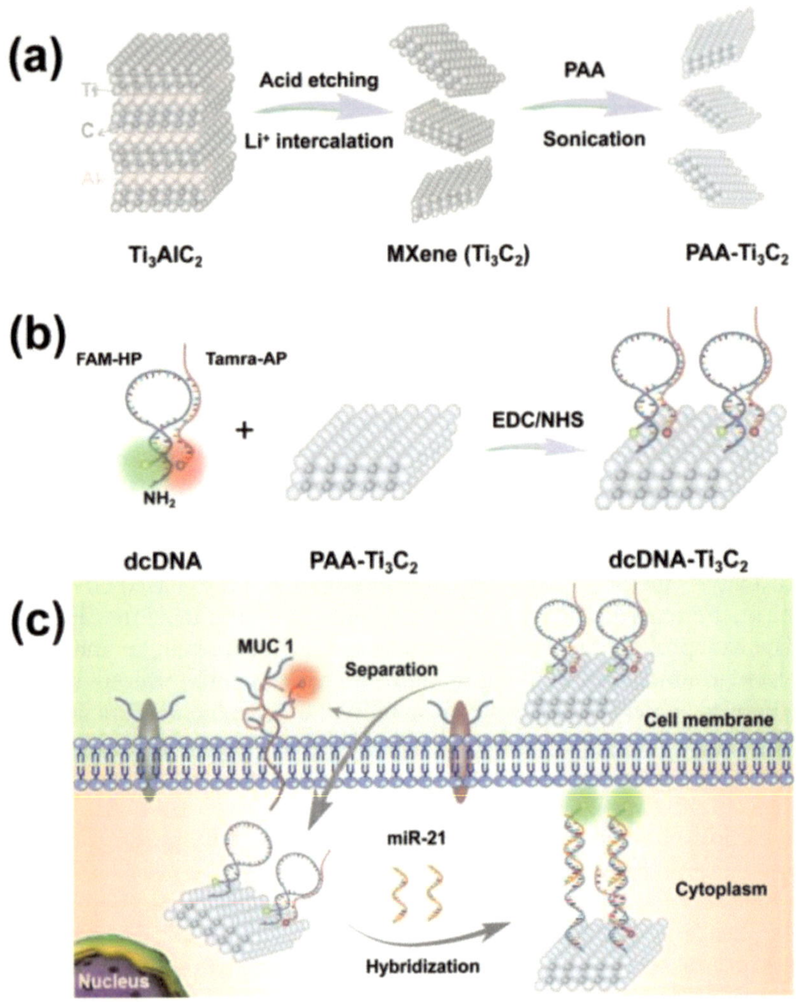

Figure 4.
(a) Illustration of the fabrication of the Ti_3C_2 MXenes and PAA-Ti_3C_2. (b) the construction of the dcDNA-Ti_3C_2 composite nanoprobe. (c) Multilayer imaging of plasma membrane glycoproteins MUC1 and cytoplasmic miR-21 using the dcDNA-Ti_3C_2 composite nanoprobe. Reprinted with permission from ref. [128]. Copyright ACS, 2019.

of 0.5 fM and 0.85 fM for miRNA-21 and miRNA-10b, respectively (a linear range from 5 fM to 100 pM) [130].

PSA was the both qualitatively and quantitatively examined through a sandwich-type immunoreaction and a photothermal measurement by applying Ti_3C_2 MXene quantum dots (QDs)-encapsulated liposome with a high photothermal efficiency [131]. Ti_3C_2 MXene QDs as the innovative photothermal signal beacons were entrapped in the liposome for the labeling of the secondary antibody on the surface. The sandwich-type assay was carried out by coupling a low-cost microplate with a homemade 3D printed device. Under NIR-laser irradiation of 808 nm, Ti_3C_2 MXene QDs converted the light energy into heat, and the shift in the temperature correlating with the analyte concentration. LOD of 0.4 ng mL^{-1} for PSA was obtained by a near-infrared (NIR) photothermal immunoassay (a linear range of 1.0 ng mL^{-1} - 50 ng mL^{-1}). The portable equipment employing a portable NIR imaging camera was able to collect the visual thermal data for semi-quantitative analysis of target PSA within 3 min [131].

Liposome, a target-responsive nanomaterial containing a bilayer of phospholipids with the spherical structure, is promising due to its superior biocompatibility, versatility of surface modification, operability of dimensional control and large-volume internal loading. The functional liposome acts as the biological signal amplifier by encapsulating numerous signal molecules and binding with biological recognition molecules like DNA, enzyme, protein and nanomaterial.

5.4 Detection of exosomes as a source of cancer biomarkers by applying 2D MXenes

Exosomes as type of endosome-derived cell-secreted vesicles with the structure of a lipid bilayer membrane are responsible for signal transduction in intercellular communication and extracellular matrix remodeling. In addition exosomes can also carry cargo affecting neighboring cells and they can form pre-metastatic niches [115]. Thus, exosomes are behind localized tumor development, progression and induction of distant tumors forming metastasis. The fact, that a substantially higher cellular activity of tumor cells results in the production of a greater number of exosomes than in normal/healthy cells, makes them hot candidates for cancer diagnostics in itself [115].

Exosomes are naturally produced biological nanoparticles, with their size usually defined in the range from ~ 30 nm up to 100 nm or sometimes up to 200 nm. Other types of extracellular vesicles (EVs) include microvesicles (50–1000 nm, which bud directly off the plasma membrane), ectosomes (vesicles assembled at and released from a plasma membrane), shedding vesicles, microparticles and apoptotic vesicles (500–2000 nm, which bud off the membrane of cells undergoing apoptosis).

Electrochemiluminescence (ECL) as an upcoming technique joining the benefits of both electrochemistry and chemiluminescence, has been widely applied for biomarker analysis thanks to its high sensitivity, short response time and low background signal [132]. A biosensor based on the application of MXene and ECL was developed for sensitive detection of exosomes [133]. First, MXene ($\zeta \sim -50$ mV) was modified by polyethyleneimine (PEI) ($\zeta \sim +55$ mV) through electrostatic interactions to prepare an MXene/PEI nanocomposite ($\zeta \sim 80$ mV). This positively charged nanocomposite was subsequently used in covalent immobilization of an aptamer against CD63 protein, which is present on the surface of the exosomes using an amine-coupling chemistry. In an effort to detect exosomes, the GCE was modified by AuNPs, which were next modified by ethylenediamine. In addition, free -NH_2 groups of ethylenediamine were activated by EDC/NHS to deposit a polymer, which was finally used for covalent immobilization of an aptamer against the EpCAM protein present on the surface of the exosomes. The signal was generated upon completion of the sandwich configuration as shown in **Figure 5**. The biosensor was most sensitive towards exosomes produced by a breast cancer cell line MCF-7, followed by a human liver cancer cell line HepG2 and a melanoma cell line B16. Exosomes released from the MCF-7 cell line were detected in the concentration range from 500 to 5×10^6 particles μL^{-1} with LOD of 125 particles μL^{-1}, which was more than 100 times lower than the conventional ELISA method. The biosensor exhibited an excellent performance by analysis of spiked serum samples with recovery indices of 95–104% [133]. In the later research of the same group, it was shown that, besides CD63 and EpCAM, other proteins

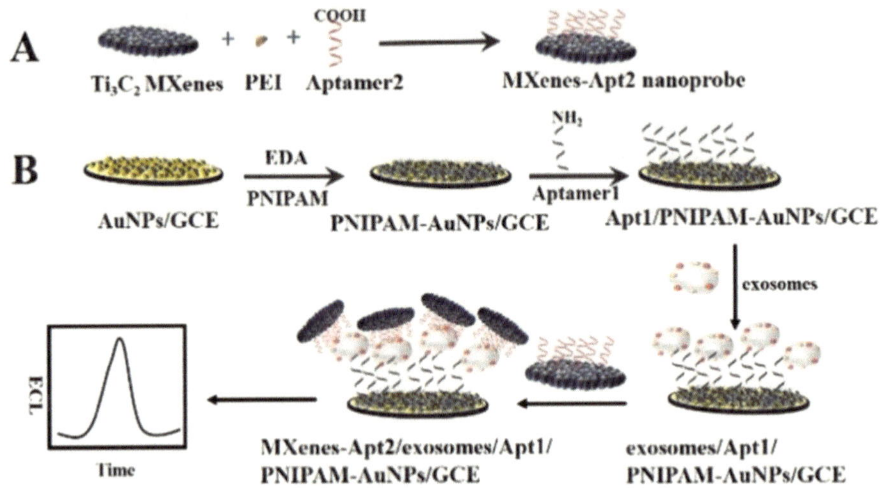

Figure 5.
The principle of the ECL biosensor for exosomes activity detection signal amplification strategy. Reprinted with permission from Ref. [134]. Copyright ACS, 2018.

present on the surface of the exosome can be targeted by DNA aptamers, including PSMA and PTK-7 [134]. Such a biosensor offered highly reproducible assays with RSD of 1.2% and 3.9% for detection of 10^8 and 10^9 exosomes mL^{-1}, respectively [134].

Another MXene-based biosensor for the detection of exosomes was prepared by Fang *et al.* [135] GCE was modified by SiNPs and ionic liquid with a final modification of the interface by EpCAM aptamers. In order to detect exosomes, a sandwich configuration was formed by a final incubation with a nanohybrid consisting of MXene modified by black phosphorus quantum dots, $Ru(bpy)_3^{2+}$ and anti-CD63 antibodies. In addition to the ECL detection of exosomes, such a configuration also made photothermal assays possible. The ECL biosensor could detect exosomes down to 37 particles μL^{-1} with a linear range of up to 5×10^7 particles μL^{-1}. The stability of the constructed biosensor was investigated by measuring 1.1×10^2 exosomes μL^{-1}. The ECL intensity kept a relatively stable value under sequential 10 cyclic scans with relative standard deviation (RSD) of 1.1% [135].

6. Conclusions

The novel 2D nanomaterial MXene has a potential to significantly influence the field of biosensing including affinity-based biosensors with expected exponential increase in related works to be published in the years to come. MXene-based biosensors offer adequate sensitivity required for detection of cancer biomarkers present in blood down to ng mL^{-1} level or better (**Table 1**). However a great deal of effort needs to be invested into finding proper decorating strategies for MXene to simultaneously allow immobilization of biomolecules, but at the same time providing resistance towards non-specific protein binding. Matching this criteria, affinity MXene-based biosensors can be applied for analysis of complex samples such as blood serum or plasma [23]. Point-of-care tests (POC) employing MXene/based devices represent promising candidates with benefits such as adaptability in different/adverse environment,

DOI: http://dx.doi.org/10.5772/intechopen.94309

Target biomarker	Biosensor architecture	Detection method	LOD	Linear range	Reference
CEA	BSA/anti-CEA/f-Ti_3C_2-MXene/GC	Electrochemical/ CV	0.000018 ng mL^{-1}	0.0001–2000 ng mL^{-1}	[77]
CEA	Ti_3C_2MXene/AuNPs/SPA/Ab_1 and MWCNTs-PDA-AgNPs/Ab_2	SPR	0.07 fM	2×10^{-16} - 2×10^{-8} M	[49]
CEA	Ab_2-conjugated SPA/HGNPs/N-Ti_3C_2-MXene	SPR	0.15 fM	0.001–1000 pM	[127]
MUC1	cDNA-Fc/MXene/Apt/Au/ GCE	Electrochemical/ DPV	0.33 pM	1.0 pM - 10 mM	[117]
miRNA-182	BSA/ssRNA/ AuNPs/ MoS_2/Ti_3C_2/GCE	Electrochemical/ DPV	0.43 fM	1 fM - 0.1 nM	[118]
miRNA-155	cDNA/$Ti_3C_2T_x$@ FePcQDs/AE	Electrochemical/ EIS	4.3 aM	0.01 fM - 10 pM	[119]
miRNA-21 and miRNA-141	ssDNAs/AuNP@ MXene/SPGE	Electrochemical/ DPV	204 aM (miRNA-21) and 138 aM (miRNA-141)	500 aM - 50 nM	[88]
PSA	BSA/anti-PSA/AuNPs-M-NTO-PEDOT/GCE	Electrochemical/ DPV	0.03 pg. L^{-1}	0.0001–20 ng mL^{-1}	[120]
PSA	HRP-Au-Ab_2-PSA-Ab_1-MXene/IDE	Electrochemical/ EIS, CV	0.031 ng mL^{-1}	0.1–50 ng mL^{-1}	[121]
$VEGF_{165}$	MB/DNA/HT/HP1/AuNPs/Ti_3C_2/$BiVO_4$/GCE	Photoelectro-chemical	3.3 fM	10 fM - 100 nM	[122]
OPN	Apt/PPy@$Ti_3C_2T_x$/ PMo_{12}/AE	Electrochemical/ EIS	0.98 fg mL^{-1}	0.05–10,000 pg. mL^{-1}	[53]
Exosomes	ECL probe - (MXenesBPQDs@Ru$(dcbpy)_3^{2+}$-PEI-Ab_{CD63} exosomes/Apt/ILs/SiO_2 NUs/GCE	ECL	37.0 particles μL^{-1}	1.1×10^2–1.1×10^7 particles μL^{-1}	[135]
Exosomes	MXenes-Apt2/exosomes/Apt1/PNIPAMAuNPs/ GCE	ECL	125 particles μL^{-1}	5×10^2–5×10^6 particles μL^{-1}	[97]
Exosomes	Cy3 labeled CD63 aptamer (Cy3-CD63 aptamer)/ Ti_3C_2 MXenes	Ratiometric fluorescence resonance	1.4×10^3 particles mL^{-1}	10^4–10^9 particles mL^{-1}	[134]
miRNA-21 and miRNA-10b	DNA-$NaYF_4$:Yb,Tm/Er UCNPs and Ti_3C_2 nanosheets	Fluorescence - fluorescence resonance energy transfer (FRET) assay	0.62 fM (miRNA-21) and 0.85 fM (miRNA-10b)	5 fM - 100 pM	[130]

Table 1.
Key characteristics of MXene-based nanobiosensors for detection of cancer biomarkers.

automation of tests, reduced cost, miniaturization, interference-free detection, *etc.* [50, 81, 136, 137].

Acknowledgements

The authors would like to acknowledge financial support from the Slovak Research and Development Agency APVV 17–0300 and from projects granted by the Ministry of Health of the Slovak Republic No. 2018/23-SAV-1 and 2019/68-CHÚSAV-1. This book chapter was supported by Qatar University Grant IRCC-2020-004. The statements made herein are solely the responsibility of the authors.

Abbreviations

Ab_1	monoclonal anti-CEA antibody, monoclonal mouse anti-human PSA capture antibody
Ab_2	polyclonal anti-CEA antibody, polyclonal rabbit anti-human PSA detection antibody
AE	bare Au electrode
AgNPs	Ag nanoparticles
anti-CEA	carcinoembryonic antibody monoclonal antibody
Apt	MUC1 aptamer;
Au, AuNPs	Au nanoparticles
BPQDs	black phosphorous quantum dots
BSA	bovine serum albumin
CEA	carcinoembryonic antigen
CV	cyclic voltammetry
cDNA	complementary deoxyribonucleic acid
cDNA-Fc	ferrocene-labeled complementary deoxyribonucleic acid
DPV	differential pulse voltammetry
EIS	electrochemical impedance spectroscopy
f-Ti_3C_2-MXene	MXene functionalized with aminosilane
FePcQDs	phthalocyanine quantum dots
GC, GCE	glassy carbon electrode
HGNPs	hollow gold nanoparticles
HP1	hairpin DNA
HRP	horseradish peroxidase
HT	hexanethiol
IDE	interdigitated microcomb electrode
IL	ionic liquid (1-carboxymehyl-3-methylimidazolium chloride)
MB	methylene blue
miRNA	microRNA
M-NTO	3D sodium titanate nanoribbons
MUC1	mucin1
MWCNTs	multi-walled carbon nanotubes
N – Ti_3C_2-MXene	amino-functionalized Ti_3C_2-MXene
PDA	polydopamine
PEDOT	poly(3,4-ethylenedioxythiophene)
PMo_{12}	phosphomolybdic acid
PNIPAM	Poly (N-isopropylacrylamide), carboxylic acid
PPy	polypyrrole
ssDNAs	single-stranded DNAs

PSA	prostate specific antigen
SiO_2 NUs	SiO_2 nanourchin
SPA	staphylococcal protein A
SPGE	screen-printed gold electrode
SPR	surface plasmon resonance
UCNPs	upconversion nanophosphors
$VEGF_{165}$	vascular endothelial growth factor 165

Author details

Lenka Lorencova[1], Kishor Kumar Sadasivuni[2], Peter Kasak[2] and Jan Tkac[1]*

1 Institute of Chemistry, Slovak Academy of Sciences, Bratislava, Slovak Republic

2 Center for Advanced Materials, Qatar University, Doha, Qatar

*Address all correspondence to: jan.tkac@savba.sk

References

[1] Novoselov, K. S.; Jiang, D.; Schedin, F.; Booth, T. J.; Khotkevich, V. V.; Morozov, S. V.; Geim, A. K. Two-dimensional atomic crystals. *Proc. Natl. Acad. Sci. U.S.A.* **2005,** *102*, 10451.

[2] Geim, A. K.; Novoselov, K. S. The rise of graphene. *Nat. Mater.* **2007,** *6*, 183–191.

[3] Duncan, C.; Roddie, H. Dendritic cell vaccines in acute leukaemia. *Best Pract. Res. Clin. Heaematol.* **2008,** *21*, 521–541.

[4] Sun, Y.; Gao, S.; Lei, F.; Xiao, C.; Xie, Y. Ultrathin Two-Dimensional Inorganic Materials: New Opportunities for Solid State Nanochemistry. *Acc. Chem. Res.* **2015,** *48*, 3–12.

[5] Chen, Y.; Tan, C.; Zhang, H.; Wang, L. Two-dimensional graphene analogues for biomedical applications. *Chem. Soc. Rev.* **2015,** *44*, 2681–2701.

[6] GÜRbÜZ, B.; Ayan, S.; Bozlar, M.; ÜStÜNdaĞ, C. B. Carbonaceous nanomaterials for phototherapy: a review. *Emergent Materials* **2020,** *3*, 479–502.

[7] Abdo, G. G.; Zagho, M. M.; Khalil, A. Recent advances in stimuli-responsive drug release and targeting concepts using mesoporous silica nanoparticles. *Emergent Materials* **2020,** *3*, 407–425.

[8] Khan, K.; Tareen, A. K.; Aslam, M.; Wang, R.; Zhang, Y.; Mahmood, A.; Ouyang, Z.; Zhang, H.; Guo, Z. Recent developments in emerging two-dimensional materials and their applications. *J. Mater. Chem. C* **2020,** *8*, 387–440.

[9] Ghidiu, M.; Lukatskaya, M. R.; Zhao, M.-Q.; Gogotsi, Y.; Barsoum, M. W. Conductive two-dimensional titanium carbide 'clay' with high volumetric capacitance. *Nature* **2014,** *516*, 78–81.

[10] Naguib, M.; Mashtalir, O.; Carle, J.; Presser, V.; Lu, J.; Hultman, L.; Gogotsi, Y.; Barsoum, M. W. Two-Dimensional Transition Metal Carbides. *ACS Nano* **2012,** *6*, 1322–1331.

[11] Naguib, M.; Gogotsi, Y. Synthesis of Two-Dimensional Materials by Selective Extraction. *Acc. Chem. Res.* **2015,** *48*, 128–135.

[12] Mashtalir, O.; Naguib, M.; Mochalin, V. N.; Dall'Agnese, Y.; Heon, M.; Barsoum, M. W.; Gogotsi, Y. Intercalation and delamination of layered carbides and carbonitrides. *Nat. Commun.* **2013,** *4*, 1716.

[13] Naguib, M.; Kurtoglu, M.; Presser, V.; Lu, J.; Niu, J.; Heon, M.; Hultman, L.; Gogotsi, Y.; Barsoum, M. W. Two-Dimensional Nanocrystals Produced by Exfoliation of Ti3AlC2. *Adv. Mater.* **2011,** *23*, 4248–4253.

[14] Pang, J.; Mendes, R. G.; Bachmatiuk, A.; Zhao, L.; Ta, H. Q.; Gemming, T.; Liu, H.; Liu, Z.; Rummeli, M. H. Applications of 2D MXenes in energy conversion and storage systems. *Chem. Soc. Rev.* **2019,** *48*, 72–133.

[15] Bu, F.; Zagho, M. M.; Ibrahim, Y. S.; Ma, B.; Elzatahry, A. A.; Zhao, D. Porous MXenes: Synthesis, structures, and applications. *Nano Today* **2020,** *30*, 100803.

[16] Wang, X.; Salari, M.; Jiang, D.-e.; Chapman Varela, J.; Anasori, B.; Wesolowski, D. J.; Dai, S.; Grinstaff, M. W.; Gogotsi, Y. Electrode material–ionic liquid coupling for electrochemical energy storage. *Nature Reviews Materials* **2020.**

[17] Shahzad, F.; Alhabeb, M.; Hatter, C. B.; Anasori, B.; Man Hong, S.; Koo, C. M.; Gogotsi, Y. Electromagnetic interference shielding with 2D

transition metal carbides (MXenes). *Science* **2016,** *353*, 1137.

[18] Iqbal, A.; Shahzad, F.; Hantanasirisakul, K.; Kim, M.-K.; Kwon, J.; Hong, J.; Kim, H.; Kim, D.; Gogotsi, Y.; Koo, C. M. Anomalous absorption of electromagnetic waves by 2D transition metal carbonitride Ti3CNTx (MXene). *Science* **2020,** *369*, 446.

[19] Rasool, K.; Pandey, R. P.; Rasheed, P. A.; Buczek, S.; Gogotsi, Y.; Mahmoud, K. A. Water treatment and environmental remediation applications of two-dimensional metal carbides (MXenes). *Mater. Today* **2019,** *30*, 80–102.

[20] Ibrahim, Y.; Kassab, A.; Eid, K.; M Abdullah, A.; Ozoemena, K. I.; Elzatahry, A. Unveiling Fabrication and Environmental Remediation of MXene-Based Nanoarchitectures in Toxic Metals Removal from Wastewater: Strategy and Mechanism. *Nanomaterials (Basel)* **2020,** *10*, 885.

[21] Sundaram, A.; Ponraj, J. S.; Wang, C.; Peng, W. K.; Manavalan, R. K.; Dhanabalan, S. C.; Zhang, H.; Gaspar, J. Engineering of 2D transition metal carbides and nitrides MXenes for cancer therapeutics and diagnostics. *J. Mater. Chem. B* **2020,** *8*, 4990–5013.

[22] Kalambate, P. K.; Gadhari, N. S.; Li, X.; Rao, Z.; Navale, S. T.; Shen, Y.; Patil, V. R.; Huang, Y. Recent advances in MXene–based electrochemical sensors and biosensors. *Trends Anal. Chem.* **2019,** *120*, 115643.

[23] Lorencova, L.; Gajdosova, V.; Hroncekova, S.; Bertok, T.; Blahutova, J.; Vikartovska, A.; Parrakova, L.; Gemeiner, P.; Kasak, P.; Tkac, J. 2D MXenes as Perspective Immobilization Platforms for Design of Electrochemical Nanobiosensors. *Electroanal.* **2019,** *31*, 1833–1844.

[24] Naguib, M.; Mochalin, V. N.; Barsoum, M. W.; Gogotsi, Y. Two-Dimensional Materials: 25th Anniversary Article: MXenes: A New Family of Two-Dimensional Materials. *Adv. Mater.* **2014,** *26*, 982–982.

[25] Michael, J.; Qifeng, Z.; Danling, W. Titanium carbide MXene: Synthesis, electrical and optical properties and their applications in sensors and energy storage devices. *Nanomater. Nanotechnol.* **2019,** *9*, 1847980418824470.

[26] Gund, G. S.; Park, J. H.; Harpalsinh, R.; Kota, M.; Shin, J. H.; Kim, T.-i.; Gogotsi, Y.; Park, H. S. MXene/Polymer Hybrid Materials for Flexible AC-Filtering Electrochemical Capacitors. *Joule* **2019,** *3*, 164–176.

[27] Er, D.; Li, J.; Naguib, M.; Gogotsi, Y.; Shenoy, V. B. Ti3C2 MXene as a High Capacity Electrode Material for Metal (Li, Na, K, Ca) Ion Batteries. *ACS Appl. Mater. Interf.* **2014,** *6*, 11173–11179.

[28] Come, J.; Black, J. M.; Lukatskaya, M. R.; Naguib, M.; Beidaghi, M.; Rondinone, A. J.; Kalinin, S. V.; Wesolowski, D. J.; Gogotsi, Y.; Balke, N. Controlling the actuation properties of MXene paper electrodes upon cation intercalation. *Nano Energy* **2015,** *17*, 27–35.

[29] Lukatskaya, M. R.; Mashtalir, O.; Ren, C. E.; Dall'Agnese, Y.; Rozier, P.; Taberna, P. L.; Naguib, M.; Simon, P.; Barsoum, M. W.; Gogotsi, Y. Cation Intercalation and High Volumetric Capacitance of Two-Dimensional Titanium Carbide. *Science* **2013,** *341*, 1502.

[30] Li, Q.; Zhou, J.; Li, F.; Sun, Z. Novel MXene-based hierarchically porous composite as superior electrodes for Li-ion storage. *Applied Surface Science* **2020,** *530*.

[31] Jia, X.; Shen, B.; Zhang, L.; Zheng, W. Construction of shape-memory

carbon foam composites for adjustable EMI shielding under self-fixable mechanical deformation. *Chemical Engineering Journal* **2021,** *405*.

[32] Rajavel, K.; Hu, Y.; Zhu, P.; Sun, R.; Wong, C. MXene/metal oxides-Ag ternary nanostructures for electromagnetic interference shielding. *Chemical Engineering Journal* **2020,** *399*.

[33] Li, Y.; Lu, Z.; Xin, B.; Liu, Y.; Cui, Y.; Hu, Y. All-solid-state flexible supercapacitor of Carbonized MXene/Cotton fabric for wearable energy storage. *Applied Surface Science* **2020,** *528*.

[34] Miao, J.; Zhu, Q.; Li, K.; Zhang, P.; Zhao, Q.; Xu, B. Self-propagating fabrication of 3D porous MXene-rGO film electrode for high-performance supercapacitors. *Journal of Energy Chemistry* **2021,** *52*, 243–250.

[35] Luo, Y.; Tian, Y.; Tang, Y.; Yin, X.; Que, W. 2D hierarchical nickel cobalt sulfides coupled with ultrathin titanium carbide (MXene) nanosheets for hybrid supercapacitors. *Journal of Power Sources* **2021,** *482*.

[36] Yang, Z.; Jiang, L.; Wang, J.; Liu, F.; He, J.; Liu, A.; Lv, S.; You, R.; Yan, X.; Sun, P.; Wang, C.; Duan, Y.; Lu, G. Flexible resistive NO2 gas sensor of three-dimensional crumpled MXene Ti3C2Tx/ZnO spheres for room temperature application. *Sensors and Actuators B: Chemical* **2021,** *326*, 128828.

[37] Deshmukh, K.; Kovářík, T.; Khadheer Pasha, S. K. State of the art recent progress in two dimensional MXenes based gas sensors and biosensors: A comprehensive review. *Coordination Chemistry Reviews* **2020,** *424*, 213514.

[38] Wang, D.; Wang, L.; Lou, Z.; Zheng, Y.; Wang, K.; Zhao, L.; Han, W.; Jiang, K.; Shen, G. Biomimetic, biocompatible and robust silk Fibroin-MXene film with stable 3D cross-link structure for flexible pressure sensors. *Nano Energy* **2020,** *78*.

[39] Wang, L.; Zhang, M.; Yang, B.; Tan, J.; Ding, X. Highly Compressible, Thermally Stable, Light-Weight, and Robust Aramid Nanofibers/Ti3AlC2 MXene Composite Aerogel for Sensitive Pressure Sensor. *ACS nano* **2020,** *14*, 10633–10647.

[40] Wang, X.; Li, M.; Yang, S.; Shan, J. A novel electrochemical sensor based on TiO2–Ti3C2TX/CTAB/chitosan composite for the detection of nitrite. *Electrochimica Acta* **2020,** *359*.

[41] Abdul Rasheed, P.; Pandey, R. P.; Gomez, T.; Jabbar, K. A.; Prenger, K.; Naguib, M.; Aïssa, B.; Mahmoud, K. A. Nb-based MXenes for efficient electrochemical sensing of small biomolecules in the anodic potential. *Electrochemistry Communications* **2020,** *119*, 106811.

[42] Tu, X.; Gao, F.; Ma, X.; Zou, J.; Yu, Y.; Li, M.; Qu, F.; Huang, X.; Lu, L. Mxene/carbon nanohorn/β-cyclodextrin-Metal-organic frameworks as high-performance electrochemical sensing platform for sensitive detection of carbendazim pesticide. *Journal of Hazardous Materials* **2020,** *396*.

[43] Kalambate, P. K.; Dhanjai; Sinha, A.; Li, Y.; Shen, Y.; Huang, Y. An electrochemical sensor for ifosfamide, acetaminophen, domperidone, and sumatriptan based on self-assembled MXene/MWCNT/chitosan nanocomposite thin film. *Microchimica Acta* **2020,** *187*.

[44] Huang, H.; Jiang, R.; Feng, Y.; Ouyang, H.; Zhou, N.; Zhang, X.; Wei, Y. Recent development and prospects of surface modification and biomedical applications of MXenes. *Nanoscale* **2020,** *12*, 1325–1338.

[45] Rafieerad, A.; Yan, W.; Amiri, A.; Dhingra, S. Bioactive and trackable

MXene quantum dots for subcellular nanomedicine applications. *Materials and Design* **2020,** *196*.

[46] Nie, Y.; Huang, J.; Ma, S.; Li, Z.; Shi, Y.; Yang, X.; Fang, X.; Zeng, J.; Bi, P.; Qi, J.; Wang, S.; Xia, Y.; Jiao, T.; Li, D.; Cao, M. MXene-hybridized silane films for metal anticorrosion and antibacterial applications. *Applied Surface Science* **2020,** *527*.

[47] Wang, W.; Feng, H.; Liu, J.; Zhang, M.; Liu, S.; Feng, C.; Chen, S. A photo catalyst of cuprous oxide anchored MXene nanosheet for dramatic enhancement of synergistic antibacterial ability. *Chemical Engineering Journal* **2020,** *386*.

[48] Zheng, K.; Li, S.; Jing, L.; Chen, P. Y.; Xie, J. Synergistic Antimicrobial Titanium Carbide (MXene) Conjugated with Gold Nanoclusters. *Advanced Healthcare Materials* **2020.**

[49] Wu, Q.; Li, N.; Wang, Y.; liu, Y.; Xu, Y.; Wei, S.; Wu, J.; Jia, G.; Fang, X.; Chen, F.; Cui, X. A 2D transition metal carbide MXene-based SPR biosensor for ultrasensitive carcinoembryonic antigen detection. *Biosens. Bioelectron.* **2019,** *144*.

[50] Liu, J.; Jiang, X.; Zhang, R.; Zhang, Y.; Wu, L.; Lu, W.; Li, J.; Li, Y.; Zhang, H. MXene-Enabled Electrochemical Microfluidic Biosensor: Applications toward Multicomponent Continuous Monitoring in Whole Blood. *Adv. Funct. Mater.* **2019,** *29*, 1807326.

[51] Shang, L.; Wang, X.; Zhang, W.; Jia, L. P.; Ma, R. N.; Jia, W. L.; Wang, H. S. A dual-potential electrochemiluminescence sensor for ratiometric detection of carcinoembryonic antigen based on single luminophor. *Sensors and Actuators, B: Chemical* **2020,** *325*.

[52] Medetalibeyoglu, H.; Kotan, G.; Atar, N.; Yola, M. L. A novel and ultrasensitive sandwich-type electrochemical immunosensor based on delaminated MXene@AuNPs as signal amplification for prostate specific antigen (PSA) detection and immunosensor validation. *Talanta* **2020,** *220*, 121403.

[53] Zhou, S.; Gu, C.; Li, Z.; Yang, L.; He, L.; Wang, M.; Huang, X.; Zhou, N.; Zhang, Z. Ti3C2Tx MXene and polyoxometalate nanohybrid embedded with polypyrrole: Ultra-sensitive platform for the detection of osteopontin. *Appl. Surf. Sci.* **2019,** *498*.

[54] Soomro, R. A.; Jawaid, S.; Kalawar, N. H.; Tunesi, M.; Karakuş, S.; Kilislioğlu, A.; Willander, M. In-situ engineered MXene-TiO2/BiVO4 hybrid as an efficient photoelectrochemical platform for sensitive detection of soluble CD44 proteins. *Biosensors and Bioelectronics* **2020,** *166*, 112439.

[55] Yang, X.; Feng, M.; Xia, J.; Zhang, F.; Wang, Z. An electrochemical biosensor based on AuNPs/Ti3C2 MXene three-dimensional nanocomposite for microRNA-155 detection by exonuclease III-aided cascade target recycling. *Journal of Electroanalytical Chemistry* **2020,** *878*.

[56] Barsoum, M.; Radovic, M. Mechanical Properties of the MAX Phases. *Annu. Rev. Mater. Res.* **2011,** *41*, 195–227.

[57] Halim, J.; Lukatskaya, M. R.; Cook, K. M.; Lu, J.; Smith, C. R.; Näslund, L.-Å.; May, S. J.; Hultman, L.; Gogotsi, Y.; Eklund, P.; Barsoum, M. W. Transparent Conductive Two-Dimensional Titanium Carbide Epitaxial Thin Films. *Chem. Mater.* **2014,** *26*, 2374–2381.

[58] Tang, H.; Hu, Q.; Zheng, M.; Chi, Y.; Qin, X.; Pang, H.; Xu, Q. MXene–2D layered electrode materials for energy storage. *Prog. Nat. Sci: Mater. International.* **2018,** *28*, 133–147.

[59] Hope, M. A.; Forse, A. C.; Griffith, K. J.; Lukatskaya, M. R.; Ghidiu, M.; Gogotsi, Y.; Grey, C. P. NMR reveals the surface functionalisation of Ti3C2 MXene. *Phys. Chem. Chem. Phys.* **2016,** *18*, 5099–5102.

[60] Gajdosova, V.; Lorencova, L.; Prochazka, M.; Omastova, M.; Micusik, M.; Prochazkova, S.; Kveton, F.; Jerigova, M.; Velic, D.; Kasak, P.; Tkac, J. Remarkable differences in the voltammetric response towards hydrogen peroxide, oxygen and Ru (NH3)63+ of electrode interfaces modified with HF or LiF-HCl etched Ti3C2Tx MXene. *Microchim. Acta* **2019,** *187*, 52.

[61] Lipatov, A.; Alhabeb, M.; Lukatskaya, M. R.; Boson, A.; Gogotsi, Y.; Sinitskii, A. Effect of Synthesis on Quality, Electronic Properties and Environmental Stability of Individual Monolayer Ti3C2 MXene Flakes. *Adv. Electron. Mater.* **2016,** *2*, 1600255.

[62] Overbury, S. H.; Kolesnikov, A. I.; Brown, G. M.; Zhang, Z.; Nair, G. S.; Sacci, R. L.; Lotfi, R.; van Duin, A. C. T.; Naguib, M. Complexity of Intercalation in MXenes: Destabilization of Urea by Two-Dimensional Titanium Carbide. *J. Am. Chem. Soc.* **2018,** *140*, 10305–10314.

[63] Mashtalir, O.; Lukatskaya, M. R.; Zhao, M.-Q.; Barsoum, M. W.; Gogotsi, Y. Amine-Assisted Delamination of Nb2C MXene for Li-Ion Energy Storage Devices. *Adv. Mater.* **2015,** *27*, 3501–3506.

[64] Alhabeb, M.; Maleski, K.; Mathis, T. S.; Sarycheva, A.; Hatter, C. B.; Uzun, S.; Levitt, A.; Gogotsi, Y. Selective Etching of Silicon from Ti3SiC2 (MAX) To Obtain 2D Titanium Carbide (MXene). *Angew. Chem. Int. Ed.* **2018,** *57*, 5444–5448.

[65] Khazaei, M.; Mishra, A.; Venkataramanan, N. S.; Singh, A. K.; Yunoki, S. Recent advances in MXenes: From fundamentals to applications. *Current Opinion in Solid State and Materials Science* **2019,** *23*, 164–178.

[66] Alhabeb, M.; Maleski, K.; Anasori, B.; Lelyukh, P.; Clark, L.; Sin, S.; Gogotsi, Y. Guidelines for Synthesis and Processing of Two-Dimensional Titanium Carbide (Ti3C2Tx MXene). *Chemistry of Materials* **2017,** *29*, 7633–7644.

[67] Zhang, J.; Kong, N.; Hegh, D.; Usman, K. A. S.; Guan, G.; Qin, S.; Jurewicz, I.; Yang, W.; Razal, J. M. Freezing Titanium Carbide Aqueous Dispersions for Ultra-long-term Storage. *ACS Applied Materials & Interfaces* **2020,** *12*, 34032–34040.

[68] Lorencova, L.; Bertok, T.; Dosekova, E.; Holazova, A.; Paprckova, D.; Vikartovska, A.; Sasinkova, V.; Filip, J.; Kasak, P.; Jerigova, M.; Velic, D.; Mahmoud, K. A.; Tkac, J. Electrochemical performance of Ti3C2Tx MXene in aqueous media: towards ultrasensitive H2O2 sensing. *Electrochim. Acta* **2017,** *235*, 471–479.

[69] Filip, J.; Zavahir, S.; Lorencova, L.; Bertok, T.; Yousaf, A. B.; Mahmoud, K. A.; Tkac, J.; Kasak, P. Tailoring electrocatalytic properties of Pt nanoparticles grown on Ti3C2Tx MXene surface. *J. Electrochem. Soc.* **2019,** *166*, H54-H62.

[70] Limbu, T. B.; Chitara, B.; Garcia Cervantes, M. Y.; Zhou, Y.; Huang, S.; Tang, Y.; Yan, F. Unravelling the Thickness Dependence and Mechanism of Surface-Enhanced Raman Scattering on Ti3C2TX MXene Nanosheets. *The Journal of Physical Chemistry C* **2020,** *124*, 17772–17782.

[71] Naguib, M.; Mashtalir, O.; Lukatskaya, M. R.; Dyatkin, B.; Zhang, C.; Presser, V.; Gogotsi, Y.; Barsoum, M. W. One-step synthesis of nanocrystalline transition metal oxides on thin sheets of disordered graphitic

carbon by oxidation of MXenes. *Chemical Communications* **2014,** *50*, 7420–7423.

[72] Lorencova, L.; Bertok, T.; Filip, J.; Jerigova, M.; Velic, D.; Kasak, P.; Mahmoud, K. A.; Tkac, J. Highly stable Ti3C2Tx (MXene)/Pt nanoparticles-modified glassy carbon electrode for H2O2 and small molecules sensing applications. *Sens. Actuat. B: Chem.* **2018,** *263*, 360–368.

[73] Lorencova, L.; Gajdosova, V.; Hroncekova, S.; Bertok, T.; Jerigova, M.; Velic, D.; Sobolciak, P.; Krupa, I.; Kasak, P.; Tkac, J. Electrochemical Investigation of Interfacial Properties of Ti3C2Tx MXene Modified by Aryldiazonium Betaine Derivatives. *Front. Chem.* **2020,** *8*, 553, DOI: 10.3389/fchem.2020.00553.

[74] Ji, J.; Zhao, L.; Shen, Y.; Liu, S.; Zhang, Y. Covalent stabilization and functionalization of MXene via silylation reactions with improved surface properties. *FlatChem* **2019,** *17*, 100128.

[75] Cao, W.-T.; Feng, W.; Jiang, Y.-Y.; Ma, C.; Zhou, Z.-F.; Ma, M.-G.; Chen, Y.; Chen, F. Two-dimensional MXene-reinforced robust surface superhydrophobicity with self-cleaning and photothermal-actuating binary effects. *Mater. Horizons* **2019,** *6*, 1057–1065.

[76] Lim, S.; Park, H.; Yang, J.; Kwak, C.; Lee, J. Stable colloidal dispersion of octylated Ti3C2-MXenes in a nonpolar solvent. *Colloids Surf., A* **2019,** *579*, 123648.

[77] Kumar, S.; Lei, Y.; Alshareef, N. H.; Quevedo-Lopez, M. A.; Salama, K. N. Biofunctionalized two-dimensional Ti3C2 MXenes for ultrasensitive detection of cancer biomarker. *Biosens. Bioelectron.* **2018,** *121*, 243–249.

[78] Bertok, T.; Lorencova, L.; Hroncekova, S.; Gajdosova, V.; Jane, E.; Hires, M.; Kasak, P.; Kaman, O.; Sokol, R.; Bella, V.; Eckstein, A. A.; Mosnacek, J.; Vikartovska, A.; Tkac, J. Advanced impedimetric biosensor configuration and assay protocol for glycoprofiling of a prostate oncomarker using Au nanoshells with a magnetic core. *Biosens. Bioelectron.* **2019,** *131*, 24–29.

[79] Bertok, T.; Dosekova, E.; Belicky, S.; Holazova, A.; Lorencova, L.; Mislovicova, D.; Paprckova, D.; Vikartovska, A.; Plicka, R.; Krejci, J.; Ilcikova, M.; Kasak, P.; Tkac, J. Mixed Zwitterion-Based Self-Assembled Monolayer Interface for Impedimetric Glycomic Analyses of Human IgG Samples in an Array Format. *Langmuir* **2016,** *32*, 7070–7078.

[80] Hetemi, D.; Noël, V.; Pinson, J. Grafting of diazonium salts on surfaces: Application to biosensors. *Biosensors* **2020,** *10*.

[81] Xu, B.; Zhi, C.; Shi, P. Latest advances in MXene biosensors. *Journal of Physics: Materials* **2020,** *3*, 031001.

[82] Kamysbayev, V.; Filatov, A. S.; Hu, H.; Rui, X.; Lagunas, F.; Wang, D.; Klie, R. F.; Talapin, D. V. Covalent surface modifications and superconductivity of two-dimensional metal carbide MXenes. *Science* **2020**, eaba8311.

[83] Liu, P.; Yao, Z.; Ng, V. M. H.; Zhou, J.; Kong, L. B.; Yue, K. Facile synthesis of ultrasmall Fe3O4 nanoparticles on MXenes for high microwave absorption performance. *Compos. Part A-Appl. Sci.* **2018,** *115*, 371–382.

[84] Rakhi, R. B.; Nayak, P.; Xia, C.; Alshareef, H. N. Novel amperometric glucose biosensor based on MXene nanocomposite. *Sci. Rep.* **2016,** *6*, 36422.

[85] Satheeshkumar, E.; Makaryan, T.; Melikyan, A.; Minassian, H.; Gogotsi, Y.; Yoshimura, M. One-step Solution Processing of Ag, Au and Pd@MXene Hybrids for SERS. *Sci. Rep.* **2016,** *6*, 32049.

[86] Jiang, Y.; Zhang, X.; Pei, L.; Yue, S.; Ma, L.; Zhou, L.; Huang, Z.; He, Y.; Gao, J. Silver nanoparticles modified two-dimensional transition metal carbides as nanocarriers to fabricate acetycholinesterase-based electrochemical biosensor. *Chem. Eng. J.* **2018,** *339*, 547–556.

[87] Song, D.; Jiang, X.; Li, Y.; Lu, X.; Luan, S.; Wang, Y.; Li, Y.; Gao, F. Metal–organic frameworks-derived MnO2/Mn3O4 microcuboids with hierarchically ordered nanosheets and Ti3C2 MXene/Au NPs composites for electrochemical pesticide detection. *J. Hazard. Mater.* **2019,** *373*, 367–376.

[88] Mohammadniaei, M.; Koyappayil, A.; Sun, Y.; Min, J.; Lee, M.-H. Gold nanoparticle/MXene for multiple and sensitive detection of oncomiRs based on synergetic signal amplification. *Biosens. Bioelectron.* **2020,** *159*, 112208.

[89] Zheng, J.; Diao, J.; Jin, Y.; Ding, A.; Wang, B.; Wu, L.; Weng, B.; Chen, J. An Inkjet Printed Ti3C2-GO Electrode for the Electrochemical Sensing of Hydrogen Peroxide. *J. Electrochem. Soc.* **2018,** *165*.

[90] Hroncekova, S.; Bertók, T.; Hires, M.; Jane, E.; Lorencova, L.; Vikartovská, A.; Tanvir, A.; Kasák, P.; Tkac, J. Ultrasensitive Ti3C2TX MXene/ Chitosan Nanocomposite-Based Amperometric Biosensor for Detection of Potential Prostate Cancer Marker in Urine Samples. *Processes* **2020,** *8*, 580.

[91] Tkac, J.; Ruzgas, T. Dispersion of single walled carbon nanotubes. Comparison of different dispersing strategies for preparation of modified electrodes toward hydrogen peroxide detection. *Electrochem. Commun.* **2006,** *8*, 899–903.

[92] Tkac, J.; Whittaker, J. W.; Ruzgas, T. The use of single walled carbon nanotubes dispersed in a chitosan matrix for preparation of a galactose biosensor. *Biosens. Bioelectron.* **2007,** *22*, 1820–1824.

[93] Wang, F.; Yang, C.; Duan, C.; Xiao, D.; Tang, Y.; Zhu, J. An Organ-Like Titanium Carbide Material (MXene) with Multilayer Structure Encapsulating Hemoglobin for a Mediator-Free Biosensor. *J. Electrochem. Soc.* **2014,** *162*, B16-B21.

[94] Liu, H.; Duan, C.; Yang, C.; Shen, W.; Wang, F.; Zhu, Z. A novel nitrite biosensor based on the direct electrochemistry of hemoglobin immobilized on MXene-Ti3C2. *Sens. Actuat. B: Chem.* **2015,** *218*, 60–66.

[95] Zheng, J.; Wang, B.; Jin, Y.; Weng, B.; Chen, J. Nanostructured MXene-based biomimetic enzymes for amperometric detection of superoxide anions from HepG2 cells. *Microchim. Acta* **2019,** *186*, 95.

[96] Wang, F.; Yang, C.; Duan, M.; Tang, Y.; Zhu, J. TiO2 nanoparticle modified organ-like Ti3C2 MXene nanocomposite encapsulating hemoglobin for a mediator-free biosensor with excellent performances. *Biosens. Bioelectron.* **2015,** *74*, 1022–1028.

[97] Zhang, H.; Wang, Z.; Zhang, Q.; Wang, F.; Liu, Y. Ti3C2 MXenes nanosheets catalyzed highly efficient electrogenerated chemiluminescence biosensor for the detection of exosomes. *Biosens. Bioelectron.* **2019,** *124–125*, 184–190.

[98] Zheng, J.; Wang, B.; Ding, A.; Weng, B.; Chen, J. Synthesis of MXene/ DNA/Pd/Pt nanocomposite for sensitive detection of dopamine. *J. Electroanal. Chem.* **2018,** *816*, 189–194.

[99] Zhang, Y.; Jiang, X.; Zhang, J.; Zhang, H.; Li, Y. Simultaneous voltammetric determination of acetaminophen and isoniazid using MXene modified screen-printed electrode. *Biosens. Bioelectron.* **2019,** *130*, 315–321.

[100] Xiang, H.; Lin, H.; Yu, L.; Chen, Y. Hypoxia-Irrelevant Photonic Thermodynamic Cancer Nanomedicine. *ACS Nano* **2019,** *13*, 2223–2235.

[101] Chae, A.; Jang, H.; Koh, D.-Y.; Yang, C.-M.; Kim, Y.-K. Exfoliated MXene as a mediator for efficient laser desorption/ionization mass spectrometry analysis of various analytes. *Talanta* **2020,** *209*, 120531.

[102] Jiang, Y.; Sun, J.; Cui, Y.; Liu, H.; Zhang, X.; Jiang, Y.; Nie, Z. Ti3C2 MXene as a novel substrate provides rapid differentiation and quantitation of glycan isomers with LDI-MS. *Chem. Commun.* **2019,** *55*, 10619–10622.

[103] Zamora-Gálvez, A.; Morales-Narváez, E.; Mayorga-Martinez, C. C.; Merkoçi, A. Nanomaterials connected to antibodies and molecularly imprinted polymers as bio/receptors for bio/sensor applications. *Applied Materials Today* **2017,** *9*, 387–401.

[104] Sinha, A.; Dhanjai; Mugo, S. M.; Chen, J.; Lokesh, K. S. MXene-based sensors and biosensors: Next-generation detection platforms. In *Handbook of Nanomaterials in Analytical Chemistry: Modern Trends in Analysis*, 2019, pp 361–372.

[105] Grieshaber, D.; MacKenzie, R.; Vörös, J.; Reimhult, E. Electrochemical Biosensors - Sensor Principles and Architectures. *Sensors* **2008,** *8*, 1400–1458.

[106] Jastrzębska, A. M.; Szuplewska, A.; Rozmysłowska-Wojciechowska, A.; Chudy, M.; Olszyna, A.; Birowska, M.; Popielski, M.; Majewski, J. A.; Scheibe, B.; Natu, V.; Barsoum, M. W. On tuning the cytotoxicity of Ti3C2 (MXene) flakes to cancerous and benign cells by post-delamination surface modifications. *2D Materials* **2020,** *7*.

[107] Gajdosova, V.; Lorencova, L.; Kasak, P.; Tkac, J. Electrochemical Nanobiosensors for Detection of Breast Cancer Biomarkers. *Sensors* **2020,** *20*, 4022.

[108] Siegel, R. L.; Miller, K. D.; Jemal, A. Cancer statistics, 2019. *CA Cancer J Clin* **2019,** *69*, 7–34.

[109] Atkinson, A. J., Jr.; Colburn, W. A.; DeGruttola, V. G.; DeMets, D. L.; Downing, G. J.; Hoth, D. F.; Oates, J. A.; Peck, C. C.; Schooley, R. T.; Spilker, B. A.; Woodcock, J.; Zeger, S. L. Biomarkers and surrogate endpoints: Preferred definitions and conceptual framework. *Clin. Pharmacol. Ther.* **2001,** *69*, 89–95.

[110] Tzitzikos, G.; Saridi, M.; Filippopoulou, T.; Makri, A.; Goulioti, A.; Stavropoulos, T.; Stamatiou, K. Measurement of tumor markers in chronic hemodialysis patients. *Saudi J. Kidney Dis. Transplant.* **2010,** *21*, 50–53.

[111] Núñez, C. Blood-based protein biomarkers in breast cancer. *Clin. Chim. Acta* **2019,** *490*, 113–127.

[112] Estakhri, R.; Ghahramanzade, A.; Vahedi, A.; Nourazarian, A. Serum levels of CA15–3, AFP, CA19–9 and CEA tumor markers in cancer care and treatment of patients with impaired renal function on hemodialysis. *Asian Pacific journal of cancer prevention : APJCP* **2013,** *14*, 1597–9.

[113] Gebrehiwot, A. G.; Melka, D. S.; Kassaye, Y. M.; Gemechu, T.; Lako, W.; Hinou, H.; Nishimura, S.-I. Exploring serum and immunoglobulin G N-glycome as diagnostic biomarkers for early detection of breast cancer in Ethiopian women. *BMC Cancer* **2019,** *19*, 588.

[114] Bertok, T.; Lorencova, L.; Chocholova, E.; Jane, E.; Vikartovska, A.; Kasak, P.; Tkac, J. Electrochemical Impedance Spectroscopy Based Biosensors: Mechanistic Principles, Analytical Examples and Challenges

towards Commercialization for Assays of Protein Cancer Biomarkers. *ChemElectroChem* **2019,** *6*, 989–1003.

[115] Lorencova, L.; Bertok, T.; Bertokova, A.; Gajdosova, V.; Hroncekova, S.; Vikartovska, A.; Kasak, P.; Tkac, J. Exosomes as a Source of Cancer Biomarkers: Advances in Electrochemical Biosensing of Exosomes. *ChemElectroChem* **2020,** *7*, 1956–1973.

[116] Reddy, K. K.; Bandal, H.; Satyanarayana, M.; Goud, K. Y.; Gobi, K. V.; Jayaramudu, T.; Amalraj, J.; Kim, H. Recent Trends in Electrochemical Sensors for Vital Biomedical Markers Using Hybrid Nanostructured Materials. *Adv. Sci.* **2020,** *7*, 1902980.

[117] Wang, H.; Sun, J.; Lu, L.; Yang, X.; Xia, J.; Zhang, F.; Wang, Z. Competitive electrochemical aptasensor based on a cDNA-ferrocene/MXene probe for detection of breast cancer marker Mucin1. *Anal. Chim. Acta* **2020,** *1094*, 18–25.

[118] Liu, L.; Wei, Y.; Jiao, S.; Zhu, S.; Liu, X. A novel label-free strategy for the ultrasensitive miRNA-182 detection based on MoS2/Ti3C2 nanohybrids. *Biosens. Bioelectron.* **2019,** *137*, 45–51.

[119] Duan, F.; Guo, C.; Hu, M.; Song, Y.; Wang, M.; He, L.; Zhang, Z.; Pettinari, R.; Zhou, L. Construction of the 0D/2D heterojunction of Ti3C2Tx MXene nanosheets and iron phthalocyanine quantum dots for the impedimetric aptasensing of microRNA-155. *Sens. Actuat. B: Chem.* **2020,** *310*, 127844.

[120] Xu, Q.; Xu, J.; Jia, H.; Tian, Q.; Liu, P.; Chen, S.; Cai, Y.; Lu, X.; Duan, X.; Lu, L. Hierarchical Ti3C2 MXene-derived sodium titanate nanoribbons/PEDOT for signal amplified electrochemical immunoassay of prostate specific antigen. *J. Electroanal. Chem.* **2020,** *860*, 113869.

[121] Chen, J.; Tong, P.; Huang, L.; Yu, Z.; Tang, D. Ti3C2 MXene nanosheet-based capacitance immunoassay with tyramine-enzyme repeats to detect prostate-specific antigen on interdigitated micro-comb electrode. *Electrochim. Acta* **2019,** *319*, 375–381.

[122] Liu, Y.; Zeng, H.; Chai, Y.; Yuan, R.; Liu, H. Ti3C2/BiVO4 Schottky junction as a signal indicator for ultrasensitive photoelectrochemical detection of VEGF165. *Chem. Commun.* **2019,** *55*, 13729–13732.

[123] Kumar, R.; Pal, S.; Verma, A.; Prajapati, Y. K.; Saini, J. P. Effect of silicon on sensitivity of SPR biosensor using hybrid nanostructure of black phosphorus and MXene. *Superlattices Microstruct.* **2020,** *145*, 106591.

[124] Wu, L.; You, Q.; Shan, Y.; Gan, S.; Zhao, Y.; Dai, X.; Xiang, Y. Few-layer Ti3C2Tx MXene: A promising surface plasmon resonance biosensing material to enhance the sensitivity. *Sens. Actuat. B: Chem.* **2018,** *277*, 210–215.

[125] Srivastava, A.; Verma, A.; Das, R.; Prajapati, Y. K. A theoretical approach to improve the performance of SPR biosensor using MXene and black phosphorus. *Optik* **2020,** *203*, 163430.

[126] Ding, X.; Li, C.; Wang, L.; Feng, L.; Han, D.; Wang, W. Fabrication of hierarchical g-C3N4/MXene-AgNPs nanocomposites with enhanced photocatalytic performances. *Mater. Lett.* **2019,** *247*, 174–177.

[127] Wu, Q.; Li, N.; Wang, Y.; Xu, Y.; Wu, J.; Jia, G.; Ji, F.; Fang, X.; Chen, F.; Cui, X. Ultrasensitive and Selective Determination of Carcinoembryonic Antigen Using Multifunctional Ultrathin Amino-Functionalized Ti3C2-MXene Nanosheets. *Anal. Chem.* **2020,** *92*, 3354–3360.

[128] Wang, S.; Wei, S.; Wang, S.; Zhu, X.; Lei, C.; Huang, Y.; Nie, Z.; Yao, S.

Chimeric DNA-Functionalized Titanium Carbide MXenes for Simultaneous Mapping of Dual Cancer Biomarkers in Living Cells. *Anal. Chem.* **2019,** *91*, 1651–1658.

[129] Guo, Z.; Zhu, X.; Wang, S.; Lei, C.; Huang, Y.; Nie, Z.; Yao, S. Fluorescent Ti3C2 MXene quantum dots for an alkaline phosphatase assay and embryonic stem cell identification based on the inner filter effect. *Nanoscale* **2018,** *10*, 19579–19585.

[130] Chen, F.; Lu, Q.; Zhang, Y.; Yao, S. Strand displacement dual amplification miRNAs strategy with FRET between NaYF4:Yb,Tm/Er upconversion nanoparticles and Ti3C2 nanosheets. *Sens. Actuat. B: Chem.* **2019,** *297*.

[131] Cai, G.; Yu, Z.; Tong, P.; Tang, D. Ti3C2 MXene quantum dot-encapsulated liposomes for photothermal immunoassays using a portable near-infrared imaging camera on a smartphone. *Nanoscale* **2019,** *11*, 15659–15667.

[132] Fang, D.; Zhang, S.; Dai, H.; Lin, Y. An ultrasensitive ratiometric electrochemiluminescence immunosensor combining photothermal amplification for ovarian cancer marker detection. *Biosens. Bioelectron.* **2019,** *146*.

[133] Zhang, H.; Wang, Z.; Zhang, Q.; Wang, F.; Liu, Y. Ti 3 C 2 MXenes nanosheets catalyzed highly efficient electrogenerated chemiluminescence biosensor for the detection of exosomes. *Biosens. Bioelectron.* **2019,** *124–125*, 184–190.

[134] Zhang, Q.; Wang, F.; Zhang, H.; Zhang, Y.; Liu, M.; Liu, Y. Universal Ti3C2 MXenes Based Self-Standard Ratiometric Fluorescence Resonance Energy Transfer Platform for Highly Sensitive Detection of Exosomes. *Anal. Chem.* **2018,** *90*, 12737–12744.

[135] Fang, D.; Zhao, D.; Zhang, S.; Huang, Y.; Dai, H.; Lin, Y. Black phosphorus quantum dots functionalized MXenes as the enhanced dual-mode probe for exosomes sensing. *Sens. Actuat. B: Chem.* **2020,** *305*, 127544.

[136] Mohammadniaei, M.; Nguyen, H. V.; Tieu, M. V.; Lee, M.-H. 2D Materials in Development of Electrochemical Point-of-Care Cancer Screening Devices. *Micromachines* **2019,** *10*, 662.

[137] Lin, H.; Chen, Y.; Shi, J. Insights into 2D MXenes for Versatile Biomedical Applications: Current Advances and Challenges Ahead. *Adv. Sci.* **2018,** *5*, 1800518.

Section 4

Biomedical and Environmental Applications of Nanomaterials

Chapter 12

Green Synthesis of Metal Nanoparticles for Antimicrobial Activity

Jerushka S. Moodley, Suresh Babu Naidu Krishna, Karen Pillay and Patrick Govender

Abstract

The development and extensive spread of multi-drug resistant bacteria are considered as a major public health concern. Failures to control severe infections due to antibiotic resistance have augmented healthcare costs as well as patient morbidity and mortality. Presently, natural product-based therapeutics are gaining significant attention both for their antimicrobial effectiveness and for not persuading drug resistance. Furthermore, recent developments in nanoscience on new drug delivery systems built on nanostructured materials from plants and microbes have emerged which focus on targeted delivery and controlled release of therapeutic agents. This review examines the recent investigations on the biological activities of plant and bacterial biological material for silver nanoparticle (AgNP) synthesis. Also, the underlying mechanism of antimicrobial activities of silver nanoparticles against human pathogens will be discussed. A fact of the biological activities and/or chemical responses of plants is required, not only for the discovery of new therapeutic agents, but because such evidence may be of value in disclosing new sources of already known biologically active compounds.

Keywords: antimicrobial resistance, biological activities, drug delivery, green synthesis, health, silver nanoparticles

1. Introduction

The antimicrobial potential of silver (Ag) and Ag-based solutions has long been established, however, their application was considered obsolete upon the discovery of antibiotics [1, 2]. In recent years, the developing crisis of multi-drug resistant pathogenic infections has led to the resurgence in this metal, however, with the use of nanotechnology to generate its nanoparticle form. For this reason, tremendous efforts have been extended in nanotechnology, particularly in the development of green synthetic strategies for silver nanoparticle (AgNPs) production to facilitate their use in antimicrobial therapeutic applications [3].

The interest in silver nanoparticles (AgNPs) as an alternative to current antibiotics has increased profoundly over the last few years. This is owed to the cumulative incidence of microbial drug-resistant infections and the lack of appropriate treatment thereof [4]. The World Health Organisation report of 2014 highlighted the probability of a post-antibiotic era in which common infections and minor injuries could potentially result in fatalities [5]. Accordingly, concerted efforts have been extended by global pharmaceuticals to formulate new or improved antibiotics. However, despite high research cost-intensive investment in the last decade or so only two new classes of antibiotics have been introduced into the market [6, 7]. The imperative need for the uncovering of novel antimicrobial scaffolds has led to the resurgence of silver, however, in its nano-particulate form [8].

The antimicrobial activities of AgNPs are well established and currently researchers are striving to develop greener synthetic strategies for their production [1, 9]. The use of nanotechnology for the synthesis of AgNPs from environmentally compatible biomaterials is evolving into an important branch of science and technology [10]. To this end, a variety of biological extracts have been explored for the bottom-up synthesis of AgNPs [11]. However, there is an ongoing search to identify novel capping structures to produce AgNPs with increased bio-efficacies. In this context, this chapter points to highlight the use of plants as an alternative green technology for nanoparticle synthesis and their biomedical applications as potential biofactories for antibacterial, antifungal and anti-cancer agents.

2. Preparation of nanosilver

2.1 Conventional nanoparticle synthetic strategies

Established technologies for AgNP synthesis and other metal preparations can be categorised distinctly into two approaches, namely: "top to bottom", which is normally employed by physicists and "bottom to up", a construction favourite of chemists [12, 13]. Both approaches converge at the nanodimension but vary drastically in the synthetic technology. "Top to bottom" approaches apply various physical methods such as grinding, milling, sputtering, evaporation-condensation and thermal/laser ablation to break down bulk solid materials to their nanoparticulate form. "Bottom to up" approaches entail various chemical and biological methods to synthesise nanoparticles by the self-assembly of atoms such as Ag^+ into nuclei that further develop into nano-sized particles [9].

Important physical "top to bottom" methods for nanoparticle preparation include evaporation-condensation and laser ablation techniques [14]. Evaporation-condensation applies a tube furnace at atmospheric temperature wherein primary material (metal Ag) contained in a boat; is centred in the furnace and vaporised into a carrier gas [9]. Several inadequacies have been identified with this technique, for example, the furnace occupies a large space, requires high energy input whilst raising the environmental temperature around the source material and requires long durations to achieve thermal stability. Additionally, a major drawback to this type of synthesis is the resulting imperfections in the surface structure of the derived nanoparticles which can ultimately alter their physical properties [9, 15]. In laser ablation, irradiation is used to remove material from a bulk metal in solution. The efficacy of this technique and characteristics of nascent particles is largely dependent on a number of parameters including the wavelength of the laser, duration of laser pulses, laser fluence, ablation duration and the effective liquid medium with or without surfactants [16, 17]. An important advantage of laser ablation for AgNP preparation is the absence of chemicals in solution which could potentially contaminate the nanoparticle preparation [18].

Regarding "bottom to up" approaches, wet chemical reduction is the most frequently practiced method for nanoparticle preparation [15] although, several other methods have been reported [19–22]. As the name suggests, wet chemical reduction involves the reduction of a metal salt precursor in aqueous or organic solution. Various organic and inorganic compounds successfully utilised as reducing agents in the synthesis of AgNPs include: ascorbate; borohydride; citrate; elemental hydrogen; formaldehyde; N-N-dimethyl formamide (DMF); Tollen's reagent; and polyethylene glycol blocks [15, 23, 24]. In addition to reducing agents, protective stabilising agents are also included in the reaction solution to prevent agglomeration of nascent nanoparticles [25, 26]. With stability achieved, this method can be useful to produce high nanoparticle yields with low preparation costs [27]. However, the efficacy of this method is challenged by the potential contamination of nascent nanoparticles by precursor chemicals, the use of toxic solvents and the generation of hazardous by-products [13, 28].

Evidently, the aforementioned physical and chemical methods have certain limitations that restrict their use in the preparation of nanoparticles for biological applications [29]. In this regard, concerted efforts have been extended to develop nanoparticle synthetic strategies that are environmentally sound. Essentially, this would entail the use of benign, biotechnological tools and has given rise to the concept of green technology. This technology can best be described as the use of biological routes such as plants and microorganisms or their byproducts in the synthesis of nanoparticles [29–31]. These bio-inspired methods **(Figure 1)** are not only environmentally welcoming but are cost effective and can be easily up-scaled for large productions [32].

2.2 Biological nanoparticle synthetic strategies

As previously eluded, biological approaches for AgNP synthesis employ the use of living organisms or their extracts as capping/reducing agents in a synthetic reaction. To date, a variety of biological entities have been explored for their Ag^+ reducing abilities and include viruses, bacteria, plants, algae, fungi, yeast and mammalian cells [11, 13, 34–36]. Biological synthesis can be divided into two strategies, specifically: bioreduction and biosorption. Bioreduction occurs when metal ions undergo chemical reduction into biologically stable complexes. Many organisms have displayed dissimilatory metal reduction involving the coupling of reduction with oxidation of an enzyme. The resulting stable, inert nanoparticles can then be safely extracted from the reaction mixture. Alternatively, biosorption involves the attachment of metal ions onto an organism itself, such as on the cell wall. Various bacteria, fungi and plant species express peptides or possess modified cell wall structures that are capable of binding metal ions, thereby forming stable complexes in the form of nanoparticles [36].

In this review, the use of plant and bacterial biological material for AgNP synthesis will be discussed. For a review on the use of alternative biological entities as AgNP factories, studies by the following authors are recommended [11, 36, 37].

2.3 AgNP synthesis from plants

Plants have shown the capacity to hyper-accumulate metals as a means to protect themselves from insects and herbivores. This observation has paved way for the technology known as phytoextraction, wherein plants are employed to extract minerals from various groundwater and soil sediments. Major applications of phytoextraction include the mining of precious metals from unfeasible ground sites (phytomining), stabilisation or recovery of non-naturally occurring contaminants (phytoremediation) and the addition of essential metals to growing crops. Interestingly, studies have unveiled that metals accumulated by

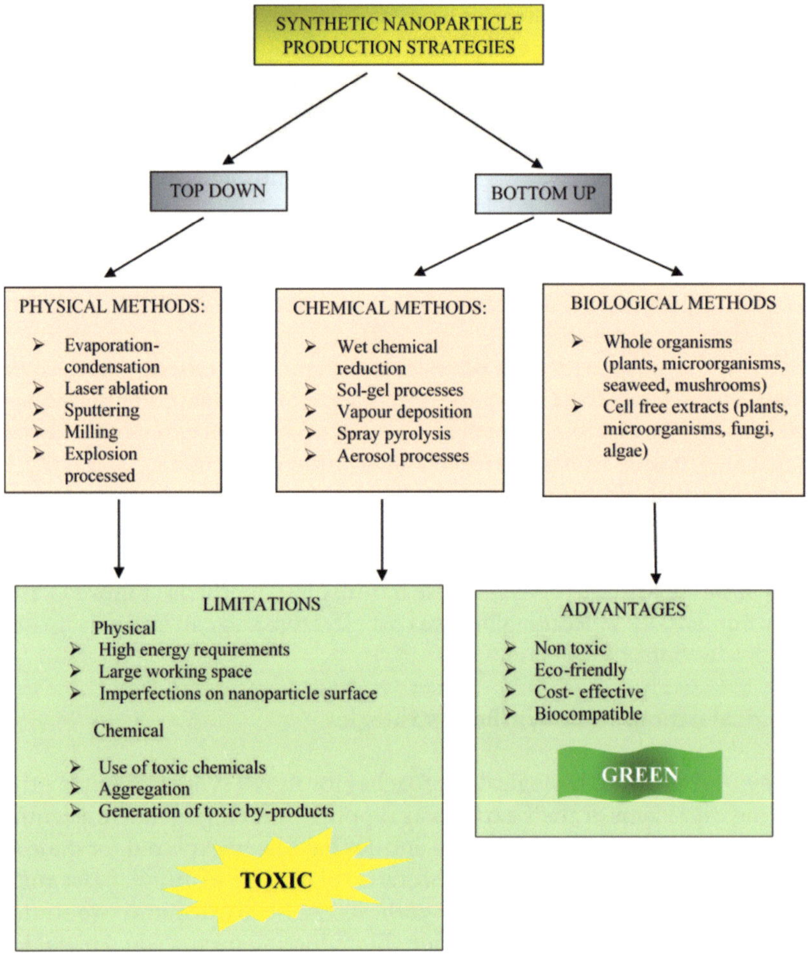

Figure 1.
Different approaches for AgNP synthesis. Adapted from [9, 33].

the plant are usually deposited in the form of nanoparticles. This has stimulated interest for the use of plants as factories for nanoparticle synthesis [35]. Whole plants have been explored for the synthesis of nanoparticles when grown on the appropriate metal enriched substrates. Species such as *Brassica juncae* (mustard greens) and *Medicago sativa* (alfalfa) have demonstrated the ability to accumulate AgNPs. For example, 50 nm sized AgNPs, at a high yield (13.6% of total plant weight) were reported for *M. sativa* when grown on silver nitrate ($AgNO_3$) [38]. Additionally, icosahedral gold nanoparticles of 4 nm size were observed in *M. sativa* and semi-spherical copper nanoparticles of 2 nm size were observed in *Iris pseudacorus* when the plants were grown on gold and copper salt enriched substrates, respectively [39, 40].

Although whole plants can potentially serve as factories for nanoparticle synthesis, several disadvantages have been identified with this technology especially when up-scaling for industrial applications. For example, physical attributes of nanoparticles such as size and shape vary upon the localisation of the particles in the plant due to the differences in metal ion content in different plant tissues and the possibility of nanoparticle movement and penetration [39]. This heterogeneity of important bioactivity-determinants such as size and shape [41, 42] limit the use of these nanoparticles and especially in applications where mono-dispersed nanoparticle preparations are required. Furthermore, recovery of nanoparticles from living plants entails laborious extraction, isolation and purification procedures and may potentially result in low yields [35].

The use of plant broths/extracts in nanoparticle synthesis was introduced by Shankar *et al*., (2003). In their study, compounds responsible for the reduction of metal ions were extracted and used as reducing agents in a synthetic reaction mixture, resulting in the extracellular production of nanoparticles [43]. This strategy tentatively offers several advantages compared to the use of whole plants. For example, nanoparticle formation occurs considerably faster as opposed to whole plants which require diffusion of metal ions throughout the plant body. Additionally, the use of extracts would be more economical due to the ease of purification [35].

This *in vitro* approach has been actively developed and applied to a variety of plant flora for the synthesis of AgNPs [28]. Various organ extracts: stem, root, leaf, bark, fruit and fruit peel have demonstrated the ability to reduce Ag^+. Particularly, biomolecules (**Figure 2**) such as proteins, amino acids, enzymes, polysaccharides, alkaloids, tannins, phenolics, saponins, terpenoids and vitamins present in the extracts act as both reducing and stabilising agents [9].

Terpenoids are a class of diverse organic polymers manufactured in plants from five-carbon isoprene units and display strong antioxidant activities. In a previous study by Shankar *et al*., involving gold nanoparticle synthesis from geranium leaf extracts, it was suggested that these polymers were actively involved in the reduction of gold ions into stable nanoparticles [44]. Later Singh

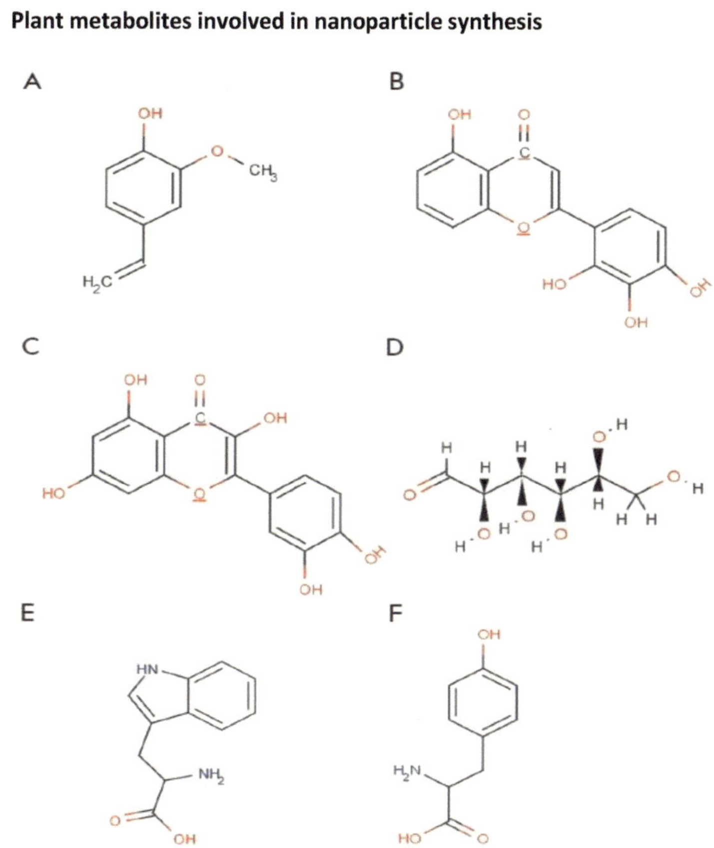

Figure 2.
Major plant metabolites involved in the synthesis of metal nanoparticles: (A)-terpenoids (eugenol); (B & C)-flavonoids (luteolin, quercetin); (D)-a reducing hexose with the open chain form; (E & F)-amino acids (tryptophan, tyrosine). Adapted from [35].

et al. reported that eugenol, the main terpenoid found in *Szyygium aromaticum* (clove), played an important role in reducing $AgNO_3$ and $HAuCL_4$. The Fourier transform infrared (FTIR) spectroscopy analysis of their study suggests that the dissociation of the proton from the OH group in eugenol leads to the formation of intermediate resonance structures which can undergo further oxidation. This latter reaction may be coupled to the reduction of Ag^+ and subsequent formation of stable AgNPs [45].

Flavonoids are made up of a large group of polyphenolic compounds containing various classes such as anthocyanins, isoflavonoids, flavonols, chalcones, flavones and flavanones. There are several functional groups present on flavonoid compounds that can participate in nanoparticle formation. It has been hypothesised that the tautomerization of flavonoids from the enol to keto form releases a reactive hydrogen atom that can participate in the reduction of metal ions. For example, studies involving AgNP synthesis from *Ocimum sanctum* extracts indicate that synthesis is likely to be the result of tautomerization of the flavonoids luteolin and rosmarinic acid [46]. Additionally, some flavonoids can chelate metal ions with their carbonyl groups or π-electron. Quercetin is an example of a flavonoid with strong chelating activity [35]. These mechanisms may explain the prevalence of flavonoid groups adsorbed on to the surface of AgNPs derived in previous studies [47, 48]. Further indication of flavonoid involvement in nanoparticle synthesis is provided by a study using *Lawsonia inermis,* in which the flavonoid apiin was extracted and successfully employed in the synthesis of gold and Ag nanoparticles [49].

Sugars contained in plant extracts are also capable of inducing nanoparticle formation. It is known that monosaccharides in the linear form containing an aldehyde (e.g. glucose), are capable reducing agents [35]. Monosaccharides harbouring a keto-group may act as antioxidants upon tautomeric transformation from a ketone to an aldehyde (e.g. fructose). In this regard, glucose is reportedly more efficient at metal ion reduction than fructose due to the kinetics of tautomerism from a ketone to an aldehyde which limits the reducing potential of fructose. Disaccharides and polysaccharides may also participate in the reduction of metal ions however, this is largely dependent on the ability of their monosaccharide components to take on an open chain configuration within an oligomer. Examples include lactose and maltose. In contrast, sucrose is unable to participate in metal ion reduction because the linkage of its glucose and fructose monomers restrict the formation of open chains. However, when sucrose was placed in tetrachloroauric and tetrachloroplatinic acids, nanoparticle formation proceeded [50]. This may be due to the acidic hydrolysis of sucrose yielding glucose and fructose. In general, it is suggested that nanoparticle formation by sugars occurs by the oxidation of an aldehyde group into a carbonyl group which subsequently leads to the reduction of metal ions and nanoparticle formation [44].

FTIR analysis of plant derived metal nanoparticles have revealed the presence of proteins on their surface, suggesting that proteins may also possess metal ion reducing ability. However, amino acids have displayed differences in their potential for metal ion reducing and binding efficiencies. For example, lysine, cysteine, arginine and methionine have been shown to bind Ag^+. In a separate study, aspartate was used to reduce tetrachloroauric acid forming nanoparticles, whilst valine and lysine did not possess this ability. Amino acids capable of binding metal ions are thought to do so through their amino or carboxyl groups or through side chain groups: carboxyl groups of aspartic and glutamic acid, imidazole ring of histidine, thiol of cysteine, thioether of methionine, hydroxyl group of serine; threonine and tyrosine, carbonyl groups of asparagine and glutamine [35].

Linkage of amino acids in a peptide chain may also affect the ability of individual amino acids to bind and reduce metal ions. For example, the R-carbon of amines and carboxylic acids in a peptide bond are inaccessible for association with metal ions. However, the free side chains of individual amino acids can still participate in binding and reduction of metal ions although, this is largely dependent on the amino acid sequence. Tan *et al.* demonstrated that synthesised peptides derived from amino acids with strong binding abilities and high reducing activities displayed lower reduction than expected [51]. A previous study suggested that protein molecules capable of nanoparticle formation display a strong attraction of metal ions to the regions on the molecule responsible for reduction however, their chelating activity is limited [52]. It was also suggested that the amino acid sequence of a protein can influence the size, shape and yield of derived nanoparticles. For example, the synthetic peptide GASLWWSEKL was found to rapidly reduce metal ions forming a large number of small nanoparticles (˂10 nm), however, replacement of the N- and C- terminal residues forming the peptide SEKLWWGASL led to slower reduction and formation of larger nanospheres and nanotriangles (40 nm). These findings seemingly suggest that peptides and proteins present in plant extracts probably play a vital role in determining nanoparticle size and shape and potentially affect the overall yield of the nanoparticles [51].

2.4 AgNP synthesis from bacteria

There exists a vast array of literature pertaining to the use of bacteria as factories for nanoparticle synthesis [53, 54]. Bacteria have a marked advantage over other microbial systems such as fungi due to their abundance, rapid growth rate, cheap cultivation and the relative ease of their manipulation [55]. Their ubiquitous nature has led to their exposure and proliferation in many environmental extremes and ultimately depends on the natural defence mechanisms of these microorganisms to resist the effects posed by environmental stresses [56]. Bacteria have demonstrated these defence mechanisms in a few non-optimal growth conditions including environments contaminated with metal ions.

AgNP synthesis by bacteria can occur intracellularly or by the use of their extracts [53]. Several studies have reported intracellular synthesis by a variety of bacterial species and as similarly reported for the use of whole plants, this technology is associated with long duration periods for nanoparticle synthesis. For example, Pugazhenthiran *et al.* reported an incubation time of 7 days for AgNP synthesis from *Bacillus* sp. [57]. Kalimuthu *et al.* reported a reaction time of 24 hours for AgNP synthesis by *Bacillus licheniformis* [58]. Although this reaction time was more industrially significant, the authors reported an additional extraction to acquire the derived nanoparticles. Synthesis of AgNPs by the use of bacterial cell free supernatant (CFS) extracts was reported by Shahverdi *et al.*, (2007). Interestingly, nanoparticle synthesis occurred within five minutes of Ag^+ coming into contact with the CFS [59]. Thus, this method presents the greatest potential for industrial production of AgNPs from bacteria. Several other studies have reported on the production of AgNPs from bacterial CFS extracts but not at the previously stated formation rate [60, 61]. This seemingly suggests that bacterial extracts differ in their metal ion reducing abilities and may require an external energy source to accelerate nanoparticle formation.

2.4.1 Bacterial metabolites involved in nanoparticle synthesis

As previously stated, metal nanoparticle synthesis in bacteria may potentially occur through resistance mechanisms attained by these organisms to overcome the toxic effects of metals. These strategies include redox state changes, efflux systems, intracellular precipitation, metal accumulation and extracellular formation of complexes **(Figure 3)** [56]. In an early study, Slawson *et al.* observed that the Ag resistant strain *Pseudomonas stutzeri* AG259, was capable of accumulating AgNPs (35–46 nm) within its periplasmic space. The formation of these nanoparticles was thought to have occurred by a mechanism involving the NADH-dependent reductase enzyme which undergoes oxidation to form NAD^+. The lost free electron may potentially reduce Ag^+ to AgNPs [62]. Later, He *et al.* reported that the NADH-dependent reductase enzyme may similarly participate in the extracellular formation of gold nanoparticles by the bacterium *Rhodopseudomonas capsulata* [63]. Other studies have reported nanoparticle formation without the use of biological enzymes. Non-enzymatic nanoparticle synthesis by a *Corynebacterium* sp. was reported by Sneha *et al.* [64]. Organic functional groups present at the cell wall were thought to induce metal ion reduction [64]. Sintubin *et al.* proposed a two-step mechanism for AgNP formation by several lactic acid bacteria, involving biosorption of Ag^+ on the cell wall which is coupled to the subsequent reduction of these ions to form the nanoparticles [65]. Parikh *et al.* identified a gene homologue in a Ag-resistant *Morganella* strain with a 99% nucleotide sequence similarity to a periplasmic Ag-binding protein-encoding gene [66]. Johnston *et al.* further reported the production of a small non-ribosomal peptide, delftibactin by *Delftia acidovorans* which they believed to be associated with a resistance mechanism. By producing inert gold nanoparticles bound to delftibactin, gold ions no longer caused toxicity to the cells [67].

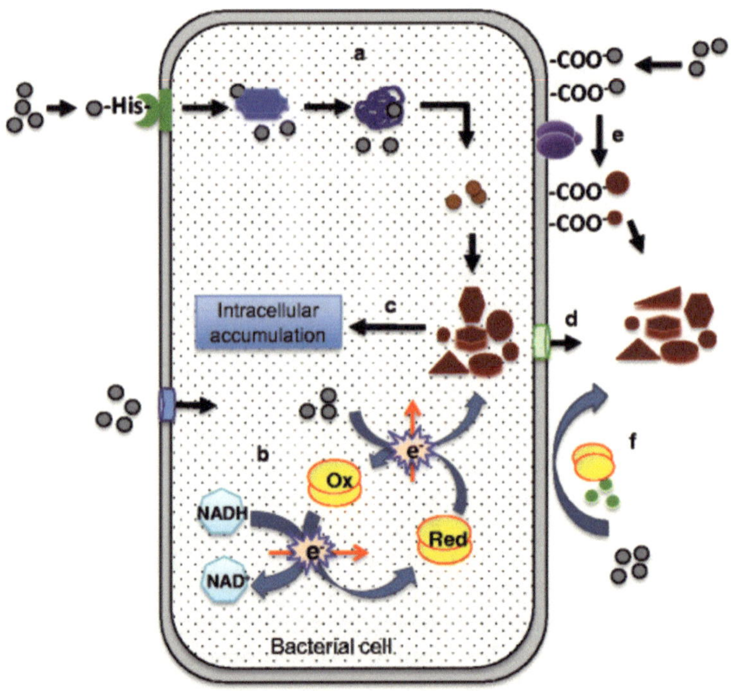

Figure 3.
Metabolites and mechanisms involved in AgNP synthesis in bacteria: (a)-uptake of Ag^+ and activation of reduction machinery; (b)-electron shuttle system involving various cofactors and enzymes; (c & d)- intra or extracellular localisation of AgNPs; (e)-electrostatic interaction between Ag^+ and cell wall peptides/proteins & (f)-extracellular reduction by enzymes or other metabolites released in solution. Adapted from [53].

3. Mechanism of nanoparticle synthesis using plants and microbes

There are three main phases in the synthesis of metal nanoparticles from plants and plant extracts. Initially, an activation phase takes place during which metal ions are reduced from mono or divalent oxidation states to zero-valent states, followed by nucleation of the reduced atoms. This step is immediately followed by a growth phase where small neighbouring nanoparticles coalesce into larger particles with greater thermodynamic stability while further biological reduction occurs. As growth proceeds nanoparticles aggregate to form various shapes such as: cubes, spheres, triangles, hexagons, pentagons, rods and wires [68]. Lastly, a termination phase follows in which nanoparticles acquire the most energetically favourable conformation, which ultimately determines the final shape of the particles (**Figure 4**) [69]. This step is largely influenced by the ability of the plant extract to stabilise the resulting nanoparticles. For example, the high surface energy of nanotriangles results in their decreased stability. Such nanoparticles would then acquire a more stable morphology such as a truncated triangle to minimise Gibbs free energy unless the stability is supported by the given extracts. It can be tentatively suggested that a similar mechanism occurs by the use of bacterial extracts since proteins and metabolites may also participate in Ag^+ reduction as previously stated.

Several controlling factors affect the synthesis and morphology of derived nanoparticles. Several researchers have associated these variations with the choice of adsorbate and catalyst used in the synthetic process [29, 70]. However, reaction parameters have also been shown to strongly affect the synthesis of nanoparticles from biological extracts.

Studies have revealed that the pH of a reaction solution strongly influences the formation of the produced nanoparticles. Variances in reaction pH tend to induce variability in the shape and size of the produced nanoparticles. Lower acidic pH values tend to produce larger particles when compared to higher pH values. In a study employing *Avena sativa* (oat) biomass for the production of gold nanoparticles, larger particles (25–85 nm) where formed at pH 2 whilst smaller particles (5–20 nm) were formed at pH 3 and 4 [71]. The researchers suggested that at pH 2, fewer functional groups were available for particle nucleation resulting in aggregation of the particles. A similar finding was observed in the synthesis of gold nanoparticles from the bacterium *Rhodopseudomonas capsulate.* At an increased pH of 7, spherical particles in the range of 10–20 nm in size were observed. In contrast, lowering the reaction pH to 4 resulted in the formation of nanoplates [63].

Temperature is an important factor in any synthesis. With respect to nanoparticle formulation with the use of biological entities, temperature elevation has demonstrated catalytic behaviour by increasing the reaction rate and efficiency of nanoparticle formation. For example, a study on the influence of reaction temperature in the synthesis of AgNPs from neem leaf extracts suggested that temperature elevation (10–50°C) was correlated with enhanced reduction of Ag^+ [72]. It was also noted that smaller sized AgNPs were produced at 50°C, similar to the finding of Kaviya *et al.* in the production of AgNPs from *Citrus sinensis* peel extracts using varying temperatures [73]. Similarly, this trend was observed in the production of AgNPs from the spent culture supernatants of *Escherichia coli* [61]. The authors tentatively suggested that the increased reaction rate might be because of temperature on a key enzyme participating in nanoparticle synthesis. However, the study importantly revealed that temperature elevation above 60°C contrastingly favoured the production of larger sized particles. The reason for this observation was reported as follows: at high temperatures, kinetic energy of the molecules increase resulting in rapid reduction of Ag^+ (facilitating reduction and nucleation), to the

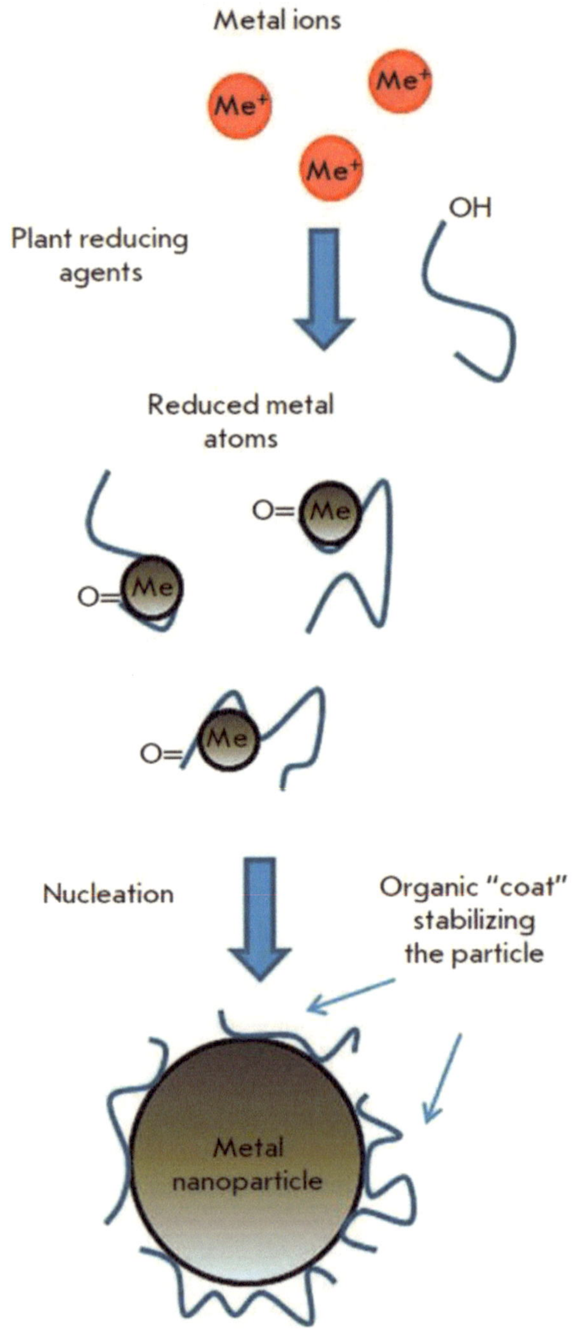

Figure 4.
Schematic representation of nanoparticle synthesis using a plant extract. Adapted from [35].

detriment of secondary reduction on the surface of nascent particles in the growth phase. However, higher temperatures beyond the optimum are thought to increase the growth of the crystal around the nucleus, resulting in the production of larger particles [48, 61].

Temperature has also been demonstrated to affect the structural form of nanoparticles. For example, AgNP synthesis using *Cassia fistula* extracts resulted in the formation of Ag nanoribbons at room temperature whilst spherical AgNPs were formed at temperatures above 60°C [74]. High temperatures in the study were thought to alter the interaction of plant biomolecules with the faces of Ag, inhibiting the coalescence of adjacent nanoparticles.

Sunlight irradiation, a recently reported primary energy source for nanoparticle formation, has been observed to derive AgNPs with desired physical attributes. Recent studies on sunlight driven AgNP synthesis using *Allium sativum* (garlic extract) and *Andrachnea chordifolia* ethanol leaf extract revealed that sunlight rapidly enhanced nanoparticle formation to produce spherical AgNPs with average diameters of 7.3 nm and 3.4 nm, respectively [75, 76]. In addition, this use of sunlight has also been used in AgNP synthesis from *Bacillus amyloliquefaciens* CFS to produce circular and triangular crystalline AgNPs with an average diameter of 14.6 nm [77].

A variety of literature reports on the synthesis of AgNPs with differing morphologies. Understanding the effects of these morphological characteristics on bioactivity is therefore an important consideration when deriving nanoparticles for therapeutic purposes. Characteristically, AgNPs are small (1–100 nm) and therefore possess a large surface area that facilitates their interaction with bacterial cell membranes [41, 78]. However, it has been suggested that within this confined size range, AgNPs present a size-dependent inhibition spectrum. Martinez-Castanon *et al.* reported that AgNPs of 7 nm in size had minimum inhibitory concentration (MIC) values of 6.25 μg ml^{-1} and 7.5 μg ml^{-1} for *E. coli* and *Staphylococcus aureus*, respectively. In contrast, larger nanoparticles (29 nm) capped with the same reducing agent displayed higher MIC values for the respective strains [79]. These results are in accordance with other studies that report nanoparticles of $<$ 10 nm in size display improved bactericidal activities [42, 80].

The interaction of AgNPs of varying shapes with *E. coli* cells has unveiled that shape plays an important factor in bioactivity. Pal *et al.* reported that at a low Ag content of 1 μg, truncated triangular nanoparticles showed nearly complete inhibition of *E. coli* cells, whilst spherical nanoparticles with a total silver content above 12.5 μg displayed a reduction in colony forming units. Rod-shaped particles and $AgNO_3$ presented inferior activities when compared to truncated triangular and spherically shaped AgNPs [41].

Considering these factors and the aforementioned factors affecting synthesis of nanoparticles, it can tentatively be suggested that the fine tuning of reaction parameters such as pH or temperature may be applied in producing AgNPs with these desired physical attributes. However, the use of sunlight irradiation provides a promising alternative in this regard.

3.1 Anti-microbial properties of silver nanoparticles

There exists an abundance of literature reporting on antimicrobial activities of biologically derived AgNPs [81–84]. Most of these studies utilise the disc diffusion assay [85] or agar well diffusion assay [86] to establish inhibitory effects. Positive indication of inhibitory activities are visualised by zones of inhibition on a microbial lawn. Veersamy *et al.* reported zones of inhibition of *S. aureus* and *E. coli* to be 15 mm and 20 mm respectively for AgNPs (20 μg ml^{-1}) derived from mangosteen leaf extracts [48]. Similarly, Logeswari *et al.* reported zones of inhibition of AgNPs synthesised from various plant extracts against several bacterial strains [81]. Although diffusion techniques are preferred amongst researchers, they seem to be labour-intensive. In addition, many researchers do not establish the initial concentration of AgNP solution prior to antimicrobial evaluation [82, 87]. Such disparities make comparison between published data inapplicable [88].

Determination of minimum inhibitory concentration (MIC) by the broth microdilution or macrodilution method [89, 90] is easy to access and provides accurate information with respect to microbial susceptibility. Moreover, MIC values are reported in various concentration units such as μg ml^{-1}, μg l^{-1} or ppm thereby facilitating comparison between publications [53]. These methods are therefore attractive for AgNP bioactivity analysis. Furthermore, determination of MICs is an important consideration for any therapeutic agent in development to assess their toxicity at the specified concentration range. As previously mentioned, the antimicrobial effects of AgNPs are well established. However, a relatively confined amount of studies has been conducted to elucidate their mechanisms of antimicrobial action. These mechanisms are poorly understood and have failed to achieve consensus amongst researchers. Despite this, three common mechanisms of bactericidal activity have been proposed by various studies. These include the uptake of Ag^+ (1), generation of reactive oxygen species (ROS) (2) and cell membrane disruption (3) (**Figure 5**) [91].

Since Ag^+ are known to possess antibacterial activities, their release from AgNPs may potentially aid to the bioactivity of the nanoparticles. It is therefore fitting to consider the mechanistic action of Ag^+ on bacterial cells.

The NADH–ubiquinone reductase has been established as one of the major targets for Ag^+. Specifically, the binding of Ag^+ to this enzyme may be responsible for their bactericidal effect even at minute concentrations [92]. Later, Dibrov *et al.* reported the binding of Ag^+ to transport proteins leads to the leakage of protons and ultimately induces the collapse of the proton motive force [93]. Such interactions with transport proteins may be attributed to the strong affinity of Ag^+ to thiol groups found on cysteine residues of these molecules [94]. Ag^+ has also been reported to inhibit phosphate uptake and additionally causes an efflux of intracellular phosphate [95]. It has also been hypothesised that the antimicrobial effect of Ag^+ is correlated with the disruption of DNA replication. DNA molecules in a relaxed conformation can be replicated effectively. However, when Ag^+ are present in bacterial cells, DNA molecules enter a condensed form and replicating ability diminishes which ultimately leads to cell death [8].

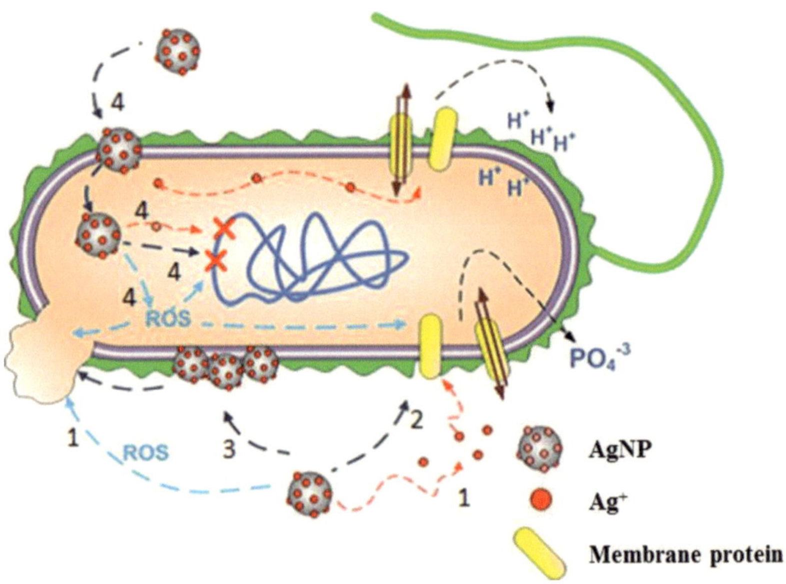

Figure 5.
Interactions of AgNPs with bacterial cells: (1) release of Ag^+ and generation of ROS; (2) interaction with cell membrane proteins; (3) accumulation in cell membrane and disruption of permeability; (4) entry into the cell and release of Ag^+, leading to generation of ROS and damage of cellular DNA. In turn, generated ROS may affect DNA, cell membrane and membrane proteins whilst released Ag^+ may affect cell membrane proteins and DNA. Adapted from [91].

The exposure of bacterial cells to AgNPs leads to the generation of ROS [96]. Naturally, ROS are metabolic by-products of respiring beings. Whist low levels of these species are skilfully controlled by various antioxidant defence mechanisms, high levels of ROS results in oxidative stress which is detrimental to any living organism. Metals can serve as catalysts and produce ROS in an oxygen containing environment [97]. AgNPs are therefore likely to catalyse reactions with oxygen leading to the production of excess free radicals. Kim *et al*. demonstrated the generation of free radicals from AgNPs by means of spin resonance measurements. Toxicity of AgNPs and $AgNO_3$ diminished upon addition of an antioxidant suggesting that the mechanism of action against bacterial strains was associated with the formation of free radicals from AgNPs. The generation of excess free radicals attack membrane lipids resulting in the breakdown of the membrane and cause damage to DNA [1].

The release of Ag^+ from nanoparticles attached to the membrane and nanoparticles inside the cell also play a role in the generation of ROS. Ag^+ released on the membrane are capable of ROS generation by acting as electron acceptors whilst those present inside the cell more likely to interact with thiol groups of respiratory chain enzymes as previously stated, or scavenging superoxide dismutase enzymes [98]. The effect of ROS scavengers on *E. coli* cells was reported by Inoue *et al*.. Specifically, ROS such as superoxide anions, hydroxyl radicals, hydrogen peroxide and singlet oxygen contributed to the bactericidal activity against *E. coli* [99]. According to literature, the bactericidal effect of AgNPs may also be the result of damage to the outer membrane of bacterial cells. Previous studies by Sondi and Salopek-Sondi suggested that treatment of *E. coli* cells with AgNPs induced changes in the membrane morphology (**Figure 6a**). This resulted in increased membrane permeability and shifts in normal transport through the plasma membrane [100]. Morones *et al*. hypothesised that these mechanisms could explain the number of nanoparticles found inside *E. coli cells* (**Figure 6b**). AgNPs with oxidised surfaces were also reported to induce the formation of holes on the surface of *E. coli* cells and portions of the cellular surface were observed to be eaten away [101]. The attachment and penetration of AgNPs has also been observed in *P. aeruginosa* (**Figure 6c**), *V. cholera* and *S. typhus* [80].

The mechanism of AgNP adhesion and penetration of bacterial cell membranes remains to be elucidated. Literature reports indicate that electrostatic interactions between positively charged particles and negatively charged cell membranes is essential for the bioactivity of these particles [102, 103]. However, this strategy does

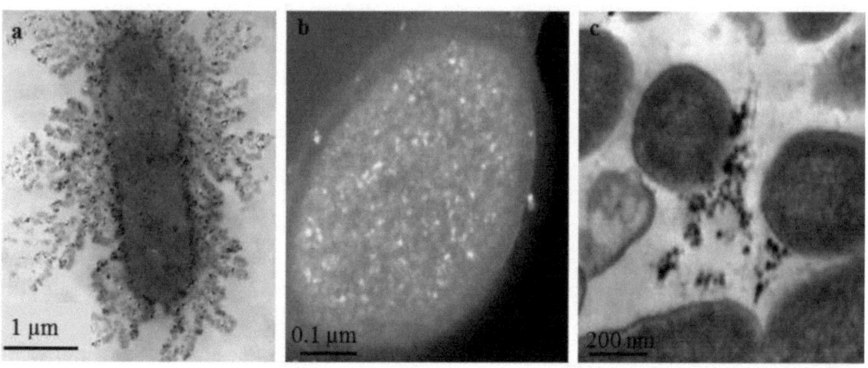

Figure 6.
Transmission electron micrographs of (a) E. coli cell after 1 h treatment with 50 µg cm^{-3} AgNPs; (b) E. coli cell after 30 min treatment with 100 µg ml^{-1} AgNPs (c) P. aeruginosa cells after 30 min treatment with 100 µg ml^{-1} AgNPs [80, 100].

not validate the adhesion and penetration abilities of negatively charged nanoparticles [104]. The researchers argued that although the particles were negatively charged, interactions between the particles and building elements of the membrane are likely to have occurred causing structural changes and degradation of the membrane. Morones *et al.* proposed that the interaction of AgNPs and bacterial membranes could be attributed to the strong affinity of the particles to sulphur containing proteins present on the membrane [80]. These interactions are thought to be conserved in the interaction of Ag^+ and thiol groups on respiratory enzymes and transport proteins [80, 91].

Sondi and Salopek-Sondi [104] further reported that damage to *E. coli* cell membranes might also occur due to the incorporation of AgNPs into their membrane structure. Scanning electron microscopy revealed the formation of "pits" on the surface of the membrane [100]. Similar findings were observed by [102]. Amro *et al.* [109] additionally reported the formation of irregularly shaped "pits" on the outer membrane of *E. coli* cells through the progressive release of lipopolysaccharide molecules. This release of LPS molecules was induced by metal depletion in the cells [105]. A membrane with such morphological changes would display a high increase in permeability, rendering the cell incapable of regulating proper transport through the membrane as previously described.

Although these studies have been conducted on Gram-negative bacteria, AgNPs have also been reported to exert inhibitory activities against Gram-positive bacteria which differ from their counterparts based on differences in cell wall structure [106]. It can be tentatively suggested that AgNPs may form interactions with Gram-positive bacteria through surface proteins present on the cell wall. Once penetrated, the mechanisms of bacterial activity are conserved with that of Gram-negative bacteria.

A relatively confined amount of literature focuses on the mechanisms of antifungal activity exerted by AgNPs. However, based on the studies that have been reported, it seems that inhibition of fungal growth by AgNPs may be the result of damage to fungal cellular membranes. Kim *et al.* demonstrated the effect of AgNPs on *Candida albicans*. Transmission electron microscopy (TEM) analysis revealed that the treatment of cells with AgNPs lead to the formation of "pits" on the cell membrane which ultimately disrupts membrane potential [107]. A similar finding was made by Nasrollahi *et al.* who reported that AgNP incubation with *C. albicans* led to damage of the cell membrane [108]. Endo *et al.* reported that disruption of membrane integrity inhibits the normal budding process of daughter cells. Therefore, the authors suggested that AgNPs exert their inhibitory activity by inhibiting the budding of daughter cells due to the destruction of the cell membrane [109].

AgNPs may also disrupt antioxidant defences in fungal cells. Eukaryotic cell studies suggest that AgNPs directly interact with gluthathione, gluthathione reductase or enzymes responsible for maintaining proper levels of gluthathione [110]. With respect to fungal cells, it has been hypothesised that Ag^+ largely affect the function of membrane bound enzymes such as those in the respiratory chain. It has also been reported that exposure of fungal cells to Ag^+ led to the loss of DNA replication ability. This results in the deactivation of ribosomal subunit protein expression and synthesis of non-functional enzymes and cellular proteins [111].

From these findings it can be tentatively suggested that bactericidal mechanisms of AgNPs are conserved in their inhibition of fungal cells. In summary, AgNPs exert their antimicrobial effects by releasing Ag^+, disrupting the cell membrane/wall, generating ROS and inhibiting proper DNA replication.

3.2 Cytotoxicity of silver nanoparticles

The unique physico-chemical and biological properties of AgNPs have extremely promising industrial and medical applications, as previously mentioned. However, there exists a dearth of knowledge regarding the effects of prolonged exposures to nanoparticles on human health and the environment [112]. It is therefore imperative to establish the *in vitro* and *in vivo* cytotoxic effect of AgNPs in mind for therapeutic purposes.

3.2.1 In vivo *studies*

Human contact with nanoparticles occurs in the form of intravenous injection, oral administration, inhalation and dermal contact [113]. Injection of AgNPs *in vivo* results in short circulation times and broad tissue distribution. Target sites often include the liver (main target), spleen, lungs and kidneys [114]. Inhalation studies suggest that AgNPs become deposited in the olfactory mucosa and olfactory nerves which can potentially induce impairment and dysfunction of brain cells [115] in addition to immunotoxicity [116]. With regard to oral administration, migration of AgNPs to the gastrointestinal tract promotes dissolution of the particles which subsequently releases Ag^+ [117]. A recent study on oral exposure to Ag^+ indicated that these ions interact with sulphur leading to the formation of sulphur containing Ag granules in the intestinal epithelium [118]. The authors suggested that during intestinal digestion, Ag^+ give rise to particle formation, possibly in the form of Ag_2S or AgCl salt. They further added that this formation might influence their uptake and reduce the toxic effects of Ag^+, however the effects of Ag salts on the intestine are yet to be elucidated [118, 119]. Reports on the exposure of workers to low doses of Ag dust indicated no significant changes in health status.

3.2.2 In vitro *studies*

Many researchers have demonstrated the cytotoxic effects of AgNPs *in vitro*, however there is still a lack of consistent and reliable data amongst publications. For example, in a recent review, Kim and Ryu (2013) attributed oxidative stress, apoptosis and genotoxicity to be the main *in vitro* outcome of AgNP exposure [120]. Later, Gliga *et al.* identified a major drawback of this review, highlighting that the AgNPs were different in each study, *i.e.* synthesised by different techniques, of varying size distributions and coatings, tested on different cell lines under different cell culture conditions and often without the use of appropriate controls [121]. Additionally, Hackenberg *et al.* reported cytotoxicity of human mesenchymal stem cells at a concentration of 10 $\mu g\ ml^{-1}$ AgNPs (˂50 nm), whereas Samberg *et al.* reported no toxicity of progenitor human adipose-derived stem cells at concentrations up to 100 $\mu g\ ml^{-1}$ AgNPs (10–20 nm) [122, 123]. To determine the effect of size on cytotoxicity, Liu *et al.* compared the cytotoxicity of AgNPs ranging in size from 5 to 50 nm on four different cell lines (A549, HepG2, MCF-7 and CGC-7901) and reported that 5 nm AgNPs were most toxic [124]. On the contrary, Kim *et al.* reported the enhanced release of lactate dehydrogenase (LDH) and reduced cell viability in the presence of 100 nm sized AgNPs when compared to smaller AgNPs (10–50 nm) [125]. It can be noted that the variation in parameters in these studies makes it difficult to observe trends and come to accurate assumptions. To achieve some consensus in this regard, Gliga *et al.* studied the cytotoxic effect of varying sized AgNPs capped by various agents on the normal bronchial epithelial cell line (BEAS-2B). They reported that 10 nm sized AgNPs induced cytotoxicity irrespective of the capping agents, at high concentrations (20–50 μg ml^{-1}), whilst larger AgNPs did not display significant cytotoxic effects at all tested concentrations. The group additionally reported that at non-cytotoxic concentrations (10 μg ml^{-1}), significant DNA damage was observed for all AgNPs independent of size and coating. In contrast, panda *et al*. reported no genotoxicity of AgNPs capped with protein at 20–80 μg ml^{-1} for 24–55 nm sized particles [126].

Overall, it is difficult to establish the cytotoxic effect of AgNPs due to the differences in nanoparticle synthetic methods, their various sizes and capping agents and lastly the diverse evaluation tests used to determine toxicity. In fact, by using different organisms and/or culture cells there is no conclusive evaluation of AgNP toxicity [127]. However, bearing in mind the results presented in this review, it can be tentatively suggested that smaller sized AgNPs are more cytotoxic than larger sized particles at higher concentrations.

4. Potential applications of biologically derived nanoparticles

The physiochemical characteristics of metal nanoparticles render them applicable across a genre of multi-disciplinary fields for a variety of uses including catalysis [128]; micro-electronics [129]; solar energy conversion [130] amongst many others [131]. They have also been recognised for their potential in a number of medical applications [132]. However, the use of nanoparticles derived from physical and chemical synthetic routes raises health and toxicity concerns due to the nature of the reaction conditions which may ultimately affect the properties of the derived particles [133].

Biologically derived nanoparticles provide a greener alternative to nanoparticles derived from the aforementioned routes since, the synthesis methods used to derive these particles are clean and non-toxic [9]. As a result, they are suitable for a number of biomedical applications (**Table 1**) including: cancer therapy; drug delivery; tumour detection; genetic disorder diagnosis; tissue repair; cell labelling; antimicrobial development; targeting and immunoassays and yet to be discovered applications [37, 114, 132, 134–136].

With respect to biologically derived AgNPs, their major exploitation exists in the development of antimicrobial agents due to their renowned microbial inhibitory activities and with the current status on antimicrobial drug resistance, these particles are being extensively sought after as possible alternatives to antibiotics [1, 8].

Plant	Applications	Reference
Moringa oleifera	Anti-microbial	[137]
Eclipta prostrata	Anti-protozoal	[138]
Gelidiella acerosa	Anti-fungal	[139]
Melia azedarach	Anti-cancer	[140]
Lampranthus coccineus	Anti-viral	[141]
Malephora lutea	Anti-Alzheimer	[142]
Melia azedarach	Wound healing	[143]
Ocimum sanctum	Anti-diabetic	[144]
Allium sativum	Antioxidant	[145]

Table 1.
Selective applications of silver nanoparticles synthesised using plant extracts.

5. Conclusion

In conclusion, it can be established that green synthetic strategies using plant and bacterial based extracts are promising alternatives to produce AgNPs. However, to produce AgNPs with enhanced bioactivities, morphological characteristics such as size and shape need to be finely tuned. Furthermore, the use of extracts with known medical value provides with attractive capping substrates that may potentially enhance the bioactivities of the produced particles.

Acknowledgements

This study was made possible through financial support from the National Research Foundation. The research facilities were provided by the University of KwaZulu-Natal.

Conflict of interest

Authors declare no conflict of interest.

Author details

Jerushka S. Moodley[1], Suresh Babu Naidu Krishna[2], Karen Pillay[1] and Patrick Govender[1]*

1 Department of Biochemistry, School of Life Sciences, University of KwaZulu Natal, Private Bag X54001, Durban, South Africa

2 Department of Biomedical and Clinical Technology, Durban University of Technology, Durban, South Africa

*Address all correspondence to: govenderpt@ukzn.ac.za

References

[1] Kim, J.S., et al., *Antimicrobial effects of silver nanoparticles.* Nanomedicine: Nanotechnology, Biology and Medicine, 2007. **3**(1): p. 95-101.

[2] Alexander, J.W., *History of the medical use of silver.* Surgical infections, 2009. **10**(3): p. 289-292.

[3] Peng, G., et al., *Nanosilver at the interface of biomedical applications, toxicology, and synthetic strategies*, in *Metal Nanoparticles for Drug Delivery and Diagnostic Applications*. 2020, Elsevier. p. 119-139.

[4] Ventola, C.L., *The antibiotic resistance crisis: part 1: causes and threats.* Pharmacy and Therapeutics, 2015. **40**(4): p. 277.

[5] Organization., W.H., *Antimicrobial resistance global report on surveillance: 2014 summary*. 2014, World Health Organization.

[6] Walsh, C., *Where will new antibiotics come from?* Nature Reviews Microbiology, 2003. **1**(1): p. 65-70.

[7] Projan, S.J., *Why is big Pharma getting out of antibacterial drug discovery?* Current Opinion in Microbiology, 2003. **6**(5): p. 427-430.

[8] Rai, M., A. Yadav, and A. Gade, *Silver nanoparticles as a new generation of antimicrobials.* Biotechnology Advances, 2009. **27**(1): p. 76-83.

[9] Ahmed, S., et al., *A review on plants extract mediated synthesis of silver nanoparticles for antimicrobial applications: A green expertise.* Journal of Advanced Research, 2016. **7**(1): p. 17-28.

[10] Ponarulselvam, S., et al., *Synthesis of silver nanoparticles using leaves of Catharanthus roseus Linn. G. Don and their antiplasmodial activities.* Asian Pacific Journal of Tropical Biomedicine, 2012. **2**(7): p. 574-580.

[11] Velusamy, P., et al., *Bio-inspired green nanoparticles: synthesis, mechanism, and antibacterial application.* Toxicological Research, 2016. **32**(2): p. 95.

[12] Biswas, A., et al., *Advances in top–down and bottom–up surface nanofabrication: Techniques, applications & future prospects.* Advances in colloid and interface science, 2012. **170**(1): p. 2-27.

[13] Thakkar, K.N., S.S. Mhatre, and R.Y. Parikh, *Biological synthesis of metallic nanoparticles.* Nanomedicine: Nanotechnology, Biology and Medicine, 2010. **6**(2): p. 257-262.

[14] El-Nour, K.M.A., et al., *Synthesis and applications of silver nanoparticles.* Arabian journal of chemistry, 2010. **3**(3): p. 135-140.

[15] Iravani, S., et al., *Synthesis of silver nanoparticles: chemical, physical and biological methods.* Research in pharmaceutical sciences, 2014. **9**(6): p. 385.

[16] Kim, S., et al., *Catalytic effect of laser ablated Ni nanoparticles in the oxidative addition reaction for a coupling reagent of benzylchloride and bromoacetonitrile.* Journal of Molecular Catalysis A: Chemical, 2005. **226**(2): p. 231-234.

[17] Chen, Y.-H. and C.-S. Yeh, *Laser ablation method: use of surfactants to form the dispersed Ag nanoparticles.* Colloids and Surfaces A: Physicochemical and Engineering Aspects, 2002. **197**(1): p. 133-139.

[18] Tsuji, T., et al., *Preparation of silver nanoparticles by laser ablation in solution: influence of laser wavelength on particle size.* Applied Surface Science, 2002. **202**(1): p. 80-85.

[19] Amin, S.A., M. Pazouki, and A. Hosseinnia, *Synthesis of TiO 2–Ag*

nanocomposite with sol–gel method and investigation of its antibacterial activity against E. coli. Powder Technology, 2009. **196**(3): p. 241-245.

[20] Yang, J. and J. Pan, *Hydrothermal synthesis of silver nanoparticles by sodium alginate and their applications in surface-enhanced Raman scattering and catalysis*. Acta Materialia, 2012. **60**(12): p. 4753-4758.

[21] Yin, B., et al., *Electrochemical synthesis of silver nanoparticles under protection of poly (N-vinylpyrrolidone)*. The Journal of Physical Chemistry B, 2003. **107**(34): p. 8898-8904.

[22] Saion, E., E. Gharibshahi, and K. Naghavi, *Size-controlled and optical properties of monodispersed silver nanoparticles synthesized by the radiolytic reduction method*. International journal of molecular sciences, 2013. **14**(4): p. 7880-7896.

[23] Pillai, Z.S. and P.V. Kamat, *What factors control the size and shape of silver nanoparticles in the citrate ion reduction method?* The Journal of Physical Chemistry B, 2004. **108**(3): p. 945-951.

[24] Chou, K.-S. and C.-Y. Ren, *Synthesis of nanosized silver particles by chemical reduction method*. Materials Chemistry and Physics, 2000. **64**(3): p. 241-246.

[25] Kapoor, S., et al., *Reduction and aggregation of silver ions in aqueous gelatin solutions*. Langmuir, 1994. **10**(9): p. 3018-3022.

[26] Bai, J., et al., *One-pot synthesis of polyacrylamide-gold nanocomposite*. Materials Chemistry and Physics, 2007. **106**(2): p. 412-415.

[27] Song, K.C., et al., *Preparation of colloidal silver nanoparticles by chemical reduction method*. Korean Journal of Chemical Engineering, 2009. **26**(1): p. 153-155.

[28] Iravani, S., *Green synthesis of metal nanoparticles using plants*. Green Chemistry, 2011. **13**(10): p. 2638-2650.

[29] Patra, J.K. and K.-H. Baek, *Green nanobiotechnology: factors affecting synthesis and characterization techniques*. Journal of Nanomaterials, 2014. **2014**: p. 219.

[30] Dauthal, P. and M. Mukhopadhyay, *Noble metal nanoparticles: Plant-mediated synthesis, mechanistic aspects of synthesis, and applications*. Industrial & Engineering Chemistry Research, 2016. **55**(36): p. 9557-9577.

[31] Rajput, S., et al., *Fungal isolate optimized for biogenesis of silver nanoparticles with enhanced colloidal stability*. Langmuir, 2016. **32**(34): p. 8688-8697.

[32] Dhuper, S., D. Panda, and P. Nayak, *Green synthesis and characterization of zero valent iron nanoparticles from the leaf extract of Mangifera indica*. Nano Trends: J Nanotech App, 2012. **13**(2): p. 16-22.

[33] Mittal, A.K., Y. Chisti, and U.C. Banerjee, *Synthesis of metallic nanoparticles using plant extracts*. Biotechnology advances, 2013. **31**(2): p. 346-356.

[34] Mohanpuria, P., N.K. Rana, and S.K. Yadav, *Biosynthesis of nanoparticles: technological concepts and future applications*. Journal of Nanoparticle Research, 2008. **10**(3): p. 507-517.

[35] Makarov, V., et al., *"Green" nanotechnologies: synthesis of metal nanoparticles using plants*. Acta Naturae (англоязычная версия), 2014. **6**(1 (20)).

[36] Pantidos, N. and L.E. Horsfall, *Biological synthesis of metallic nanoparticles by bacteria, fungi and plants*. Journal of Nanomedicine & Nanotechnology, 2014. **5**(5): p. 1.

[37] Singh, P., et al., *Biological synthesis of nanoparticles from plants and microorganisms.* Trends in biotechnology, 2016.

[38] Harris, A.T. and R. Bali, *On the formation and extent of uptake of silver nanoparticles by live plants.* Journal of Nanoparticle Research, 2008. **10**(4): p. 691-695.

[39] Gardea-Torresdey, J., et al., *Formation and growth of Au nanoparticles inside live alfalfa plants.* Nano letters, 2002. **2**(4): p. 397-401.

[40] Manceau, A., et al., *Formation of metallic copper nanoparticles at the soil-root interface.* Environmental science & technology, 2008. **42**(5): p. 1766-1772.

[41] Pal, S., Y.K. Tak, and J.M. Song, *Does the antibacterial activity of silver nanoparticles depend on the shape of the nanoparticle? A study of the gram-negative bacterium Escherichia coli.* Applied and environmental microbiology, 2007. **73**(6): p. 1712-1720.

[42] Jeong, Y., D.W. Lim, and J. Choi, *Assessment of size-dependent antimicrobial and cytotoxic properties of silver nanoparticles.* Advances in Materials Science and Engineering, 2014. **2014**.

[43] Shankar, S.S., A. Ahmad, and M. Sastry, *Geranium leaf assisted biosynthesis of silver nanoparticles.* Biotechnology progress, 2003. **19**(6): p. 1627-1631.

[44] Shankar, S.S., et al., *Bioreduction of chloroaurate ions by geranium leaves and its endophytic fungus yields gold nanoparticles of different shapes.* Journal of Materials Chemistry, 2003. **13**(7): p. 1822-1826.

[45] Singh, A.K., et al., *Biosynthesis of gold and silver nanoparticles by natural precursor clove and their functionalization with amine group.* Journal of Nanoparticle Research, 2010. **12**(5): p. 1667-1675.

[46] Ahmad, N., et al., *Rapid synthesis of silver nanoparticles using dried medicinal plant of basil.* Colloids and Surfaces B: Biointerfaces, 2010. **81**(1): p. 81-86.

[47] Gannimani, R., et al., *Synthesis and antibacterial activity of silver and gold nanoparticles produced using aqueous seed extract of Protorhus longifolia as a reducing agent.* 2014.

[48] Veerasamy, R., et al., *Biosynthesis of silver nanoparticles using mangosteen leaf extract and evaluation of their antimicrobial activities.* Journal of Saudi Chemical Society, 2011. **15**(2): p. 113-120.

[49] Kasthuri, J., S. Veerapandian, and N. Rajendiran, *Biological synthesis of silver and gold nanoparticles using apiin as reducing agent.* Colloids and Surfaces B: Biointerfaces, 2009. **68**(1): p. 55-60.

[50] Panigrahi, S., et al., *General method of synthesis for metal nanoparticles.* Journal of Nanoparticle Research, 2004. **6**(4): p. 411-414.

[51] Tan, Y.N., J.Y. Lee, and D.I. Wang, *Uncovering the design rules for peptide synthesis of metal nanoparticles.* Journal of the American Chemical Society, 2010. **132**(16): p. 5677-5686.

[52] Glusker, J., A. Katz, and C. Bock, *Metal ions in biological systems.* The Rigaku Journal, 1999. **16**(2): p. 8-16.

[53] Singh, R., et al., *Bacteriagenic silver nanoparticles: synthesis, mechanism, and applications.* Applied microbiology and biotechnology, 2015. **99**(11): p. 4579-4593.

[54] Iravani, S., *Bacteria in nanoparticle synthesis: current status and future prospects.* International Scholarly Research Notices, 2014. **2014**.

[55] Pantidos, N. and L.E. Horsfall, *Biological synthesis of metallic nanoparticles by bacteria, fungi and plants.* Journal of Nanomedicine & Nanotechnology, 2014. **2014**.

[56] Shah, M., et al., *Green synthesis of metallic nanoparticles via biological entities.* Materials, 2015. **8**(11): p. 7278-7308.

[57] Pugazhenthiran, N., et al., *Microbial synthesis of silver nanoparticles by Bacillus sp.* Journal of Nanoparticle Research, 2009. **11**(7): p. 1811-1815.

[58] Kalimuthu, K., et al., *Biosynthesis of silver nanocrystals by Bacillus licheniformis.* Colloids and Surfaces B: Biointerfaces, 2008. **65**(1): p. 150-153.

[59] Shahverdi, A.R., et al., *Rapid synthesis of silver nanoparticles using culture supernatants of Enterobacteria: a novel biological approach.* Process Biochemistry, 2007. **42**(5): p. 919-923.

[60] Saifuddin, N., C. Wong, and A. Yasumira, *Rapid biosynthesis of silver nanoparticles using culture supernatant of bacteria with microwave irradiation.* Journal of Chemistry, 2009. **6**(1): p. 61-70.

[61] Gurunathan, S., et al., *Biosynthesis, purification and characterization of silver nanoparticles using Escherichia coli.* Colloids and Surfaces B: Biointerfaces, 2009. **74**(1): p. 328-335.

[62] Slawson, R.M., et al., *Germanium and silver resistance, accumulation, and toxicity in microorganisms.* Plasmid, 1992. **27**(1): p. 72-79.

[63] He, S., et al., *Biosynthesis of gold nanoparticles using the bacteria Rhodopseudomonas capsulata.* Materials Letters, 2007. **61**(18): p. 3984-3987.

[64] Sneha, K., et al., *Corynebacterium glutamicum-mediated crystallization of silver ions through sorption and reduction processes.* Chemical Engineering Journal, 2010. **162**(3): p. 989-996.

[65] Sintubin, L., et al., *Lactic acid bacteria as reducing and capping agent for the fast and efficient production of silver nanoparticles.* Applied microbiology and biotechnology, 2009. **84**(4): p. 741-749.

[66] Parikh, R.Y., et al., *Extracellular synthesis of crystalline silver nanoparticles and molecular evidence of silver resistance from Morganella sp.: towards understanding biochemical synthesis mechanism.* ChemBioChem, 2008. **9**(9): p. 1415-1422.

[67] Johnston, C.W., et al., *Gold biomineralization by a metallophore from a gold-associated microbe.* Nature chemical biology, 2013. **9**(4): p. 241-243.

[68] Kim, J., et al., *Peptide-mediated shape-and size-tunable synthesis of gold nanostructures.* Acta biomaterialia, 2010. **6**(7): p. 2681-2689.

[69] Si, S. and T.K. Mandal, *Tryptophan-Based Peptides to Synthesize Gold and Silver Nanoparticles: A Mechanistic and Kinetic Study.* Chemistry–A European Journal, 2007. **13**(11): p. 3160-3168.

[70] Ajayan, P.M., *Nanotechnology: How does a nanofibre grow?* Nature, 2004. **427**(6973): p. 402-403.

[71] Armendariz, V., et al., *Size controlled gold nanoparticle formation by Avena sativa biomass: use of plants in nanobiotechnology.* Journal of Nanoparticle Research, 2004. **6**(4): p. 377-382.

[72] Verma, A. and M.S. Mehata, *Controllable synthesis of silver nanoparticles using Neem leaves and their antimicrobial activity.* Journal of Radiation Research and Applied Sciences, 2016. **9**(1): p. 109-115.

[73] Kaviya, S., et al., *Biosynthesis of silver nanoparticles using Citrus sinensis peel extract and its antibacterial activity.* Spectrochimica Acta Part A: Molecular and Biomolecular Spectroscopy, 2011. **79**(3): p. 594-598.

[74] Lin, L., et al., *Nature factory of silver nanowires: Plant-mediated synthesis using broth of Cassia fistula leaf.* Chemical Engineering Journal, 2010. **162**(2): p. 852-858.

[75] Karimi Zarchi, A., et al., *A sunlight-induced method for rapid biosynthesis of silver nanoparticles using an Andrachnea chordifolia ethanol extract.* Applied Physics A: Materials Science & Processing, 2011. **103**(2): p. 349-353.

[76] Rastogi, L. and J. Arunachalam, *Sunlight based irradiation strategy for rapid green synthesis of highly stable silver nanoparticles using aqueous garlic (Allium sativum) extract and their antibacterial potential.* Materials Chemistry and Physics, 2011. **129**(1): p. 558-563.

[77] Wei, X., et al., *Synthesis of silver nanoparticles by solar irradiation of cell-free Bacillus amyloliquefaciens extracts and AgNO 3.* Bioresource Technology, 2012. **103**(1): p. 273-278.

[78] Christian, P., et al., *Nanoparticles: structure, properties, preparation and behaviour in environmental media.* Ecotoxicology, 2008. **17**(5): p. 326-343.

[79] Martinez-Castanon, G., et al., *Synthesis and antibacterial activity of silver nanoparticles with different sizes.* Journal of Nanoparticle Research, 2008. **10**(8): p. 1343-1348.

[80] Morones, J.R., et al., *The bactericidal effect of silver nanoparticles.* Nanotechnology, 2005. **16**(10): p. 2346.

[81] Logeswari, P., S. Silambarasan, and J. Abraham, *Synthesis of silver nanoparticles using plants extract and analysis of their antimicrobial property.* Journal of Saudi Chemical Society, 2012. **19**(3): p. 311-317.

[82] Prasad, T. and E. Elumalai, *Biofabrication of Ag nanoparticles using Moringa oleifera leaf extract and their antimicrobial activity.* Asian Pacific Journal of Tropical Biomedicine, 2011. **1**(6): p. 439-442.

[83] Awwad, A.M., N.M. Salem, and A.O. Abdeen, *Green synthesis of silver nanoparticles using carob leaf extract and its antibacterial activity.* International Journal of Industrial Chemistry, 2013. **4**(1): p. 29.

[84] Sharma, V.K., R.A. Yngard, and Y. Lin, *Silver nanoparticles: green synthesis and their antimicrobial activities.* Advances in colloid and interface science, 2009. **145**(1): p. 83-96.

[85] Biemer, J.J., *Antimicrobial susceptibility testing by the Kirby-Bauer disc diffusion method.* Annals of Clinical & Laboratory Science, 1973. **3**(2): p. 135-140.

[86] Perez, C., M. Pauli, and P. Bazerque, *An antibiotic assay by the agar well diffusion method.* Acta Biol Med Exp, 1990. **15**(1): p. 113-5.

[87] Jagtap, U.B. and V.A. Bapat, *Green synthesis of silver nanoparticles using Artocarpus heterophyllus Lam. seed extract and its antibacterial activity.* Industrial Crops and Products, 2013. **46**: p. 132-137.

[88] Allahverdiyev, A.M., et al., *Coping with antibiotic resistance: combining nanoparticles with antibiotics and other antimicrobial agents.* Expert review of anti-infective therapy, 2011. **9**(11): p. 1035-1052.

[89] Eloff, J., *A sensitive and quick microplate method to determine the minimal inhibitory concentration of plant extracts for bacteria.* Planta medica, 1998. **64**(08): p. 711-713.

[90] Schwalbe, R., L. Steele-Moore, and A.C. Goodwin, *Antimicrobial susceptibility testing protocols*. 2007: Crc Press.

[91] Marambio-Jones, C. and E.M. Hoek, *A review of the antibacterial effects of silver nanomaterials and potential implications for human health and the environment.* Journal of Nanoparticle Research, 2010. **12**(5): p. 1531-1551.

[92] Semeykina, A.L. and V.P. Skulachev, *Submicromolar Ag+ increases passive Na+ permeability and inhibits the respiration-supported formation of Na+ gradient in Bacillus FTU vesicles.* FEBS letters, 1990. **269**(1): p. 69-72.

[93] Dibrov, P., et al., *Chemiosmotic mechanism of antimicrobial activity of Ag+ in Vibrio cholerae.* Antimicrobial agents and chemotherapy, 2002. **46**(8): p. 2668-2670.

[94] Liau, S., et al., *Interaction of silver nitrate with readily identifiable groups: relationship to the antibacterialaction of silver ions.* Letters in applied microbiology, 1997. **25**(4): p. 279-283.

[95] Schreurs, W. and H. Rosenberg, *Effect of silver ions on transport and retention of phosphate by Escherichia coli.* Journal of Bacteriology, 1982. **152**(1): p. 7-13.

[96] Ninganagouda, S., et al., *Growth kinetics and mechanistic action of reactive oxygen species released by silver nanoparticles from Aspergillus niger on Escherichia coli.* BioMed research international, 2014. **2014**.

[97] Stohs, S. and D. Bagchi, *Oxidative mechanisms in the toxicity of metal ions.* Free radical biology and medicine, 1995. **18**(2): p. 321-336.

[98] Park, H.-J., et al., *Silver-ion-mediated reactive oxygen species generation affecting bactericidal activity.* Water research, 2009. **43**(4): p. 1027-1032.

[99] Inoue, Y., et al., *Bactericidal activity of Ag–zeolite mediated by reactive oxygen species under aerated conditions.* Journal of inorganic biochemistry, 2002. **92**(1): p. 37-42.

[100] Sondi, I. and B. Salopek-Sondi, *Silver nanoparticles as antimicrobial agent: a case study on E. coli as a model for Gram-negative bacteria.* Journal of colloid and interface science, 2004. **275**(1): p. 177-182.

[101] Smetana, A.B., et al., *Biocidal activity of nanocrystalline silver powders and particles.* Langmuir, 2008. **24**(14): p. 7457-7464.

[102] Stoimenov, P.K., et al., *Metal oxide nanoparticles as bactericidal agents.* Langmuir, 2002. **18**(17): p. 6679-6686.

[103] Hamouda, T. and J. Baker, *Antimicrobial mechanism of action of surfactant lipid preparations in enteric Gram-negative bacilli.* Journal of applied microbiology, 2000. **89**(3): p. 397-403.

[104] Sondi, I., D.V. Goia, and E. Matijević, *Preparation of highly concentrated stable dispersions of uniform silver nanoparticles.* Journal of colloid and interface science, 2003. **260**(1): p. 75-81.

[105] Amro, N.A., et al., *High-resolution atomic force microscopy studies of the Escherichia coli outer membrane: structural basis for permeability.* Langmuir, 2000. **16**(6): p. 2789-2796.

[106] Dipankar, C. and S. Murugan, *The green synthesis, characterization and evaluation of the biological activities of silver nanoparticles synthesized from Iresine herbstii leaf aqueous extracts.* Colloids and Surfaces B: Biointerfaces, 2012. **98**: p. 112-119.

[107] Kim, K.-J., et al., *Antifungal activity and mode of action of silver nano-particles on Candida albicans.* Biometals, 2009. **22**(2): p. 235-242.

[108] Nasrollahi, A., K. Pourshamsian, and P. Mansourkiaee, *Antifungal activity of silver nanoparticles on some of fungi.* International Journal of Nano Dimension, 2011. **1**(3): p. 233-239.

[109] Endo, M., et al., *Fungicidal action of aureobasidin A, a cyclic depsipeptide antifungal antibiotic, against Saccharomyces cerevisiae.* Antimicrobial agents and chemotherapy, 1997. **41**(3): p. 672-676.

[110] Carlson, C., et al., *Unique cellular interaction of silver nanoparticles: size-dependent generation of reactive oxygen species.* The journal of physical chemistry B, 2008. **112**(43): p. 13608-13619.

[111] Elgorban, A.M., et al., *Antifungal silver nanoparticles: synthesis, characterization and biological evaluation.* Biotechnology & Biotechnological Equipment, 2016. **30**(1): p. 56-62.

[112] Ravishankar Rai, V. and A. Jamuna Bai, *Nanoparticles and their potential application as antimicrobials.* Science against microbial pathogens, communicating current research and technological advances. Formatex, Badajoz, 2011: p. 197-209.

[113] Chen, X. and H. Schluesener, *Nanosilver: a nanoproduct in medical application.* Toxicology letters, 2008. **176**(1): p. 1-12.

[114] Ahamed, M., M.S. AlSalhi, and M. Siddiqui, *Silver nanoparticle applications and human health.* Clinica chimica acta, 2010. **411**(23): p. 1841-1848.

[115] Oberdörster, G., et al., *Translocation of inhaled ultrafine particles to the brain.* Inhalation toxicology, 2004. **16**(6-7): p. 437-445.

[116] Rahman, M., et al., *Expression of genes related to oxidative stress in the mouse brain after exposure to silver-25 nanoparticles.* Toxicology letters, 2009. **187**(1): p. 15-21.

[117] Zhang, T., et al., *Cytotoxic potential of silver nanoparticles.* Yonsei medical journal, 2014. **55**(2): p. 283-291.

[118] Loeschner, K., et al., *Distribution of silver in rats following 28 days of repeated oral exposure to silver nanoparticles or silver acetate.* Particle and fibre toxicology, 2011. **8**(1): p. 18.

[119] Walczak, A.P., et al., *Behaviour of silver nanoparticles and silver ions in an in vitro human gastrointestinal digestion model.* Nanotoxicology, 2012. **7**(7): p. 1198-1210.

[120] Kim, S. and D.Y. Ryu, *Silver nanoparticle-induced oxidative stress, genotoxicity and apoptosis in cultured cells and animal tissues.* Journal of Applied Toxicology, 2013. **33**(2): p. 78-89.

[121] Gliga, A.R., et al., *Size-dependent cytotoxicity of silver nanoparticles in human lung cells: the role of cellular uptake, agglomeration and Ag release.* Particle and fibre toxicology, 2014. **11**(1): p. 11.

[122] Hackenberg, S., et al., *Silver nanoparticles: evaluation of DNA damage, toxicity and functional impairment in human mesenchymal stem cells.* Toxicology letters, 2011. **201**(1): p. 27-33.

[123] Samberg, M.E., et al., *Silver nanoparticles do not influence stem cell differentiation but cause minimal toxicity.* Nanomedicine, 2012. **7**(8): p. 1197-1209.

[124] Liu, W., et al., *Impact of silver nanoparticles on human cells: effect of particle size.* Nanotoxicology, 2010. **4**(3): p. 319-330.

[125] Kim, T.H., et al., *Size-dependent cellular toxicity of silver nanoparticles.*

Journal of Biomedical Materials Research Part A, 2012. **100**(4): p. 1033-1043.

[126] Panda, K.K., et al., *In vitro biosynthesis and genotoxicity bioassay of silver nanoparticles using plants.* Toxicology in vitro, 2011. **25**(5): p. 1097-1105.

[127] Lima, R., A.B. Seabra, and N. Durán, *Silver nanoparticles: a brief review of cytotoxicity and genotoxicity of chemically and biogenically synthesized nanoparticles.* Journal of Applied Toxicology, 2012. **32**(11): p. 867-879.

[128] Crooks, R.M., et al., *Dendrimer-encapsulated metals and semiconductors: synthesis, characterization, and applications*, in *Dendrimers III*. 2001, Springer. p. 81-135.

[129] Gittins, D.I., et al., *A nanometre-scale electronic switch consisting of a metal cluster and redox-addressable groups.* Nature, 2000. **408**(6808): p. 67-69.

[130] Kamat, P.V., *Meeting the clean energy demand: nanostructure architectures for solar energy conversion.* The Journal of Physical Chemistry C, 2007. **111**(7): p. 2834-2860.

[131] Fedlheim, D.L. and C.A. Foss, *Metal nanoparticles: synthesis, characterization, and applications*. 2001: CRC press.

[132] Rai, M., et al., *Strategic role of selected noble metal nanoparticles in medicine.* Critical reviews in microbiology, 2016. **42**(5): p. 696-719.

[133] Baker, S., et al., *Plants: emerging as nanofactories towards facile route in synthesis of nanoparticles.* BioImpacts: BI, 2013. **3**(3): p. 111.

[134] Khlebtsov, N. and L. Dykman, *Biodistribution and toxicity of engineered gold nanoparticles: a review of in vitro and in vivo studies.* Chemical Society Reviews, 2011. **40**(3): p. 1647-1671.

[135] Huang, X., et al., *Gold nanoparticles: interesting optical properties and recent applications in cancer diagnostics and therapy.* Nanomedicine, 2007. **2**(5): p. 681-693.

[136] Bhadra, M.P., B. Sreedhar, and C.R. Patra, *Potential theranostics application of bio-synthesized silver nanoparticles (4-in-1 system).* 2014.

[137] Prasad, T. and E. Elumalai, *Biofabrication of Ag nanoparticles using Moringa oleifera leaf extract and their antimicrobial activity.* Asian Pacific Journal of Tropical Biomedicine, 2011. **1**(6): p. 439.

[138] Rajakumar, G. and A.A. Rahuman, *Larvicidal activity of synthesized silver nanoparticles using Eclipta prostrata leaf extract against filariasis and malaria vectors.* Acta tropica, 2011. **118**(3): p. 196-203.

[139] Vivek, M., et al., *Biogenic silver nanoparticles by Gelidiella acerosa extract and their antifungal effects.* Avicenna Journal of Medical Biotechnology, 2011. **3**(3): p. 143.

[140] Sukirtha, R., et al., *Cytotoxic effect of Green synthesized silver nanoparticles using Melia azedarach against in vitro HeLa cell lines and lymphoma mice model.* Process Biochemistry, 2012. **47**(2): p. 273-279.

[141] Haggag, E.G., et al., *Antiviral potential of green synthesized silver nanoparticles of Lampranthus coccineus and Malephora lutea.* International journal of nanomedicine, 2019. **14**: p. 6217-6229.

[142] Youssif, K.A., et al., *Anti-Alzheimer potential, metabolomic profiling and molecular docking of green synthesized silver nanoparticles of Lampranthus*

coccineus and Malephora lutea aqueous extracts. PLoS One, 2019. **14**(11): p. e0223781.

[143] Chinnasamy, G., S. Chandrasekharan, and S. Bhatnagar, *Biosynthesis of Silver Nanoparticles from Melia azedarach: Enhancement of Antibacterial, Wound Healing, Antidiabetic and Antioxidant Activities.* Int J Nanomedicine, 2019. **14**: p. 9823-9836.

[144] Malapermal, V., et al., *Enhancing antidiabetic and antimicrobial performance of Ocimum basilicum, and Ocimum sanctum (L.) using silver nanoparticles.* Saudi Journal of Biological Sciences, 2017. **24**(6): p. 1294-1305.

[145] Otunola, G.A., et al., *Characterization, Antibacterial and Antioxidant Properties of Silver Nanoparticles Synthesized from Aqueous Extracts of Allium sativum, Zingiber officinale, and Capsicum frutescens.* Pharmacognosy magazine, 2017. **13**(Suppl 2): p. S201-S208.

Chapter 13

Preparation, Structural Characterization, and Biomedical Applications of Gypsum-Based Nanocomposite Bone Cements

Hesham F. El-Maghraby and Yaser E. Greish

Abstract

Hard tissues are natural nanocomposites comprising collagen nanofibers that are interlocked with hydroxyapatite (HAp) nanocrystallites. This mechanical interlocking at the nanoscale provides the unique properties of hard tissues (bone and teeth). Upon fracture, cements are usually used for treatment of simple fractures or as an adhesive for the treatment of complicated fractures that require the use of metallic implants. Most of the commercially available bone cements are polymer-based, and lack the required bioactivity for a successful cementation. Besides calcium phosphate cements, gypsum is one of the early recognized and used biomaterials as a basi for a self-setting cementation. It is based on the controlled hydration of plaster of Paris at room temperature and its subsequent conversion to a self-setting solid gypsum product. In our work, we have taken this process further towards the development of a set of nanocomposites that have enhanced bioactivity and mechanical properties. This chapter will outline the formation, characterization, and properties of gypsum-based nanocomposites for bone cement applications. These modified cements can be formulated at room temperature and have been shown to possess a high degree of bioactivity, and are considered potential candidates for bone fracture and defect treatment.

Keywords: gypsum, bone cement, nanocomposite, biomimetic, apatite-forming ability

1. Introduction

Hard tissues, such as bone and teeth, are natural composites consisting of two types of material. The first material is an organic extracellular matrix that contains collagen, accounts for approximately 30–35% of the dry weight of bone, and is responsible for providing flexibility and resilience to the bone. The second material consists primarily of calcium and phosphorous salts, especially hydroxyapatite [$Ca_{10}(PO_4)_6(OH)_2$; HAp], accounts for approximately 65–70% of the dry weight of bone, and contributes to the hardness and rigidity of the bone [1]. Hard tissues, in general, are responsible for providing support for the whole body, for the attachment to ligaments and tendons, and protect vital organs. Based on the composition of hard tissues, they are also considered as a reservoir for the minerals in addition to iron that maintains the process of hemostasis.

IntechOpen

Bone can be classified macroscopically as cortical tissue and cancellous (trabecular) tissue [1]. Both types are morphologically lamellar bone. Cortical tissue relies on osteons for cell communication. Because trabecular width is small, the canaliculi can communicate directly with blood vessels in the medullary canal. The basic differences between cortical tissue and cancellous tissue relate to porosity and apparent density. The porosity of cortical tissue typically ranges from 5–30%, and that of cancellous tissue ranges from 30–90%. The apparent density of cortical tissue is approximately 1.8 g/cm^3, and that of cancellous tissue typically ranges from 0.1 to 1.0 g/cm^3. The distinction between cortical tissue and cancellous tissue is arbitrary. However, in biomechanical terms, the two tissues are often considered one material with a specific range of porosity and density.. The organization of cortical and cancellous tissue in bone allows adaptation to function. Cortical tissue always surrounds cancellous tissue, but the relative quantity of each type of tissue varies with the bone's functional requirements. In long bones, the diaphysis's cortical tissue is arranged as a hollow cylinder to best resist bending. The metaphyseal region of the long bones flares to increase the bone volume and surface area in a manner that minimizes the stress of joint contact. The cancellous tissue in this region provides an intricate network that distributes weight-bearing forces and joint reaction forces into the bulk of the bone tissue [2–4]. **Figure 1** shows a schematic diagram of a typical human bone, showing its detailed composition as well as the various types of bone.

If fractured or damaged, hard tissues are either left to self-heal, depending on the extent of the fracture or replaced partially or totally by an implant. **Figure 2** shows the mechanism of self-healing of fractured bone. However, surgical intervening is mostly required to ensure proper fixation of the fractured bone and avoid future health consequences. In this regard, natural or synthetic materials, also known as biomaterials, are often used to partially or totally fix the fractured bone. The term "biomaterials" is used to indicate materials that constitute parts of medical implants, extracorporeal devices, and disposables utilized in medicine, surgery, dentistry, veterinary medicine, and in every aspect of patient health care [5, 6].

Biomaterials can be generally divided into three main categories that are regulated by tissue reaction. In basic context, inert materials (more narrowly, almost inert) preclude no or minimal tissue reaction. Active materials enhance bonding to surround tissue by promoting, for example, new bone formation. In the underlying tissue, degradable or resorbable materials are introduced or can even dissolve completely over a period. Commonly, metals are inert, ceramics may be inert, active or resorbable, and polymers may be resorbable or inert. The main property required of a biomaterial is that it does not elicit an adverse reaction when placed into service. In addition to biomaterials used for partial or total fixation of the fractured bone, an additional class of materials known as "bone cements" is also used for stand-alone

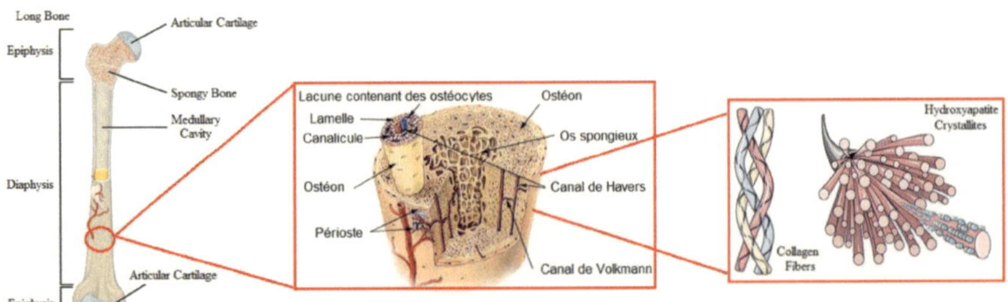

Figure 1.
Detailed structure of a typical human bone.

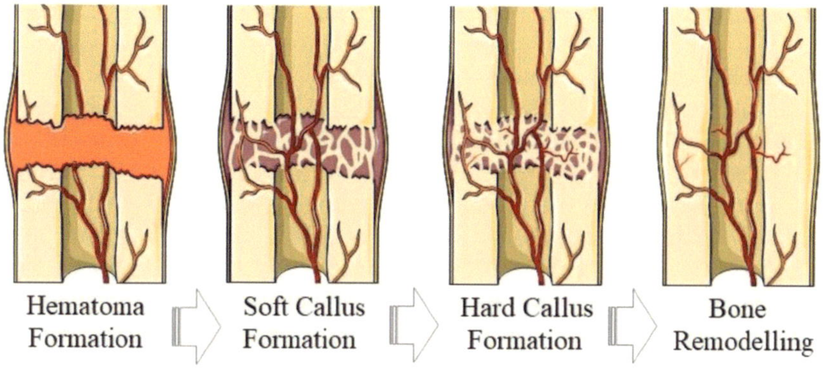

Figure 2.
Self-healing process involved in a fractured human bone.

fixation of the fractures bone or as an auxiliary material with other biomaterials to help in the fixation process. More details about bone cements will be discussed in the following sections.

2. Bone cements

By definition, bone cements are biomaterials that are obtained by mixing a powder phase with a liquid phase, forming a paste that can solidify into a final set product upon implantation within the body [7]. Bone cements are highly characterized by the ability to be injected in the body, extending their application to minimally invasive surgical applications [7]. The first known bone cement is poly(methyl methacrylate) (PMMA), which was used in 1940s as a bone filler to close fractures and defects in the skull [8]. PMMA was also used to fix total hip arthroplasty replacement components, similar to a total hip replacement [9]. This process takes place through the in-situ polymerization of the PMMA; hence a set product is formed.

Biocompatible polymeric bone cements are further classified as bioinert, bioactive, or biodegradable depending on their composition. However, none of them possesses a similar chemical composition to that of natural hard tissues. Accordingly, non-polymeric bone cements were investigated. These include calcium phosphate cements, as well as gypsum-based cements [10]. These cements are based on formulations that are composed of the mineral powder and an aqueous liquid. Their blending results in the hydration of the powder and the instant chemical transformation to the final set product [10].

2.1 Gypsum bone cements

Gypsum is one of the first used biomaterials. It is a common non-metallic mineral consisting of hydrated calcium sulfate ($CaS0_4{\cdot}2H_20$). It crystallizes in the monoclinic system in white or colorless crystals; Gypsum is formed through the hydration of its hemihydrate precursor (Plaster of Paris; POP) according to Eq. (1):

$$CaSO_4{\cdot}½H_2O + 1½\,H_2O \rightarrow CaS0_4{\cdot}2H_20 \qquad (1)$$

This reaction can take place at room temperature. A two-phase suspension of hemihydrate particles in a saturated aqueous solution is formed as the hemihydrate dissolves. Crystals nucleate in the suspension as the solution becomes super-saturated with dihydrate and form a precipitate. Until the solution is no longer saturated, nucleation and crystal formation proceed, leading to the further dissolution of the hemihydrate. Alternative dissolution and precipitation keep, with the growth of existing crystals or nucleation of new crystals [11].

The transformation of POP to gypsum is often followed by simple characterization techniques, such as x-ray diffraction (XRD), infrared spectroscopy (IR), and scanning electron microscopy (SEM). **Figure 3** shows a comparison between the XRD patterns (3a), IR spectra (3b), and SEM micrographs of POP and gypsum phases. POP is characterized by its XRD peaks at 2Θ values of 14.79, 25.48, 29.34, 29.77, 31.84, 33.03, 38.46, 39.77, 41.34, 42.31, 45.36, 47.64, 48.66, and 49.32°; as shown in **Figure 3a**. Gypsum, on the other hand, is characterized by its XRD peaks at 2Θ values of 11.64, 20.76, 23.43, 25.48, 28.19, 29.17, 31.13, 32.16, 33.35, 34.61, 36.02, 36.611, 37.38, 40.69, 42.20, 43.40, 44.27, 45.57, 46.49, 47.06, and 48.51°; as shown in **Figure 3a**. These peaks were in accordance with their standard XRD cards 01–0999, and 33–0311 for gypsum and POP phases, respectively.

Figure 3b shows the IR spectra of POP and gypsum solid powders. Both POP and gypsum have water of crystallization, ½ H_2O in POP, and $2H_2O$ in gypsum. This was shown in the IR spectra as relatively broad bands with medium-strong intensities. Both appeared as a doublet at 3549.7 and 3607 cm^{-1} [12]. In the IR spectrum of gypsum, more broadness of this assembly was observed, extending its absorption to have a broad shoulder at 3416 cm^{-1} [12]. These differences are attributed to the difference in the proportion of water of crystallization in the two phases. Sulfate ions are shown in these phases' patterns at four places; a sharp strong intensity doublet at 604 and 656 cm^{-1}, a relatively broad but very strong band with a peak at 1143.9 cm^{-1}, and a medium intensity sharp band at 1630.9 cm^{-1} [12]. All four bands appeared with almost the same intensity in the two IR spectra of POP and gypsum.

The microstructure of POP and gypsum powders is shown in **Figure 3c,d**. POP phase is known to exist in the form of irregular shaped crystals with a polydisperse particle size distribution, as shown in **Figure 3c**. Upon reaction with water and formation of gypsum, the dissolution-precipitation mechanism by which reaction proceeds results in the formation of smaller crystallites with a more uniform size distribution, as shown in **Figure 3d**.

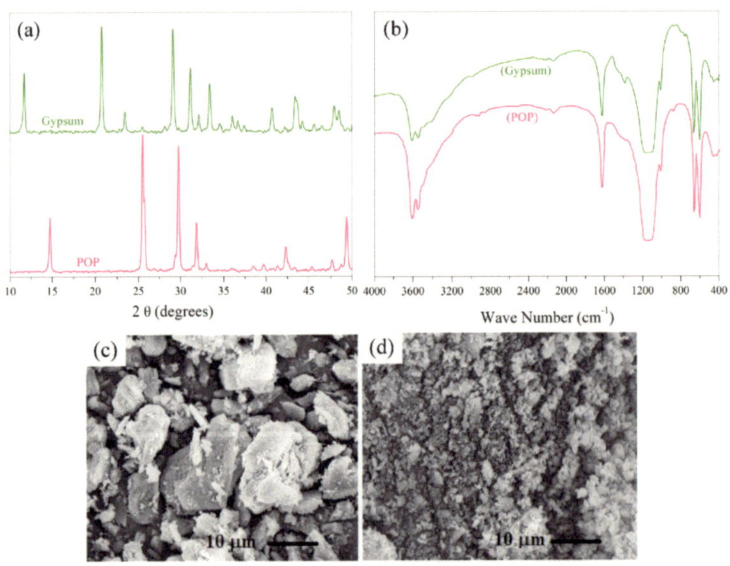

Figure 3.
Physico-chemical characterization of gypsum and POP: (a) X-ray diffraction analysis, (b): Fourier-transformed infrared spectroscopy, Scanning electron micrographs of (c) POP and (d) gypsum powders.

There are two varieties of the gypsum hemihydrate form, α- and β-, which differ in crystal size, surface area, and lattice imperfections. While these materials are chemically similar, their physical properties vary considerably. The dental stone from which diagnostic casts is the α-hemihydrate type. When compared to the β-hemihydrate, it is very difficult and relatively insoluble. The β-hemihydrate is distinguished by an aggregate of interstitial capillary pores of abnormal crystals, while the α-hemihydrate comprises cleavage fragments and rod and prism-shaped crystals [13]. The α-form needs much less water than the β-form (0.3 versus 0.6 g/g of hemihydrate, respectively) due to their various particle characteristics. As a result, the α-form results in a super thick, heavier and less soluble dihydrate than the β-form [11]. The formed gypsum exhibits a layered structure with the water molecules alternating with calcium sulfate layers explaining this mineral's easy cleavage [14], as shown in **Figure 4**. The hemihydrate has the water molecules arranged in channels between chains of calcium sulfate. It explains the ease with which this compound can lose most of its coordinated water without disrupting the structure.

Gypsum as one of the oldest known construction materials today is manufactured in a huge amounts for renders, plasters, indoor finishing, retardants for cement, ceramics, and medical supplements or implants [15–20]. The wide applications of gypsum plaster are primarily based on its unique properties, e.g., setting time, suitable workability, and volume stability. However, some disadvantages appear when neat gypsum plaster is used, where a relatively high amount of water is required for mixing. The setting time may accordingly be too long, and as a result, it adversely affects the set product's mechanical properties. These properties are merely developed by introducing other ingredients to form what is known as gypsum composites as an attempt to improve the plaster properties.

Gypsum is a highly biocompatible material that is one of the simplest synthetic bone graft materials with the longest clinical history, spanning more than 100 years [21]. It is classified as a bioresorbable material. It has been used effectively for the treatment of periodontal disorders, endodontic lesions, alveolar bone loss and augmentation of the maxillary sinus [21]. It has also been used to promote healing and stop failure of the grafting material as a binder. In addition, it is tissue compliant and does not interfere with the process of healing [22]. At rates as high as 1 mm per week, Ricci et al. [23] observed a rapid rate of dissolution of both in vitro and in vivo gypsum from the outer surface inwards. Despite this dissolution, they reported

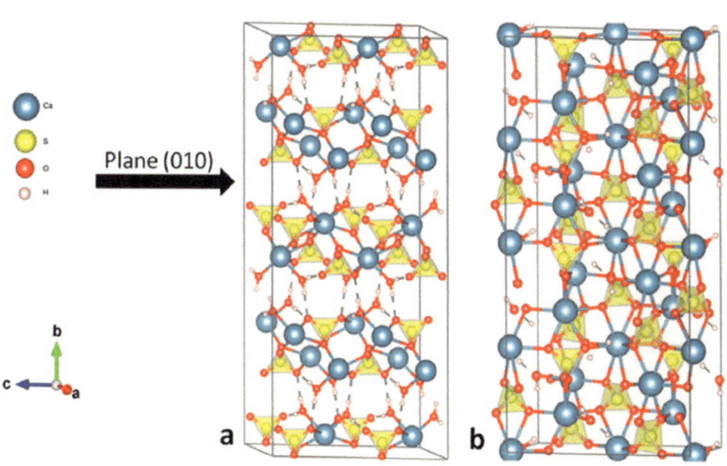

Figure 4.
Crystal structure of (a) gypsum, and (b) POP phases [14].

that while bone was not observed to come into direct contact with gypsum, gypsum stimulated new bone formation. The formed bone took the shape of concentric rings in surrounding tissue, and histologically, these deposits stain like bone mineral and often showed attachment of osteoid and new bone [23].

2.2 Biomedical applications of gypsum bone cements

Gypsum and its precursor; POP, have been used as bone void fillers for more than a century [24]. Kelly et al. in 2001 used Osteosets (surgical grade calcium sulfate, Wright Medical Technology, Arlington TN) to fill bone defects caused by benign bone tumors, trauma, cyst, etc. [25]. They concluded that surgical grade calcium sulfate is reliable, convenient, safe, and readily available bone graft substitute that yields consistent results [25]. Osteosets, Borrelli et al. [26] successfully treated non-union osseous defects caused by trauma with a combination of autogenous iliac bone and Osteoset using the same surgical-grade gypsum. In 2005, Chen et al. concluded that, as an artificial bone expander with a fair fusion rate, surgeons may use calcium sulfate combined with locally harvested morselized bone [27].

Lazary et al. showed that MC3T3-E1 osteoblastic mouse cells placed on a gypsum disc expressed genes that are relevant in the formation of new bones in a way that is different and better than poly(methyl methacrylate) (PMMA), which is a typical bone void filler [28]. They also concluded that a more effective bone repair environment was created by gypsum. Because of its unique crystal structure and high calcium content, it is also chosen to have the potential for osteoinductivity [27, 28]. In addition to the above-mentioned characteristics, Rohmiller et al. suggested gypsum as a cement for lumbar pedicle screw fixation [29]. They found it more promising than PMMA, in particular because, unlike the exothermic hardening of PMMA [29], no heat forms during its framework. In a parallel study, it was found that cells plated on gypsum disc express genes important in new bone formation with different expression ratios compared to PMMA, generally used as a bone void filler, suggesting that gypsum provides a more efficient environment for bone repair. These findings indicate that gypsum possess a potential Osteoinductivity, as a result of its special crystal structure and high calcium content [30].

Properties and applications of gypsum-based bone cements were also enhanced through the inclusion of various types of additives; natural or synthetic and inorganic or organic. One of these additives' main objectives is to improve the bioactive and mechanical properties of gypsum as a bone cement so that its biomedical applications are further extended. One of the first reports about the use of additives to gypsum was shown by Sanad et al. in 1982, where a combination of calcium oxide (CaO) and gum Arabic was found to enhance the mechanical properties of gypsum [31]. This was related to lime (CaO) precipitation in the hydration and setting processes of POP, while the gum Arabic acted as a gluing matrix [31].

One of the effective approaches to modulating the mechanical properties of gypsum [32–35] has been considered to be the blending of gypsum with polymers. The presence of such functional groups, such as hydroxyl and carboxyl groups, on these polymers is often preferred. These were found to connect with the gypsum products set together with the calcium sites. In general, polymers that do not have these groups are passive during the gypsum setting reaction. However, the mechanical interlocking of these polymers with the gypsum crystals collected improves the overall mechanical performance of the composites made. For biomedical applications, polymers used with gypsum should be biocompatible to avoid rejection by the human immune system. Different polymers could be used in this regard, ranging from bioactive to bioinert, depending on the application type and site. El-Maghraby et al. evaluated gypsum composites with poly(vinyl alcohol) and its copolymers with vinyl acetate and itaconic acid [36]. They concluded that the presence of carboxylic acid-groups along the polymeric ingredients enhanced the mineralization ability of the produced composites. In another study, a novel injectable, photo-cross-linkable PCL/calcium sulfate system has been developed to overcome calcium sulfate brittleness and fast resorption rate enhancing its performance in bone regeneration techniques [37]. Moreover, the addition of viscous polymers, such as Carboxymethylcellulose (CMC), and hyaluronan, improved the handling characteristics and mechanical properties of the CS [38, 39]. Gelatin was also used as a natural additive to a mixture of gypsum and POP, which result in the formation of a porous scaffold upon setting [40]. This novel composite system showed high potential in tissue engineering applications [40]. Mineralization of a gypsum-PLLA composite bone cement indicated that the material may be osteoconductive. Both gypsum and PLLA are classified as bioresorbable biomaterials. Therefore, their composite bone cement can be designed in such a way that its overall degradation can be controlled [41, 42].

2.3 Resorption of gypsum bone cement

Gypsum is classified as a resorbable biomaterial. This characteristic does not interfere with its wide application as a bone cement, where the resorption of gypsum cement after implantation could be controlled to match the fixation of the fractured bone and the subsequent bone formation. The resorption of gypsum has been studied by various authors. Randolph et al. patented the ideas of having it in the form of pellets that will release certain ingredients upon dissolution [43–45]. Besides, Hanker et al. patented the gypsum resorption concept for the repair of damaged bone [46]. Doadrio et al. developed a gypsum-based cement and used it to deliver an antibiotic; cephalexin [47]. The inclusion of this drug was shown not to affect the structural composition or the physicochemical properties of gypsum [47].

Despite the benefit of gypsum bioresorption, which makes it an appealing candidate for particular applications, its relatively low mechanical properties have limited its scope of use as a bone replacement implant or even as bone cement. To boost its mechanical properties, various materials were mixed with gypsum; what is referred to as biphasic composites [48]. POP has often been mixed with various ceramic and polymeric ingredients to form gypsum composites there from.

On the other hand, the expedited sorption rate of gypsum can be reduced significantly through the use of various additives, particle/bead size, and sintering techniques of the ready-made gypsum made biomaterials. It was shown that the degradation rate of sintered CS specimens can be adjusted through the introduction of pores. Through various techniques, the introduction of spherical pores with amounts ranging from 6.7 to 68% into sintered CS specimens was reflected on a corresponding variable degradation rate in Hank's solution in the range of 1.9–7.7%/day [49]. Upon using different-sized calcium sulfate beads, their ability to elute multiple antibiotics in vitro was observed, as a possible method to improve the therapeutic delivery in patients [50].

2.4 Intrinsic bioactivity of gypsum bone cements

Preliminary evaluation of biomaterials, implants, and cements is intended to be in a closer contact with hard tissues, occurs through the study of the bone-like apatite formation and growth ability of the potential biomaterials. This process is also termed "biomimetic", where bone-like apatite formation and growth resembles the natural mineralization of collagen in nature, which takes place during the continued bone formation process with time. This process was discussed in details in a previous work by Hafiz Uddin et al. [51] and was first recognized and established by Kokubo et al. in 1981 and reported in 1990 [52].

The hypothesis behind this approach is to mimic the biomineralization process where Ca^{2+} and PO_4^{3-} ions deposit in the form of apatite nanocrystallites onto certain nucleation sites along the collagen nanofibers. These nucleation sites are the functional groups along with the various amino acids within the collagen fibrils, such as -OH, -COOH, and $-NH_2$ groups. Accordingly, biomaterials that possess any of these groups upon the immersion of a simulated body fluid (SBF), will undergo a similar mechanism resulting in the deposition of bone-like apatite spherolites, which were shown to contain apatite nanocrystallites. Protein-free SBF media contain ions with concentrations similar to those in the human blood plasma, and are supersaturated with respect to stoichiometric HAp (Ca/P 1.67). **Table 1** shows a detailed composition of a typical SBF solution.

The mechanism of biomimetic deposition of bone-like apatite takes place at a physiologic pH and temperature through the instant gravitation and binding of the Ca^{2+} ions from the solution onto the surface functional groups, followed by the attachment of the PO_4^{3-} ions, hence nuclei of bone-like HAp are formed. Upon continued immersion in SBF, these nuclei further grow and adopt a typical bone-like apatite morphology. **Figure 5** shows a schematic representation of the apatite formation's biomimetic process onto OH-carrying biomaterial surfaces, such as silicate-containing implants.. Similarly, biomaterial surfaces that provide these functional groups were widely explored for the biomimetic formation of nanostructured bone-like apatite.

Gypsum, as a final set product, was also investigated in our laboratories for its affinity towards the biomimetic deposition of bone-like HAp [36, 53]. In a typical experiment, gypsum samples were immersed in a freshly prepared SBF medium and maintained at a physiologic temperature of 37.4°C for up to 2 weeks. Aliquots were collected at various time intervals to study the variation of the most relevant ions; Ca^{2+}, PO_4^{3-}, and SO_4^{2-}, in the medium as a function of time. Dry samples at the end of the 15-days experiment were collected and studied by scanning electron microscopy for the variation of its surface morphology. **Figure 6** shows typical SEM micrographs of a SBF-treated gypsum sample (**b**), compared to a pristine gypsum sample (**a**). Upon further treatment of the SBF-treated gypsum sample in a more concentrated SBF solution (1.5X), further growth of the deposited apatite nanostructured spherolites were shown with a more detailed bone-like morphology as shown in **Figure 6c**. Elemental analysis of the nanostructured spherolites indicated the presence of Ca and P peaks that confirmed the identity of the deposited apatite [36, 53].

In a typical SBF experiment, both Ca^{2+} and PO_4^{3-} ions are leached out from solution as a result of initially being supersaturated with respect to these ions. This process results in the precipitation of the corresponding nanostructured bone-like apatite spherolites. Chemical analysis of these ions with time usually confirms these findings. However, the immersion of pure gypsum in SBF for up to 15 days

Type of solution	Concentration (mM)							
	Na^+	K^+	Ca^{2+}	Mg^{2+}	HCO_3^-	Cl^-	HPO_4^{2-}	SO_4^{2-}
SBF	142.0	5.0	2.5	1.5	4.2	148.0	1.0	0.5
Blood plasma	142.0	5.0	2.5	1.5	27.0	103.0	1.0	0.5

Table 1.
A detailed composition of a typical SBF solution, compared with a blood plasma [52].

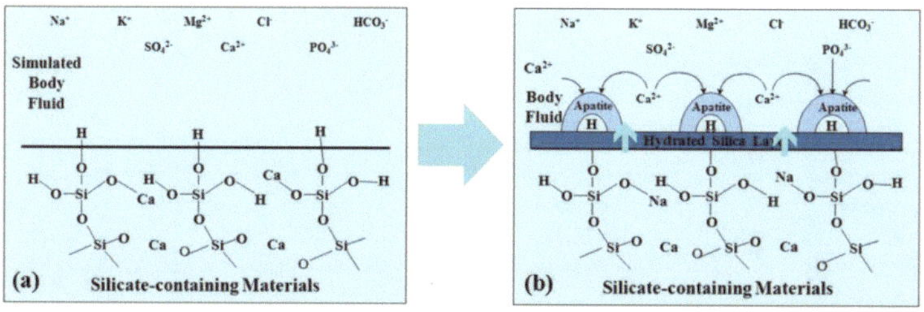

Figure 5.
A schematic representation of the mechanism of biomimetic deposition of bone-like apatite spherolites on a typical silicate-based biomaterial; before (a), and after (b) immersion in a simulated body fluid.

Figure 6.
Scanning electron micrographs of (a) set gypsum surface and nanotextured, bone-like apatite formation onto the surfaces of a set gypsum cement sample after immersion in (b) SBF and (c) 1.5 SBF media at 37°C.

showed the decrease of the concentration of the PO_4^{3-} ions with time, with an unexpected increase in the concentration of the Ca^{2+} ions, as shown in **Figure 7** [53]. To explain this behavior, the concentration of SO_4^{2-} ions were also analyzed and was also proven to show a continued increase with time. The combined increase in the concentrations of Ca^{2+} and SO_4^{2-} ions with time was attributed to gypsum's bioresorbable nature. It is believed that the process of biomimetic mineralization of gypsum takes place through a combined mechanism of dissolution (via resorption) of gypsum crystal surfaces and the subsequent release of Ca^{2+} and SO_4^{2-} ions into solution. This is followed by an abrupt increase in the supersaturation of the SBF medium with respect to Ca^{2+} and PO_4^{3-} ions and their eventual precipitation in the form of the nanostructured bone-like apatite spherolites. Gypsum bone cement with biomimetically grown nano-textured apatite was successfully used as a drug delivery vehicle in addition to its cementation effect. This is attributed to the add-value of the nanotextured apatite spherolites that can be used as drug carriers, where the sorption of gypsum cement results in the slow release of the drugs. An example was shown in the sustained release cephalexin and gentamicin from both pure calcium sulfate and nanocomposite cements into SBF, which was demonstrated in the work of Hesaraki et al. [54].

2.5 Nano-textured gypsum bone cements with enhanced bioactivity

In addition to the intrinsic bioactivity of pristine gypsum bone cements and its ability to form bone-like nano-textured apatite on its surfaces, these bioactivity indicators were further enhanced through the inclusion of bioactive fillers. These include calcium phosphates, which are known for their structural similarity to bone apatite, and calcium silicates, which were the first candidates to show an enhanced affinity to grow bone-like nanotextured apatite spherolites in SBF media [55–62]. In addition, POP-based biomaterials have also exhibited promise as grafts in a pre-clinical repair model of intrabony periodontal defects, as well as in clinical reports for sinus augmentation and treatments of femoral shaft non-unions [60–62].

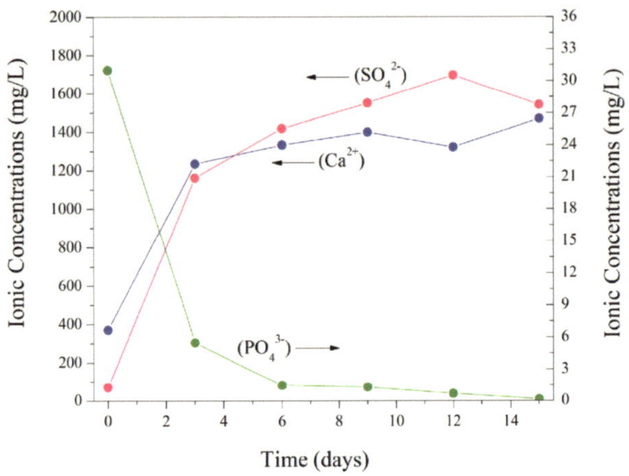

Figure 7.
Variation of the ionic concentrations of Ca^{2+}, PO_4^{3-}, and SO_4^{2-} ions in SBF media with time as a result of immersion of gypsum samples for up to 15 days at 37°C.

2.6 Calcium phosphate-doped gypsum bone cements

POP was applied to enhance the setting reactions of a biodegradable calcium phosphate cement that was composed of β-tricalcium phosphate (β-$Ca_3(PO_4)_2$; β-TCP) and monocalcium phosphate monohydrate ($Ca(H_2PO_4)_2.H_2O$; MCPM) [56] due to its relatively faster setting process. These composites of cement have been shown to be osteo-conductive [56]. The setting time, workability, and porosity of a cement composed of an aqueous slurry of β –TCP [57] were also modulated using POP. When combined with β-TCP, Nilsson et al. [58] realized the essential function of gypsum, helping to create pores in the implanted material, thereby ensuring the growth of new bone tissue. The use of β-TCP in a CS matrix produced significantly more vital new bone fill and preserved bone dimensions than β-TCP alone [59]. Composite bone cement was also made of gypsum and granules of β-TCP, and dicalcium phosphate dihydrate, and were shown to have a greater bone filling ability than did pure gypsum cement [60]. Moreover, the physical properties of the bone regenerated with the CS/CP composite were similar to or greater than native bone [60].

Combinations of ready-made HAp particles with gypsum were also evaluated. Sato et al. indicated the promising characteristics of gypsum after mixing it with HAp particles, based on the relatively fast absorption of gypsum without interfering with bone healing [48]. Cabanas et al. concluded that for a paste of calcium sulfate-calcium phosphate cements to be injected percutaneously using a syringe or implanted operatively, it was essential to control parameters such as working time, setting time, or thermodynamic behavior [61]. It was also shown by Guo et al. that a calcium phosphate-doped gypsum bone cement has a controlled setting time within the range of 5–20 min., and the material can easily be molded before setting [62]. A novel HAp-gypsum-POP cement system was successfully used as a reservoir for the delivery of growth factor-beta 1 (TGF-1) and vascular endothelial growth factor (VEGF) [63].

Two novel dual-setting apatite-gypsum bone cements were studied, in which the setting reaction of POP and its transformation to gypsum was combined with the setting reaction of apatitic calcium phosphate precursors [64, 65]. These precursors normally react with water at room temperature through an acid–base reaction leading to the formation of calcium phosphate bone cements with various degrees of stoichiometry, as shown in Eqs. (2) and (3):

$$2Ca_4(PO_4)_2O + 2CaHPO_4.2H_2O \rightarrow Ca_{10}(PO_4)_6(OH)_2\,(\mathbf{SHAp}) \quad (2)$$

$$3Ca_4(PO_4)_2O + 6CaHPO_4.2H_2O \rightarrow 2Ca_9(HPO_4)(PO_4)_5(OH)(\mathbf{CD-HAp}) \quad (3)$$

As observed in both cases, the findings showed a remarkable delay in the growth kinetics of gypsum, with different extensions depending on the starting Ca/P molar ratios of the apatite and its proportion in the composites. The synthetic version of Ca-def apatite's bioresorbability also revealed its retardation effect on the formation of gypsum. An overall near-physiological pH regimen was obtained by mixing POP in their solutions with each of the Ca-def apatites [64]. However, variations in the pH of the studied solutions showed the effect of POP on decreasing the alkalinity of the media containing SHAp precursors [65].

Fabrication of osteoconductive scaffold with osteoinductive capability and appropriate resorption rate was also achieved through the introduction of Strontium in the crystal structure of gypsum as well as the combined effect of HAp. Together, Sr-gypsum/HAp was shown to promote bone formation by recruiting and stimulating osteogenic differentiation of BMSCs [66].

2.7 Calcium silicate-doped gypsum bone cements

Silicate-based materials have been proven to possess an excellent affinity towards the biomimetic deposition of nano-textured bone-like apatite spherolites in SBF media. Accordingly, they were considered bioactive, and their inclusion in the formation of composite biomaterials is highly believed to lead to enhancement of the bioactivity of the biomaterials produced thereof. Accordingly, silicates were considered potential candidates to be added to gypsum bone cements to further enhance their bioactivity. Silicate-based materials; either crystalline or amorphous, could be used in this regard as they share the mechanism of biomimetic deposition of apatite when exposed to SBF media. In the work of Greish et al. [53], highly crystalline wollastonite fibers were added to gypsum bone cements in order to study their effect on the mechanical properties and bioactivity of the produced composite cement systems. Due to the silicate composition of the wollastonite fibers and the gypsum matrix's intrinsic bioactivity, both phases were shown to develop nano-structured apatite coatings onto their surfaces; **Figure 8a**. Furthermore, alkali-treated wollastonite fibers were shown to have an extensive formation of these apatitic coatings on its surfaces, as shown in **Figure 8b**. These novel cements provide a dual mechanism of biomimetic deposition of the bone-like spherolites, hence show an enhanced bioactivity as compared with pure gypsum cements [53]. Tricalcium silicate was also added to gypsum in the work of Huan et al. [67] to improve its handling and physical properties, in addition to improved mechanical properties. Furthermore, the composite made thereafter was also shown to have enhanced bioactivity and favorable resorption characteristics as compared with pure gypsum cement [67]. A ternary organic–inorganic composite bone cements of tricalcium silicate/sodium alginate/POP (C_3S/SA/POP) were successfully shown to exhibit good proliferation, excellent attachment, enhanced alkaline phosphatase activity, increased calcium deposition, and osteogenic-related gene expressions with growing calcium sulfate component [68].

In the course of 3D printing, POP was also supplemented with silicate materials. In order to boost the scaffold efficiency, a combined cement was integrated functionally

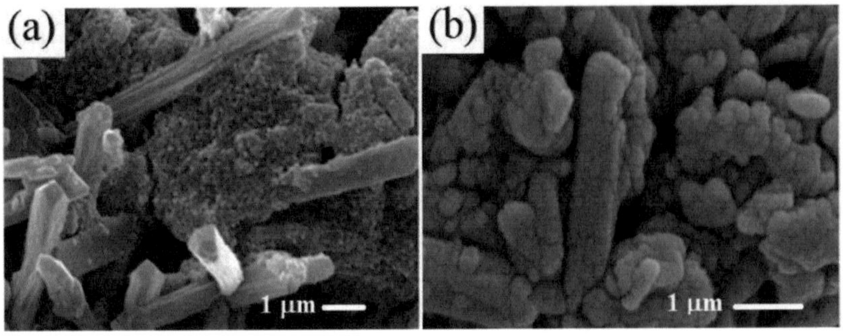

Figure 8.
Scanning electron micrographs of gypsum cements containing a) pristine, and b) alkali-treated wollastonite fibers, after immersion in SBF media for 15 days at 37°C.

into mesoporous calcium silicate (MCS) via a 3D printing technique. Compared to printed MCS scaffolds, the characteristics showed that 20 percent CSH integration over 4 weeks of hydration increased their compressive strength by 2 times. Moreover, MCS component in the composite scaffolds exhibited sustained release behavior of dexamethasone drugs to assist bone regeneration [69]. Similarly, scaffolds made of POP and mesoporous bioactive glass (MBG) scaffolds successfully fabricated using a 3D printing technique, which had a regular and uniform square macroporous structure, high porosity and excellent apatite mineralization ability. Thus 3D printed POP/MBG scaffolds would be promising candidates for promoting bone regeneration [70].

3. Conclusions

Bone cements are widely used for the treatment of bone defects and fractures. Gypsum-based bone cements are characterized by their ease of preparation and affordability. Gypsum has been classified as a bioresorbable material. Moreover, our ongoing research has clearly shown a string evidence of its bioactivity where gypsum is mineralized in SBF media in a manner similar to that was initially proposed for the biomimetic growth of bone-like apatite on the surfaces of bioactive materials. Moreover, the inclusion of other biocompatible fillers further enhance its bioactivity. The biomimetically-grown bone-like apatite adopts the morphology of nanostructured spherolites which are made of apatite nanocrystallites. The overall assembly of nanostructured gypsum-based bone cements, therefore, represents a potential modality for the treatment of fractured bone with an enhanced bioactivity. Moreover, an added value of the improvement in the mechanical properties of these composite cements is highly believed to extend the applications of these cements to be used as bone implants for non-load bearing applications.

Acknowledgements

The authors would like to extend their acknowledgement to the continuous financial support of the UAE University.

Conflict of interest

The authors hereby declare the absence of any conflict of interest with other researchers and research entities.

Author details

Hesham F. El-Maghraby[1,2] and Yaser E. Greish[1,2]*

1 Department of Chemistry, UAE University, Al Ain, UAE

2 Department of Ceramics, National Research Centre, Cairo, Egypt

*Address all correspondence to: y.afifi@uaeu.ac.ae

References

[1] Buckwalter JA, Glimcher MJ, Cooper RR, Recker R. Bone biology. I: Structure, blood supply, cells, matrix, and mineralization. Instructional Course Lectures. 1996a;**45**:371-386

[2] Skinner HB. Current Diagnosis and Treatment in Orthopedic, 4th ed, McGraw Hill, 2006

[3] Gao C, Peng S, Feng P, Shuai C. Bone biomaterials and interactions with stem cells Bone Research 2017; 5:17059. DOI:10.1038/boneres.2017.59

[4] Beamer B, Hettrich C, Lane J. Vascular endothelial growth factor: An essential component of angiogenesis and fracture healing. HSSJ. 2010;**6**:85-94. DOI: 10.1007/s11420-009-9129-**4**

[5] Boretos JW, Murray E. Contemporary biomaterials: Material and host response, clinical applications, new technology, and legal aspects. Conference Proceedings, Biomedical materials; Congress, 1984.

[6] Dee KC, Puleo DA, Bizios R. An Introduction To Tissue-Biomaterial Interactions, John Wiley & Sons, Chapter 1: Biomaterials. 2002; 1-2. DOI:10.1002/0471270598

[7] Ginebra M-P, Montufar EB. Cements as bone repair materials. In: Pawelec, KM, Planell, JA (eds) Bone Repair Biomaterials (Second Edition). Cambridge, UK: Woodhead Publishing, 2019;233-271. https://doi.org/10.1016/B978-0-08-102451-5.00009-3

[8] Woolf JI, Walker AE. Cranioplasty: Collective review. Int Abs Surg. 1945;81:1-23.

[9] Webb JCJ, Spencer RF. The role of polymethylmethacrylate bone cement in modern orthopaedic surgery. Journal of Bone and Joint Surgery. 2007;**89-B**:851-857. DOI: https://doi.org/10.1302/0301-620X.89B7.19148

[10] Graça MPF, Gavinho SR. Calcium phosphate cements in tissue engineering. IntechOpen, 2020;1-27. DOI: 10.5772/intechopen.89131.

[11] Thomas MV, Puleo DA. Calcium sulfate: Properties and clinical applications-review. J Biomed Mater Res Part B: Appl Biomater. 2009; 88B: 597-610. DOI: 10.1002/jbm.b.31269.

[12] Nyan M, Sato D, Oda M, Machida T, Kobayashi H, Nakamura T, et al. Bone formation with the combination of simvastatin and calcium sulfate in critical-sized rat calvarial defect. Journal of Pharmacological Sciences. 2007;**104**:384-386. DOI: 10.1254/jphs.SC0070184

[13] Anusavice KJ. Gypsum products. In: Anusavice KJ, editor. Phillips' Science of Dental Materials. St. Louis, MO: Saunders; 2003; 255p.

[14] Carreno-Marquez IJA, Menendez-Me´ndez E, Esparza-Ponce HE, Fuentes-Cobas L, García-Roves R,´Castillo-Sandoval I, Luna-Porres M, Frutos-Vaquerizo J, Montero-Cabrera ME. Naica's Giant Crystals: Deterioration Scenarios. Cryst. Growth Des. 2018; 18: 4611-4620

[15] Arikan M, Sobolev K. The optimization of gypsum based composite material. Cement and Concrete Research. 2002;**32**:1725-1728. DOI: 10.1016/S0008-8846(02)00858-X

[16] Karni J, Karni E. Gypsium in construction: Origin and properties. Materials and Structures. 1995;**28**:92-100

[17] Melo LGN, Nagata MJH, Basco AF, Ribeiro LLG, Leite CM. Bone Healing and Surgically Created

Defect treated with Either Bioactive Glass Particles, a Calcium Sulfate Barrier, or a Combination of both materials. Clin, Oral. Impl. Res.2005; 16: 683-691. https://doi.org/10.1111/j.1600-0501.2005.01090.x

[18] Pagageorgiou A, Tzouvalas G, Tsimas S. Use of inorganic setting retarders in cement, industry. Cem. Concr. Res. 2005;27:183-189. 10.1016/j.cemconcomp.2004.02.005

[19] Peter CP, Hines JL, Bachus KN, Craig MA, Bloebaum RD. Biological effect of calcium sulfate as bone graft substitute in ovine metaphyseal defects. Journal of Biomedical Materials Research. Part A. 2005;**76**(3):456-462. DOI: https://doi.org/10.1002/jbm.a.30569

[20] Sievert T, Wolter W, Singh NB. Hydration of anhydrite of gypsum (CaSO4.II) in a ball mill. Cem. Concr. Res. 2005;**35**:623-630. DOI: 10.1016/j.cemconres.2004.02.010

[21] Orsini G, Ricci J, Scarano A, Pecora G, Petrone G, Lezzi G. Bone-defect healing with calcium-sulfate particles and cement: An experimental study in rabbit. J. Biomed. Mater. Res. Part B: Appl. Biomater. 2004;**68B**:199-208. DOI: https://doi.org/10.1002/jbm.b.20012

[22] Maragos P, Bissada NF, Wang R, Cole RP. Comparison of three methods using calcium sulfate as a graft/barrier material for the treatment of class II mandibular molar furcation defects. Int. J. Periodont. Restor. Dent. 2002;**22**:493-501

[23] Ricci JL, Weiner MJ, Iorio DD, Mamidwar S, Alexander H. Evaluation of timed release calcium sulfate (CS-TR) bone graft substitutes. Microsc microanal. 2005;11(2):1256-1257. DOI. DOI: https://doi.org/10.1017/S1431927605506585

[24] Peltier LF. The use of plaster of Paris to fill defects in bone. Clinical Orthopaedics. 1961;**21**:1-31

[25] Kelly CM, Wilkins RM, Gitelis S, Hartjen C, Watson JT, Kim PT. The use of a surgical grade calcium sulfate as a bone graft substitute: Results of a multicenter trial. Clinical Orthopaedics and Related Research. 2001;**382**:42-50. DOI: 10.1097/00003086-200101000-00008

[26] Borrelli JJ, Prickett WD, Ricci WM. Treatment of nonunions and osseous defects with bone graft and calcium sulfate. Clinical Orthopaedics and Related Research. 2003;**411**:245-254. DOI: 10.1097/01.blo.0000069893.31220.6f

[27] Chen WJ, Tsai TT, Chen LH, Niu CC, Lai PL, Fu TS. The fusion rate of calcium sulfate with local autograft bone compared with autologous iliac bone graft for instrumented short-segment spinal fusion. Spine. 2005;**30**(20):2293-2297. DOI: 10.1097/01.brs.0000182087.35335.05

[28] Zhu T, Ren H, Li A, Liu B, Cui C, Dong Y, et al. Novel bioactive glass based injectable bone cement with improved osteoinductivity and its in vivo evaluation. Scientific Reports. 2017;**7**:3622. DOI: https://doi.org/10.1038/s41598-017-03207-9

[29] Rohmiller MT, Schwalm D, Glattes C, Elalayli TG, Spengler DM. Evaluation of calcium sulfate paste for augmentation of lumbar pedicle screw pullout strength. The Spine Journal. 2002;2:255-260. https://doi.org/10.1016/S1529-9430(02)00207-3

[30] Lazary A'r, Balla B. Ja'nos P. Ko'sa, Krisztia'n Ba'csi, Zsolt Nagy, Istva'n Taka'cs, Pe'ter P. Varga, Ga'bor Speer, Pe'ter Lakatos. Effect of gypsum on proliferation and differentiation of MC3T3-E1 mouse osteoblastic cells. Biomaterials. 2007;**28**:393-399

[31] Sanad ME, Combe EC, Grant AA. The use of additives to improve the mechanical properties of gypsum products. Journal of Dental Research. 1982;**61**:808-810. DOI: https://doi.org/10.1177/00220345820610063201

[32] Olak AC. "Characteristics of acrylic latex-modified and partially epoxyimpregnated gypsum", Cement and Concrete Research 2001;31:1539-1547. https://doi.org/10.1016/S0008-8846(01)00575-0

[33] Eve S, Gomina M, Hamel J, Orange G. Investigation of the setting of polyamid fiber/latex filled plaster composites. Journal of the European Ceramic Society. 2006;**26**:2541-2546. DOI: https://doi.org/10.1016/j.jeurceramsoc.2005.07.063

[34] Olak AC. Physical and mechanical properties of polymer–plaster composites. Materials Letters. 2006;**60**:1977-1982. DOI: https://doi.org/10.1016/j.matlet.2005.12.062

[35] El-Maghraby HF, Gedeon O, Khalil AA. Formation and characterization of poly(vinyl alcohol-co-vinyl acetate-co-itaconic acid)/plaster composites, part II: Composite formation and characteristics. Ceramics-Silikaty. 2007;**51**(3):168-172

[36] El-Maghraby HF, Gedeon O, Rohanova D, Greish YE. Compressive strength and preliminary in vitro evaluation of gypsum and gypsum-polymer compostes in protein-free SBF at 37°C. ceramics International. 2010;36:1561-1569. https://doi.org/10.1016/j.ceramint.2010.02.023

[37] La Gatta A, De Rosa A, Laurienzo P, Malinconico M, De Rosa M, Schiraldi C. A novel injectable poly(e-caprolactone)/calcium sulfate system for bone regeneration: Synthesis and characterization. Macromolecular Bioscience. 2005;**5**:1108-1117. DOI: https://doi.org/10.1002/mabi.200500114

[38] Lewis KN, Thomas MV, Puleo DA. Mechanical and degradation behavior of polymer-calcium sulfate composites.J Mater Sci Mater Med 2006;17:531-537. https://doi.org/10.1007/s10856-006-8936-0

[39] Urban RM, Turner TM, Hall DJ, Infanger SI, Cheema N, Lim TH, et al. An injectable calcium sulfate-based bone graft putty using hydroxypropylmethylcellulose as the plasticizer. Orthopedics. 2004;**27**(1):s155-s159. DOI: https://doi.org/10.3928/0147-7447-20040102-16

[40] Gao C, Huo S, Li X, You X, Zhang Y, Gao J. Characteristics of calcium sulfate/gelatin composite biomaterials for bone repair. Journal of Biomaterials Science. Polymer Edition. 2007;**18**:799-824. DOI: https://doi.org/10.1163/156856207781367710

[41] Mamidwar SS, Arena C, Kelly S, Alexander H, Ricci J. In vitro characterization of a calcium sulfate/PLLA composite for use as a bone graft material. Journal of Biomedical Materials Research. Part B, Applied Biomaterials. 2007;**81**:57-65. DOI: https://doi.org/10.1002/jbm.b.30636

[42] Mamidwar S, Weiner M, Alexander H, Ricci J. In vivo bone response to calcium sulfate/poly L-lactic acid composite. Implant Dentistry. 2008;**17**:208-216. DOI: 10.1097/ID.0b013e3181776b7d

[43] Randolph DA, Nagri JL, Devine TR, Gitelis S. Controlled Dissolution Pellet Containing Calcium Sulfate. USA 5,614,206.

[44] Randolph DA, Nagri JL, Devine TR, Gitelis S. Calcium Sulfate Controlled Release Matrix. USA 6,030,636.

[45] Randolph DA, Nagri JL, Devine TR, Gitelis S. Calcium Sulfate Controlled Release Matrix. USA 5,807,567.

[46] Hanker JS, Terry BC, Ambrose WW, Lupton CR. Plaster of Paris as A bioresorbable Scaffold in Implants for Bone Repair. USA 4,619,655.

[47] Doadrio JC, Arcos D. Cabanas MV. Vallet-Regi M. calcium sulphate-based containing cephalexin. Biomaterials. 2004;**25**:2629-2635. DOI: https://doi.org/10.1016/j.biomaterials.2003.09.037

[48] Sato S, Koshino T, SaitoT. Osteogenic response of rabbit tibia to hydroxyapatite particle–plaster of Paris mixture. Biomaterials 1998;19:1895-1900. https://doi.org/10.1016/S0142-9612(98)00102-1

[49] Kuo ST, Wu HW, Tuan WH, Tsai YY, Wang SF, Sakka Y. Porous calcium sulfate ceramics with tunable degradation rate. Journal of Materials Science: Materials in Medicine. 2012;**23**:2437-2443. DOI: https://doi.org/10.1007/s10856-012-4704-5

[50] Randy Roberts, Stephen J. McConoughey, Jason H. Calhoun. Size and composition of synthetic calcium sulfate beads influence dissolution and elution rates in vitro. J. Biomed Mater Res Part B: Appl Biomater, 102B: 667-673, 2014

[51] Uddin MH, Matsumoto T, Okazaki M, Nakahira A, Sohmura T. Biomimetic Fabrication of Apatite Related Biomaterials, Biomimetics Learning from Nature, Amitava Mukherjee, IntechOpen, (March 1st 2010). DOI: 10.5772/8777

[52] Kokubo T, Kushitani H, Sakka S. Kitsugi T, Yamamuro T. Solutions able to reproduce in vivo surface structure changes in bioactive glass-ceramic A-W. Journal of Biomedical Materials Research 1990; 24:721-734. https://doi.org/10.1002/jbm.820240607

[53] Greish Y, Attia N, Mourad AH. Formation and biomimetic deposition of bone-like hydroxyapatite on wollastonite-gypsum composites. Journal of Bionic Engineering. 2018;**15**:894-906. DOI: https://doi.org/10.1007/s42235-018-0076-y

[54] Hesaraki S, Moztarzadeh F, Nemati R, Nezafati N. Preparation and characterization of calcium sulfate–biomimetic apatite nanocomposites for controlled release of antibiotics. J Biomed Mater Res Part B: Appl Biomater. 2009;**91B**:651-661. DOI: https://doi.org/10.1002/jbm.b.31441

[55] Greish YE, Brown PW, Bender JD, Allcock HR, Lakshmi S, Laurencin CT. Hydroxyapatite-polyphosphazene composites prepared at low temperatures. Journal of the American Ceramic Society. 2007;**90**:2728-2734. DOI: https://doi.org/10.1111/j.1551-2916.2007.01780.x

[56] Lemaitre J, Munting E, Mirtchi A. Setting, hardening, and resorption of calcium phosphate hydraulic cements. STOMOTO-91 Proceedings, COM. 1991:1-5p.

[57] Ferna'ndez E, Vlad M, Gel M, Lo'pez J, Torres R, Cauich J, Bohner M. Modulation of porosity in apatitic cements by the use of α-tricalcium phosphate—calcium sulphate dihydrate mixtures. Biomaterials. 2005;26:3395-3404. https://doi.org/10.1016/j.biomaterials.2004.09.023

[58] Nilsson M, Fernandez E, Sarda S, Lidgren L, Planell JA. Characterization of a novel calcium phosphate/sulphate bone cement. Journal of Biomedical Materials Research. 2002;**61**:600-607. DOI: https://doi.org/10.1002/jbm.10268

[59] Podaropoulos L, Veis AA, Papadimitriou S, Alexandridis C, Kalyvas D. Bone regeneration using b-tricalcium phosphate in a calcium sulfate matrix. Journal of Oral Implantology. 2009;**35**(1):28-36. DOI: https://doi.org/10.1563/1548-1336-35.1.28

[60] Urban RM, Turner TM, Hall DJ, Inoue N, Gitelis S. Increased bone formation using calcium sulfate-calcium phosphate composite graft. Clinical Orthopaedics and Related Research. 2007;**459**:110-117. DOI: 10.1097/BLO.0b013e318059b902

[61] Cabanas MV, Rodriguez-Lorenzo LM, Vallet-Regi M. Setting behaviour and in vitro bioactivity of hydroxyapatite/calcium sulfate cements. Chemistry of Materials. 2002;**14**:3550-3555. DOI: https://doi.org/10.1021/cm021121w

[62] Guo H, Wei J, Liu CS. Development of a degradable cement of calcium phosphate and calcium sulfate composite for bone reconstruction. Biomedical Materials. 2006;**1**:193-197. DOI: https://doi.org/10.1088/1748-6041/1/4/003

[63] Chiang YC, Chang HH, Wong CC, Wang YP, Wang YL, Huang WH, et al. Nanocrystalline calcium sulfate/hydroxyapatitebiphasic compound as a TGF-1/VEGF reservoir forvital pulp therapy. Dental Mater. 2016;**32**:1197-1208. DOI: https://doi.org/10.1016/j.dental.2016.06.013

[64] Greish YE. Phase evolution during the low temperature formation of stoichiometric hydroxyapatite-gypsum composites. Ceramics International. 2011;**37**:715-723. DOI: https://doi.org/10.1016/j.ceramint.2010.09.046

[65] Greish YE. Phase evolution during the low temperature formation of calcium-deficient hydroxyapatite–gypsum composites. Ceramics International. 2011;**37**:1493-1500. DOI: https://doi.org/10.1016/j.ceramint.2011.01.007

[66] Chang H, Xiang H, Yao Z, Yang S, Tu M, Zhang X, et al. Strontium-substituted calcium sulfate hemihydrate/hydroxyapatite scaffold enhances bone regeneration by recruiting bone mesenchymal stromal cells. Journal of Biomaterials Applications. 2020;**35**(1):97-107. DOI: https://doi.org/10.1177/0885328220915816

[67] Huan Z, Chang J. Self-setting properties and in vitro bioactivity of calcium sulfate hemihydrate-tricalcium silicate composite bone cements. Acta Biomaterialia. 2007;**3**:952-960. DOI: https://doi.org/10.1016/j.actbio.2007.05.003

[68] Ji M, Chen H, Yan Y, Ding Z, Ren H, Zhong Y. Effects of tricalcium silicate/sodium alginate/calcium sulfate hemihydrate composite cements on osteogenic performances in vitro and in vivo. Journal of Biomaterials Applications. 2020;**34**(10):1422-1436. DOI: https://doi.org/10.1177/0885328220907784

[69] Pei P, Wei D, Zhu M, Du X, Zhu Y. The effect of calcium sulfate incorporation on physiochemical and biological properties of 3D-printed mesoporous calcium silicate cement scaffolds. Microporous and Mesoporous Materials. 2017;**241**:11-20. DOI: https://doi.org/10.1016/j.micromeso.2016.11.031

[70] Qi X, Pei P, Zhu M, Du X, Xin C, Zhao S, et al. Three dimensional printing of calcium sulfate and mesoporous bioactive glass scaffolds for improving bone regeneration in vitro and in vivo. Scientific Reports. 2017;**7**:42556. DOI: 10.1038/srep42556

Chapter 14

Nanoengineered Polysaccharide-Based Adsorbents as Green Alternatives for Dye Removal from Wastewater

Hugues Kamdem Paumo, Lebogang Katata-Seru, Tshepiso Moremedi, Mpitloane Joseph Hato, Soumen Sardar and Abhijit Bandyopadhyay

Abstract

Water represents one of the essential resources on earth, and all living organisms rely on it for survival. However, freshwater systems are directly under serious threat by human activities. A recent World Health Organization report has estimated that 2 billion people use contaminated water sources. The major organic contaminants in water bodies include organic dyes. These are directly related to the spread of diseases owing to their allergenic, mutagenic, and carcinogenic characteristics. Dyes occur in the environment through untreated industrial effluents. Also, the advancement in human civilisation cannot be considered without the development of modern industries. However, an attempt to limit the resulting impacts of coloured effluents on global freshwater quality has become the focus of recent research. For this reason, the use of efficient, simple, and low-cost methods for the treatment of dyes-containing industrial wastewater could serve as a useful tool. An effort to meet the demand for "green chemistry" and sustainable development has led to several investigations on polysaccharide-based adsorbents. This chapter exposes a critical discussion on the literature concerning the biopolymeric xanthan gum and *kappa*-carrageenan polysaccharides as adsorbents for dye removal in an aqueous medium. Functionalisation of these biopolymers through graft copolymerisation and inclusion of inorganic nanoparticles for improved adsorption performance is systematically elaborated.

Keywords: xanthan gum, *kappa*-carrageenan, grafted polysaccharides, nanocomposite, dyes, adsorption

1. Introduction

Water is a precious resource indispensable for the sustainability of life on earth. It is vital to human well-being and economic development. However, safe water remains a significant issue around the world [1]. According to a recent report by the World Health Organization (WHO 2019), approximately 785 million people lack clean water for their basic needs. Supplying water to an increasing global population necessitates concurrent consideration of several societal sectors that compete for limited resources [2]. For that reason, a vast majority are dependent on surface water and groundwater for drinking purpose. The WHO report also highlights the fact that more than two million people die every year from the use of unsafe drinking water. This is well-reported to spread sickness and waterborne diseases such as typhoid and cholera. This situation emerges from the occurrence of pollutants as a result of the environmental disposal of untreated effluent released by human activities [3]. Indeed, it is inconceivable to achieve progress in human civilisation without industrialisation. However, its exponential development in a competitive era and an increasing global population has seriously impaired the quality of freshwater systems.

Among an assortment of environmental pollution, waterbody contamination owing to the discharge of untreated water-containing organic species has attracted significant consideration in recent years [4, 5]. Organic dye-containing wastewater from industries such as textiles, petrochemical, cosmetics, papers and plastics, for example, has been described for their carcinogenic and mutagenic nature [6]. Also, the organic dyes are oxygen-sequestering agents capable of reducing light penetration in the water systems and thereby restraining the photosynthesis of aquatic vegetation [7, 8]. **Figure 1** illustrates the fate of organic dyes in the natural environment. Thought the textile industry had played an enormous role in the development of the South African economy [9], this has also significantly impacted the water resources. The dye-containing effluents disposed to the environment without proper treatment can be highly toxic even at a concentration lower than one ppm [10]. Therefore, the removal of toxic organic dyes from contaminated effluents before being discharged into the environment has evoked considerable attention.

From the commercial and environmental viewpoint, the focus of this chapter is to provide a comprehensive discussion on the ability of natural polymers to perform as adsorbents for industrial wastewater remediation. The hybridisation of these

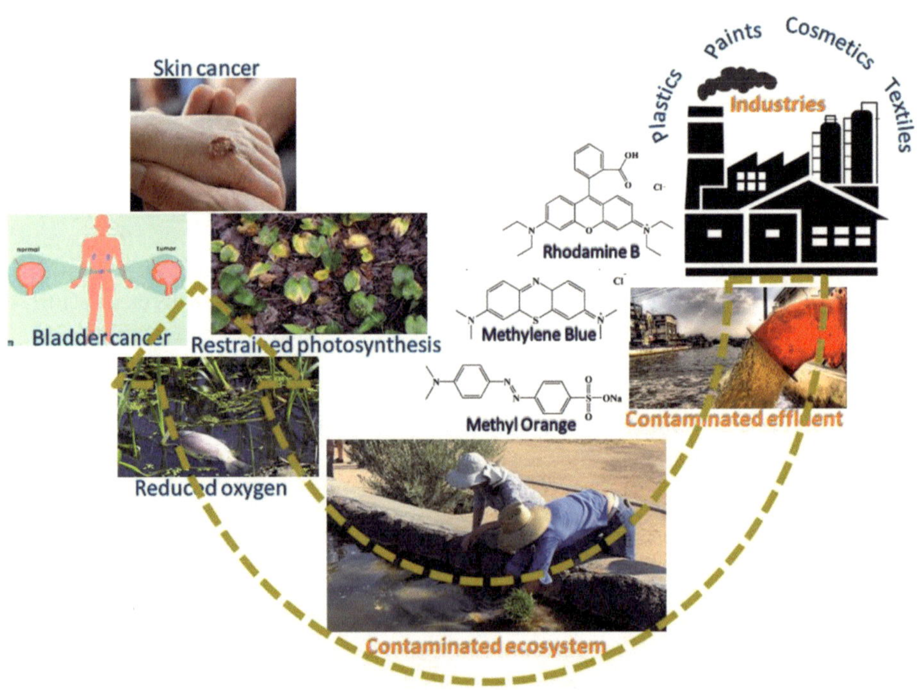

Figure 1.
The fate of the dye-contaminated effluent in the aquatic environment.

biomaterials by grafting of segments and combination with inorganic nanoparticles is presented as an ideal strategy to improve their inherent characteristics and sorption capacity.

2. Treatment of dye-contaminated water in adsorption procedure

Several techniques have been reported to be able of alleviating the occurrence of organic dye in contaminated industrial effluents, and these include flocculation, membrane filtration, biodegradation, chemical degradation, and photochemical treatment [11]. Although these treatment procedures have been applied successfully, they have also displayed shortcomings like high operational costs, low selectivity and toxic sludge production. Adsorption procedure, on the other hand, has frequently been described as a promising approach for wastewater treatment because of the reasons like simplicity in design, low operational cost and comparatively high efficiency [12, 13]. Furthermore, this treatment procedure can be applied on a large scale, as it can handle large flow rates and yields a high quality of water without producing sludge and residuals [14].

Numerous adsorbents are reported in the literature for the removal of contaminants in wastewater, and these include activated carbon, clay minerals, zeolites, industrial wastes, metal oxides, and polymeric materials [15–20]. Among these adsorbents activated carbon is the most widely employed material for the removal of dye in aqueous solution due to its high surface area [21]. However, activation of carbon sources is energy-intensive and requires the use of expensive chemicals and equipments. The low regeneration of spent activated carbon also restricts its practical application. A large majority of the reported adsorbents are also associated with drawbacks like non-biodegradability and low efficiency. These limitations can be overcome using naturally occurring "green" polysaccharide adsorbents owing to their cost-effectiveness, ease availability, and biodegradability.

3. Polysaccharide adsorbents for the removal of organic dye in aqueous solution

Polysaccharides are hydrophilic, non-toxic, and relatively cheap polymers consisting of repeating subunits of sugar linked with glycosidic bonds. The composition and sources of polysaccharides that are generally investigated for their potential to act as adsorbents in aqueous media are presented in **Table 1** [22–25]. The adsorption ability of these biomaterials is due to the occurrence of functionalities such as hydroxyl (-OH), sulphonic acid (-SO_3H), the carboxylic acid (-COOH), amino (-NH_2) and amide (-$CONH_2$) groups which can serve as binding sites [26]. This feature, complemented with the porous nature, make polysaccharides good candidates for water treatment applications. Some of the commercially available polysaccharides include cellulose, starch, guar gum, chitosan, xanthan gum and carrageenan. Among these, eco-friendly xanthan gum and carrageenan (**Figure 2**) with the ability to form gel have been gaining considerable attractions recently [27, 28]. So far, these have been described to be among the most effective adsorbents for the removal of toxic dyes in aqueous solution due to their tunable surface chemistry and feasible regeneration [29, 30].

Among these, eco-friendly xanthan gum and carrageenan (**Figure 2**) with the ability to form gel have been gaining considerable attractions recently [27, 28]. So far, these have been described to be among the most effective adsorbents for the removal of toxic dyes in aqueous solution due to their tunable surface chemistry and feasible regeneration [29, 30].

Name	Composition	Sources
Cellulose	Anhydroglucose units linked by the β-(1 → 4) glycosidic bonds	Fruit, nuts, grains, and vegetables
Chitin	Glycosidically β-(1 → 4)-linked 2-acetamido-2-deoxy-D-glucose	Shrimp shell
Chitosan	Copolymer of *N*-glucosamine and *N*-acetylglucosamine units	Deacetylation of chitin
Alginate	β-(1 → 4)-Linked D-mannuronic acid and α-(1 → 4)-linked L-guluronic acid	Algae, Bacteria
Starch	(1 → 4)-Attached α-D-glucopyranosyl units with α-(1 → 6) branch linkages	Cereal grains, tuber crops
Guar gum	Glycosidically β-(1 → 4)-linked D-mannose subunits and glycosidically (1 → 6)-linked D-galactose subunits	Guar seed
Xanthan gum	(1 → 4)-Attached D-glucose D-mannosyl, D-glucuronyl acid, O-acetyl and pyruvyl residues	*Xanthomonas* bacteria
Carrageenan	D-galactose and 3,6-anhydrogalactose connected by α-(1 → 3)- and β-(1 → 4)-glycosidic linkages	Marine algae

Table 1.
Polysaccharides used as adsorbents for wastewater treatment.

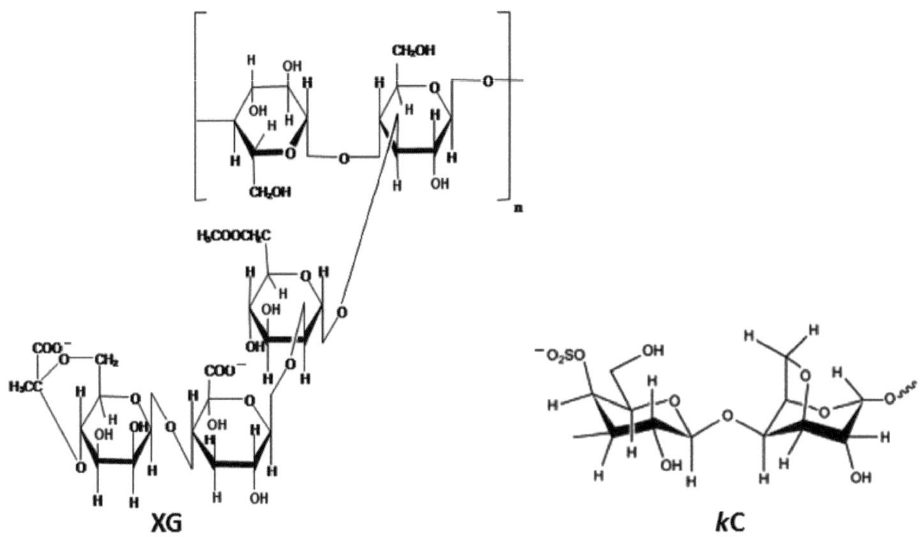

Figure 2.
The molecular structure of xanthan gum (XG) and kappa-*carrageenan (*kC*) polysaccharides.*

Xanthan gum (XG) is an anionic polymer obtained by *Xanthmonous campestris* bacterial fermentation of carbohydrate source. This biopolymer is recovered through precipitation in ethanol, isopropyl alcohol or *tert*-butanol [31]. It is a high molecular weight polysaccharide, which provides suspension, thickness, and stabilisation of the combined material. The negative charge of XG biopolymer is mainly ascribed to the presence of carboxylic acid group within its backbone, thereby indicating its affinity for cationic species. XG main chain is made of repeating cellulosic units with side chains involving an α-D-mannose with an acetyl group, a β-D-glucuronic residues and a terminal β-D mannose bearing the pyruvate substituent [32]. The physicochemical characteristics of XG include higher stability in solution under a wide range of pH (2–12), higher viscosity at low concentration (1% or less), and higher shelf life. The presence of hydroxyl and carboxylic groups that are pH tunable also confers a polyionic character to XG. Thus this adsorbent can be used for binding various ions through electrostatic interaction. Moreover, the –OH groups are also capable of interacting with the electron clouds of the aromatic rings of organic dye molecules through hydrogen-bonding. Thus, adsorption of toxic dye onto the biodegradable polymeric adsorbent can be examined using Fourier transformed infrared spectroscopy (FTIR). For example, the as-described mechanism was reported by Lozano-Álvarez et al. during adsorption of disperse yellow 54 dye in aqueous solution onto XG [33].

Carrageenans are sulfated polysaccharides obtained through alkaline extraction from some red marine algae. Depending on the extraction method and the algae species from which this anionic polysaccharide is obtained, there are several types of carrageenans with different solubility. The polymer chains of carrageenans consist of alternate units of D-galactose and 3,6-anhydrogalactose joined by α-1,3- and β-1,4-glycosidic linkages. The main types are lambda (λ), kappa (k), and iota (i). k-Carrageenan is less soluble than the others owing to the hydrophobic 3.6-anhydro-D-galactose group, which form part of its repeating unit and the relatively lesser number of sulphate groups (one sulfate functional group for each disaccharide unit).

Although these renewable materials show unique properties for water treatment in the adsorption process, they exhibit poor specific surface area and mechanical properties which limits their applicability as lasting adsorbents [34, 35]. In general, the effectiveness of an adsorbent strongly depends on its chemical and mechanical stability, which determines the suitability for application under harsh conditions.

4. Functionalised polysaccharide adsorbents for the efficient removal of organic dye in aqueous solution

The polysaccharide surface modification can improve their physicochemical properties and mechanical characteristics [26]. This strategy has also been reported to avoid leaching of organic substances and improve the adsorption potential of the nanoengineered adsorbent. The polysaccharide surface modification can be accomplished through graft copolymerisation and/or incorporation of specific nanoscale inorganic particles. Deposition of inorganic nanoparticles such as SiO_2, Fe_3O_4, TiO_2, and carbon nanotube onto polymeric supports has been reported to enhance their chemical, mechanical and thermal stabilities [36, 37].

4.1 Polysaccharide functionalised by graft copolymerisation

Graft copolymerisation of vinyl monomers onto polysaccharides is a well-reported and versatile technique that allows for the increased potential applicability of biopolymers. This procedure usually involves the attachment of a vinyl monomer to the polysaccharide backbone. In general, the vinyl monomer will undergo polymerisation in the presence of polysaccharide chains to generate a copolymeric network. The vinyl monomers that are frequently reported for the modification of gum-based polysaccharides involve the acrylamide (AAm), methyl methacrylate (MAA), acrylic acid (AA), *N*-vinyl imidazole (VI), and acrylonitrile (AN). The synthesis of polysaccharide graft copolymers is achieved by amending the biopolymer molecules *via* formation of branches of synthetic polymers. This can be accomplished through "grafting onto" or "grafting from" methodology [38]. The "grafting from" process entails the development of polymer chains from initiating sites onto the polysaccharide framework. In the "grafting onto" method, on the other hand, the pre-formed polymer bearing a reactive end-functionality interacts with the functional groups that are located on the polysaccharide backbone. However, the latter presents inherent shortcomings, including the crowding of chains at the polysaccharide surface and a limited number of attachments [39].

Many techniques have been employed for the attachment of monomers onto polysaccharide surface, and these involve free radical graft copolymerisation, living radical polymerisation, and ionic polymerisation. Grafting using free radical approach requires an initiator (chemicals, photoirradiation, plasma ions or gamma rays exposure) to generate a free radical on the polysaccharide backbone. In ionic polymerisation technique, on the other hand, the chemical initiator generates cationic or anionic active centers which participate in the grafting process. Nevertheless, more than 60% of all the reported polymers are still synthesised by free radical polymerisation method. It is worth noting that the solubility, wettability, glass transition temperature, and elasticity of polysaccharides are tailored through grafting with synthetic monomer. For example, grafting of chitosan with vinyl monomer *N*-acryloylglycine in the presence of 2,2-dimethoxy-2-phenyl acetophenone initiator was reported to yield a material with relatively decreased solubility and wettability profiles [40]. The polysaccharide affinity to water limits its adsorption property owing to competitive phenomenon between the water molecules and pollutants in the aquatic milieu. Characterisation of the graft copolymerised polysaccharides can be ascertained using FTIR spectra analysis. Data obtained from this technique indicate the formation of covalent bonds. Considering the involvement of hydroxyl groups during graft copolymerisation, a shift and change in intensity of the band corresponding to O–H vibration can be evidenced.

In a recent study by our research team, the surface-modified XG polysaccharide obtained through grafting with acrylamide and acrylic acid monomers to afford XG grafted poly(AAm/AA) [XG-g-P(AAm/AA)], was used as an adsorbent for the removal of rhodamine B and methylene blue (MB) in aqueous solution. The UV irradiation method in the presence of benzophenone as initiator was employed to effect the attachment of monomers. UV irradiation approach is of interest because of its mild reaction conditions and less adverse effect to change bulk properties [41]. Moreover, lower radiation energy may be applied for the modification to proceed. The structural change from XG to XG-g-P(AAm/AA) was elucidated using FTIR spectroscopy (**Figure 3**).

The spectra of pristine XG showed absorption bands at 3263, 2915, 1712, 1653, 1416 and 1019 cm^{-1} attributed to O–H stretching vibration, C–H stretching vibration, C–O stretching, O–H bending, the symmetrical stretching of –CCO– group of glucuronic acid and C–O–C of the ether group, respectively [42, 43]. The formation of XG-g-P(AAm/AA) was evidenced with the change in intensity of the band around 3263 cm^{-1}, the strong vibrational bands at 1593, and the shift of band at 1416 cm^{-1}. After dye adsorption, the FTIR spectrum of loaded XG-g-P(AAm/AA) was characterised by a shift in absorption bands, indicating a dye–adsorbent interaction. However, the peak of 2895 cm^{-1} attributed to C–H stretching did not experience a significant shift, suggesting that the dye adsorption took place through electrostatic and hydrogen-bonding interactions with the XG-g-P(AAm/AA) polysaccharide (**Figure 4**). Electrostatic interactions presumably occur between the nucleophilic functional groups ($-COO^-$) and the positively charged centres of the organic dye molecules (S and N). Also, hydrogen-bonding interactions are expected to take place between the hydroxyl groups of XG and aromatic π-electrons and/or nitrogen lone pair of electrons of the dye molecules.

The adsorption kinetics investigation for the removal of rhodamine B and methylene blue dyes in synthetic solutions using our prepared XG-g-P(AAm/AA)

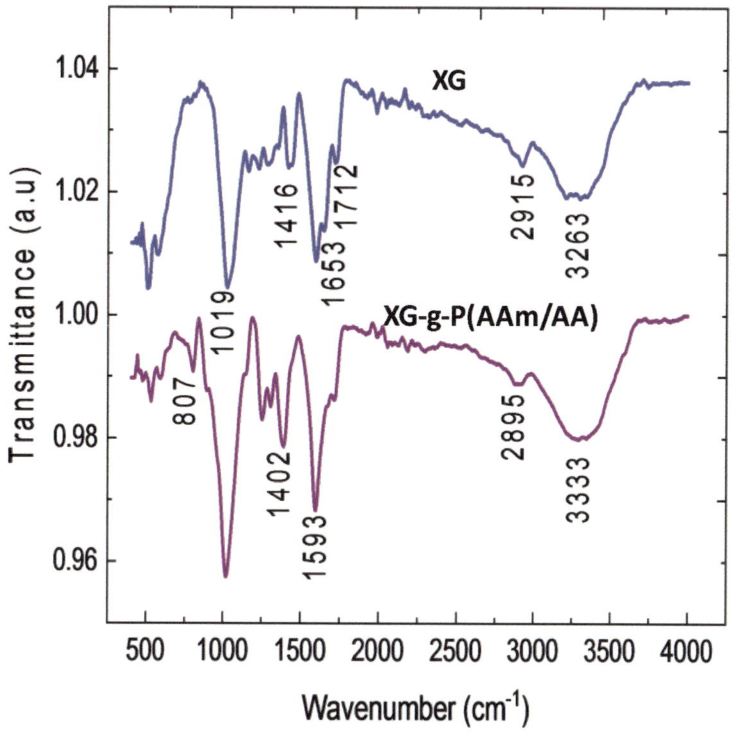

Figure 3.
FTIR spectra of XG and XG-g-P(AAm/AA).

material demonstrated that the procedure followed a pseudo-second-order model. On the other hand, the isothermal study indicated multilayer adsorption behaviour. In a different study, Elella et al. reported the synthesis of XG-g-poly(*N*-vinyl imidazole) copolymers using *N,N′*-methylene bisacrylamide as cross-linker [44]. This modified XG polysaccharide was obtained by a free radical technique using potassium persulfate initiator and then evaluated as an adsorbent for the removal of crystal violet dye from synthetic water samples. The maximum dye uptake onto this cross-linked grafted XG (0.04 g) was determined to be 625 mg/g in 50 ml crystal violet dye solution (500 mg/l, pH 7).

Interestingly, the analogous XG-g-poly(*N*-vinyl imidazole) derivatives without linker have also been found to exhibit antibacterial activity against *Staphyloccus aureus* and *Escherichiacoli* [45].

4.2 Polysaccharide functionalised by incorporation of inorganic nanoparticles

Incorporation of inorganic NPs with the higher surface area has also been described as a fascinating strategy for improved adsorption procedure. This results from a strong synergistic outcome between the organic polysaccharide matrixes and embedded inorganic NPs. This methodology also affects the modification in mechanical properties of the hybrid adsorbent. Moreover, the polysaccharide moiety is anticipated to stabilise the nanoparticles and prevent aggregation. Metal oxide nanoparticles having desirable attributes as adsorbents often suffer tendency to agglomerate in aqueous solution due to their higher surface energy, leading to lowered efficiency [46].

XG grafted polyacrylamide XG-g-PAAm incorporated with nanosilica (SiO_2 NPs), for example, was obtained *in situ* through hydrolysis and condensation of

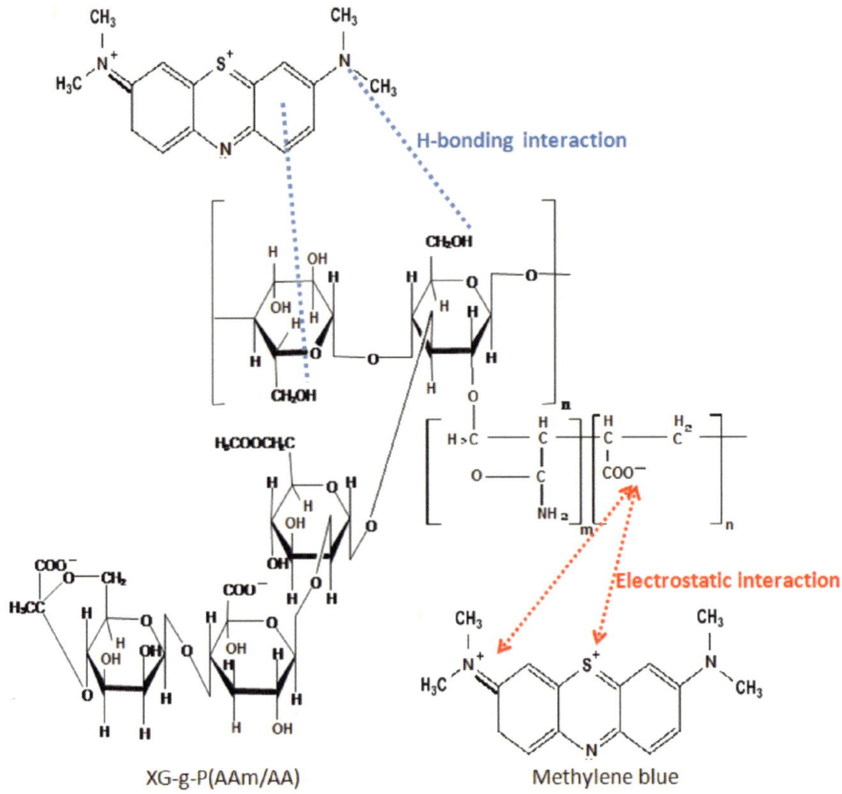

Figure 4.
The mechanism for dye adsorption onto XG-g-P(AAm/AA).

tetraethylorthosilicate [$Si(OC_2H_5)_4$] in the presence of ammonia at the copolymer surface [10]. This synthetic approach exploits the occurrence of hydrophilic groups on the polysaccharide backbone to control the hydrolysis, condensation and nucleation growth of metal oxide nanoparticles [47]. The functionalised SiO_2 NPs@XG-g-PAAm composite was described as a highly improved adsorbent for the removal of anionic Congo red dye in aqueous solution. Grafting of polyacrylamide chain on XG was accomplished *via* free radical polymerisation technique in the presence of potassium persulphate initiator. The branched XG gum matrix acted as a template for the formation, growth, and stabilisation of SiO_2 NPs through hydrogen bonding interaction between the –OH bonded surface group of SiO_2 NPs with oxygen atom of the polysaccharide –COOH surface group [48].

Decoration of XG-g-P(AAm/AA) with SiO_2 NPs (diameters range 2–7 nm) was also reported by Ghorai and coworkers [34]. These authors subjected the XG-g-PAAm material to hydrolysis in the presence of NaOH to generate the graft copolymer of XG [XG-g-P(AAm/AA)] (**Figure 5**). Subsequently, functionalisation of the later with SiO_2 NPs was achieved *in situ* through hydrolysis and condensation of silica sol–gel $Si(OC_2H_5)_4$ precursor (**Figure 6**). The resultant nanocomposite was employed as an adsorbent for the removal of cationic dyes in the aqueous milieu. The adsorbent exhibited good adsorption efficiency of 497.5 mg/g and 378.8 mg/g towards MB at pH 8 (adsorbent dose: 0.03 g/25 mL solution, contact time: 20 min, and temperature: 50°C) and methyl violet at pH 9 (adsorbent dose: 0.04 g/25 mL solution, contact time: 15 min, and temperature: 40°C), respectively.

*k*C modified with carbon nanotubes (10–20 nm diameter), and Fe_3O_4 (10–25 nm) was also synthesised following polymer grafting on the surface of multiwall carbon

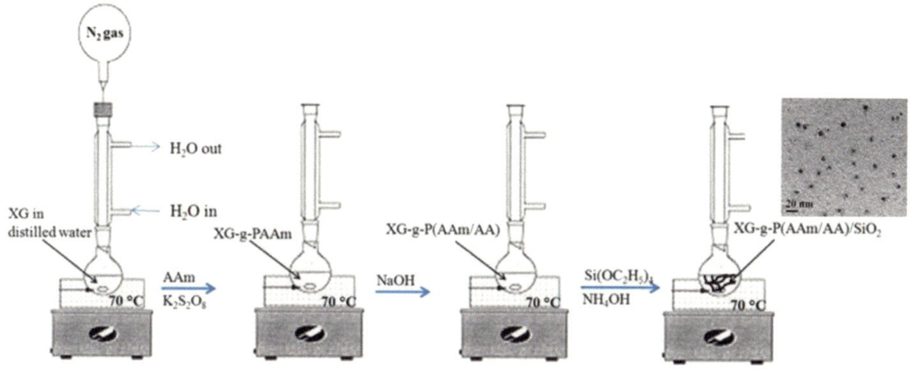

Figure 5.
Fabrication of SiO_2 NPs@XG-g-P(AAm/AA) composite and TEM image.

$$Si(OC_2H_5)_4 + 4H_2O \longrightarrow Si(OH)_4 + 4C_2H_5OH$$

$$Si(OH)_4 \longrightarrow SiO_2 + 2H_2O$$

Figure 6.
Hydrolysis of $Si(OC_2H_5)_4$ and condensation of the silicic acid intermediate.

nanotubes and suspension in a solution of iron precursors Fe(III)/Fe(II) under nitrogen atmosphere [49]. The Brunauer–Emmett–Teller (BET) technique indicated an increase in specific surface area after attachment of carbon nanotubes (S_{BET}: 238 m^2/g) and deposition of magnetic Fe_3O_4 nanoparticles (S_{BET}: 55 m^2/g). The occurrence of this metal oxide was evidenced by transmission electron microscopy (TEM) pictures analysis. Magnetism is an exclusive physical property that has been demonstrated to ease the water treatment procedure in adsorption technique. This property allows for the removal of spent adsorbent from the aqueous solution by a simple application of a magnet. The prepared magnetic *k*C-carbon nanotubes-Fe_3O_4 composite exhibited high adsorption toward MB dye in aqueous solution. The maximum dye uptake onto 0.4 g/L of this biofunctional nanocomposite was determined to be 1.24×10^{-4} mol/g^{-1} at pH 6.5. In aqueous solution, the adsorption potential of metal oxide (MOH) is regulated by the pH-dependent formation of complex ions MOH_2^+ and MO^- [12, 50, 51]. Therefore, dye adsorption onto metal oxide surfaces arise from the Coulombic interaction between MOH_2^+ and MO^-, and the negatively charged (e.g. methyl orange $R–SO_3^-$) and positively charged (e.g. methylene blue $R = N(CH_3)^+$) centers, respectively. In a cognate investigation, Mahdavinia et al. described the modification of *k*C with polyvinyl alcohol (PVA) and Fe_3O_4 nanoparticles [52]. This nanocomposite was engineered using the *in-situ* chemical co-precipitation of Fe(II) and Fe(III) salts in the presence of PVA and ionic *k*C biopolymer under basic condition (**Figure** 7). Cross-linking of the occurring Fe_3O_4 nanoparticles and polymer matrixes was achieved by the freezing–thawing technique and K^+/*k*-carrageenan–SO_3^- interaction. Freezing–thawing plays a crucial role in dispersing nanoparticles into gelatin environment and limiting their agglomeration. This technique has been reported to afford bio-related materials with better thermal stability and mechanical properties [53]. The magnetic behaviour of this cationic dye adsorbent was assessed with the vibrating sample magnetometer (VSM) standard method. The hysteresis loop revealed an S-shape with an estimated value of saturation magnetisation 3.4 emu/g, suggesting that the spent adsorbent can be removed easily with an external magnet.

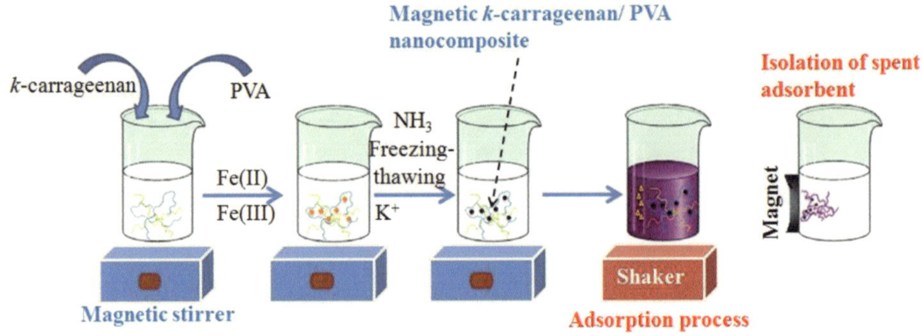

Figure 7.
Fabrication of magnetic kC/PVA *nanocomposite as an adsorbent for cationic dye removal in aqueous solution.*

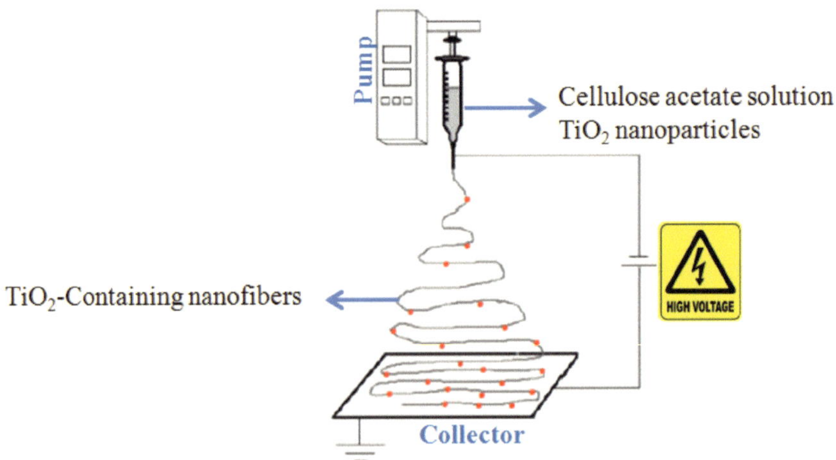

Figure 8.
Experimental setup for electrospinning of TiO_2 nanoparticles in polysaccharide solution.

FTIR, X-ray photoelectron spectroscopy (XPS) spectra analysis and assessment of the pH at the point of zero charge (pH_{pzc}) can be used to clarify the adsorption mechanism at the surface of metal oxide in aqueous milieu,. Thus, this additional interaction is anticipated to improve the adsorption capacity of polysaccharide adsorbents functionalised with metal oxides. The –OH groups of polysaccharides are highly reactive in encouraging polycondensation, or in interacting with cations or hydroxylated cations, capable of undergoing nucleation and growth processes. Polysaccharide matrix also provides a cavity that is capable of immobilising the developing inorganic entities and controlling their growth. Furthermore, carbon nanotubes ability to abstract organic dyes from wastewater has been reported to take place through π–π stacking, hydrogen bonding, hydrophobic, Coulombic, and/or van der Waals interactions [54–56]. The occurrence of defects and active centers, and the morphology of the carbon nanotubes play a key role in their dye adsorption capability. The carbon nanotubes are rolled-up graphene or graphitic sheets of single-layer carbon atoms. These are π-conjugative structures possessing a hydrophobic surface. The dynamic mechanical investigation of polymer adsorbents and their inorganic particle-functionalised derivatives has uncovered relatively better properties of the latter through restriction of the mobility of polymer macromolecular chains [57].

Cai and coworkers also described the cellulose nanofibers (average diameter 237–443 nm) modified with TiO_2 nanoparticles as a precursor for the synthesis of composite hydrogels [58]. Graft copolymerisation of these nanofibers using AAm, AA, and N,N′-methylene bisacrylamide in the presence of ammonium persulfate afforded the polysaccharide-based hydrogels with good MB dye adsorption capacity. The nanofibers material was fabricated through electrospinning of cellulose acetate solution containing TiO_2 nanoparticles of average diameter 25 nm, followed by deacetylation under basic condition. The voltage power and flow rate applied in this investigation were 20 kV and 1 mL/h, respectively. Electrospinning is a versatile and efficient method for the fabrication of nanofibers. This technique utilises high voltage to charge the surface of a polymer solution and initiates the ejection of fluid jets through a small hole (**Figure 8**). Solidification of these thin jets yields nanofibers. The size of nanofibers obtained using this procedure depends on parameters like solution physical properties, voltage, hydrostatic pressure, size of a hole, and distance hole-collector [59]. The light-driven catalytic activity of TiO_2 has also been described to improve the MB removal performance of TiO_2-containing cellulose nanofibers. TiO_2, also known as titania, is a low-cost and environmentally benign oxide that has gained commercial success in beauty, cosmetic, and personal care applications. Moreover, interest in TiO_2 nanoparticles for the degradation of organic pollutants in wastewater has been tremendous since the early report by Frank and Bard [60]. This is attributed to their unique electronic structure, impressive UV-light absorption properties, prolonged excited-state lifetimes and enhanced charge transport features. The high-energy photons emitted by UV-light initiates the excitation of TiO_2 valence band electrons and formation of $^{\bullet}OH/O_2^{\bullet-}$ radicals. These species are responsible for the degradation of organic pollutants in aqueous solution.

Composite hydrogel of NH_2–silica functionalised TiO_2 NPs@*k*C-g-PAA was developed by Pourjavadi et al. for the removal of malachite green (MG) in synthetic water samples [61]. The functionalised TiO_2 NPs were synthesised by the hydrolysis of $TiCl_4$ at 90°C in the presence of HNO_3 followed by the treatment of hydrated TiO_2 NPs with 3-aminopropyltriethoxysilane. Graft copolymerisation of *k*C with AA monomer in the presence of ammonium persulfate initiator, methylenebisacrylamide as a crosslinking agent, and the pre-synthesised NH_2–silica functionalised TiO_2 NPs yielded an adsorbent with impressive potential for the removal of cationic dye. MG adsorption onto NH_2–silica functionalised TiO_2 NPs@*k*C-g-PAA hydrogel followed the pseudo-second-order rate model, and best fitted the Langmuir isotherm, with the maximum capacity of 666 mg/g.

5. Conclusion

Herein, we present a well-elaborated discussion on the developed polysaccharide-based materials for the removal of highly toxic organic dyes from contaminated water using adsorption procedure. Naturally occurring, non-toxic, and biodegradable xanthan gum and kappa-carrageenan matrixes were used as representatives for effective dye remediation owing to their surface charged functionalities that serve as active binding sites. The polysaccharide surface modification through graft copolymerisation with monomers and/or incorporation of nano-sized inorganic particles having high surface areas like metal oxides and carbon nanotubes has been found to yield composites with improved mechanical stability and impressive adsorption capacities.

Author details

Hugues Kamdem Paumo[1], Lebogang Katata-Seru[1*], Tshepiso Moremedi[1], Mpitloane Joseph Hato[2,3], Soumen Sardar[4] and Abhijit Bandyopadhyay[4]

1 Department of Chemistry, School of Physical and Chemical Sciences, Faculty of Natural and Agricultural Sciences, North-West University, Private Bag X2046, Mmabatho, South Africa

2 Department of Chemistry, School of Physical and Mineral Sciences, Faculty of Sciences and Agriculture, University of Limpopo (Turfloop), South Africa

3 Department of Environmental Sciences, College of Agriculture and Environmental Sciences, University of South Africa (UNISA), Florida Science Campus, Johannesburg, South Africa

4 Department of Polymer Science and Technology, University of Calcutta 92, Kolkata, India

*Address all correspondence to: lebo.seru@nwu.ac.za

References

[1] Boelee E, Geerling G, der Zaan B, Blauw A, Vethaak AD. Water and health: From environmental pressures to integrated responses. Acta Tropica. 2019;**193**:217-226. DOI: 10.1016/j.actatropica.2019.03.011

[2] World Health Organization. fact-sheets, detail, drinking-water. 2019

[3] Santhi T, Manonmani S, Vasantha VS, Chang YT. A new alternative adsorbent for the removal of cationic dyes from aqueous solution. Arabian Journal of Chemistry. 2016;**9**(Supplement 1):S466-S474. DOI: 10.1016/j.arabjc.2011.06.004

[4] Zhang H, Chen D, Lv X, Wang Y, Chang H, Li J. Energy-Efficient Photodegradation of Azo Dyes with TiO_2 Nanoparticles Based on Photoisomerization and Alternate UV–Visible Light. Environmental Science & Technology. 2010;**44**:1107-1111. DOI: 10.1021/es9029123

[5] Ai L, Yue H, Jiang J. Sacrificial template-directed synthesis of mesoporous manganese oxide architectures with superior performance for organic dye adsorption. Nanoscale. 2012;**4**:5401-5408. DOI: 10.1039/c2nr31333b

[6] Gupta VK, Kumar R, Nayak A, Saleh TA, Barakat M. Adsorptive removal of dyes from aqueous solution onto carbon nanotubes: a review. Advances in colloid and interface science. 2013;**193**:24-34. DOI: 10.1016/j.cis.2013.03.003

[7] Das R, Sypu VS, Paumo HK, Bhaumik M, Maharaj V, Maity A. Silver decorated magnetic nanocomposite (Fe3O4@PPy-MAA/Ag) as highly active catalyst towards reduction of 4-nitrophenol and toxic organic dyes. Applied Catalysis B: Environmental. 2019;**244**:546-558. DOI: 10.1016/j.apcatb.2018.11.073

[8] Hai FI, Yamamoto K, Fukushi K. Hybrid Treatment Systems for Dye Wastewater. Critical Reviews in Environmental Science and Technology. 2007;**37**:315-377. DOI: 10.1080/10643380601174723

[9] van Zyl G, Matswalela K. A comparative analysis of the level of competitiveness of the South African clothing and textile industry. Journal of Economic and Financial Sciences. 2016;**9**:370-391

[10] Ghorai S, Sarkar AK, Panda AB, Pal S. Effective removal of Congo red dye from aqueous solution using modified xanthan gum/silica hybrid nanocomposite as adsorbent. Bioresource Technology. 2013;**144**:485-491. DOI: 10.1016/j.biortech.2013.06.108

[11] Sarkar AK, Pal A, Ghorai S, Mandre N, Pal S. Efficient removal of malachite green dye using biodegradable graft copolymer derived from amylopectin and poly (acrylic acid). Carbohydrate polymers. 2014;**111**:108-115. DOI: 10.1016/j.carbpol.2014.04.042

[12] Chigondo M, Paumo HK, Bhaumik M, Pillay K, Maity A. Hydrous CeO_2-Fe_3O_4 decorated polyaniline fibers nanocomposite for effective defluoridation of drinking water. Journal of Colloid and Interface Science. 2018;**532**:500-516. DOI: 10.1016/j.jcis.2018.07.134

[13] Chigondo M, Paumo HK, Bhaumik M, Pillay K, Maity A. Magnetic arginine-functionalized polypyrrole with improved and selective chromium(VI) ions removal from water. Journal of Molecular Liquids. 2019;**275**:778-791. DOI: 10.1016/j.molliq.2018.11.032

[14] Badruddoza AZM, Shawon ZBZ, Tay WJD, Hidajat K, Uddin MS. Fe_3O_4/cyclodextrin polymer nanocomposites for selective heavy metals removal from industrial wastewater. Carbohydrate polymers. 2013;**91**:322-332. DOI: 10.1016/j.carbpol.2012.08.030

[15] Sud D, Mahajan G, Kaur MP. Agricultural waste material as potential adsorbent for sequestering heavy metal ions from aqueous solutions–A review. Bioresource Technology. 2008;**99**:6017-6027. DOI: 10.1016/j.biortech.2007.11.064

[16] Nguyen TA, Ngo HH, Guo WS, Zhang J, Liang S, Yue QY, et al. Applicability of agricultural waste and by-products for adsorptive removal of heavy metals from wastewater. Bioresource Technology. 2013;**148**:574-585. DOI: 10.1016/j.biortech.2013.08.124

[17] Ahmed MJ, Ahmaruzzaman M. A review on potential usage of industrial waste materials for binding heavy metal ions from aqueous solutions. Journal of Water Process Engineering. 2016;**10**:39-47. DOI: 10.1016/j.jwpe.2016.01.014

[18] Wong S, Ngadi N, Inuwa IM, Hassan O. Recent advances in applications of activated carbon from biowaste for wastewater treatment: A short review. Journal of Cleaner Production. 2018;**175**:361-375. DOI: 10.1016/j.jclepro.2017.12.059

[19] Uddin MK. A review on the adsorption of heavy metals by clay minerals, with special focus on the past decade. Chemical Engineering Journal. 2017;**308**:438-462. DOI: 10.1016/j.cej.2016.09.029

[20] Zare EN, Motahari A, Sillanpää M. Nanoadsorbents based on conducting polymer nanocomposites with main focus on polyaniline and its derivatives for removal of heavy metal ions/dyes: A review. Environmental research. 2018;**162**:173-195. DOI: 10.1016/j.envres.2017.12.025

[21] Mezohegyi G, van der Zee FP, Font J, Fortuny A, Azael Fabregat A. Towards advanced aqueous dye removal processes: A short review on theversatile role of activated carbon. Journal of Environmental Management. 2012;**102**:148-164. DOI: 10.1016/j.jenvman.2012.02.021

[22] Gomez-Maldonado D, Vega Erramuspe IB, Peresin MS. Natural polymers as alternative adsorbents and treatment agents for water remediation. BioResources. 2019;**14**:10093-10160

[23] Kumar A, Rao KM, Han SS. Application of xanthan gum as polysaccharide in tissue engineering: A review. Carbohydrate Polymers. 2018;**180**:128-144. DOI: 10.1016/j.carbpol.2017.10.009

[24] Sedayu BB, Cran MJ, Bigger SW. A review of property enhancement techniques for carrageenan-based films and coatings. Carbohydrate Polymers. 2019;**216**:287-302. DOI: 10.1016/j.carbpol.2019.04.021

[25] Thakur S, Sharma B, Verma A, Chaudhary J, Tamulevicius S, Thakur VK. Recent approaches in guar gum hydrogel synthesis for water purification. International Journal of Polymer Analysis and Characterization. 2018;**23**:621-632. DOI: 10.1080/1023666X.2018.1488661

[26] Mittal H, Ray SS, Okamoto M. Recent progress on the design and applications of polysaccharide-based graft copolymer hydrogels as adsorbents for wastewater purification. Macromolecular Materials and Engineering. 2016;**301**:496-522. DOI: 10.1002/mame.201500399

[27] Rosalam S, England R. Review of xanthan gum production from unmodified starches by *Xanthomonas*

comprestris sp. Enzyme and Microbial Technology. 2006;**39**:197-207. DOI: 10.1016/j.enzmictec.2005.10.019

[28] Necas J, Bartosikova L. Carrageenan: a review. Veterinární Medicína. 2013;**58**:187-205

[29] Salgueiro AN, Daniel-da-Silva AL, Girão AV, Pinheiro PC, Trindade T. Unusual dye adsorption behavior of *k*-carrageenan coated superparamagnetic nanoparticles. Chemical Engineering Journal. 2013;**229**:276-284. DOI: 10.1016/j.cej.2013.06.015

[30] Patel J, Maji B, Narayana Moorthy NSH, Maiti S. Xanthan gum derivatives: review of synthesis, properties and diverse applications. RSC Advances. 2020;**10**:27103-27136. DOI: 10.1039/d0ra04366d

[31] Flahive JJ III, Foufopoulos A. Etzel MR, Alcohol precipitation of xanthan gum from pure solutions and fermentation broths. Separation Science and Technology. 1994; 29:1673-1687. DOI: 10.1080/01496399408002164

[32] Khouryieh H, Puli G, Williams K, Aramouni F. Effects of xanthan–locust bean gum mixtures on the physicochemical properties and oxidative stability of whey protein stabilised oil-in-water emulsions. Food Chemistry. 2015;**167**:340-348. DOI: 10.1016/j.foodchem.2014.07.009

[33] Lozano-Álvarez JA, Jáuregui-Rincón J, Mendoza-Díaz G, Rodríguez-Vázquez R. Frausto-Reyes C. Study of Sorption Equilibrium of Biopolymers Alginic Acid and Xanthan with C.I. Disperse Yellow 54. Journal of the Mexican Chemical Society. 2009;**53**:59-70

[34] Ghorai S, Sarkar A, Raoufi M, Panda AB, Holger Schonherr H, Pal S. Enhanced Removal of Methylene Blue and Methyl Violet Dyes from Aqueous Solution Using a Nanocomposite of Hydrolyzed Polyacrylamide Grafted Xanthan Gum and Incorporated Nanosilica. ACS Applied Materials & Interfaces. 2014;**6**:4766-4777. DOI: 10.1021/am4055657

[35] M-m W, Wang L. Synthesis and characterisation of carboxymethyl cellulose/organic montmorillonite nanocomposites and its adsorption behavior for Congo Red dye. Water Science and Engineering. 2013;**6**:272-282. DOI: 10.3882/j.issn.1674-2370.2013.03.004

[36] Ng LY, Mohammad AW, Leo CP, Hilal N. Polymeric membranes incorporated with metal/metal oxide nanoparticles: A comprehensive review. Desalination. 2013;**308**:15-33. DOI: 10.1016/j.desal.2010.11.033

[37] Zare EN, Motahari A, Sillanpää M. Nanoadsorbents based on conducting polymer nanocomposites with main focus on polyaniline and its derivatives for removal of heavy metal ions/dyes: A review. Environmental research. 2018;**162**:173-195. DOI: 10.1016/j.envres.2017.12.025

[38] Ito S, Goseki R, Ishizone T, Hirao A. Synthesis of well-controlled graft polymers by living anionic polymerisation towards exact graft polymers. Polymer Chemistry. 2014;**5**:5523-5534. DOI: 10.1039/c4py00584h

[39] Roy D, Semsarilar M, Guthrie JT, Perrier S. Cellulose modification by polymer grafting: a review. Chemical Society Reviews. 2009;**38**:2046-2064. DOI: 10.1039/b808639g

[40] El-Sherbiny IM. Synthesis, characterisation and metal uptake capacity of a new carboxymethyl chitosan derivative. European Polymer Journal. 2009;**45**:199-210. DOI: 10.1016/j.eurpolymj.2008.10.042

[41] Liao Y, Zheng H, Qian L, Sun Y, Dai L, Xue W. UV-initiated polymerisation of hydrophobically associating cationic polyacrylamide modified by a surface-active monomer: a comparative study of synthesis, characterisation, and sludge dewatering performance. Industrial & Engineering Chemistry Research. 2014;**53**:11193-11203. DOI: 10.1021/ie5016987

[42] Pal S, Ghorai S, Das C, Samrat S, Ghosh A, Panda AB. Carboxymethyl tamarind-g-poly (acrylamide)/ silica: A high performance hybrid nanocomposite for adsorption of methylene blue dye. Industrial & Engineering Chemistry Research. 2012;**51**:15546-15556. DOI: 10.1021/ ie301134a

[43] Thakur S, Pandey S, Arotiba OA. Development of a sodium alginate-based organic/inorganic superabsorbent composite hydrogel for adsorption of methylene blue. Carbohydrate Polymers. 2016;**153**:34-46. DOI: 10.1016/j.carbpol.2016.06.104

[44] Elella MHA, Sabaa MW, ElHafeez EA, Mohamed RR. Crystal violet dye removal using crosslinked grafted xanthan gum. International Journal of Biological Macromolecules. 2019;**137**:1086-1101. DOI: 10.1016/j. ijbiomac.2019.06.243

[45] Elella MHA, Mohamed RR, ElHafeez EA, Sabaa MW. Synthesis of novel biodegradable antibacterial grafted xanthan gum. Carbohydrate Polymers. 2017;**173**:305-311. DOI: 10.1016/j.carbpol.2017.05.058

[46] Ali A, Zafar H, Zia M, ul Haq I, Phull AR, Ali JS, et al. Synthesis, characterisation, applications, and challenges of iron oxide nanoparticles. Nanotechnology, Science and Applications. 2016;**9**:49-67. DOI: 10.2147/NSA.S99986

[47] Boury B, Plumejeau S. Metal oxides and polysaccharides: an efficient hybrid association for materials chemistry. Green Chemistry. 2015;**17**:72-88. DOI: 10.1039/c4gc00957f

[48] Mittal H, Maity A, Ray SS. Synthesis of co-polymer-grafted gum karaya and silica hybrid organic–inorganic hydrogel nanocomposite for the highly effective removal of methylene blue. Chemical Engineering Journal. 2015;**279**:166-179. DOI: 10.1016/j.cej.2015.05.002

[49] Duman O, Tunc S,, Polat TG, Bozoglan BK. Synthesis of magnetic oxidised multiwalled carbon nanotube-*k*-carrageenan-Fe_3O_4 nanocomposite adsorbent and its application in cationic Methylene Blue dye adsorption. Carbohydrate Polymers. 2016;**147**:79-88. DOI: 10.1016/j.carbpol.2016.03.099

[50] Bulut E, Ozacar M, Sengil IA. Adsorption of malachite green onto bentonite: Equilibrium and kinetic studies and process design. Microporous and Mesoporous Materials. 2008;**115**:234-246. DOI: 10.1016/j. micromeso.2008.01.039

[51] Kumar KY, Muralidhara HB, Nayaka YA, Balasubramanyam J, Hanumanthappa H. Low-cost synthesis of metal oxide nanoparticles and their application in adsorption of commercial dye and heavy metal ion in aqueous solution. Powder Technology. 2013;**246**:125-136. DOI: 10.1016/j. powtec.2013.05.017

[52] Mahdavinia GR, Massoudi A, Baghban A, Shokri E. Study of adsorption of cationic dye on magnetic kappa-carrageenan/ PVA nanocomposite hydrogels. Journal of Environmental Chemical Engineering. 2014;**2**:1578-1587. DOI: 10.1016/j. jece.2014.05.020

[53] Guo J, Li X, Mu C, Zhang H, Qin P, Li D. Freezingethawing effects on the

properties of dialdehyde carboxymethyl cellulose crosslinked gelatin-MMT composite films. Food Hydrocolloids. 2013;**33**:273-279. DOI: 10.1016/j.foodhyd.2013.04.004

[54] Yu J-G, Zhao X-H, Yang H, Chen X-H, Yang Q, Yu L-Y, et al. Aqueous adsorption and removal of organic contaminants by carbon nanotubes. Science of the Total Environment. 2014;**482-483**:241-251. DOI: 10.1016/j.scitotenv.2014.02.129

[55] Rajabi M, Mahanpoora K, Moradi O. Removal of dye molecules from aqueous solution by carbon nanotubes and carbon nanotube functional groups: critical review. RSC Advances. 2017;7:47083-47090. DOI: 10.1039/c7ra09377b

[56] Gupta VK, Kumar R, Nayak A, Saleh TA, Barakat MA. Adsorptive removal of dyes from aqueous solution onto carbon nanotubes: A review. Advances in Colloid and Interface Science. 2013;**193-194**:24-34. DOI: 10.1016/j.cis.2013.03.003

[57] Kontou E, Niaounakis M. Thermo-mechanical properties of LLDPE/SiO_2 nanocomposites. Polymer. 2006;**47**:1267-1280. DOI: 10.1016/j.polymer.2005.12.039

[58] Cai J, Zhang D, Xu W, Ding W, Zhu Z, He J, et al. Polysaccharide-Based Hydrogels Derived from Cellulose: The Architecture Change from Nanofibers to Hydrogels for a Putative Dual Function in Dye Wastewater Treatment. Journal of Agricultural and Food Chemistry. 2020;**68**:9725-9732. DOI: 10.1021/acs.jafc.0c03054

[59] Huang Z-M, Zhang Y-Z, Kotaki M, Ramakrishna S. A review on polymer nanofibers by electrospinning and their applications in nanocomposites. Composites Science and Technology. 2003;**63**:2223-2253. DOI: 10.1016/S0266-3538(03)00178-7

[60] Frank SN, Bard AJ. Semiconductor electrodes. 12. Photoassisted oxidations and photoelectrosynthesis at polycrystalline titanium dioxide electrodes. Journal of the American Chemical Society. 1977;**99**:4667-4675. DOI: 10.1021/ja00456a024

[61] Pourjavadi A, Doulabi M, Doroudian M. Adsorption characteristics of malachite green dye onto novel kappa-carrageenan-g-polyacrylic acid/TiO_2–NH_2 hydrogel nanocomposite. Journal of the Iranian Chemical Society. 2014;**11**:1057-1065. DOI: 10.1007/s13738-013-0374-6

Section 5

Applications of Nanomaterials in Coating and Building Industries

Chapter 15

Nanostructured Materials for the Development of Superhydrophobic Coatings

Jeyasubramanian Kadarkaraithangam, Thangaiyanadar Suyambulingam Gokul Raja, Silambuselvan Parani Bramma Nayagi and Karthikeyan Krishnamoorthy

Abstract

This chapter describes the results of developing superhydrophobic coatings using porous ZnO nanostructures impregnated metal stearates and their applications. The porous ZnO nanostructures with a surface area of 9.7 m^2/g and pores in the range from 200 to 400 nm have been prepared via precipitation cum calcination route. The superhydrophobic coatings comprising ZnO/metal stearate film have been deposited using a spray coating method. The developed superhydrophobic films possess a water contact angle of 161° that can be explained using the Cassie-Baxter model. The prepared films exhibited excellent floating properties and high load-bearing characteristics over a prolonged time. Additionally, the self-cleaning properties of the developed superhydrophobic films towards dust removal and self-cleaning urinary coatings are also demonstrated. This chapter collectively presented the novel applications of superhydrophobic coating in the development of biomedical coatings and applications in water surveillance and underwater robotics.

Keywords: nanomaterials, superhydrophobic, Cassie-Baxter equation, functional coatings, nanoscale roughness

1. Introduction

Superhydrophobic surfaces received many research interest from academic to industrial sectors due to their intriguing self-cleaning properties [1–4]. The study on superhydrophobic surfaces was initiated during early 1907 when Ollivier observed a contact angle of ~180° for surfaces modified with soot, lycopodium powder, and arsenic trioxide [1]. Later, Coghill and Anderson (in 1923) studied the surface modification of galena via deposition of stearic acids leading to achieve a water contact angle (WCA) of 160° [1]. Until the 1990s, minimal research work has been carried out on superhydrophobic surfaces. In contrast, it was reactivated in 1997 when Neinhuis and Barthlott discussed the origin of superhydrophobicity via the principle of "lotus effect" [5]. Many researchers have focused on superhydrophobic surfaces by mimicking nature and fabricating similar structures via artificial methods through surface modification. Besides fundamental understanding of the superhydrophobic phenomenon, researchers currently focuses on extending the application of superhydrophobic surfaces like self-cleaning coatings, friction reduction coatings on ship hulls, corrosion prevention oil–water separation, and antibacterial textiles/bandages [6–9]. Generally, roughness plays a vital role in the hydrophobicity of a hydrophobic solid. Flat solids display almost 100° to 120° contact angles for water and reach up to 160° to 175° if they are rough or micro/nanotextured. Such improved superhydrophobic behavior is not only due to the solids' surface chemistry alone [10]. There are two distinct processes involved in the enhancement of superhydrophobic properties that can be explained based on the (i) Wenzel model and (ii) Cassie-Baxter model [11, 12]. The Wenzel model relies on the surface roughness induced high surface area of the solid resulting in superhydrophobic nature. The Cassie model revealed that the air trapped below the drop could also result in stable superhydrophobic properties. In both models, an apparent contact angle θ^* of a drop on the surface of a rough substrate/solid will be formed via reducing the surface energy of a drop to Young's contact angle θ (determined on a flat surface of the same) [10].

Naturally, the water repellent properties (superhydrophobic surfaces) can be seen in many plant surfaces (lotus leaves), and animal furs, so on [13]. For instance, lotus leaves are considered as one of the finest examples of superhydrophobic surfaces. The WCA of the lotus leaf is about ~162° with a hysteresis of 2°. Barthlott and Neihuis study the micro-structure of lotus leaf using a scanning electron microscope (SEM) [14]. They demonstrated the presence of two different ranges of roughness present in lotus leaf viz. (i) one with 10 μm (rough structure) and (ii) other with 100 nm (fine structure). These studies confirmed micro- and nano-textured surfaces on the lotus leaf that lead to the origin of their self-cleaning properties [15]. To date, there are different methods used for the creation of superhydrophobic surfaces that can be broadly classified as (i) Top-down and (ii) bottom-up methods. Top-down approaches include the costly lithographic process, template-assisted fabrication methods, and delicate surfaces' plasma treatment [16–18]. Bottom-up approaches include self-assembled layers/films, chemical deposition, and layer-by-layer (LBL) deposition, etc., [19–21]. The available methods from the bottom-up approaches are considered cost-effective and scalable compared to the top-down methods. More importantly, synergistic methods involving both top-down and bottom-up methods are also in practice. Most of them rely on solution casting, phase separation, electrospinning/spraying, and nanostructure impregnated polymer composite films [22, 23]. The superhydrophobic surfaces/coatings possess numerous applications in automobile windows, optical windows for electronic devices, eyeglasses, fluidic drag reduction, enhanced water supporting force, water corrosion prevention, and humidity proof coatings, anti-biofouling, self-cleaning textiles, and oil–water separation, etc. [24–27]. Additionally, the superhydrophobic materials as electrodes for batteries and fuel cells result in their extended shelf-life time as experimentally demonstrated by Lifton et al. [28]. Recently, studies demonstrated that the superhydrophobic coatings could also be used for applications such as anti-icing, anti-fogging, and anti-frosting sectors.

The role of nanostructured materials in superhydrophobic coatings is rapidly rising mainly due to their exceptional physical/chemical properties [29]. The choice of nanostructures over their bulk counterparts for superhydrophobic coatings is due to the increased surface area and high roughness on the surfaces. The nanostructured materials are mostly used as fillers in polymers to modify their surface roughness and porosity of the polymer surfaces, leading to superhydrophobic properties. Additionally, the choice of nanomaterials in superhydrophobic coatings also enhanced their durability and focused on biomedical applications such as creating antibacterial textiles, medical implants, etc. [29–31]. This chapter discusses the recent trends in the development of nanostructured materials based on superhydrophobic coatings and their applications.

2. Methods

2.1 Preparation of porous ZnO nanostructures

A facile precipitation cum calcination route was used for the preparation of porous ZnO nanoparticles [32]. Briefly, appropriate amount of zinc nitrate (4 g) was dissolved in distilled water, followed by the addition of polyethylene glycol (PEG) (3 g) and hexamine (3 g). The entire precursor solution is subjected to a vigorous stirring process using a magnetic stirrer for 30 minutes at a temperature of 60 °C. Following this, the ammonia solution was added to the precursor solution until their pH reached 9. After that, the solution is allowed to gelation process by placing them in a hot plate at the temperature of 90 °C, which results in the formation of a dark brown gel. After 4 hours, the dried brownish gel was placed in a silica crucible and calcined at a temperature of 500 °C for 2 h that finally lead to the formation of white-colored porous ZnO nanostructures.

2.2 Synthesis of copper stearate and magnesium stearate

The metal (copper/magnesium stearate powders) were prepared via a precipitation method using metal salts, stearic acid, and ammonia, as reported in our recent works [32, 33].

2.3 Fabrication of superhydrophobic films via spray coating process

A spray coating process is used to fabricate the superhydrophobic films comprising ZnO/metal stearate with the various weight ratio of ZnO in our recent study [32]. The coating thickness was varied by spraying the solution from different time intervals ranging from 30 seconds to 30 minutes.

3. Results and discussion

3.1 Characterization of porous ZnO nanostructures

Figure 1(A) shows the X-ray diffraction (XRD) pattern of ZnO nanostructures prepared via the calcination route explained in section 2.1. High intense sharp diffraction peaks are seen in **Figure 1(A)** matched with the wurtzite structure of ZnO ((JCPDS No. 89–7102) [34]. The Raman spectrum of the prepared ZnO nanostructures (given in **Figure 1(B)**) indicated a sharp band located at 437 cm^{-1} as a result of E_2 (high) mode vibrations [35]. **Figure 1(C, D)** represents the deconvoluted X-ray photoelectron spectroscopy of Zn and O states present in the ZnO nanostructures. The Zn 2p states (given in **Figure 1(C)**) showed the presence of two peaks corresponding to the Zn $2p_{3/2}$ (at 1022 eV) and Zn $2p_{1/2}$ (at 1045 eV), respectively. A value of 23 eV is obtained for the difference between the peak positions of Zn $2p_{3/2}$, and Zn $2p_{1/2}$ states that matched with the reported ones and indicates that Zn possesses an oxidation state of +2 in the synthesized ZnO nanostructures [32]. The O 1 s spectrum evidences the broad peak centered at 532 eV (given in **Figure 1(D)**) arises from the oxygen content present in the wurtzite ZnO [35]. The field emission scanning electron micrographs (FE-SEM) of the as prepared ZnO nanostructures (given in **Figure 1(E)** and **(F)**) showed the presence of nanoparticles with a high amount of pores. The high magnification micrograph (**Figure 1(F)**) evidences the honeycomb-like porous ZnO nanostructures with pore sizes ranging from 200 to

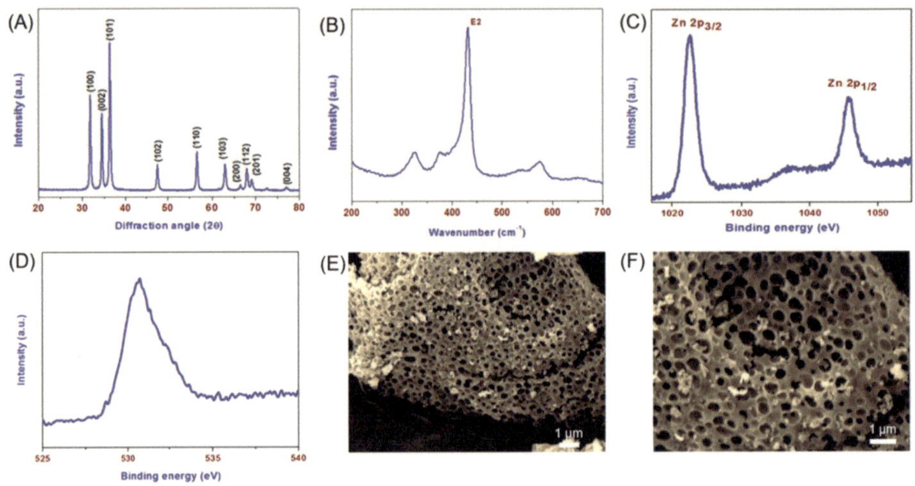

Figure 1.
(A) X-ray diffraction pattern, (B) laser Raman spectrum, (C) Zn 2p core-level X-ray photoelectron spectrum, and (D) O 1 score-level photoelectron spectrum of porous ZnO nanostructures and scanning electron micrographs of porous ZnO nanoparticles with (E) low and (F) high magnification.

400 nm. The Brunauer-Emmet-Teller (BET) analysis using the Barrett–Joyner–Halenda (BJH) method of the ZnO nanostructures revealed the Type IV isotherm with a hysteresis (data not shown) highlighting the presence of mesoporous nature [36]. The prepared ZnO nanostructures possess a pore volume and surface area of about 0.028 cm^3/g, and 9.7 m^2/g. The obtained high surface area of the ZnO nanostructures is due to the role of PEG in the preparation. The guest-host chemistry of metal ions with polymer matrix (i.e., Zn^{2+} ions as a guest in the PEG host matrix) results in the formation of mesoporous ZnO nanostructures via decomposition of PEG during the calcination process.

3.2 Characterization of copper stearate

The formation of copper stearate via precipitation method was studied using Fourier transformed infra-red spectroscopy (FT-IR) and XRD analysis (given in **Figure 2(A,B)**). During the formation of copper stearate from stearic acid using copper salts, the carboxyl group's and hydrogen atom in the stearic acid is replaced with the Cu ions. **Figure 2(A)** compares the FT-IR spectra of copper stearate to that of stearic acid. The presence of vibration due to the COO- group in stearic acid is noticed via its characteristic peak at 1700 cm^{-1} and is disappeared/shifted towards a lower peak position at 1583 cm^{-1}, respectively [37]. The sharp bands observed at 3352 (-OH) and 1046 cm^{-1} (C-O stretching) in the spectrum of stearic acid were disappeared in the FT-IR spectrum of copper stearate. Additionally, the FT-IR spectrum of copper stearate revealed the presence of characteristic bands at 2847, 2917, 1441, 720, and 880 cm^{-1} raised from the CH_2, C-H, C-C, CH_2 (rocking), and CH_3 rocking vibrations, respectively [38, 39]. The presence of characteristic diffraction peaks from 5° to 20° in the XRD pattern (given in **Figure 2(B)**) confirmed the formation of copper stearate [32, 38].

3.3 Characterization of ZnO/copper stearate composite films

Figure 2(C) shows the FT-IR spectrum of bare ZnO nanostructures and ZnO/copper stearate composite films. The presence of vibration bands centered at 518 and

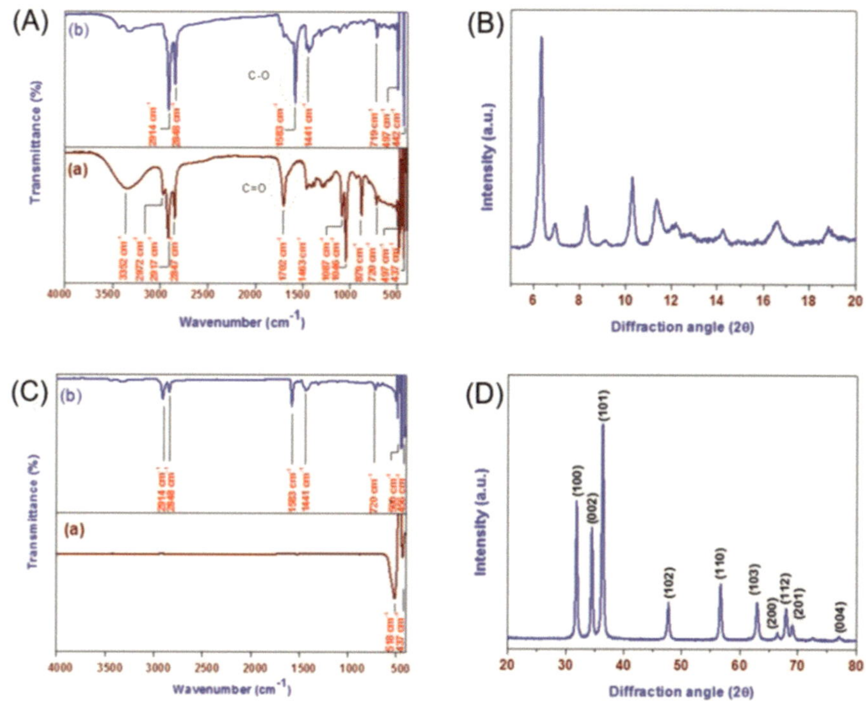

Figure 2.
(A) Fourier transform infrared spectrum and (B) X-ray diffraction pattern of copper stearate, (C) FT-IR spectrum of porous ZnO and ZnO/copper stearate coatings, (D) X-ray diffraction pattern of ZnO/copper stearate coatings.

437 cm^{-1} was raised from the Zn-O vibrations present in ZnO nanostructures [40]. The FT-IR spectrum of ZnO/copper stearate film given in **Figure 2(C)** shows almost all the characteristics vibration bands of ZnO nanostructures and copper stearate with slight variation in their peak positions, that is due to the interaction between the ZnO nanostructures and copper stearate via guest-host chemistry. **Figure 2(D)** shows the XRD pattern of the ZnO/copper stearate coatings. The peaks corresponding to the crystalline ZnO nanostructures were visible in the XRD pattern, whereas the peaks due to the copper stearate were diminished/not observed. This is due to the low crystallinity of copper stearate compared to ZnO nanostructures' highly crystalline nature. **Figure 3(A)** depicts the SEM image of spray-coated ZnO/copper stearate films. It is quite challenging to distinguish ZnO nanostructures in the spray-coated films due to the low weight percentage of ZnO to that of copper stearate and better dispersibility of ZnO nanostructures in the copper stearate due to the ultrasonication process. The size of the pores present in the ZnO/copper stearate coatings was found to be in the range from 100 to 300 nm determined using ImageJ software [41]. Here, it is noteworthy that micron-sized ZnO in similar coatings resulted in irregular surface formation due to the low dispersion index of micron-sized ZnO. **Figure 3(B-D)** shows the elemental maps of Zn, O, and Cu components present in the ZnO/copper stearate coatings. The Zn map of the spray-coated films (shown in **Figure 3(B)**) indicated the presence of well-dispersed ZnO nanostructures in the copper stearate matrix. **Figure 3(C)** presents the oxygen map indicating the existence of array-like arrangement due to the chain-like structures of copper stearate in addition to the ZnO components of the coating. The mapping of copper elements in the spray coated films (given in **Figure 3(D)**) shows that the copper elements are randomly distributed in the films. Since the copper elements are attached

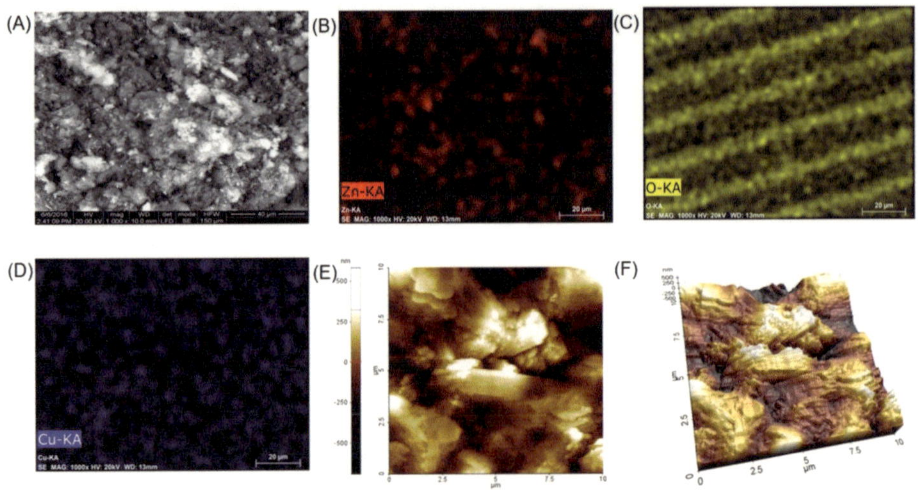

Figure 3.
Field emission scanning electron micrograph of (A) ZnO/copper stearate and (B-D) shows the elemental maps of Zn, O, and Cu present in these coatings and (E-F) 2D and 3D Atomic force micrograph of ZnO/copper stearate coatings.

to the end of the chain-like structure of copper stearate. More probably, the copper ends were attached to the glass substrates due to their hydrophilic nature leaving the hydrophobic methyl group on the exterior surfaces. The 2D and 3D topographic analysis of the spray coated ZnO/copper stearate films analyzed by atomic force micrograph is shown in **Figure 3(E, F)**. These studies demonstrated the existence of porous and rough surfaces nature of the spray-coated ZnO/copper stearate films.

3.4 Superhydrophobic properties of ZnO/copper stearate coatings

The water contact angle measurement was carried out for determining the superhydrophobic properties of the ZnO/copper stearate coatings with various loading ratios of ZnO nanostructures in the copper stearate matrix. The plain glass substrates possess a WCA of 27.4°, indicating their hydrophilic surfaces [42]. The bare copper stearate coatings (with low surface energy) on glass substrates possess a WCA of 152.4°, demonstrating their hydrophobic properties. It is expected that the inclusion of porous ZnO nanostructures in the copper stearate matrix might alter their roughness and thus improves the hydrophobicity. The effect of porous ZnO nanostructures loading ratio on the water repellent properties of the ZnO/copper stearate coatings is summarized in **Figure 4(A)**. It displayed that the superhydrophobic effect was not obtained in the composite films with a loading ratio of ZnO nanostructures until 0.01 g. The ZnO/copper stearate coatings displayed the superhydrophobic properties with a WCA of 161° when the ZnO weight percentage is increased up to 0.14 g. This can be due to the improvements in the films' roughness by the inclusion of highly-porous ZnO nanostructures. The mechanism of superhydrophobic effect achieved in the spray-coated films can be described via the Cassie-Baxter model using the following relation [43]:

$$\cos\theta* = \varphi_s \cos\theta + \varphi_s - 1\ldots \quad (1)$$

Here θ^*, φ_s and θ represents the apparent contact angle, substantial fraction in contact with liquid, and Young's contact angle, respectively. The substantial fraction of the bare copper stearate coating is about 0.197, and these values decrease with

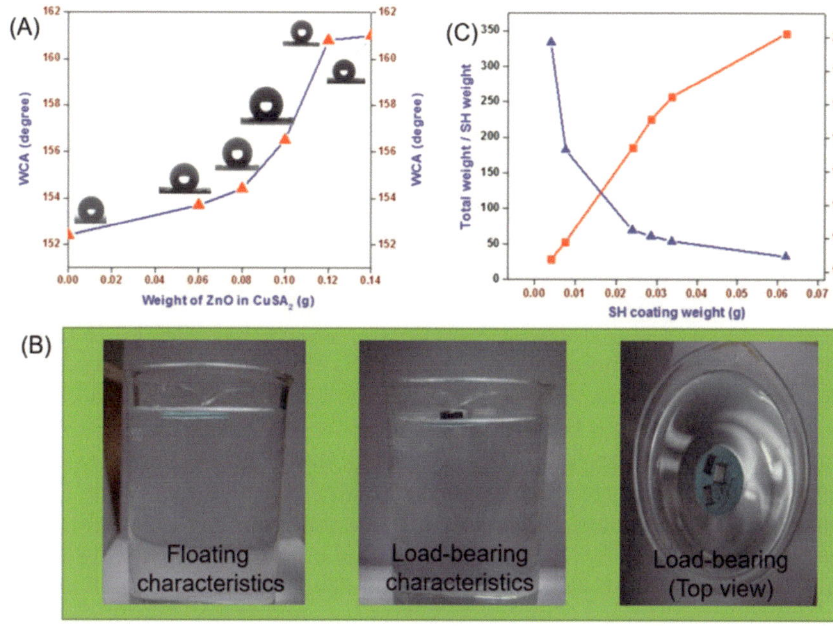

Figure 4.
(A) Water contact angle, (B) floating characteristics, and (C) load-bearing properties of ZnO/copper stearate coatings.

an increase in the loading ratio of porous ZnO. A low substantial fraction of about 0.094 was reached for the ZnO/copper stearate coatings with a ZnO loading of about 0.14 g. The observed decrease in solid fraction values indicated the enhancement in the surface roughness due to porous ZnO nanostructures' impregnation. A similar effect is not observed in films coated using non-porous ZnO or micron-sized ZnO powders. This further substantiates porous ZnO nanostructures' significance for modulating the roughness of copper stearate films to obtain superhydrophobic properties.

3.5 Floating and load-bearing characteristics of superhydrophobic ZnO/copper stearate coatings

Floating and load-bearing characteristics are major applications where superhydrophobic coatings for underwater robotics, monitoring water pollution, water quality analysis, and surveillance applications can be developed [44–46]. Based on the Archimedes principle, it is well known that any object (with density higher than water) will sink in water. However, water strider possesses the ability to float in water and stride freely on the water surface due to the hierarchical fibrous architecture of strider leg exhibiting superhydrophobic effect. The bare glass substrates immediately drown after placing in water solution, which can be explained using the Archimedes principle [47]. It is expected that superhydrophobic coatings can possess a floating nature. The floating characteristics of the spray-coated ZnO/copper stearate coatings (laid down) in the water solution is shown in **Figure 4(B)**. The fundamental mechanism of an object's floating nature will be achieved if the buoyancy force exerted by the object is higher than the drown force acts on it. The ZnO/copper stearate coatings on glass substrates float on water solution for more than a week without any noticeable sinking effects. Here, in addition to the buoyancy force, a curvature force acts on the surface of superhydrophobic coatings, acting against the drown force that enables them to float over a prolonged time. The superhydrophobic ZnO/

copper stearate coatings possess a trapped air film on their exterior surfaces leading to an excessive displaced volume of water. The entire phenomenon can be termed as super-buoyancy [48]. These studies demonstrated the plastron effect's vital role in the ZnO/copper stearate coatings' floating properties. The ZnO/copper stearate coatings' load-bearing characteristics are investigated by loading known mass (stapler pins) on their top surface during floating, as shown in **Figure 4(B)**. It showed that after loading a few stapler pins, there are no signs of sinking for the ZnO/copper stearate coatings. The effect of thickness of superhydrophobic ZnO/copper stearate coatings via different deposition times (180 to 1800 seconds) on their load-bearing properties is provided in **Figure 4(C)**. It showed that ZnO/copper stearate coatings with a high thickness (weight of about 0.062 g) remain floating and can bear about 0.3667 g (19 stapler pins) without sinking issues. **Figure 4(C)** illustrated that the ratio of net floating weight to the weight of ZnO/copper stearate coating decreases with increasing thickness. This can be explained based on the more inert/dead layers present in the coatings with high thickness. On the other hand, the capability of load-bearing is superior for thin coatings (deposition at 30 seconds) that can bear 52 pins (~333 times higher than their net weight). The findings on the floating and load-bearing characteristics of the superhydrophobic ZnO/copper stearate coatings can be applied for water floating micro-robots and surveillance applications.

3.6 Applications of superhydrophobic coatings towards self-cleaning urinary coatings

The superhydrophobic coatings' self-cleaning properties can be used for developing water-free urinaries, which doesn't require any systematic cleaning process to mollify unpleasant odor [49, 50]. Since the surface tension of water (71.2 mN/m) and urine (70 mN/m) are of similar values, it is expected that a water-repellent surface possesses a tendency to repel urine as well [51, 52]. The ZnO/magnesium stearate coatings showed excellent dye and urine-repellant properties, as shown in **Figure 5(a, b)**. It is observed that the droplets of urine samples roll off over the surface of ZnO/magnesium stearate coatings due to their extreme repellent

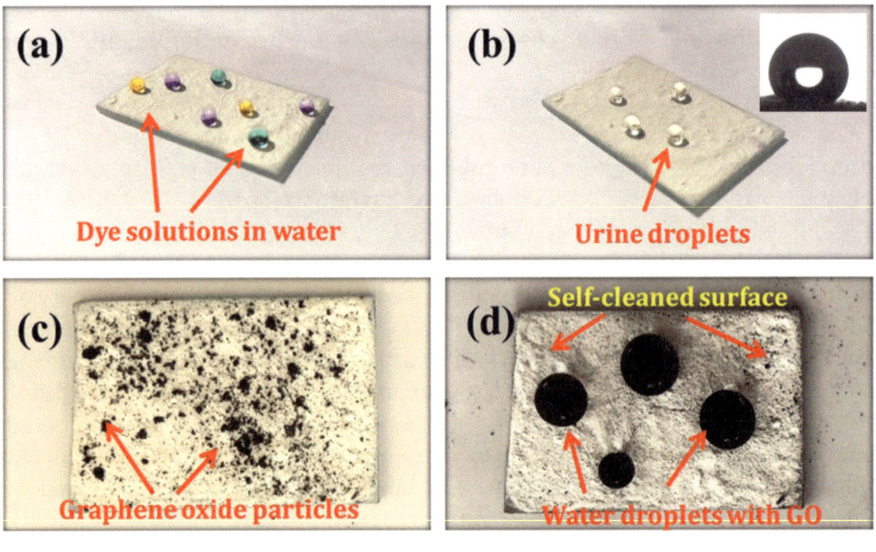

Figure 5.
Superhydrophobic applications of ZnO/magnesium stearate (a) dye repellency, (b) urine repellency, (c) spreading of graphene oxide (GO) dust on coatings, the (d) self-cleaning ability of ZnO/magnesium stearate to remove the GO dust.

properties [33]. The FT-IR spectrum measured after urine droplet tests on the ZnO/magnesium stearate showed the absence of vibration bands due to urine, which showed their excellent superhydrophobic effects [33]. Additionally, these superhydrophobic films' self-cleaning properties are also examined using graphene oxide as a dust spread on their surfaces. After adding few water droplets on their surface, the water droplets cleaned all the dust (graphene oxide) on their surfaces, as shown in **Figure 5(c, d)**, and ensures their self-cleaning properties.

4. Conclusions

In conclusion, the significant role of porous nanostructures for developing superhydrophobic coatings is demonstrated in this chapter. A cost-effective sonochemical approach combined with a spray coating process for the fabrication of superhydrophobic coatings is presented. The impact of nanostructures on the coatings' porosity is explained with the experimental findings like AFM micrograph and water contact angle measurements. The exciting applications of superhydrophobic coatings like water-floating properties, load-bearing applications, dust removal, self-cleaning urinary coatings are illustrated in this chapter.

Acknowledgements

The authors thank the Management and the Principal, Mepco Schlenk Engineering College, Sivakasi, for the constant encouragement and support.

Conflict of interest

"The authors declare no conflict of interest."

Author details

Jeyasubramanian Kadarkaraithangam[1*],
Thangaiyanadar Suyambulingam Gokul Raja[1],
Silambuselvan Parani Bramma Nayagi[1] and Karthikeyan Krishnamoorthy[2*]

1 Department of Mechanical Engineering, Centre for Nano Science and Technology, Mepco Schlenk Engineering College, Sivakasi, India

2 Nanomaterials and System Laboratory, Major of Mechatronics Engineering, Faculty of Applied Energy System, Jeju National University, Jeju, South Korea

*Address all correspondence to: kjeya@mepcoeng.ac.in
and karthi.nanotech@gmail.com

References

[1] X. Zhang, F. Shi, J. Niu, Y. Jiang, Z. Wang, Superhydrophobic surfaces: From structural control to functional application, J. Mater. Chem. 18 (2008) 621-633. doi:10.1039/b711226b.

[2] S. S. Latthe, R. S. Sutar, V. S. Kodag, A. K. Bhosale, A. MadhanKumar, K. K. Sadasivuni, R. Xing, S. Liu, Self – cleaning superhydrophobic coatings: Potential industrial applications, Prog. Org. Coat. 128 (2019) 52-28. 10.1016/j.porgcoat.2018.12.008. LECTACTA.2017.01.044.

[3] T. Yang, M. Wang, X. Wang, X. Di, C. Wang, Y. Li, Fabrication of a waterborne, superhydrophobic, self-cleaning, highly transparent and stable surface, Soft Matter 16 (2020) 3678-3685. doi: 10.1039/C9SM02473E.

[4] X. Dong, J. Meng, Y. Hu, X. Wei, X. Luan, H. Zhou, Fabrication of Self-Cleaning Superhydrophobic Surfaces with Improved Corrosion Resistance on 6061 Aluminum Alloys, Micromachines (Basel). 11 (2020) 159. doi: 10.3390/mi11020159.

[5] C. Neinhuis, Characterization and Distribution of Water-repellent, Self-cleaning Plant Surfaces, Ann. Bot. 79 (1997) 667-677. doi:10.1006/anbo.1997.0400.

[6] K. Watanabe, Yanuar, H. Udagawa, Drag reduction of Newtonian fluid in a circular pipe with a highly water-repellent wall, J. Fluid Mech. 381 (1999) 225-238. doi:10.1017/S0022112098003747.

[7] T. Liu, Y. Yin, S. Chen, X. Chang, S. Cheng, Super-hydrophobic surfaces improve corrosion resistance of copper in seawater, Electrochim. Acta. 52 (2007) 3709-3713. doi:10.1016/j.electacta.2006.10.059.

[8] M. Ma, R.M. Hill, J.L. Lowery, S. V. Fridrikh, G.C. Rutledge, Electrospun poly(styrene-block-dimethylsiloxane) block copolymer fibers exhibiting superhydrophobicity, Langmuir. 21 (2005) 5549-5554. doi:10.1021/la047064y.

[9] J. Kadarkaraithangam, M. Anthony Raja, K. Krishnamoorthy, S. Natarajan, Fabrication of superhydrophobic surfaces using CuO nanoneedles blended polymer nanocomposite film, Nanosci. Nanotechnol. Lett. 5 (2013) 558-562. doi:10.1166/nnl.2013.1579.

[10] A. Lafuma, D. Quéré, Superhydrophobic states, Nat. Mater. 2 (2003) 457-460. doi:10.1038/nmat924.

[11] A.B.D. Cassie, S. Baxter, Wettability of porous surfaces, Trans. Faraday Soc. 40 (1944) 546-551. doi:10.1039/tf9444000546.

[12] R.N. Wenzel, Resistance of solid surfaces to wetting by water, Ind. Eng. Chem. 28 (1936) 988-994. doi:10.1021/ie50320a024.

[13] X.M. Li, D. Reinhoudt, M. Crego-Calama, What do we need for a superhydrophobic surface? A review on the recent progress in the preparation of superhydrophobic surfaces, Chem. Soc. Rev. 36 (2007) 1350-1368. doi:10.1039/b602486f.

[14] A. Otten, S. Herminghaus, How Plants Keep Dry: A Physicist's Point of View, Langmuir. 20 (2004) 2405-2408. doi:10.1021/la034961d.

[15] A. Scardino, R. De Nys, O. Ison, W. O'Connor, P. Steinberg, Microtopography and antifouling properties of the shell surface of the bivalve molluscs *Mytilus galloprovincialis* and *Pinctada imbricata*, in: Biofouling, Biofouling, 2003: pp. 221-230. doi:10.1080/0892701021000057882.

[16] K Y. Suh, S. Jon, Control over Wettability of Polyethylene Glycol

Surfaces Using Capillary Lithography, Langmuir 21 (2005) 6836-6841. doi:10.1021/LA050878+.

[17] D. Öner, T.J. McCarthy, Ultrahydrophobic surfaces. Effects of topography length scales on wettability, Langmuir. 16 (2000) 7777-7782. doi:10.1021/la000598o.

[18] J. Fresnais, L. Benyahia, F. Poncin-Epaillard, Dynamic (de)wetting properties of superhydrophobic plasma-treated polyethylene surfaces, Surf. Interface Anal. 38 (2006) 144-149. doi:10.1002/sia.2235.

[19] Y. Wu, M. Bekke, Y. Inoue, H. Sugimura, H. Kitaguchi, C. Liu, O. Takai, Mechanical durability of ultra-water-repellent thin film by microwave plasma-enhanced CVD, in: Thin Solid Films, Elsevier, 2004: pp. 122-127. doi:10.1016/j.tsf.2003.12.007.

[20] J.T. Han, Y. Zheng, J.H. Cho, X. Xu, K. Cho, Stable superhydrophobic organic-inorganic hybrid films by electrostatic self-assembly, J. Phys. Chem. B. 109 (2005) 20773-20778. doi:10.1021/jp052691x.

[21] G. Zhang, D. Wang, Z.Z. Gu, H. Mchwald, Fabrication of superhydrophobic surfaces from binary colloidal assembly, Langmuir. 21 (2005) 9143-9148. doi:10.1021/la0511945.

[22] H. Yabu, M. Shimomura, Single-step fabrication of transparent superhydrophobic porous polymer films, Chem. Mater. 17 (2005) 5231-5234. doi:10.1021/cm051281i.

[23] K. Acatay, E. Simsek, C. Ow-Yang, Y.Z. Menceloglu, Tunable, superhydrophobically stable polymeric surfaces by electrospinning, Angew. Chemie - Int. Ed. 43 (2004) 5210-5213. doi:10.1002/anie.200461092.

[24] A. Kavitha Sri, P. Deeksha, G. Deepika, J. Nishanthini, G.S. Hikku, S. Antinate Shilpa, K. Jeyasubramanian, R. Murugesan, Super-hydrophobicity: Mechanism, fabrication and its application in medical implants to prevent biomaterial associated infections, J. Ind. Eng. Chem. 92 (2020) 1-17. doi:10.1016/j.jiec.2020.08.008.

[25] R. Liao, C. Li, Y. Yuan, Y. Duan, A. Zhuang, Anti-icing performance of ZnO/SiO_2/PTFE sandwich-nanostructure superhydrophobic film on glass prepared via RF magnetron sputtering, Mater. Lett. 206 (2017) 109-112. doi:10.1016/j.matlet.2017.06.127.

[26] Z. Li, B.L. Nguyen, Y.C. Cheng, J. Xue, G. MacLaren, C.H. Yap, Durable, flexible, superhydrophobic and blood-repelling surfaces for use in medical blood pumps, J. Mater. Chem. B. 6 (2018) 6225-6233. doi:10.1039/C8TB01547C.

[27] Y. Wu, J. Wen, P. Zhang, Application of AOI light source modes in multi-chip modules inspection, in: Proc. - 2018 19th Int. Conf. Electron. Packag. Technol. ICEPT 2018, Institute of Electrical and Electronics Engineers Inc., 2018: pp. 141-143. doi:10.1109/ICEPT.2018.8480769.

[28] V.A. Lifton, S. Simon, R.E. Frahm, Reserve battery architecture based on superhydrophobic nanostructured surfaces, Bell Labs Tech. J. 10 (2005) 81-85. doi:10.1002/bltj.20105.

[29] Y. Si, Z. Guo, Superhydrophobic nanocoatings: From materials to fabrications and to applications, Nanoscale. 7 (2015) 5922-5946. doi:10.1039/c4nr07554d.

[30] M. Shateri Khalil-Abad, M.E. Yazdanshenas, Superhydrophobic antibacterial cotton textiles, J. Colloid Interface Sci. 351 (2010) 293-298. doi:10.1016/j.jcis.2010.07.049.

[31] K. Jeyasubramanian, G.S. Hikku, A.V.M. Preethi, V.S. Benitha, N.

Selvakumar, Fabrication of water repellent cotton fabric by coating nano particle impregnated hydrophobic additives and its characterization, J. Ind. Eng. Chem. 37 (2016) 180-189. doi:10.1016/j.jiec.2016.03.023.

[32] G.R.T. Suyambulingam, K. Jeyasubramanian, V.K. Mariappan, P. Veluswamy, H. Ikeda, K. Krishnamoorthy, Excellent floating and load bearing properties of superhydrophobic ZnO/copper stearate nanocoating, Chem. Eng. J. 320 (2017). doi:10.1016/j.cej.2017.03.052.

[33] T.S.G. Raja, K. Jeyasubramanian, Tuning the superhydrophobicity of magnesium stearate decorated ZnO porous structures for self-cleaning urinary coatings, Appl. Surf. Sci. 423 (2017) 293-304. doi:10.1016/j.apsusc.2017.06.188.

[34] H.E. Yong, K. Krishnamoorthy, K.T. Hyun, S.J. Kim, Preparation of ZnO nanopaint for marine antifouling applications, J. Ind. Eng. Chem. 29 (2015) 39-42. doi:10.1016/j.jiec.2015.04.020.

[35] S. Thangavel, K. Krishnamoorthy, V. Krishnaswamy, N. Raju, S.J. Kim, G. Venugopal, Graphdiyne-ZnO Nanohybrids as an Advanced Photocatalytic Material, J. Phys. Chem. C. 119 (2015) 22057-22065. doi:10.1021/acs.jpcc.5b06138.

[36] A.U. Dogan, M. Dogan, M. Omal, Y. Sarikaya, A. Aburub, D.E. Wurster, Baseline studies of The Clay Minerals Society source clays: Specific surface area by Brunauer Emmett Teller (BET) method, Clays Clay Miner. 54 (2006) 62-66. doi:10.1346/CCMN.2006.0540108.

[37] J. Li, X. Liu, Y. Ye, H. Zhou, J. Chen, A facile solution-immersion process for the fabrication of superhydrophobic surfaces with high water adhesion, Mater. Lett. 66 (2012) 321-323. doi:10.1016/j.matlet.2011.08.044.

[38] Y. Huang, D.K. Sarkar, X.G. Chen, A one-step process to engineer superhydrophobic copper surfaces, Mater. Lett. 64 (2010) 2722-2724. doi:10.1016/j.matlet.2010.09.010.

[39] N. Xu, D.K. Sarkar, X. Grant Chen, H. Zhang, W. Tong, Superhydrophobic copper stearate/copper oxide thin films by a simple one-step electrochemical process and their corrosion resistance properties, RSC Adv. 6 (2016) 35466-35478. doi:10.1039/c6ra01944g.

[40] Y. Zhang, P. Ju, C. Zhao, X. Qian, In-situ Grown of MoS2/RGO/MoS2@Mo Nanocomposite and Its supercapacitor Performance, Electrochim. Acta. 219 (2016) 693-700. doi:10.1016/j.electacta.2016.10.072.

[41] C.A. Schneider, W.S. Rasband, K.W. Eliceiri, NIH Image to ImageJ: 25 years of image analysis, Nat. Methods. 9 (2012) 671-675. doi:10.1038/nmeth.2089.

[42] S.A. Mahadik, F. Pedraza, R.S. Vhatkar, Silica based superhydrophobic coating for long-term industrial and domestic applications, J. Alloys Compd. 663 (2016) 487-493. doi:10.1016/j.jallcom.2015.12.016.

[43] B. V. Ramana, A. Das, S. Dhara, S. Amirthapandian, A.K. Tyagi, Synthesis and surface functionalization of SnO2 nanoparticles and their superhydrophobic coatings, Sci. Adv. Mater. 5 (2013) 865-872. doi:10.1166/sam.2013.1532.

[44] X.Q. Kong, J.L. Liu, W.J. Zhang, Y.D. Qu, Load-bearing ability of the mosquito tarsus on water surfaces arising from its flexibility, AIP Adv. 5 (2015) 037101. doi:10.1063/1.4908027.

[45] Q. Pan, M. Wang, Miniature boats with striking loading capacity fabricated

from superhydrophobic copper meshes, ACS Appl. Mater. Interfaces. 1 (2009) 420-423. doi:10.1021/am800116d.

[46] X. Zhang, J. Zhao, Q. Zhu, N. Chen, M. Zhang, Q. Pan, Bioinspired aquatic microrobot capable of walking on water surface like a water strider, ACS Appl. Mater. Interfaces. 3 (2011) 2630-2636. doi:10.1021/am200382g.

[47] J.L. Liu, X.Q. Feng, G.F. Wang, Buoyant force and sinking conditions of a hydrophobic thin rod floating on water, Phys. Rev. E - Stat. Nonlinear, Soft Matter Phys. 76 (2007) 066103. doi:10.1103/PhysRevE.76.066103.

[48] G. He, S. Lu, W. Xu, S. Szunerits, R. Boukherroub, H. Zhang, Controllable growth of durable superhydrophobic coatings on a copper substrate via electrodeposition, Phys. Chem. Chem. Phys. 17 (2015) 10871-10880. doi:10.1039/c5cp00059a.

[49] Self-cleaning flow shut off valve, 2010.

[50] Toilet bowl cleaning and/or deodorizing device, 2011. www.dripirrigation.com, (accessed November 7, 2020).

[51] N.B. Vargaftik, B.N. Volkov, L.D. Voljak, International Tables of the Surface Tension of Water, J. Phys. Chem. Ref. Data. 12 (1983) 817-820. doi:10.1063/1.555688.

[52] C.O. Mills, E. Ellas, G.H.B. Martin, M.T.C. Woo, A.F. Winder, Surface Tension Properties of Human Urine: Relationship with Bile Salt Concentration1), Clin. Chem. Lab. Med. 26 (1988) 187-194. doi:10.1515/cclm.1988.26.4.187.

Chapter 16

Novel Applications of Nanoparticles in Nature and Building Materials

Juyoung Ha

Abstract

Nanoparticles are assemblies of atoms in the size range less than 100 nanometers. At these length scales, the properties of particles may deviate significantly from those of the equivalent bulk material indicating that changes in physical and chemical properties of materials depend on the dimensions of the particle. The presence of mineral nanoparticles has been reported in a range of natural environments. Such nanoparticles can arise from a variety of mechanisms, including chemical weathering processes, precipitation from relatively saturated solutions in hydothermal and acid mine drainage environments, evaporation of aqueous solutions in soils, and biological formation by a variety of different microorganisms. Furthermore, recent increased applications of nanoparticles in different types of industries, including construction and building material manufacturing, have caused prevalent occurrences of different types of synthetic nanoparticles in the environment. In this chapter, a comprehensive reviews on occurrences and observations of naturally and anthropogeniccally generated nanoparticles in the environment and their characterization techniques will be discussed along with directions and suggestions for the future research topics and areas for nanomaterials.

Keywords: nanoparticle, nanotoxicity, geopolymer, Scanning transmission X-ray microscopy (STXM), X-ray absorption fine structure (XAFS)

1. Introduction

Nanoparticles (NPs) are assemblies of atoms in the size range less than 100 nanometers (i.e., one nm = billionth of a meter, which is roughly ten times the size of an individual atom). Apart from size, NPs may also be classified by of their physical parameters, such as electrical charge, chemical characteristics, shapes, and origin (natural vs. artificial). Regardless of how NPs are classified and defined, at these length scales, the properties of particles may deviate significantly from those of the equivalent bulk material indicating that changes in physical and chemical properties of materials depend on the dimensions of the particle. For example, at the surface of nanoparticles, significantly greater fractions of atoms are expected to be exposed and disordered rather than confined in the bulk crystalline structure for nanoparticles. This structural modification causes the nanoparticles to be more reactive and prone to greater dissolution than the bulk materials. In addition, thermodynamic analyses illustrate that adsorption on smaller particles decrease the free energy and the activation energy barrier of the system to a greater degree than adsorption on larger particles; thus, the driving force for adsorption onto smaller particles is larger and more favored. As a result, NPs possess higher surface reactivities than larger particles of the same phase, subsequently affecting the metal and organic ion availability and sequestration much more significantly than the larger particles of the same phase. However, more work still is needed to fully understand the chemical reactivity dependence on the particle size and the impact of the nano-sized particles in the environment.

The presence of mineral nanoparticles has been reported in a range of natural environments. Such nanoparticles can arise from a variety of mechanisms, including chemical weathering processes, precipitation from relatively saturated solutions in hydothermal and acid mine drainage environments, evaporation of aqueous solutions in soils, and biological formation by a variety of different microorganisms [1–9]. When compared with larger particles of the same material, mineral nanoparticles possess a number of unique and potentially important physical, chemical, and magnetic properties. Interestingly, while a number of these unique features of mineral nanoparticles have been extensively studied with respect to their applications in the medicine, pharmacotherapies, semiconductor, microelectronics, and catalysis industries, comparatively little is known about the properties of nanoparticles with respect to their potential importance in natural environments (e.g., enhanced adsorption coefficients and chemical reactivities). Recent studies have reported a ubiquitous presence of different types of nanoparticles in virtually all water domains, including the oceans, surface waters, groundwater, atmospheric water, and even treated drinking water [10–16]. Wigginton et al. [17] reported these naturally occurring environmental nanoparticles can play a critical role in determining an important chemical characteristics and the overall quality of natural and engineered waters. Moreover, aquatic nanoparticles have the ability to influence environmental and engineered water chemistry and processes in a much different way than similar materials of larger sizes. Zhu et al. [18] reported toxic effects of a range of metal oxide nanoparticles on zebra fish while Li et al. [19] investigated ecotoxicological impacts of metal oxide nanoparticles released to aquatic environments on *Ceriodaphnia dubia*, a species of water flea. They reported chronic exposure of nanoparticles induced a significant increase of severe stress response. Pakrashi et al. [20] also reported that the aluminum oxide nanoparticles play a significant role in the cytotoxicity towards freshwater algae. They identified that the surface charge driven interaction between the aluminum oxide nanoparticles and the cell surface functional groups are the dominating reaction mechanism resulting in the cell membrane damage and increased oxidative stress. Their study also elucidated the dissolution of the nanoparticles and release of Al^{+3} ions into the solution caused enhanced cytotoxicity. The particle aggregation of NPs and ion release from the nanoparticles will significantly alter the solution phase dynamics and the subsequent change will cause challenging ecological problems in understanding the impact of NPs in an environmental matrices.

At a broader Earth ecosystem scale, naturally occurring iron nanoparticles have been found to significantly impact global biogeochemical of various metals and metalloid ions [21]. Iron nanoparticles are of particular interest since iron oxide nanoparticle surface chemistry and the subsequent reactions within the interface between the water and iron nanoparticles determine the long-term fate and transport behavior for nutrients and pollutants in natural systems [22–24]. Iron oxide nanoparticles have been shown to be an effective agent for hazardous waste site remediation [25]. Tagliabue et al. [26] have shown that iron nanoparticles can drive primary productivity of the ocean where its biogeochemical systems rely primarily on iron and suggested that iron nanoparticles can play an integral role in global ocean biogeochemistry. Nowack and Bucheli [27] also provided evidence of ubiquitous presence of nanoparticles including those engineered ones in the aquatic environments and presented comprehensive results from ecotoxicological studies to show that only certain nanoparticles have effects on organisms under environmental conditions, though mostly at elevated concentrations. They argued that the assessment of the risks posed by nanoparticles in the environment should be re-considered and re-evaluated especially when the current available technology and analytical tools lack to measure materials in the nanometer scale. Recent article by Hyden et al. [21] reviewed the current analytical approaches that can be used to characterize natural Fe nanoparticles using a synchrotron-based X-ray spectro-microscopic techniques (more details in section. They measured suspended Fe nanoparticles collected from fluvial, marine, and lacustrine surface waters. They successfully identified different oxidative state Fe into ferrous, ferric and magnetite classes of Fe nanoparticles (10–100 nm). The heterogeneity of iron oxidation state within the collected samples was attributed to the possible presence of nanoparticle aggregates, and to the low degrees of crystallinity and ubiquitous presence of impurities in natural samples. Their results provided an important baseline for natural nanoparticle speciation in pristine aquatic systems and elucidated the importance of inter-particle variability, which should be considered as an important variable for making accurate biogeochemical models. Furthermore, their study suggests that the fate of released engineered or natural Fe nanoparticles must be considered as a time dependent kinetic reactions as they evolve and transform in natural systems.

This review covers recent advances made in identifying nanoparticles in aqueous phase from a variety of sources, and advances in understanding their very interesting properties and reactivity that affect the chemical characteristics and behavior of natural water and soil. More specifically, an overview of recent scientific advances enhancing the understanding of the (i) sources and (ii) fate of nanoparticles, (iii) the effects of nanoparticles in simplified studies, and (iv) how nanoparticles interact with biota in a more complex environment will be discussed.

2. Naturally occurring nanoparticles

From an environmental perspective, one of the most important features of mineral nanoparticles is their high characteristic surface area, which potentially allows them to act as powerful sink of contaminant ions through sorption processes. Higher mobility of nanomaterials in the environment is expected due to its colloidal properties and it implies a greater potential of exposure and persistence for nanomaterials in the environment. Classsical thermodynamic forces such as attractive London-van der Waals and attractive or repulsive electrical double-layer forces (e.g., the classic Derjaguin, Landau, Verwey and Overbeek or DLVO forces) that are known to influence particle attachment deviate significantly when particle size gets small in the nanometer range as shown in **Figure 1**. The stability of colloidal nanoparticles is greatly influenced by the presence of adsorbed natural organic matter. Without any functionalized surface modification in water, colloidal nanoparticles will tend to grow to become larger particles through aggregation and flocculation in order to stabilize the disordered structure of surfaces and to reduce the excess surface energy. Hence, understanding the surface reactions on nanoparticles are of particularly important and deserve more investigations as current models and predictions do not apply and cannot accurately predict the fate and transport of the nanoparticles in the environment.

This review attempts to investigate the environmental impacts of naturally occurring nanoparticles by studying their unique properties and sorption reactions

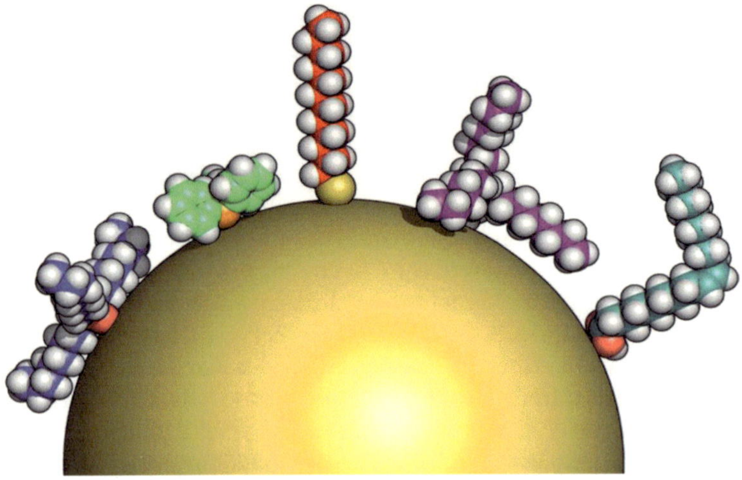

Figure 1.
A nanoparticle of 5 nm core diameter with different organic molecules drawn to scale.

with various environmental constituents and by comparing those results of the nanoparticles to that of microparticles to elucidate the role of nanoparticles in the environment. A wide range of nanoparticle phases has been reported in various environmental conditions as shown in **Table 1**. Of particular, the dependence of aquatic chemical reactivity of iron oxides on particle size is important because iron oxide plays a key role in the geocycling of elements, rock weathering and soil formation, as well as in the transport of aqueous metal species and contaminants. Iron oxide clusters are also found in living organisms (e.g., plants, bacteria, molluscs, fish, birds, and humans). They are also applied to (nano)technologies, including alternative energy, catalysis, electronics, optoelectronics, memory devices, corrosion protection, cleaning of waters and control of acid mine drainage, radioactive waste storage and disposal, flotation, pigments, magnetocaloric refrigeration, colour imaging, biochemical engineering, sensors, and other surface-based

Solid Phase	Typical size (nm)	Shape	Environments	Reference
Ferrihydrite	2–10	Irregular and spherical	Oxic surface water and soils, bacterial surfaces	[28]
Goethite	A few nm to micrometer	Acicular	Soils, water	[28]
Hematite	A few nm to micrometer	Rounded or platy	Oxic water, bacterial surfaces	[28, 29]
Magnetite	Several nm	Octahedron	Anoxic soils and sediments	[28]
Glauconite	A few nm to 100 nm	Acicular	Anoxic soils and sediments, bacterial surfaces	[28, 30]
Gold	< 200 nm	Platy	Groundwater, mining pits	[31]
Aluminum Oxide	< 200 nm	Rounded or platy		[32]
Manganese Oxide	A few to hundreds nm	Rounded or platy	Bacterial surfaces, anoxic soil and water	[33, 34]

Table 1.
Examples of naturally occurring nanoparticles in various environments.

applications. Such wide diversity in occurrence and application of iron oxides stems from the richness of their physical, chemical, and structural properties with continuous or sudden change between them, which in turn originates from the transition character of iron and the complex crystal and electronic structures of its compounds.

2.1 Iron oxide nanoparticles

Hematite (α-Fe_2O_3) is one of the most naturally abundant iron oxide mineral phases [12, 22–24, 28]. It is also a commonly present in a nanoparticle form, occurring in soils, acid mine drainage effluent, and on bacterial surfaces as well as in atmospheric dusts (**Figure 2**). Iron is one of the essential element in governing the biogeochemical cycling of nutrients in marines and sedimentary environments. During the iron cycling process in those environment, various nanostructures of iron oxides and oxyhydroxides form and persist under certain conditions, especially at redox and pH interfaces [29]. Recent surveys on the global budget of naturally occurring iron oxide nanostructures suggest 105 Tg (teragram) of iron oxide including hematite phase is introduced annually in soil by mass [30]. The ubiquitous existence of hematite nanoparticles has a significant implication on fate on toxic heavy metal contaminants. For example, heavy metals such as Cr and U that are introduced to environments by anthropogenic activities such as mining and spills can be effectively sequestered by iron oxide nanostructures through sorption and precipitation reactions due to the thermodynamically active nanomorphologies and crystallinity of the nanparticulate phase. It has been shown that the adsorption capacities of Cr and U by ferrihydrite decrease remarkably with either increasing

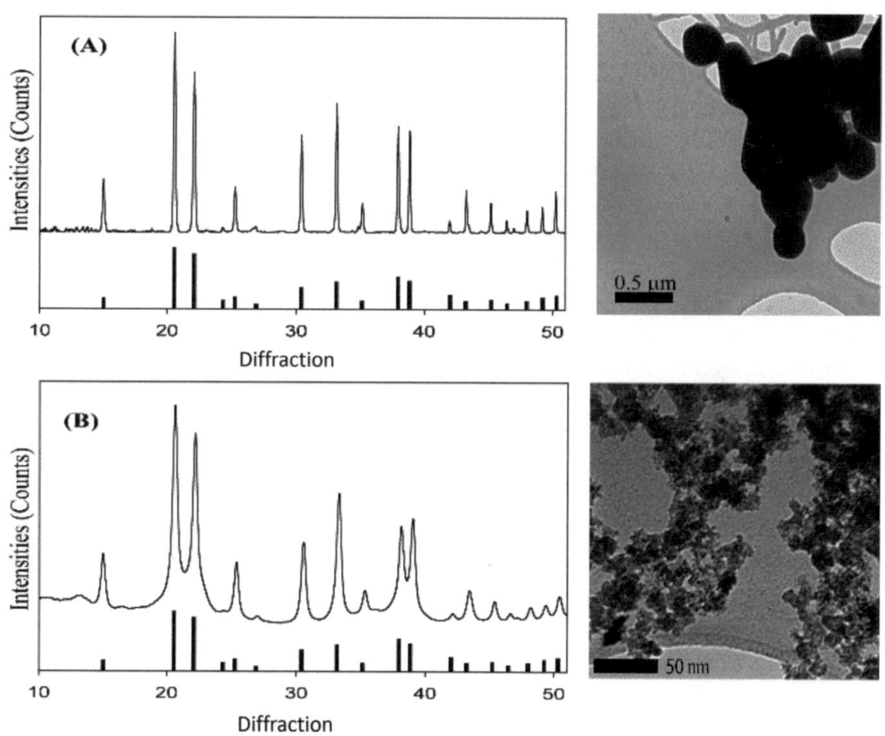

Figure 2.
(A) Powder X-ray diffraction pattern of hematite microparticles (HM) compared with the reference XRD pattern of hematite46 and the TEM image of HM; (B) Powder X-ray diffraction pattern of hematite nanoparticles (HN) compared with the reference XRD pattern of hematite.

crystallinity or transformation to more crystalline phases (such as goethite or hematite) further confirming the effectiveness of the nanoparticle phases in contaminant removal [31–33].

Moreover, hematite nanoparticles recently has received much attention from industry for different applications due to their unique properties, such as extremely small size, high surface-area-to-volume ratio, surface modifiability, excellent magnetic properties and great biocompatibility. A range of environmental clean-up technologies have been proposed in wastewater treatment which applied iron oxide nanomaterials as nanosorbents and photocatalysts [34]. Nanoscale zero-valent iron (nZVI) is an example of extensively applied iron nanomaterials for groundwater and hazardous waste treatment. Over the past decade, nZVI synthesis and application have been comprehensively investigated for its remediation applications focusing on enhanced sequestration of a wide spectrum of contaminants in addition to the well-documented chlorinated solvents both in the laboratory and field experiments [25, 32, 35–37]. At least 50 successful field applications of nZVI for in-situ groundwater and soil cleanup worldwide were reported by recent reviews by Karn et al. [38] and Mueller et al. [39]. As the application of the nZVI gained more attention, colloidal stability and transport properties of nZVI in porous media, and the effects of nZVI amendment on the biogeochemical environment were also studied in order to understand the impacts of the nZVI in the environment once released [25, 35, 40]. Dong et al. [40] observed the presence of humic acid increases the stability of nZVI in the aqueous phase due to enhanced the electrosteric repulsion effect but it can also cause coagulation among nZVI particules via bridging effect if too much humic acid is present. Hence, the nanoparticle stability and transport behavior depends on the concentrations of other environmental constituents, especially organic acids, both in aqueous and soil environments. Further studies are needed to enhance the colloidal stability and transport properties of nZVI in porous media to fully understand effects of nZVI on the biogeochemical environment.

Lastly, iron oxide nanoparticles have received a great attention recently in biomedical applications due to their non-toxic role in the biological systems [41–43]. Iron oxide nanoparticles have both magnetic behavior and semiconductor property, which lead to multifunctional biomedical imaging applications. According to Chen et al. [41] gold coated nanoparticle of iron oxide has enhanced magnetic properties compared to the same phase of larger particles. Cheong et al. [43] similarly reported greater cellular MRI contrast enhancement of nanoparticles compared to iron oxides of a bigger size without increase in cytotoxicity. Iron oxide nanoparticles also became popular for its application in biomedical fields as antibacterial, antifungal and anticancer agents as well as bone marrow treatments and cell labelling activities for its unique biocompatibility, biodegradability, ease of synthesis and different magnetic behaviors [44].

In addition to the above reported laboratory synthesized nanoparticles, iron oxide nanoparticles exist in nature at low-temperature environmental conditions and places that have high degrees of supersaturation. The supersaturation condition is typically created by changing the physical and chemical conditions, such as influx of Fe(II)-rich hydrothermal vent fluids, mixing of highly acidic solutions with neutral pH water, and the evaporation of soil solutions [12, 22–24, 45, 46]. Once formed, iron oxide and oxyhydroxide nanoparticles are redistributed by rivers, glaciers, winds, and ocean currents into various ecosystems, where they can undergo continuous phase transformations, dissolution, and morphology changes. One potentially important role played by naturally occurring hematite nanoparticles is their interaction with various types of organic acids. In many natural environmental settings, ubiquitous presence of naturally occurring low-molecular weight (LMW) organic compounds and nanoparticles often controls the fate and transport behavior of many heavy metal contaminants by sorption and/or co-precipitation processes. For example, mobile nanoparticles can serve as carriers for strongly sorbed contaminants and thereby facilitate contaminant transport in soils, groundwater aquifers, and fractured rocks [21, 25]. The colloidal stability of nanoparticles is greatly influenced by the presence of adsorbed natural organic matter. Recent studies have suggested that sorption of organic acids can dramatically enhance the particle-based transport of heavy metal contaminants by physically stabilizing contaminant-containing nanoparticles in aquatic environments [47, 48]. In addition, the organic matters typically bind strongly to common iron- and aluminum-(oxy)hydroxide minerals as well as to heavy metal contaminants (e.g., Pb, Hg, Cr, and Zn) under circumneutral to moderately acidic pH conditions [47, 49, 50]. Therefore, the nature of bonding between organic species and nanoparticle surfaces can substantially alter the properties of mineral nanoparticle-water interfaces and thereby affect the geochemical cycling of metals.

Ha et al. [51] studied the interaction of the L-lactate ion (L-$CH_3CH(OH)COO^-$, $Lact^{-1}$) with hematite (α-Fe_2O_3) nanoparticles (average diameter 11 nm) in the presence of bulk water at pH 5. Their combined dissolution and ATR-FTIR spectroscopy data suggested different hydrogen bonding environments was found as $Lact^{-1}$ surface coverage on hematite nanoparticle surfaces increases which resulted in a concomitant increase in Fe(III) dissolution from the hematite nanoparticles due to the inner-sphere complex formation as shown in **Figure 3**.

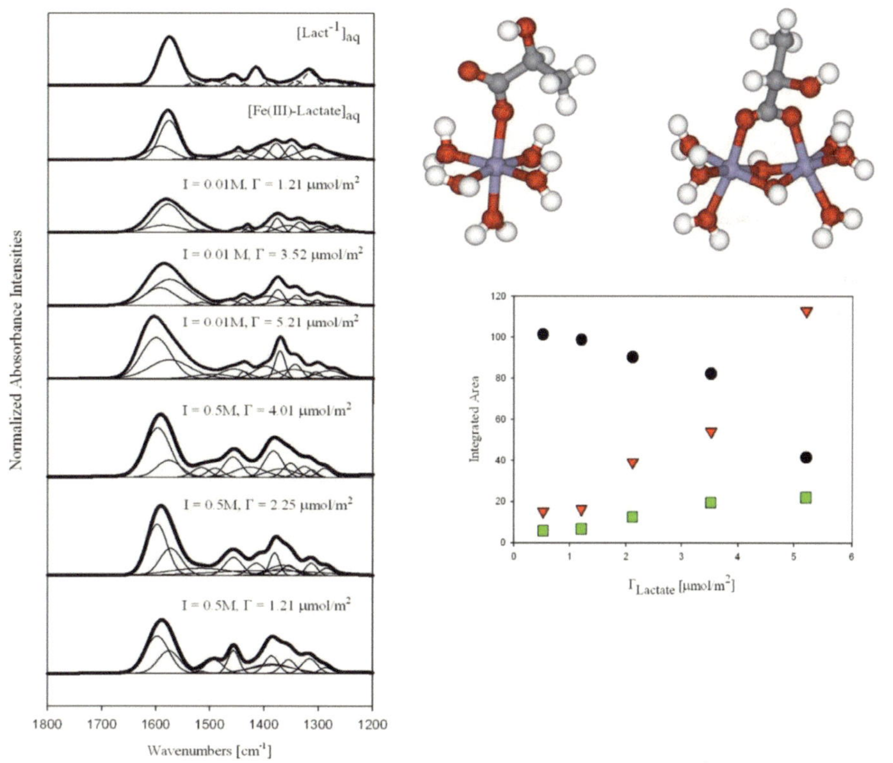

Figure 3.
(Left) Fit of ATR-FTIR spectra of aqueous deprotonated lactate species, aqueous Fe(III)-lactate complex, and Lact-1 species at the hematite/water interface at 250C, pH 5.0 and 0.01 M and 0.5M of NaCl for different Lact-1 surface coverages. (Top Right) Molecular model for Lact-1 sorption on monomer and dimetric Fe cluster unit. (Bottom Right) Fit of ATR-FTIR spectra of aqueous deprotonated lactate species, aqueous Fe(III)-lactate complex, and Lact-1 species at the hematite/water interface at 250C, pH 5.0 and 0.01 M and 0.5M of NaCl for different Lact-1 surface coverages demonstrating different surface complexation

Lenhart et al. [52] also reported the impacts of fumaric acid and maleic acid, which are naturally occurring dicarboxylic acids, on aggregation kinetics of nano-sized hematite. Interestingly, they found that the structure and orientation of the adsorbed dianion at the hematite surface, not the adsorption mechanism, defined the resulting effect. Maleate, which directed both carboxyl groups to the surface in the form of inner- and outer-sphere surface complexes, enhanced colloidal stability. Fumarate, however, which bounded to the hematite surface as an outer-sphere complex with just one carboxyl group only slightly influenced particle stability. Their research outcome suggested that subtle differences in the structure of adsorbed acids produced important differences in the physicochemical behavior of particles in dilute aquatic systems. Another study by Dickson et al. [47] similarly suggested that the surface chemical reactions and dissolution of the iron oxide nanoparticles strongly depend on the presence of organic molecules. They found significantly increased aggregation and sedimentation reactions of iron oxide nanoparticles in presence of humic acids regardless of different ionic strength of the solution. Their results suggested that understanding the effects of important environmental factors on the stability of nanoparticles is key to fully and accurately be able to predict the mobility of the nanoparticles in aquatic environment.

On the contrary, Palomino and Stoll [53] identified that iron oxide nanoparticle aggregation process significantly depends on the solution pH conditions even in the presence of fulvic acids. Dispersion and stability of the nanoparticles were maintained only at pH conditions of which the surface charge of the nanoparticles remained positive regardless of the fulvic acid concentrations. Their results showed that the examined environmentally relevant range of fulvic acid concentrations were expected to promote not only the nanoparticle stabilization but also the disaggregation of nanoparticle aggregates, but only at very low concentrations of fulvic acid and specific solution pH conditions. This finding suggests that hematite nanoparticle behavior in natural aquatic environments are much more complex and dynamic that it cannot generalized or linearly predicted based on a single parameter. A recent study by Xu et al. [50] presented laboratory experimental results to simulate more environmentally relevant conditions by investigating the hematite nanoparticles coated with naturally found peat humic acid and soil humic acid. Their model system reflects the fact that hematite nanoparticles exist naturally and ubiquitously in soil, and they are always associated with soil organic matter by forming organic–inorganic complexes. In this work, the organic coated hematite nanoparticles reacted with hydrophobic organic contaminants (HOCs) to simulate the sorption processes in soil. The sorption of HOCs on organic acid coated nanoparticles were inhibited with increasing pH values of solution due to the deprotonation reaction of the organic acid functional groups within the adsorbed humic acids. Their findings further elucidated the mechanisms involved in contaminant sorption processes by organic acids coated hematite nanoparticles are complex reactions governed both by the surface complex structures formed within the mineral-water interfaces as well as the solution chemistry of the aquatic environments. Results on sorption of a commonly occurring pollutant, Zn(II), on hematite nanoparticles in presence or absence of dicarboxylate organic compound, oxalate, also identified the complexity of the nanoparticle reactions with naturally occurring constituents [54, 55]. At higher concentrations of Zn(II), formation of surface precipitates on hematite nanoparticles was observed based on comparison of the EXAFS spectra of the sorption samples with that of zinc-bearing hydrotalcite ($Zn_6Al_2(CO_3)(OH)_{16} \cdot 4H_2O$), Cauchy wavelet analysis, and fitting of the Zn K-edge EXAFS data. On the contrary, no surface precipitate was observed on the bigger size hematite particles even at the same concentrations of Zn(II) suggesting enhanced dissolution of hematite nanoparticles was promoted by the divalent metal ions in solution (**Figure 4**). Similarly, different types of Zn(II) surface complex structures formed in the presence of oxalate on hematite nanoparticle surfaces. These studies provide a direct comparative and quantitative evidence that different surface chemical reactions on hematite nanoparticles persist.

There are still many open questions regarding the ability of surface reactions to alter the fate and toxicity of nanoparticles in the environment. Evidence so far suggests that surface coatings and complexation with organic acids affect nanoparticle fate and transport mechanisms hence consequently impact nanoparticle toxicity in the environment. However, as surfactant adsorption on silica nanoparticle studies revealed that such surface complexation reaction can be readily reversible so that desorption can be kinetic reactions whereas adsorption for high molecular weight polymers is essentially an irreversible process [56, 57]. Clearly, the fate of nanomaterials in the environment is highly dependent on their surface coatings and reactions, hence, it is imperative to understand the fate of these coatings and surface reactions of organic matter with nanoparticles in the environment.

2.2 Nanoparticles and microorganisms

Bacteria can be considered as another extension of organic compounds in the natural environment. Similar to LMW organic acids, they have been shown to exhibit a strong affinity for heavy metal contaminants and metal (oxyhydr)oxide surfaces through reactions such as sorption, bioaccumulation, and precipitation [52, 58–60]. With their estimated biomass close to the total amount of carbon in plants [61], they can potentially passivate naturally occurring mineral surfaces

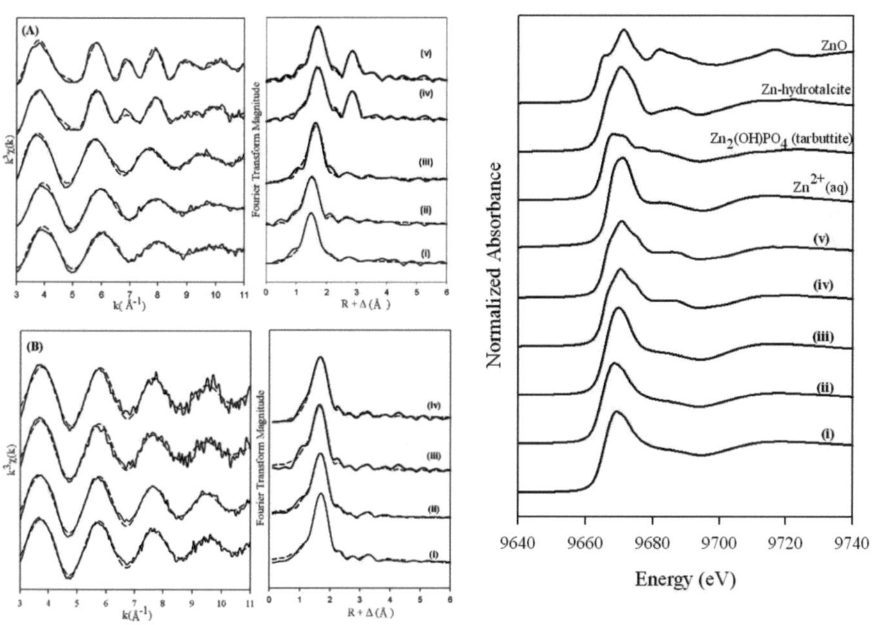

Figure 4.
(Left) (A) Background-subtracted, normalized, and k3-weighted Zn K-edge EXAFS spectra of Zn(II) sorbed on hematite nanoparticles (HN) at different Zn surface coverages, and corresponding Fourier transforms (not phase-shift corrected) of sorption samples with following surface concentration: (i) 68.35 micromol/g; (ii) 195.80 micromol/g; (iii) 384.92 micromol/g; (iv)568.49 micromol/g; (v) 640.19 micromol/g; (B) Background-subtracted, normalized, and k3-weighted Zn K-edge EXAFS spectra of Zn sorbed on hematite microparticles (HM) at similar Zn surface coverages, and corresponding Fourier transforms (not phase-shift corrected) of sorption samples. (Right) Zn K-edge XANES spectra of Zn/HN sorption samples and four reference compounds: (i) 68.35 micromol/g; (ii) 195.80 micromol/g; (iii) 384.92 micromol/g; (iv) 568.49 micromol/g; (v) 640.19 micromol/g.

either hindering or enhancing the reactivity of the underlying mineral surfaces. The resulted the cell-mineral surface interactions occurring in natural soil and water environments can retain pollutants in great extent [62, 63]. Therefore, the presence of microorganisms can result in significant modification of metal speciation or contaminant sequestration and transport, and hence, investigating the interfaces of bacteria-metal and bacteria-mineral surfaces is essential in predicting the mobility of heavy metal contaminants accurately and quantitatively.

More importantly, understanding the dynamics of the bacterial interaction with engineered and/or naturally occurring NPs have received significant attention because the fate and transport reactions and mechanisms of them in the terrestrial and aquatic environments strongly depend on the nature and extent of the bacterial sorption and reduction of the nanoparticles [64]. For example, Schwegmann et al. [65] observed iron oxide nanoparticle sorption on microorganisms (*Saccharomyces cerevisiae* and *Escherichia coli*) significantly shifted the point of zero charge for bacteria. Their results imply that overall electrostatic interaction between dissolved heavy metal contaminants and bacteria can be considerably altered due to the presence of the nanoparticles. Another study showed an enhanced removal of the heavy metals from the solution phase in conventional activated sludge wastewater treatment plants when nanoparticles were added to reactors including different types of bacteria. This study suggested that the production and transformations of the surface properties of biomass due to NP would be key factors in determining the fate of the waste toxic metals in the environment [66]. Droz et al. [67] observed biogenic MnO2 nanoparticles affected a wide range of metal fate and transport in natural and engineered systems by strongly sorbing metals ions.

Of particular importance, the dissimilatory microbial iron reduction is one of the important processes in determining the biogeochemical cycling of iron under anaerobic conditions and has significant influences on the aqueous geochemistry and mineralogy of sedimentary environments [68, 69]. Iron cycling has dramatic implications for trace element and heavy metal contaminant mobility, and the complex interplay of biological and chemical processes determine the extent and reaction pathways of iron cycling in the environment [69]. Past studies have identified many different chemical and biological factors controlling the microbial reduction of iron oxides, yet the role of nanoparticles in determining the extent and reaction pathways for dissimilatory iron reduction are still poorly understood. Despite the important roles of nanoparticles, current molecular-level understanding on the interaction of organic compounds, metals, and microorganisms with nanoparticle surfaces and the resulting impacts on pollutant speciation at nanoparticle-water interfaces are very limited. One of the main reasons for this limitation is lack of appropriate tools with high resolution and/or high sensitivity to molecular structural changes occurring on nanometer scales. In addition, ability to study the materials in situ, i.e., in presence of water with minimal alteration of the sample from its natural state, has been limited. For many environmental samples, especially those involving interactions at solid-water interfaces, the presence of water plays a critical role.

Due to the small size of particles, different mechanisms and processes have been observed when bacteria interacts with nanoparticles. For example, a greater bioavailability of iron was observed when iron nanoparticle reacted with *Pseudomonas mendocina* bacteria [70] because of the possible penetration of the nanoparticles through the cell membrane and proximity of the particles to the bacteria. Sulfate-reducing bacteria (*Desulfovibrio desulfuricans*) also showed different metabolic responses in presence of iron oxide nanoparticles and generated biogenic pyrrhotite formation suggesting a potential impact of iron oxide nanoparticles on geomagnetic field behavior of sediments. Different dissimilatory reduction of hematite nanoparticles and microparticles were also observed for *Shewanella oneidensis* bacteria [71]. Results show that proximity and encapsulation of the nanoparticles near the cell membrane allowed different iron oxide reductive mechanisms resulting in a significantly enhanced iron oxide reductive transformation rates by *Shewanella oneidensis* (**Figure** 5). Clearly different reaction pathways and microbial responses toward nano-meter materials are present compared to the larger particle size even for the same mineral phases.

Nanoparticles are also generated by a biogenic enzymatic process and used as engineered materials. The development of eco-friendly technologies in material synthesis has become important and widely applied. Synthesis of nanoparticles using different microorganisms, and their applications in many cutting-edge technological areas have been explored. Recent study shows that biogenically generated iron(III) (oxyhydr)oxide ($Fe(OH)_3$) clusters by gram positive bacteria *Clostridium* could be used to create lithium storage capacity [72]. This study provides another potential use of hematite nanoparticles as a substitute for an industrial product. Other studies also have found that many microorganisms can produce different types of inorganic nanoparticles through either intracellular or extracellular routes, and such biosynthesized nanoparticles have been used in a variety of applications including drug carriers for targeted delivery, cancer treatment, gene therapy and DNA analysis, antibacterial agents, biosensors, enhancing reaction rates, separation science, and magnetic resonance imaging (MRI) [50, 73–75]. As shown, there have been tremendous developments in the field of microorganism-produced nanoparticles and their applications over the last decade. However, much work is needed to improve the synthesis efficiency and the control of particle size and morphology.

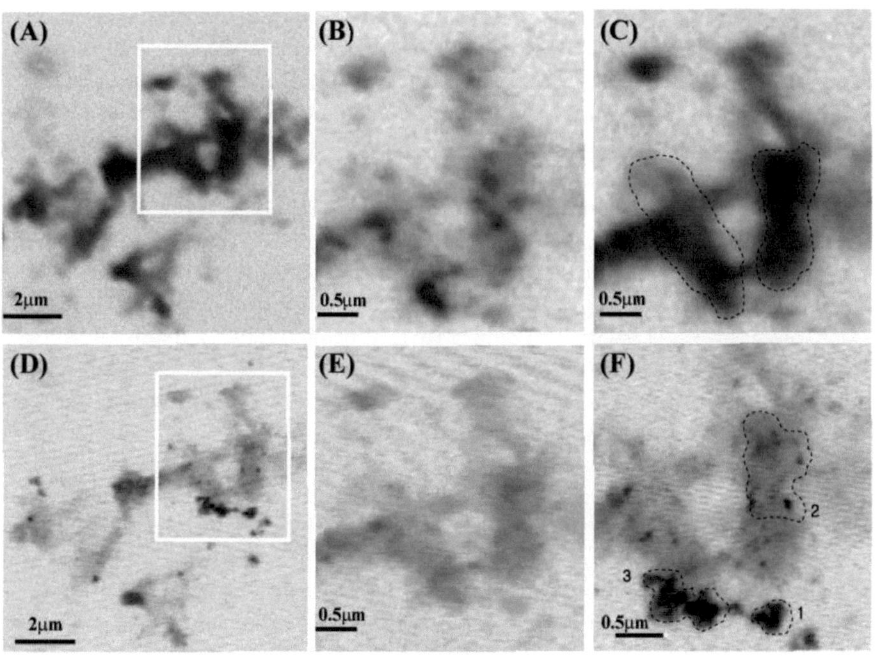

Figure 5.
STXM images of S. oneidensis WT strain and hematite nanoparticle reacted for 98 hr under anaerobic condition at initial hematite concentration of 1mM in an aqueous suspension measured at (A) C K-edge (288.2 eV) and (D) Fe L-edge (709.5 eV). The outlined areas labeled in (A) and (D) are shown in higher magnification at (B) below C K-edge (280.0 eV), (C) C K-edge (288.2 eV), (E) below Fe L-edge (700.0 eV), and (F) Fe L-edge (709.5 eV).

2.3 Spectroscopic and theoretical tools used to study nanoparticle-water interface

Synchrotron-based [32, 54, 55, 71]. X-ray absorption spectroscopy (XAS) is one of the most structurally and chemically sensitive tools probing solid-water interface [32, 54, 55, 71]. It has been used in many different scientific research fields such as material sciences, chemistry, and environmental sciences [76–79] . Because XAS is non-destructive and element sensitive, it allows to probe the interface between the solid and liquid phases in-situ to elucidate the chemical coordination of metal ions in many different matrices. Furthermore, XAS can simultaneously characterize both the amorphous and crystalline portions of samples which makes it to be one of the most suitable tools to investigate biological and environmental samples because many samples involving nanoparticles and/or environmental matrices are often either non-crystalline or in a solution matrix [80–82]. Much advances in molecular understanding of the interactions of ions in solution with nanoparticle surfaces and the structures of complexes are made since XAS has been used as a research tool. XAS spectroscopy consists of two different complimentary techniques, extended X-ray absorption fine structure (EXAFS) and X-ray absorption near edge structure (XANES) [76, 78, 80, 83–85]. XANES provides information on the oxidation state, three-dimensional geometry, and coordination environment of the element under investigation while EXAFS provides information on the coordination environment and nearest neighboring atoms to the atom of interest. Of particular, the Extended X-ray absorption fine structure (EXAFS) analyze the oscillatory variation of the X-ray absorption as a function of photon energy beyond an absorption edge. The absorption, normally expressed in terms of absorption coefficient (μ), can be determined from a measurement of the attenuation of X-rays upon their passage through a material. When the X-ray photon energy (E) is tuned to the binding energy of some core level of an atom in the material, an abrupt increase in the absorption coefficient, known as the absorption edge, occurs. For isolated atoms, the absorption coefficient decreases monotonically as a function of energy beyond the edge. For atoms either in a molecule or embedded in a condensed phase, the variation of absorption coefficient at energies above the absorption edge displays a complex fine structure called EXAFS. This is how EXAFS can be used to probe the molecular sorption complex of different metal ions on various surfaces.

In the last 15 years, X-ray absorption spectroscopy (XAS) has found wide application in determination of the local atomic and electronic structures of absorbing centers (atoms) in materials science, physics, chemistry, biology, and geophysics. With its elemental specificity and ability to determine the molecular-scale speciation in situ even at the parts per million concentration range, EXAFS has become a useful method for the analysis of environmentally relevant elements in natural sediment and soils, and in laboratory model system studies involving nanoparticles [83, 86–88]. Quantitative measures of interatomic distances and coordination numbers for the first and second coordination neighbors around specific elements can be obtained based on EXAFS analysis, and these information, especially of the second-shell coordination, provide the means to construct surface complex structures in which an adsorbate ion is bonded to a sorbent surfaces. Sheng et al. [89] studied nanoparticle zero-valent iron (NZVI) interaction with uranium using EXAFS. They found that reduction of highly toxic and mobile $UO2^{2+}$ into less toxic and mobile UO2 could be enhanced using NZVI. Their EXAFS analysis provide evidence to support the proposed detailed reaction mechanisms the iron nanoparticles sequester insoluble products like UO2, and thus more reactive sites could be used for U(VI) reduction and increase the rate and extent of the overall reductive reactions. Others also successfully characterized CuO nanoparticles and nanocomposites crystal structures and the detailed local electronic structures using X-ray absorption near edge structure (XANES) and extended X-ray absorption fine structure (EXAFS) spectroscopy [85]. Frenkel [90] also demonstrated that mono- and heteroatomic nanoparticles can be modeled using EXAFS experimental data and theory the multiple-scattering FEFF6 theory. Combined with the results of other experimental evidence such as transmission electron microscopy and electron diffraction, EXAFS analysis can be used to determine the size and shape of the nanoparticles with much higher accuracy than any other experimental tools.

Synchrotron-based scanning transmission x-ray microscopy (STXM) is another powerful spectroscopic tool capable of both imaging and spectroscopic speciation for samples with ~30 nm spatial resolution under fully hydrated conditions [91–94]. The emergence of combination of microspectroscopic and fluorescence-based techniques with imaging has permitted investigations of carbon containing materials such as microorganisms at nanometer scale. Imaging of nanoparticle samples coupled with X-ray absorption near edge structure (XANES) spectroscopy provides an excellent opportunity to identify and fingerprint the fine structures of carbon and directly image micro to nanometer sized environmental samples with nanometer spatial resolution. More recently, STXM and NEXAFS spectromicroscopic analyses has been effectively employed to investigate soil carbon and mineral associations at a nanometer scale. For example, using the STXM technique, Obst et al. [92] found that cyanobacteria produced an amorphous or nanocrystalline calcium carbonate phase with a short-range structure order. STXM results provided direct evidence that the bacteria induced the formation of the nanocrystliine as part of their metabolic activity on the extracellular polymeric substances (EPS). Because of the spectral data and mapping information provided by STXM technique, they were able to discriminate the nucleation of the amorphous aragonite-like nanoparticles were taking place within the cyanobacteria cell wall structure. Lawrence et al. [95] also reported for the first time the spatial correlation between the copper nanoparticles and natural river biofilms based on the STXM experimental data. Their results of copper nanoparticles dissolved and redistributed in the biofilm lipid and polymers have significant implications on the fate and transport mechanism of the metal particulates in the natural environments. Several other studies were also carried out successfully using STXM to elucidate the nanomaterial sorption reaction with subsequent dissolution and/or re-sorption chemodynamics with bacterial surfaces [96, 97]. Leuf et al. [97] used STXM and identified iron-reducing bacteria accumulate ferric oxyhydroxide nanoparticle aggregates that may support planktonic growth suggesting the aquatic food chain can be substantially altered due to the presence of the nanoparticles. Clearly, the advances in the spectroscopic and microscopic tools and techniques enhance our understanding of the nanoparticle dynamics with other environmental constituents and their roles in the ecosystem as well as the subsequent effects on the long term fate and transport of other metal contaminants.

3. Nanoparticle applications

Nanomaterials have gained much attention due to their unique properties. Many engineered nanoparticles, such as nano silver (nano-Ag), nano TiO2 (nano-TiO2), nano aluminum oxide (nano-Al2O3) and carbon nanotubes (CNT), have found potential commercial applications in catalysis, biomedical researches, medicinal applications, etc. Their subsequent impending effects on the ecosystem have also raised concerns over human health and environments as evidence has suggested that engineered nanomaterials are likely to present potential risks. Numerous active research and studies are carried out to analyze the health and environmental risk assessments. Although none of the analytical techniques employed provide unequivocal evidence as to the nature and impacts of the nanoparticles, advances in the analytical tools and techniques allowed to significantly enhance our understanding on the properties and roles of nanoparticles. One of the recent emerging fields for nanoparticle applications is agriculture and construction. In this review, examples of nanoparticle usages in these two fields of practices to date and the challenges associated with nanoparticle applications are introduced.

3.1 Nanoparticle and bioichar

Biochar (BC) is a carbon-rich product generated by pyrolysis of biomass such as wood, crop residues, and manure in a closed container with little or no oxygen. Creating biochar is a carbon-negative process because synthesizing BC utilizes naturally decaying organic matters to turn into useable sustainable materials while bypassing the release of CO2 into the atmosphere. Studies have shown that BC can substantially enhance physicochemical and biological properties of the soil to increase the plant and crop yields [98–101]. A number of remediation strategies using BC as sorbents are developed to remove toxic contaminants from polluted soils because of the high surface areas and presence of various carbon functional groups in BC. Furthermore, chemical structure of BC is stable and highly resistant to microbial degradation due to the recalcitrant nature of BC hence allowing BC to be an alternative growing medium or substrates for agronomic applications. For example, in 2010, the American Society of Agronomy Soil Science Society of the American Environmental Quality Division identified the agronomic applications of BC to be one of the immediate challenges and solutions to address the global food shortage. Clearly, more research on BC and its application to agronomic industry is urgently necessary for globally sustainable agricultural practices.

Recent report by Saxina et al. [102] provided evidence of biochar contains carbon nanoparticles. Presence of these carbon nanoparticles in the biochar significantly enhanced the soil fertility and nutrient retention compared to the ones without the carbon nanoparticles. Others also reported the properties of biochar containing the NPs differ from the properties of their macro-counterparts [103]. Different zeta potential, cation exchange capacity, elemental compositions, and aromaticity/polarity not only based on the type of biochars, but also the regularities in the differences between their macro- and nano-structure were observed suggesting the nanoparticles have substantial impacts on governing the overall properties of the biochar. They accounted that the larger surface area and smaller pore sizes than the corresponding macro-biochars are the main physicochemical properties. On the contrary, Chen et al. [104] reported nanoscale biochar particles may carry the inherent contaminants along the soil profile, posing a potential risk to the groundwater. Their study on the transport and retention of wood chip-derived biochar nanoparticles (NPs) in water-saturated columns packed with a paddy soil identified mobility of biochar NPs in natural soils must be taken into consideration for accurately assessing their environmental impacts. **Figure 6** illustrates the effects of different type of nanoparticle modifications on biochars to enhance the contaminant removal and transformation reactions of a wide range of environmental and anthropogenically abundant toxic matters.

In addition to the laboratory studies on biochar nanoparticles, field-scale experiments were carried and provided promising evidence on the impacts of the biochar nanoparticles to increase food production and remediation. Su et al. [105] synthesized biochar-supported zero-valent iron nanoparticles and used in the remediation of Cr(VI)-contaminated soil. They found removal of Cr(VI) was significantly

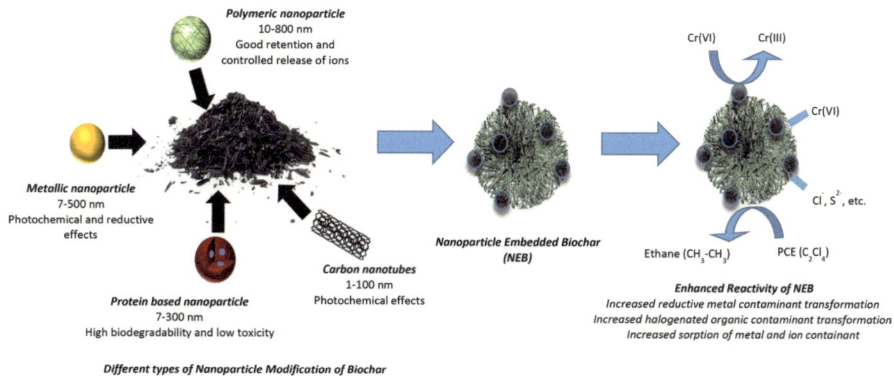

Figure 6.
Schematic representation of biochar modifications by different types of nanoparticles and the subsequent enhanced reaction mechanisms in transformation of the contaminants in the environment.

improved and observed substantially increased cabbage mustard growth as the phytotoxicity of Cr(VI) and Fe in the seedlings was effectively decreased by the biochar nanoparticle treatment in the field. Yue et al. [106] also demonstrated that application of the biochar nanoparticles on the rice plants significantly alleviated the phytotoxicity of Cd2+. However, their study reported that the nano-biochar significantly increased the contents of antioxidative enzyme activities indicating that nano-biochar could also induce oxidative stress in the rice plants. These results indicate that nano-biochar could greatly reduce the uptake and phytotoxicity of Cd2+, but its potential risk should not be overlooked during the environmental and agricultural applications of biochar. Clearly more studies are needed before fully applying nanoparticles in biochar synthesis or application to the fields especially focusing on the the interaction between biochar nanoparticles and plant roots in the rhizosphere where most of the physico-chemical reactions between the plants and nutrients occur.

3.2 Nanoparticle and cement

Portland cement based materials like concrete, mortar, fiber reinforcement and others are widely used building materials. About 4.1 billion metric tons of cement were produced in 2020 worldwide and a continuously increased demand is expected, elevating the environmental impacts related with this worldwide industry. Efforts have been made to mitigate the environmental impacts of cement production by creating different synthesis approaches and alternative methods for production. For example, geopolymeric cements are developed and commercially produced to reduce the carbon footprint of cement production, while also being highly durable and comparable to the traditional cements [107–110]. Recently, nano-engineered cement based materials is also actively studied because materials, such as nano-SiO2, nano-TiO2, nano-Fe2O3, nano-Al2O3, nano-CaCO3, nano-ZnO2, nano-cement particles of C2S and C3S phases, nano-clays, and Carbon Nanotubes, can work as effective binder to improve the cement based materials performance [111]. The addition of nanoparticles in cementitious materials can act as a filler agent, producing a dense matrix and reduce the growth of micro pores. Some nanoparticles also help in the secondary reactions forming cement composite and contribute to the strength development. Moreover, with the advancement on the analytical tools, the current knowledge of the microstructure, mechanical strength and durability of cementitious materials when incorporating different types of nanoparticles increased during the past decades [112–114].

For example, Said et al. [115] demonstrate a wide range of nanoparticles like Al_2O_3, C–S–H-phase, and quartz have the best potential as accelerators for cement hydration and therefore are most suitable for cement production based on their result of heat flow calorimetry measurements, particles size of nanomaterials and their percentage in the cement paste. Gaitero et al. [116] also identified that the addition of small amounts of silica nanoparticles to the cement paste could substantially slow down a degradation process due to the progressive dissolution of the cement paste. Nazari and the co-workers [111] also identified nano-phase Al_2O_3 particles with the average diameter of 15 nm increased flexural strength of fresh concrete and the extent of this positive effects of nanoparticles amplified as the content of Al_2O_3 nanoparticles increased. It is concluded that partial replacement of cement with nanophase Al_2O_3 particles improves the split tensile and flexural strength of concrete but decreases its setting time. Kawashima et al. [112] combined two nanomaterials, namely calcium carbonate nanoparticles and nanosilica, to synthesize cementatious materials and observed the mixture can offset the negative effects of fly ash (i.e., precursor for cements) on early-age properties to facilitate the development of a more environmentally friendly, high-volume fly ash concrete. Lastly, Land and his co-workers [113] also investigated the effects of colloidal nano-silica on concrete incorporating cement binders and found that the concentrations of nano-silica determined the micro-structural and thermal property changes. Many studies to date suggest that utilizing nanoparticles in cement synthesis can benefit the system by improving many of the structural properties. Further in-depth study on the mechanisms underlying the influence of nanoparticles on the compressive strength gain and other physical properties of cement systems need to be carried out to fully assess the impacts of nanoparticles.

Lastly, another example of where nanotechnology made a tremendous impact on commercial building materials is synthesis of nano-paint. For example, Krishnamoorthy et al. [117] developed a multifunctional graphene oxide (GO) nanopaint by incorporating GO sheets in an alkyd resin of the paint. The prepared GO nanopaint exhibited enhanced corrosion-resistant behavior in both acidic and high-salt-content solutions as well as substantial inhibition of the bacterial growth on its surface. This in-situ biofouling test results demonstrated incorporating the nanoparticles into the conventional paint materials produce significant performance benefits. Another study also concluded that preparation of titanium oxide (TiO2) nanopaint by embedding the TiO2 nanoparticles in alkyd resin matrix exhibited substantially enhanced antibacterial properties against a wide range of different types of bacterial strains [118]. Others also reported similar findings of enhanced corrosion-resistant behavior and anti-bacterial properties of nanoformulated paints [119–124]. Clearly, results from various studies demonstrate that these new coating formulation of nanopaint to be of tremendous value to researchers and industry. In conclusion the result of nanopaint characterization and performance evaluation opens up a new promising field of study and building materials for next generation.

4. Conclusion

This review highlights the diversity in naturally occurring and engineered nanoparticle in various environments. Given the production of engineered nanoparticles is expected to increase significantly in forthcoming years with more applications and productions, the information provided herein review provide an important baseline from which to interpret future environmental change. Clearly the impacts and effects of nanoparticles on the bacterial toxicity, cement materials, agricultural practices, and sequestration of contaminants cannot be generalized because the physico-chemical reactions occurring at the nanoparticle surfaces depend on a wide range of environmental conditions. Further work should continue to focus on the speciation, biogeochemical behavior and ecotoxicological impacts of both natural and engineered nanoparticles to understand the long-term effects as well as fate and transport behavior of them in the nature.

Acknowledgements

The author would like to thank beamline scientist for assistance with data collection at Lawrence Berkeley National Laboratory (Berkeley, CA, USA) and Stanford Synchrotron Radiation Lightsource (Stanford, CA, USA). The author also wishes to thank Kean University Foundation office for financial support and the anonymous reviewers and the editors for the comments which served to strengthen the manuscript.

Author details

Juyoung Ha
School of Environmental and Sustainability Sciences, Kean University, Union, New Jersey, USA

*Address all correspondence to: haj@kean.edu

References

[1] Chen, L.X., et al., *Fe2O3 nanoparticle structures investigated by X-ray absorption near-edge structure, surface modifications, and model calculations.* J. Phys. Chem. B., 2002. **106**: p. 8539.

[2] Chen, K.L. and M. Elimelech, *Relating Colloidal Stability of Fullerene (C-60) Nanoparticles to Nanoparticle Charge and Electrokinetic Properties.* Environ. Sci. Technol., 2009. **43**(19): p. 7270.

[3] Hotze, E.M., T. Phenrat, and G.V. Lowry, *Nanoparticle Aggregation: Challenges to Understanding Transport and Reactivity in the Environment.* J. Environ. Qual., 2010. **39**(6): p. 1909.

[4] Liu, W.-T., *Nanoparticles and their biological and environmental applications.* Journal of Bioscience and Bioengineering, 2006. **102**(1): p. 1-7.

[5] Chen, M., et al., *Preparation and study of polyacryamide-stabilized silver nanoparticles through a one-pot process.* J. Phys. Chem. B, 2006. **110**(23): p. 11224.

[6] Huynh, K.A. and K.L. Chen, *Aggregation Kinetics of Citrate and Polyvinylpyrrolidone Coated Silver Nanoparticles in Monovalent and Divalent Electrolyte Solutions.* Environ. Sci. Technol., 2011. **45**(13): p. 5564.

[7] Mylon, S.E., K.L. Chen, and M. Elimelech, *Influence of natural organic matter and ionic composition on the kinetics and structure of hematite colloid aggregation: Implications to iron depletion in estuaries.* Langmuir, 2004. **20**: p. 9000.

[8] Dai, M.-H. and J.-M. Martin, *First data on trace metal level and behaviour in two major Arctic river-estuarine systems (Ob and Yenisey) and in the adjacent Kara Sea, Russia.* Earth and Planetary Science Letters, 1995. **131**(3): p. 127-141.

[9] Fulaz, S., et al., *Nanoparticle–Biofilm Interactions: The Role of the EPS Matrix.* Trends in Microbiology, 2019. **27**(11): p. 915-926.

[10] Fabrega, J., et al., *Silver nanoparticles: Behaviour and effects in the aquatic environment.* Environ. Int., 2011. **37**(2): p. 517.

[11] Kaegi, R., et al., *Behavior of Metallic Silver Nanoparticles in a Pilot Wastewater Treatment Plant.* Environ. Sci. Technol., 2011. **45**(9): p. 3902.

[12] Kim, B., et al., *Discovery and characterization of silver sulfide nanoparticles in final sewage sludge products.* Environ. Sci. Technol., 2010. **44**(19): p. 7509.

[13] Levard, C., et al., *Environmental Transformations of Silver Nanoparticles: Impact on Stability and Toxicity.* Environmental Science & Technology, 2012. **46**(13): p. 6900-6914.

[14] Navarro, E., et al., *Environmental behavior and ecotoxicity of engineered nanoparticles to algae, plants, and fungi.* Ecotoxicology, 2008. **17**(5): p. 372.

[15] Roemer, I., et al., *Aggregation and dispersion of silver nanoparticles in exposure media for aquatic toxicity tests.* J. Chromatogr., A, 2011. **1218**(27): p. 4226.

[16] Yin, Y., et al., *Synthesis and characterization of stable aqueous dispersions of silver nanoparticles through the Tollens process.* J. Mater. Chem., 2002. **12**: p. 522.

[17] Wigginton, N.S., K.L. Haus, and M.F. Hochella Jr, *Aquatic environmental nanoparticles.* Journal of Environmental Monitoring, 2007. **9**(12): p. 1306-1316.

[18] Zhu, X., et al., *Comparative Toxicity of Several Metal Oxide Nanoparticle*

Aqueous Suspensions to Zebrafish (Danio rerio) Early Developmental Stage. Journal of environmental science and health. Part A, Toxic/hazardous substances & environmental engineering, 2008. **43**: p. 278-84.

[19] Li, M., K.J. Czymmek, and C.P. Huang, *Responses of Ceriodaphnia dubia to TiO2 and Al2O3 nanoparticles: a dynamic nano-toxicity assessment of energy budget distribution.* Journal of hazardous materials, 2011. **187**(1-3): p. 502-508.

[20] Pakrashi, S., et al., *Cytotoxicity of aluminium oxide nanoparticles towards fresh water algal isolate at low exposure concentrations.* Aquatic Toxicology, 2013. **132-133**: p. 34-45.

[21] von der Heyden, B., A. Roychoudhury, and S. Myneni, *Iron-Rich Nanoparticles in Natural Aquatic Environments.* Minerals, 2019. **9**(5).

[22] Hochella, M.F. and A.F. White, *Reviews in Mineralogy, Vol. 23, Mineral-Water Interface Geochemistry*. 1990. 87.

[23] Hochella, M.F., et al., *Direct observation of heavy metal-mineral association from the Clark Fork River Superfund Complex: Implications for metal transport and bioavailability.* Geochimica et Cosmochimica Acta, 2005. **69**(7): p. 1651-1663.

[24] Hochella, M.F., et al., *Nanominerals, mineral nanoparticles and earth systems.* Science, 2008. **319**: p. 1631.

[25] Yan, W., et al., *Iron nanoparticles for environmental clean-up: recent developments and future outlook.* Environ Sci Process Impacts, 2013. **15**(1): p. 63-77.

[26] Tagliabue, A., et al., *The integral role of iron in ocean biogeochemistry.* Nature, 2017. **543**(7643): p. 51-59.

[27] Nowack, B. and T.D. Bucheli, *Occurrence, behavior and effects of nanoparticles in the environment.* Environ Pollut, 2007. **150**(1): p. 5-22.

[28] Chernyshova, I.V., M.F. Hochella Jr, and A.S. Madden, *Size-dependent structural transformations of hematite nanoparticles. 1. Phase transition.* Physical Chemistry Chemical Physics, 2007. **9**(14): p. 1736-1750.

[29] Guo, H. and A.S. Barnard, *Naturally occurring iron oxide nanoparticles: morphology, surface chemistry and environmental stability.* Journal of Materials Chemistry A, 2013. **1**(1): p. 27-42.

[30] Yang, Y., et al., *Nanoparticles in road dust from impervious urban surfaces: distribution, identification, and environmental implications.* Environmental Science: Nano, 2016. **3**(3): p. 534-544.

[31] Richmond, W.R., et al., *Arsenic Removal from Aqueous Solution via Ferrihydrite Crystallization Control.* Environmental Science & Technology, 2004. **38**(8): p. 2368-2372.

[32] Hu, B., et al., *X-ray absorption fine structure study of enhanced sequestration of U(vi) and Se(iv) by montmorillonite decorated with zero-valent iron nanoparticles.* Environmental Science: Nano, 2016. **3**(6): p. 1460-1472.

[33] Yan, W., et al., *Iron nanoparticles for environmental clean-up: recent developments and future outlook.* Environmental Science: Processes & Impacts, 2013. **15**(1): p. 63-77.

[34] Xu, P., et al., *Use of iron oxide nanomaterials in wastewater treatment: A review.* Science of The Total Environment, 2012. **424**: p. 1-10.

[35] Li, P., et al., *Applying modified biochar with nZVI/nFe3O4 to immobilize*

Pb in contaminated soil. Environmental Science & Pollution Research, 2020. **27**(19): p. 24495-24506.

[36] Hasan, M.S., et al., *Enhanced Heavy Metal Removal from Synthetic Stormwater Using Nanoscale Zerovalent Iron–Modified Biochar.* Water, Air & Soil Pollution, 2020. **231**(5): p. 1-15.

[37] Stefaniuk, M., P. Oleszczuk, and Y.S. Ok, *Review on nano zerovalent iron (nZVI): From synthesis to environmental applications.* Chemical Engineering Journal, 2016. **287**: p. 618-632.

[38] Karn, B., T. Kuiken, and M. Otto, *Nanotechnology and in situ remediation: a review of the benefits and potential risks.* Environmental health perspectives, 2009. **117**(12): p. 1813-1831.

[39] Mueller, N.C., et al., *Application of nanoscale zero valent iron (NZVI) for groundwater remediation in Europe.* Environ Sci Pollut Res Int, 2012. **19**(2): p. 550-8.

[40] Dong, H. and I.M.C. Lo, *Influence of humic acid on the colloidal stability of surface-modified nano zero-valent iron.* Water Research, 2013. **47**(1): p. 419-427.

[41] Chen, M., et al., *Gold-coated iron nanoparticles for biomedical applications.* Journal of Applied Physics, 2003. **93**(10): p. 7551-7553.

[42] Tran, N. and T.J. Webster, *Magnetic nanoparticles: biomedical applications and challenges.* Journal of Materials Chemistry, 2010. **20**(40): p. 8760-8767.

[43] Cheong, S., et al., *Synthesis and Stability of Highly Crystalline and Stable Iron/Iron Oxide Core/Shell Nanoparticles for Biomedical Applications.* ChemPlusChem, 2012. 77(2): p. 135-140.

[44] Sangaiya, P. and R. Jayaprakash, *A Review on Iron Oxide Nanoparticles and Their Biomedical Applications.* Journal of Superconductivity and Novel Magnetism, 2018. **31**(11): p. 3397-3413.

[45] Madden, A.S. and M.F. Hochella, *A test of geochemical reactivity as a function of mineral size: Manganese oxidation promoted by hematite nanoparticles.* Geochim. Cosmochim. Acta, 2005. **69**: p. 389.

[46] Madden, A.S., M.F. Hochella, and T.P. Luxton, *Insights for size-dependent of hematite nanomineral surfaces through Cu2+ sorption.* Geochim. Cosmochim. Acta, 2006. **70**: p. 4095.

[47] Dickson, D., et al., *Dispersion and stability of bare hematite nanoparticles: Effect of dispersion tools, nanoparticle concentration, humic acid and ionic strength.* Science of The Total Environment, 2012. **419**: p. 170-177.

[48] Hwang, E.T., et al., *Analysis of the toxic mode of action of silver nanoparticles using stress-specific bioluminescent bacteria.* Small, 2008. **4**(6): p. 746.

[49] Pettibone, J.M., et al., *Adsorption of Organic Acids on TiO2 Nanoparticles: Effects of pH, Nanoparticle Size, and Nanoparticle Aggregation.* Langmuir, 2008. **24**(13): p. 6659-6667.

[50] Xu, B., et al., *Sorption of pentachlorophenol and phenanthrene by humic acid-coated hematite nanoparticles.* Environmental Pollution, 2019. **248**: p. 929-937.

[51] Ha, J., et al., *Adsorption of Organic Matter at Mineral/Water Interfaces: 7. ATR-FTIR and Quantum Chemical Study of Lactate Interactions with Hematite Nanoparticles.* Langmuir, 2008. **24**(13): p. 6683-6692.

[52] Lenhart, J.J., et al., *The influence of dicarboxylic acid structure on the stability of colloidal hematite.* Journal of Colloid and Interface Science, 2010. **345**(2): p. 556-560.

[53] Palomino, D. and S. Stoll, *Fulvic acids concentration and pH influence on the stability of hematite nanoparticles in aquatic systems.* Journal of Nanoparticle Research, 2013. **15**(2): p. 1428.

[54] Ha, J., et al., *Interaction of Zn(II) with hematite nanoparticles and microparticles: Part 2. ATR-FTIR and EXAFS study of the aqueous Zn(II)/ oxalate/hematite ternary system.* Langmuir, 2009. **25**(10): p. 5586.

[55] Ha, J., et al., *Interaction of Zn(II) with hematite nanoparticles and microparticles: Part 2. EXAFS study of Zn(II) adsorption and precipitation.* Langmuir, 2009. **25**(10): p. 5574.

[56] Braem, A.D., D.C. Prieve, and R.D. Tilton, *Electrostatically Tunable Coadsorption of Sodium Dodecyl Sulfate and Poly(ethylene oxide)-b-poly(propylene oxide)-b-poly(ethylene oxide) Triblock Copolymer to Silica.* Langmuir, 2001. **17**(3): p. 883-890.

[57] Burleson, D.J., M.D. Driessen, and R.L. Penn, *On the Characterization of Environmental Nanoparticles.* Journal of Environmental Science and Health, Part A, 2004. **39**(10): p. 2707-2753.

[58] Harter, R.D. and R. Naidu, *Role of Metal-Organic complexation in metal sorption by Soils*, in *Advances in Agronomy*, D.L. Sparks, Editor. 1995, Academic Press. p. 219-263.

[59] Sun, J., L. Liu, and F. Yang, *A WO3/ PPy/ACF modified electrode in electrochemical system for simultaneous removal of heavy metal ion Cu2+ and organic acid.* Journal of Hazardous Materials, 2020. **394**: p. 122534.

[60] Renella, G., L. Landi, and P. Nannipieri, *Degradation of low molecular weight organic acids complexed with heavy metals in soil.* Geoderma, 2004. **122**(2): p. 311-315.

[61] López-Bucio, J., et al., *Organic acid metabolism in plants: from adaptive physiology to transgenic varieties for cultivation in extreme soils.* Plant Science, 2000. **160**(1): p. 1-13.

[62] Li, N., et al., *Contaminant removal and microorganism response of activated sludge in sulfamethazine wastewater treatment.* International Biodeterioration & Biodegradation, 2019. **143**: p. 104705.

[63] Masciandaro, G., et al., *Organic matter–microorganism–plant in soil bioremediation: a synergic approach.* Reviews in Environmental Science and Bio/Technology, 2013. **12**(4): p. 399-419.

[64] Deschênes, L. and T. Ells, *Bacteria-nanoparticle interactions in the context of nanofouling.* Advances in Colloid and Interface Science, 2020. **277**: p. 102106.

[65] Schwegmann, H., A.J. Feitz, and F.H. Frimmel, *Influence of the zeta potential on the sorption and toxicity of iron oxide nanoparticles on S. cerevisiae and E. coli.* Journal of Colloid and Interface Science, 2010. **347**(1): p. 43-48.

[66] Kiser, M.A., et al., *Biosorption of nanoparticles to heterotrophic wastewater biomass.* Water Research, 2010. **44**(14): p. 4105-4114.

[67] Droz, B., et al., *A Comparison of the Sorption Reactivity of Bacteriogenic and Mycogenic Mn Oxide Nanoparticles.* Environmental Science & Technology, 2015. **49**(7): p. 4200-4208.

[68] Pan, W., et al., *Dissimilatory microbial iron reduction release DOC (dissolved organic carbon) from carbon-ferrihydrite association.* Soil Biology and Biochemistry, 2016. **103**: p. 232-240.

[69] Hansel, C.M., et al., *Secondary mineralization pathways induced by dissimilatory iron reduction of ferrihydrite under advective flow.* Geochimica et Cosmochimica Acta, 2003. **67**(16): p. 2977-2992.

[70] Dehner, C.A., et al., *Size-Dependent Bioavailability of Hematite (α-Fe2O3) Nanoparticles to a Common Aerobic Bacterium.* Environmental Science & Technology, 2011. **45**(3): p. 977-983.

[71] Ha, J., et al., *Role of extracellular polymeric substances in metal ion complexation on Shewanella oneidensis: Batch uptake, thermodynamic modeling, ATR-FTIR, and EXAFS study.* Geochimica et Cosmochimica Acta, 2010. **74**(1): p. 1-15.

[72] Kim, T.-Y., et al., *Biogenic Hematite from Bacteria: Facile Synthesis of Secondary Nanoclusters for Lithium Storage Capacity.* ACS Applied Materials & Interfaces, 2019. **11**(7): p. 6948-6957.

[73] Klaus, T., et al., *Silver-based crystalline nanoparticles, microbially fabricated.* Proceedings of the National Academy of Sciences, 1999. **96**(24): p. 13611.

[74] Husseiny, M.I., et al., *Biosynthesis of gold nanoparticles using Pseudomonas aeruginosa.* Spectrochimica Acta Part A: Molecular and Biomolecular Spectroscopy, 2007. **67**(3): p. 1003-1006.

[75] Philip, D., *Biosynthesis of Au, Ag and Au–Ag nanoparticles using edible mushroom extract.* Spectrochimica Acta Part A: Molecular and Biomolecular Spectroscopy, 2009. **73**(2): p. 374-381.

[76] Penner-Hahn, J.E., *X-ray absorption spectroscopy in coordination chemistry.* Coordination Chemistry Reviews, 1999. **190-192**: p. 1101-1123.

[77] Meitzner, G., *Experimental aspects of X-Ray absorption spectroscopy.* Catalysis Today, 1998. **39**(4): p. 281-291.

[78] Newville, M., *EXAFS analysis using FEFF and FEFFIT.* Journal of Synchrotron Radiation, 2001. **8**(2): p. 96-100.

[79] Koningsberger, D.C., *X-ray absorption: principles, applications, techniques of EXAFS, SEXAFS, and XANES.* 1988, United States: John Wiley and Sons.

[80] Bostick, B.C., et al., *Cesium Adsorption on Clay Minerals: An EXAFS Spectroscopic Investigation.* Environmental Science & Technology, 2002. **36**(12): p. 2670-2676.

[81] Brown, G.E., et al., *Metal oxide surfaces and their interactions with aqueous solutions and microbial organisms.* J. Colloid Interface Sci., 1999. **241**: p. 317.

[82] Brown, G.E. and G.A. Parks, *Sorption of trace elements on mineral surfaces: Modern perspectives from spectroscopic studies, and comments on sorption in the marine environment.* Int. Geol. Rev., 2001. **43**: p. 963.

[83] Cheng, G., J.D. Carter, and T. Guo, *Investigation of Co nanoparticles with EXAFS and XANES.* Chemical Physics Letters, 2004. **400**(1): p. 122-127.

[84] Peterson, M.L., et al., *Differential redox and sorption of Cr (III/VI) on natural silicate and oxide minerals: EXAFS and XANES results.* Geochimica et Cosmochimica Acta, 1997. **61**(16): p. 3399-3412.

[85] Sharma, A., et al., *XANES, EXAFS and photocatalytic investigations on copper oxide nanoparticles and nanocomposites.* RSC Advances, 2015. **5**(28): p. 21762-21771.

[86] Rockenberger, J., et al., *EXAFS Studies on the Size Dependence of Structural and Dynamic Properties of CdS Nanoparticles.* The Journal of Physical Chemistry B, 1997. **101**(14): p. 2691-2701.

[87] Sun, Y., et al., *Characterization of Palladium Nanoparticles by Using X-ray Reflectivity, EXAFS, and Electron*

Microscopy. Langmuir, 2006. **22**(2): p. 807-816.

[88] Guczi, L., et al., *AuPd bimetallic nanoparticles on TiO2: XRD, TEM, in situ EXAFS studies and catalytic activity in CO oxidation.* Journal of Molecular Catalysis A: Chemical, 2003. **204-205**: p. 545-552.

[89] Sheng, G., et al., *New insights into the primary roles of diatomite in the enhanced sequestration of UO22+ by zerovalent iron nanoparticles: An advanced approach utilizing XPS and EXAFS.* Applied Catalysis B: Environmental, 2016. **193**: p. 189-197.

[90] Frenkel, A., *Solving the structure of nanoparticles by multiple-scattering EXAFS analysis.* Journal of Synchrotron Radiation, 1999. **6**(3): p. 293-295.

[91] Bernard, S., et al., *Geochemical evolution of organic-rich shales with increasing maturity: A STXM and TEM study of the Posidonia Shale (Lower Toarcian, northern Germany).* Marine and Petroleum Geology, 2012. **31**(1): p. 70-89.

[92] Obst, M., et al., *Precipitation of amorphous CaCO3 (aragonite-like) by cyanobacteria: A STXM study of the influence of EPS on the nucleation process.* Geochimica et Cosmochimica Acta, 2009. **73**(14): p. 4180-4198.

[93] Beetz, T. and C. Jacobsen, *Soft X-ray radiation-damage studies in PMMA using a cryo-STXM.* Journal of Synchrotron Radiation, 2003. **10**(3): p. 280-283.

[94] Solomon, D., et al., *Micro- and nano-environments of carbon sequestration: Multi-element STXM–NEXAFS spectromicroscopy assessment of microbial carbon and mineral associations.* Chemical Geology, 2012. **329**: p. 53-73.

[95] Lawrence, J.R., et al., *Monitoring the fate of copper nanoparticles in river biofilms using scanning transmission X-ray microscopy (STXM).* Chemical Geology, 2012. **329**: p. 18-25.

[96] Burke, K.B., et al., *Scanning transmission x-ray microscopy of polymer nanoparticles: probing morphology on sub-10 nm length scales.* Nanotechnology, 2011. **22**(26): p. 265710.

[97] Luef, B., et al., *Iron-reducing bacteria accumulate ferric oxyhydroxide nanoparticle aggregates that may support planktonic growth.* The ISME Journal, 2013. **7**(2): p. 338-350.

[98] Baowei, Z., et al., *Effects of Soybean Stover-Derived Biochar on Microbial Community and Structure in Loess Soil.* Nature Environment & Pollution Technology, 2020. **19**(2): p. 703-710.

[99] Hong, W., et al., *Biochar Mitigates Greenhouse Gas Emissions from an Acidic Tea Soil.* Polish Journal of Environmental Studies, 2020. **29**(1): p. 323-330.

[100] Vijayaraghavan, K., *Recent advancements in biochar preparation, feedstocks, modification, characterization and future applications.* Environmental Technology Reviews, 2019. **8**(1): p. 47-64.

[101] Wang, B., et al., *Biochar addition can reduce NOx gas emissions from a calcareous soil.* Chemical Speciation & Bioavailability, 2019. **31**(1): p. 38-48.

[102] Saxena, M., S. Maity, and S. Sarkar, *Carbon nanoparticles in 'biochar' boost wheat (Triticum aestivum) plant growth.* RSC Advances, 2014. **4**(75): p. 39948-39954.

[103] Oleszczuk, P., et al., *Characterization of nanoparticles of biochars from different biomass.* Journal of Analytical and Applied Pyrolysis, 2016. **121**: p. 165-172.

[104] Chen, M., et al., *Transport and retention of biochar nanoparticles in a paddy soil under environmentally-relevant solution chemistry conditions.* Environmental Pollution, 2017. **230**: p. 540-549.

[105] Su, H., et al., *Remediation of hexavalent chromium contaminated soil by biochar-supported zero-valent iron nanoparticles.* Journal of Hazardous Materials, 2016. **318**: p. 533-540.

[106] Yue, L., et al., *The effect of biochar nanoparticles on rice plant growth and the uptake of heavy metals: Implications for agronomic benefits and potential risk.* Science of The Total Environment, 2019. **656**: p. 9-18.

[107] Duxson, P., et al., *Understanding the relationship between geopolymer composition, microstructure and mechanical properties.* Colloids and Surfaces A: Physicochemical and Engineering Aspects, 2005. **269**(1): p. 47-58.

[108] Duxson, P., et al., *Geopolymer technology: the current state of the art.* Journal of Materials Science, 2007. **42**(9): p. 2917-2933.

[109] Singh, B., et al., *Geopolymer concrete: A review of some recent developments.* Construction and Building Materials, 2015. **85**: p. 78-90.

[110] Wang, H., H. Li, and F. Yan, *Synthesis and mechanical properties of metakaolinite-based geopolymer.* Colloids and Surfaces A: Physicochemical and Engineering Aspects, 2005. **268**(1): p. 1-6.

[111] Nazari, A., et al., *Mechanical properties of cement mortar with Al2O3 nanoparticles.* Journal of American Science, 2010. **6**(4): p. 94-97.

[112] Kawashima, S., et al., *Modification of cement-based materials with nanoparticles.* Cement and Concrete Composites, 2013. **36**: p. 8-15.

[113] Land, G. and D. Stephan, *Controlling cement hydration with nanoparticles.* Cement and Concrete Composites, 2015. **57**: p. 64-67.

[114] Li, S., et al., *Effects of nano-SiO2 coated multi-walled carbon nanotubes on mechanical properties of cement-based composites.* Construction and Building Materials, 2021. **281**: p. 122577.

[115] Said, A.M., et al., *Properties of concrete incorporating nano-silica.* Construction and Building Materials, 2012. **36**: p. 838-844.

[116] Gaitero, J.J., I. Campillo, and A. Guerrero, *Reduction of the calcium leaching rate of cement paste by addition of silica nanoparticles.* Cement and Concrete Research, 2008. **38**(8): p. 1112-1118.

[117] Krishnamoorthy, K., et al., *Graphene oxide nanopaint.* Carbon, 2014. **72**: p. 328-337.

[118] Subbiah, G., et al., *Preparation of TiO2 nanopaint using ball milling process and investigation on its antibacterial properties.* Materials Express, 2014. **4**(5): p. 393-399.

[119] Benitha, V.S., K. Monisha, and K. Jeyasubramaniyan, *Formulation and evaluation of TiO2.Fe2O3 nanopaint.* Archives of Materials Science and Engineering, 2016. **Vol. 77, nr 1**: p. 40-44.

[120] Kumar, A., et al., *Silver-nanoparticle-embedded antimicrobial paints based on vegetable oil.* Nature Materials, 2008. **7**(3): p. 236-241.

[121] Stephen, A. and S. Seethalakshmi, *Phytochemical Synthesis and Preliminary Characterization of Silver Nanoparticles Using Hesperidin.* Journal of Nanoscience, 2013. **2013**: p. 126564.

[122] Krishnamoorthy, K., et al., *Nanostructured molybdenum oxide-based antibacterial paint: effective growth inhibition of various pathogenic bacteria*. 2014, Nanotechnology. p. 315101.

[123] Yong, H.E., et al., *Preparation of ZnO nanopaint for marine antifouling applications.* Journal of Industrial and Engineering Chemistry, 2015. **29**: p. 39-42.

[124] Krishnamoorthy, K., et al., *Mechanochemical Reinforcement of Graphene Sheets into Alkyd Resin Matrix for the Development of Electrically Conductive Paints.* ChemNanoMat, 2018. **4**(6): p. 568-574.